Lecture Notes in Mathematics

Edited by A. Dold and B. Eckmann

1285

I. W. Knowles Y. Saitō (Eds)

Differential Equations and Mathematical Physics

Proceedings of an International Conference held in Birmingham, Alabama, USA, March 3–8, 1986

Springer-Verlag

Berlin Heidelberg New York London Paris Tokyo

Editors

Ian W. Knowles
Yoshimi Saitō
Department of Mathematics, The University of Alabama at Birmingham
Birmingham, AL 35294, USA

Mathematics Subject Classification (1980): 34, 35, 42, 73, 76, 78, 81

ISBN 3-540-18479-1 Springer-Verlag Berlin Heidelberg New York
ISBN 0-387-18479-1 Springer-Verlag New York Berlin Heidelberg

Printed in Germany

Printing and binding: Druckhaus Beltz, Hemsbach/Bergstr.
2146/3140-543210

This volume is respectfully dedicated
to Professor Tosio Kato on the occasion
of his seventieth birthday.

PREFACE

This volume forms a permanent record of lectures given at the International Conference on Differential Equations and Mathematical Physics held at the University of Alabama at Birmingham during March 3-8, 1986.

The conference was supported by about 250 mathematicians from the following countries: Belgium, Canada, Czechoslovakia, Denmark, Egypt, Finland, France, Hungary, India, Ireland, Japan, Kuwait, Nigeria, Norway, P.R. of China, South Africa, Sweden, Switzerland, The Netherlands, the U.K., the U.S.A., and West Germany. Its main purpose was to provide a forum for the discussion of recent developments in the theory of ordinary and partial differential equations, both linear and non-linear, with particular reference to work relating to the equations of mathematical physics. Invited one-hour lectures were given by P. Deift, R. DiPerna, W.N. Everitt, C. Foias, T. Kato, S. Kotani, A. Majda, J. Mawhin, J. McLaughlin, J. McLeod, C. Morawetz, R. Newton, R. Phillips, M. Reed, I. Sigal, and B. Simon. The remainder of the program consisted of invited one-half hour lectures.

On behalf of the participants, the conference directors acknowledge, with gratitude, the generous financial support provided by the U.S. National Science Foundation, under grant number DMS-8516772, the Department of Mathematics and the Graduate School, University of Alabama at Birmingham, and the College of Arts and Sciences, the Graduate School and the Office of Academic Affairs, University of Alabama. We acknowledge also the valuable support provided by the other members of the conference committee: Robert Kauffman, Roger Lewis, and Fred Martens from UAB, and Richard Brown and James Ward from UA. As always, the committee is much indebted to the faculty, staff, and graduate students of the Department of Mathematics at UAB for their manifold contributions; here, we wish to make particular mention of Mrs. Eileen Schauer for undertaking the onerous task of typing much of the conference material, including many of the the papers appearing in this volume.

Ian W. Knowles
Yoshimi Saitō
Conference Directors

CONTENTS

LECTURES NOT APPEARING IN THE PROCEEDINGS

A Nonlinear Eigenvalue Problem in Astrophysical Magnetohydrodynamics

John A. Adam (Old Dominion University, U.S.A.)

Existence of Non-Trivial Periodic Solutions of a Certain Third-Order Non-Linear Differential Equation

Anthony Uyi Afuwape (University of Ife, NIGERIA)

Stabilization of Solutions for a Class of Degenerate Equations in Divergence Form in One Space Dimension

N. Alikakos (University of Tennessee, U.S.A.)

Spectral Properties of Indefinite Elliptic Problems

W. Allegretto (University of Alberta, CANADA)

Quasilinear Parabolic Systems

H. Amann (Universitat Zurich, SWITZERLAND)

Convergence Properties of Strongly-Damped Semilinear Wave Equations

Joel D. Avrin (University of North Carolina-Charlotte, U.S.A.)

On Smoothness of Solutions of Elliptic Equations in n-Dimensional Nonsmooth Domains

A. Azzam (Kuwait University, KUWAIT)

Singular Elliptic Operators with Discrete Spectra

J. V. Baxley (Wake Forest University, U.S.A.) *and R. O. Chapman* (Oxford University, U.K.)

Numerical Solution of Nonlinear Parabolic Variational and Quasi-Variational Inequalities

S. A. Belbas (University of Alabama, U.S.A.)

The Kolomogoroff-Arnold-Moses Theorem in Schrodinger's Equation

Jean Bellissard (California Institute of Technology, U.S.A.)

The Limiting Absorption Principle for Differential Operators with Short-Range Perturbations

Matania Ben-Artzi (University of California, Los Angeles, U.S.A.)

Random Wave Operators

Marc A. Berger (Georgia Institute of Technology, U.S.A.)

Singular Linear Differential and Difference Operators in the Complex Plane and in B^*-Algebras

C. E. Billigheimer (McMaster University, CANADA)

Resonance Regions Determined by a Projection Operator Formulation

Erkki Brandas and Erik Engdahl (Uppsala University, SWEDEN)

ANALYTICAL SOLUTIONS FOR ORDINARY AND PARTIAL DIFFERENTIAL EQUATIONS

G. Adomian[1]

Our objective is to address the need for realistic solution of the nonlinear stochastic systems of equations in space and time which arise in the modeling of frontier problems in physics. What is meant by realistic solution is solution of the problem as it is rather than forcing it into an oversimplified mold to make it easily solvable. For a wide range of problems, of course, it is adequate to use perturbation, linearization, etc., but generally assumptions of weak nonlinearity, small fluctuations, and convenient but unphysical stochastic processes may be unjustified, and we resort to them only when no other approach is possible. (For some nonlinear systems, exact linearization is possible by clever transformations of variables to make the equations linear and solvable. However, this is not generally possible and one resorts to ad hoc or perturbative methods.) In systems involving stochastic parameters, e.g., differential equations with stochastic process coefficients - the stochastic operator case - usual analyses employ perturbation or hierarchy methods which require that fluctuations be small. Another common restrictive assumption is an assumed special nature or behavior for the processes - for mathematical rather than physical reasons. The literature abounds with unrealistic unphysical assumptions and approximations such as white noise, monochromatic approximation, local independence, etc.

These limitations and assumptions are made for mathematical tractability and use of well known theory. Yet our final objective must not be simply the satisfaction of quoting theorems and stating an abstruse

[1]Center for Applied Mathematics, University of Georgia, Athens, Georgia 30602, U.S.A.

solution but finding solutions in close correspondence with actual physical behavior. Numerical results on supercomputers may lead to massive printouts which make dependences and relationships difficult to see; computers are only faster, not wiser than mathematicians. The solution we want is that of the problem at hand, not one tailored to machine computation or the use of existing theorems.

Thus we propose to solve systems of multidimensional nonlinear stochastic partial differential equations in space and time - or ordinary differential equations or integro-differential or delay-differential equations - (and special cases where equations become linear or deterministic or one-dimensional) without linearization, or discretization, or perturbation, etc. (If we have a problem of waves propagating in a random medium, we will solve the stochastic partial differential equation, without resort to a monochromatic approximation, and Helmholtz equation.) We have called the method the *decomposition method* (feeling the term is more appropriate here than in Galois theory or even in the simplification of large-scale systems to get disjoint state spaces).

This method is an approximation method. It yields a series solution. It is not less desirable as a result than a so-called closed form analytical solution which has been arrived at by forcing the problem into a linear deterministic mold. All modeling is an approximation and a solution which provides, as this does, a rapidly converging continuous analytic approximation to the nonlinear problem (rather than a so-called exact solution to a linearized problem) may very well be much more "exact."

Certainly, it is important to know that attempts to compute solutions will be successful. Mathematically this means the problem is well-set, e.g., in the sense of Hadamard - that an operator exists which uniquely and continuously takes elements in a suitable class of initial data into a class of solutions. Statement of the precise mathematical conditions unfortunately requires a complicated symbolism - but the meaning is simple enough. Solutions should exist for reasonable input data and each solution should be unique so it can serve as a physical approximation, and depend continuously without jumps on the given conditions. Also, it is reasonable to say small changes in parameters of the model should cause no more than small changes in our solutions. However, to talk of a problem being "well-set" and then to neglect the nonlinear (or stochastic) effects or to approximate them to first order means that "solution" is only a mathematical solution, not the real solution. Nonlinear equations can be very sensitive to small input changes. If one linearizes a strongly nonlinear equation in his model then precisely defines conditions under which a mathematical solution to the simplified equation is valid, the solution of the model retaining nonlinearity seems preferable even if one knows merely that $u(x,0) = g(x)$ and not that $g(x)$ belongs to a Sobolev space of L_2 functions with generalized first derivatives also in L_2. After all, g represents a physical quantity. We are dealing with physical problems and the physical system has a solution and the parameters are generally well-defined without discontinuities. This, and the fact that our general forms are for *operator* equations where the operator may be algebraic, differential, or partial differential, allows us to conceive of solving problems in wide-ranging applications.

We begin with the (deterministic) form $Fu = g(t)$ where F is a nonlinear ordinary differential operator with linear and nonlinear terms. The linear term is written $Lu + Ru$ where L is invertible. To avoid difficult integrations we choose L as the highest ordered derivative. R is the remainder of the linear operator. The nonlinear term is represented by Nu. Thus $Lu + Ru + Nu = g$ and we write

$$Lu = g - Ru - Nu$$

$$L^{-1}Lu = L^{-1}g - L^{-1}Ru - L^{-1}Nu.$$

For initial-value problems we conveniently define L^{-1} for $L = d^n/dt^n$ as the n-fold definite integration operator from 0 to t. For the operator $L = d^2/dt^2$, for example, we have $L^{-1}Lu = u - u(0) - tu'(0)$ and therefore

$$u = u(0) + tu'(0) + L^{-1}g - L^{-1}Ru - L^{-1}Nu. \tag{1}$$

For the same operator but a boundary value problem, we let L^{-1} be an indefinite integral and write $u = A + Bt$ for the first two terms and evaluate A,B from the given conditions. The first three terms of (1) are identfied as u_0 in the assumed decomposition $u = \sum_{n=0}^{\infty} u_n$. Finally we write $Nu = \sum_{n=0}^{\infty} A_n(u_0,u_1,\ldots,u_n)$ where the A_n are specially generated polynomials for the particular nonlinearity which depend only on the u_0 to u_n components. They are defined in [1] and discussed extensively in [2] (in which an alternative faster converging form is also discussed) and elsewhere. We have now

$$u = \sum_{n=0}^{\infty} u_n = u_0 - L^{-1}R \sum_{n=0}^{\infty} u_n - L^{-1} \sum_{n=0}^{\infty} A_n$$

so that

$$u_1 = - L^{-1}Ru_0 - L^{-1}A_0$$

$$u_2 = - L^{-1}Ru_1 - L^{-1}A_1$$

$$u_3 = - L^{-1}Ru_2 - L^{-1}A_2$$

etc., and all components are determinable since A_0 depends only on u_0 , A_1 depends only on u_0,u_1, etc. The practical solution will be the n-term approximation

$$\phi_n = \sum_{i=0}^{n-1} u_i$$

and $\lim_{n \to \infty} \phi_n = \sum_{i=0}^{\infty} u_i = u$ by definition.

In the linear case where Nu vanishes we have

$$u = u_0 - L^{-1}Ru_0 - L^{-1}Ru_1 - \ldots$$

$$= u_0 - L^{-1}Ru_0 + (L^{-1}R)(L^{-1}R)u_0 - \ldots$$

$$= \sum_{n=0}^{\infty} (-1)^n (L^{-1}R)^n u_0.$$

If the specified conditions vanish, $u_0 = L^{-1}g$ thus

$$u = \sum_{n=0}^{\infty} (-1)^n (L^{-1}R)^n L^{-1} g$$

thus $Fu = g$ becomes $u = F^{-1}g$ where the inverse is $F^{-1} = \sum_{n=0}^{\infty} (-1)^n (L^{-1}R)^n L^{-1}$

Suppose $Fu = g$ is a multidimensional equation such as $\nabla^2 u + u_t = g$ with $g = g(x,y,z,t)$ which we write as

$$[L_x + L_y + L_z + L_t]u = g$$

where L_x, L_y, L_z are $\partial^2/\partial x^2$, $\partial^2/\partial y^2$, $\partial^2/\partial z^2$ respectively and $L_t = \partial/\partial t$. We solve this exactly as before with one change. We must solve for each linear operator term, invert, add and divide by the number of equations to get a single equation. Thus we have

$$[L_t + L_x + L_y + L_z]u = g$$

from which we get four equations

$$L_t u = g - L_x u - L_y u - L_z u$$

$$L_x u = g - L_t u - L_y u - L_z u$$

$$L_y u = g - L_t u - L_x u - L_z u$$

$$L_z u = g - L_t u - L_x u - L_y u.$$

Applying the integrals L_t^{-1} to the first, L_x^{-1} to the second, etc., we get*

---

*With each integration, we will also get terms containing initial and boundary conditions (see [2]) so our resulting u_0 must contain all initial/boundary conditions as well as the forcing function term as shown in the following examples. We have omitted these terms here in this discussion; their inclusion considerably complicates this discussion and is left to a forthcoming research paper now in final preparation.

$$u = L_t^{-1}g - L_t^{-1}(L_x + L_y + L_z)u$$

$$u = L_x^{-1}g - L_x^{-1}(L_t + L_y + L_z)u$$

$$u = L_y^{-1}g - L_y^{-1}(L_t + L_x + L_z)u$$

$$u = L_z^{-1}g - L_z^{-1}(L_t + L_x + L_y)u.$$

Adding and dividing by four [1,2]

$$u = (1/4)\{L_t^{-1} + L_x^{-1} + L_y^{-1} + L_z^{-1}\}g$$

$$- (1/4)\{L_t^{-1}(L_x + L_y + L_z) + L_x^{-1}(L_t + L_y + L_z)$$

$$+ L_y^{-1}(L_t + L_x + L_z) + L_z^{-1}(L_t + L_x + L_y)\}u.$$

Define

$$u_0 = (1/4)\{L_t^{-1} + L_x^{-1} + L_y^{-1} + L_z^{-1}\}g^*$$

Replacing u by $\sum_{n=0}^{\infty} u_n$, identify

$$u_{n+1} = - (1/4)\{L_t^{-1}(L_x + L_y + L_z) + L_x^{-1}(L_t + L_y + L_z)$$

$$+ L_y^{-1}(L_t + L_x + L_z) + L_z^{-1}(L_t + L_x + L_y)\}u_n$$

for $n \geq 0$. Since each component depends on the preceding components, all of them can be written in terms of u_0. Thus

$$u_1 = - (1/4)\{L_t^{-1}(L_x + L_y + L_z) + L_x^{-1}(L_t + L_y + L_z)$$

$$+ L_y^{-1}(L_t + L_x + L_z) + L_z^{-1}(L_t + L_x + L_y)\}u_0$$

$$u_2 = - (1/4)\{L_t^{-1}(L_x + L_y + L_z) + L_x^{-1}(L_t + L_y + L_z)$$

$$+ L_y^{-1}(L_t + L_x + L_z) + L_z^{-1}(L_t + L_x + L_y)\}u_1$$

$$\vdots$$

$$u_n = (-1)^n (1/4)^n \{\cdot\}^n (1/4)[L_t^{-1} + L_x^{-1} + L_y^{-1} + L_z^{-1}]g$$

so that

$$u = \sum_{n=0}^{\infty} (-1)^n (1/4)^n \{\cdot\}^n (1/4) [L_t^{-1} + L_x^{-1} + L_y^{-1} + L_z^{-1}]g.$$

Consequently we can write $u = \mathcal{L}^{-1}g$ with $\mathcal{L}^{-1}$ defined by

$$\mathcal{L}^{-1} = \sum_{n=0}^{\infty} (-1)^n (1/4)^n \{\cdot\}^n (1/4) [L_t^{-1} + L_x^{-1} + L_y^{-1} + L_z^{-1}].$$

Consider the bracketed term $\{\cdot\}$ which becomes

$$L_t^{-1}L_x + L_t^{-1}L_y + L_t^{-1}L_z + L_x^{-1}L_t + L_x^{-1}L_y + L_x^{-1}L_z$$
$$+ L_y^{-1}L_t + L_y^{-1}L_x + L_y^{-1}L_z + L_z^{-1}L_t + L_z^{-1}L_x + L_z^{-1}L_y$$

or

$$(L_x^{-1}L_y + L_y^{-1}L_x) + (L_x^{-1}L_z + L_z^{-1}L_x)$$
$$+ (L_x^{-1}L_t + L_t^{-1}L_x) + (L_y^{-1}L_z + L_z^{-1}L_y) \qquad (2)$$
$$+ (L_t^{-1}L_y + L_y^{-1}L_t) + (L_t^{-1}L_z + L_z^{-1}L_t).$$

To prove $\mathcal{L}^{-1}$ is the inverse of $\mathcal{L}$, write

$$\mathcal{L}^{-1}\mathcal{L} = \sum_{n=0}^{\infty} (-1)^n (1/4)^n \{\cdot\}^n (1/4) [L_t^{-1} + L_x^{-1} + L_y^{-1} + L_z^{-1}]$$
$$\cdot [L_t + L_x + L_y + L_z] .$$

Consider the product $[L_t^{-1} + L_x^{-1} + L_y^{-1} + L_z^{-1}][L_t + L_x + L_y + L_z]$. The terms $L_t^{-1}L_t + L_x^{-1}L_x + L_y^{-1}L_y + L_z^{-1}L_z$ are equal to $4\,I$ where I is the identity operator. The remaining terms are identical to the expression (2) and therefore the bracketed term. Thus

$$\mathcal{L}^{-1}\mathcal{L} = \sum_{n=0}^{\infty} (-1)^n (1/4)^n \{\cdot\} \quad (1/4)[4I + \{\cdot\}]$$
$$= \sum_{n=0}^{\infty} (-1)^n (1/4)^n \{\cdot\} \; I - \sum_{n=0}^{\infty} (-1)^{n+1} (1/4)^{n+1} \{\cdot\}^{n+1}$$
$$= I.$$

Hence $\mathcal{L}^{-1}$ is indeed the inverse. To recapitulate, in $Fu = g$ where $F = L_t + L_x + L_y + L_z$

$$F^{-1} = \sum_{n=0}^{\infty} (-1)^n (1/4)^n \{(L_x^{-1}L_y + L_y^{-1}L_x) + (L_x^{-1}L_z + L_z^{-1}L_x)$$

$$+ (L_x^{-1}L_t + L_t^{-1}L_x) + (L_y^{-1}L_z + L_z^{-1}L_y)$$

$$+ (L_t^{-1}L_y + L_y^{-1}L_t) + (L_t^{-1}L_z + L_z^{-1}L_t)\}^n$$

$$\times (1/4)[L_t^{-1} + L_x^{-1} + L_y^{-1} + L_z^{-1}].$$

Remark: The 1/4 is a dimensionality factor. If we have only L_t, L_x, for example, it would be 1/2.

Now suppose our equations are coupled differential or partial differential equations. Suppose we have, for example, a system of equations in u and v. It is only necessary to define the pair u_0, v_0 then find u_1, v_1 in terms of u_0, v_0, etc. For n equations we have an n-vector of terms for the first component. Then an n-vector of second components is found in terms of the first. The procedure is discussed completely in [2] and we will only illustrate the procedure here by examples.

The equation $u_{xx} - u_t = 0$ with conditions $u(0,t) = t$, $u(x,0) = x^2/2$, $u_x(0,t) = 0$ leads immediately by this procedure to $u = t + (x^2/2)$. The six term approximation is within 2%. The ten term approximation is within 0.1% of the correct solution. This example is discussed in [2]. Let us consider a nonlinear example.

Example: Partial Differential Equation

Let $\nabla^2 = (\partial^2/\partial x^2) + (\partial^2/\partial y^2) + (\partial^2/\partial z^2) = L_x + L_y + L_z$ (where the notation L_x symbolizes a linear (deterministic) differential operator $\partial^2/\partial x^2$, etc.) and $N(p)$, or Np for convenience, symbolizes the nonlinear (deterministic) operator acting on p to give $\sinh p$.

We will consider the nonlinear equation $\nabla^2 p = k^2 \sinh p$ which we write as

$$[L_x + L_y + L_z]p = k^2Np$$

where $Np = \sinh p$. Hence

$$L_xp = k^2Np - L_yp - L_zp$$

$$L_yp = k^2Np - L_xp - L_zp$$

$$L_zp = k^2Np - L_xp - L_yp$$

$$L_x^{-1}L_xp = L_x^{-1}k^2Np - L_x^{-1}L_yp - L_x^{-1}L_zp$$

$$L_y^{-1}L_yp = L_y^{-1}k^2Np - L_y^{-1}L_xp - L_y^{-1}L_zp \tag{3}$$

$$L_z^{-1}L_zp = L_z^{-1}k^2Np - L_z^{-1}L_xp - L_z^{-1}L_yp.$$

But

$$L_x^{-1}L_xp = p - p(0,y,z) - x\,\frac{\partial p}{\partial x}\,(0,y,z)$$

$$L_y^{-1}L_yp = p - p(x,0,z) - y\,\frac{\partial p}{\partial y}\,(x,0,z) \tag{4}$$

$$L_z^{-1}L_zp = p - p(x,y,0) - z\,\frac{\partial p}{\partial z}\,(x,y,0).$$

Consequently, summing (3) with the substitution (4), and dividing by three

$$\begin{aligned} p = \tfrac{1}{3}\,&[p(0,y,z) + x\,\frac{\partial p}{\partial x}\,(0,y,z) \\ &+ p(x,0,z) + y\,\frac{\partial p}{\partial y}\,(x,0,z) \\ &+ p(x,y,0) + z\,\frac{\partial p}{\partial z}\,(x,y,0)] \\ &+ \tfrac{1}{3}\,[(L_x^{-1}L_y + L_y^{-1}L_z) + (L_x^{-1}L_z + L_z^{-1}L_x) \\ &\quad + (L_y^{-1}L_z + L_z^{-1}L_y)]p \\ &+ \tfrac{1}{3}\,[L_x^{-1} + L_y^{-1} + L_z^{-1}]\,\sum A_n \end{aligned}$$

where $N(u)$ has been replaced by Adomian's A_n polynomials [1,2].
We take immediately the first term of our approximation

$$p_0 = \frac{1}{3}[p(0,y,z) + x\frac{\partial p}{\partial x}(0,y,z)$$
$$+ p(x,0,z) + y\frac{\partial p}{\partial y}(x,0,z)$$
$$+ p(x,y,0) + z\frac{\partial p}{\partial z}(x,y,0)]$$

and the following terms given by

$$p_1 = \frac{1}{3}[(L_x^{-1}L_y + L_y^{-1}L_z) + (L_x^{-1}L_z + L_z^{-1}L_x)$$
$$+ (L_y^{-1}L_z + L_z^{-1}L_y)]p_0$$
$$+ \frac{1}{3}[L_x^{-1} + L_y^{-1} + L_z^{-1}]A_0$$
$$\vdots$$
$$p_n = \frac{1}{3}[L_x^{-1}L_y + L_y^{-1}L_z) + (L_x^{-1}L_z + L_z^{-1}L_x)$$
$$+ (L_y^{-1}L_z + L_z^{-1}L_y)p_{n-1}$$
$$+ \frac{1}{3}[L_x^{-1} + L_y^{-1} + L_z^{-1}]A_{n-1}\,.$$

Now we must evaluate the A_n which has been discussed elsewhere [1,3]
Then

$$A_0 = \sinh p_0$$
$$A_1 = p_1 \cosh p_0$$
$$A_2 = p_2 \cosh p_0 + \frac{1}{2}p_1^2 \sinh p_0$$
$$A_3 = p_3 \cosh p_0 + p_1 p_2 \sinh p_0 + \frac{1}{6}p_1^3 \cosh p_0$$
$$A_4 = p_4 \cosh p_0 + [\frac{1}{2}p_2^2 + p_1 p_3] \sinh p_0$$
$$+ \frac{1}{2}p_1^2 p_2 \cosh p_0 + \frac{1}{24}p_1^4 \sinh p_0$$

etc.

The A_n are easily written down by the procedures given in the references for as many terms as desired. We now have the complete decomposition $p = \sum_{n=0}^{\infty} p_i$ and hence the solution.

As another example of partial differential equations consider

$$\frac{\partial u}{\partial t} = x^2 - \frac{1}{4}\left(\frac{\partial u}{\partial x}\right)^2 \qquad u(x,0) = 0.$$

(We note that since u_x is in the nonlinear term, we have only one linear operator so we do not get several equations and the procedure is the same as for an ordinary differential equation.)

Writing $L_t = \partial/\partial t$, we have

$$L_t u = x^2 - (1/4)(u_x)^2.$$

The inverse $L_t^{-1} = \int_0^t [\cdot]dt$ hence

$$L_t^{-1} L_t u = L_t^{-1} x^2 - (1/4)\, L_t^{-1}(u_x)^2 .$$

Since the left side is $u - u(0) = u$ we have

$$\sum_{n=0}^{\infty} u_n = u_0 - (1/4)\, L_t^{-1} \sum_{n=0}^{\infty} A_n$$

where we let $u = \sum_{n=0}^{\infty} u_n$, identify $u_0 = L_t^{-1} x^2 = x^2 t$, and replace the nonlinearity $(u_x)^2$ by the A_n polynomials. The A_n polynomials for u^2 are $A_0 = u_0^2$, $A_1 = 2u_0 u_1$, $A_2 = u_1^2 + 2u_0 u_2, \ldots$. See e.g. [1]. Consequently

$$u_1 = -(1/4)\, L_t^{-1}(u_{0_x})^2 = -(1/4)\, L_t^{-1}(4x^2t^2) = -x^2t^3/3$$

$$u_2 = -(1/4)\, L_t^{-1}(2u_{0_x} u_{1_x}) = (2/15)x^2t^5$$

$$u_3 = -(1/4)\, L_t^{-1}[u_{1_x}^2 + 2u_{0_x} u_{2_x}]$$

$$\vdots$$

so that

$$u = x^2(t - \frac{t^3}{3} + \frac{2}{15}t^5 - \cdots)$$

$$u = x^2 \tanh t \qquad |t| < \pi/2$$

which is easily verified not only for $u = x^2 \tanh t$ but for the series $\sum_{i=0}^{n-1} u_i$ for any n as discussed in [2].

Systems of Nonlinear Partial Differential Equations: Consider now a system of nonlinear partial differential equations given by:

$$u_t = uu_x + vu_y$$

$$v_t = uv_x + vv_y$$

$$u(x,y,0) = f(x,y)$$

$$v(x,y,0) = g(x,y).$$

We wish to investigate the solution by the decomposition technique. Let $L_t = \partial/\partial t$, $L_x = \partial/\partial x$, $L_y = \partial/\partial y$ and write the form:

$$L_t u = uL_x u + vL_y u$$

$$L_t v = uL_x v + vL_y v$$

where $L_t^{-1} = \int_0^t [\cdot]dt$; hence

$$u = u(x,y,0) + L_t^{-1}uL_x u + L_t^{-1}vL_y u$$

$$v = v(x,y,0) + L_t^{-1}uL_x v + L_t^{-1}vL_y v.$$

Let $u = \sum_{n=0}^{\infty} u_n$ and $v = \sum_{n=0}^{\infty} v_n$ and let

$$u_0 = u(x,y,0) = f(x,y)$$

$$v_0 = v(x,y,0) = g(x,y)$$

so that the first term of u and of v are known. We now have:

$$u = u_0 + L_t^{-1} uL_x u + L_t^{-1} vL_y u$$

$$v = v_0 + L_t^{-1} uL_x v + L_t^{-1} vL_y v.$$

We can use the A_n polynomials for the nonlinear terms, thus

$$u = u_0 + L_t^{-1} \sum_{n=0}^{\infty} A_n(uL_x u) + L_t^{-1} \sum_{n=0}^{\infty} A_n(vL_y u)$$

$$v = v_0 + L_t^{-1} \sum_{n=0}^{\infty} A_n(uL_x v) + L_t^{-1} \sum_{n=0}^{\infty} A_n(vL_y v).$$

(The notation $A_n(uL_x u)$ means the A_n generated for uu_x.)

$$A_0(uL_x u) = u_0 L_x u_0$$

$$A_1(uL_x u) = u_0 L_x u_1 + u_1 L_x u_0$$

$$A_2(uL_x u) = u_0 L_x u_2 + u_1 L_x u_1 + u_2 L_x u_0$$

$$\vdots$$

etc., for the other A_n. A simple rule here is the sum of the subscripts of each term is the same as the subscript of A. Consequently we can write:

$$u_1 = L_t^{-1} u_0 L_x u_0 + L_t^{-1} v_0 L_y u_0$$

$$v_1 = L_t^{-1} u_0 L_x v_0 + L_t^{-1} v_0 L_y v_0$$

which yields the next component of u and of v. Then

then u_1 , v_1 can be calculated as

$$u_1 = L_t^{-1}u_0L_xu_0 + L_t^{-1}v_0L_yu_0$$

$$= L_t^{-1}(x+y)L_x(x+y) + L_t^{-1}(x+y)L_y(x+y)$$

$$= xt + yt + xt + yt = 2xt + 2yt$$

$$v_1 = L_t^{-1}u_0L_xv_0 + L_t^{-1}v_0L_yv_0 = 2xt + 2yt$$

and u_2 , v_2 are calculated as

$$u_2 = L_t^{-1}[(x+y)L_x(2xt + 2yt)$$

$$+ (2xt + 2yt)L_x(x+y)]$$

$$u_2 = L_t^{-1}[u_0L_xu_1 + u_1L_xu_0] + L_t^{-1}[v_0L_yu_1 + v_1L_yu_0]$$

$$v_2 = L_t^{-1}[u_0L_xv_1 + u_1L_xv_0]$$

$$+ L_t^{-1}[v_0L_yv_1 + v_1L_yv_0]$$

$$u_3 = L_t^{-1}[u_0L_xu_2 + u_1L_xu_1 + u_2L_xu_0]$$

$$+ L_t^{-1}[v_0L_yu_2 + v_1L_yu_1 + v_2L_yu_0]$$

$$v_3 = L_t^{-1}[u_0L_xv_2 + u_1L_xv_1 + u_2L_xv_0]$$

$$+ L_t^{-1}[v_0L_yv_2 + v_1L_yv_1 + v_2L_yv_0]$$

etc., up to some u_n , v_n then we have the n term approximations $\sum_{i=0}^{n-1} u_i$ for u and $\sum_{i=0}^{n-1} v_i$ for v as our approximate solutions.

Since the solution can exhibit a shock phenomenon for finite t, we select f, g such that the shock occurs for a value of t far from our region of interest. Let $f(x,y) = g(x,y) = x + y$. Therefore,

$$u_0 = v_0 = x + y$$

$$+ L_t^{-1}[(x+y)L_y(2xt + 2yt)$$
$$+ (2xt + 2yt)L_y(x+y)]$$
$$= 4t^2(x+y)$$
$$v_2 = 4t^2(x+y).$$

Thus

$$u = (x+y) + 2t(x+y) + 4t^2(x+y) + \ldots$$

$$v = (x+y) + 2t(x+y) + 4t^2(x+y) + \ldots$$

which we can write also as

$$u(x,y) = (x+y)/(1-2t)$$
$$v(x,y) = (x+y)/(1-2t).$$

The inclusion of stochastic processes is dealt with in [1,2] and elsewhere. The appoximation ϕ_n becomes a stochastic series and no statistical independence problems [1] are encountered in obtaining first- and second-order statistics from the ϕ_n.

REFERENCES

1) G. Adomian, Stochastic Systems, Academic Press, 1983.

2) G. Adomian, Nonlinear Stochastic Operator Equations, Academic Press, 1986.

ON THE RATIO OF THE FIRST TWO EIGENVALUES OF SCHRÖDINGER OPERATORS WITH POSITIVE POTENTIALS

Mark S. Ashbaugh[1]
Department of Mathematics
University of Missouri
Columbia, MO 65211

Rafael Benguria[2]
Departamento de Física
Facultad de Ciencias Físicas y Matemáticas
Universidad de Chile
Casilla 487/3
Santiago, Chile

ABSTRACT. We survey current knowledge on the ratio, λ_2/λ_1, of the first two eigenvalues of the Schrödinger operator $H_V = -\Delta + V(x)$ on the region $\Omega \subset \mathbb{R}^n$ with Dirichlet boundary conditions and nonnegative potentials. We discuss the Payne-Pólya-Weinberger conjecture for $H_0 = -\Delta$ and generalize the conjecture to Schrödinger operators. Lastly, we present our recent result giving the best possible upper bound $\lambda_2/\lambda_1 \leq 4$ for one-dimensional Schrödinger operators with nonnegative potentials and discuss some extensions of this result.

1. Introduction

We consider the Schrödinger operator

$$H_V = -\Delta + V(x) \tag{1}$$

acting on $L^2(\Omega)$ where Ω is a bounded subset of $\mathbb{R}^n$, H_V has Dirichlet boundary conditions, and the potential V is nonnegative.

---

[1]Partially supported by grants from the Research Council of the Graduate School, University of Missouri-Columbia and the Programa de las Naciones Unidas para el Desarrollo (PNUD grant CHI-84-005)

[2]Partially supported by the Departamento de Investigación y Bibliotecas de la Universidad de Chile (Grant E-1959-8522)

Letting $\lambda_1 < \lambda_2 \leq \lambda_3 \leq \ldots$ be the eigenvalues of H_V we consider upper bounds on the ratio λ_2/λ_1 that hold for all operators H_V for fixed dimension n, i.e. the bounds should hold for all $V \geq 0$ and all bounded $\Omega \subset \mathbb{R}^n$, n fixed. Observe that $|\Omega|$ does not affect the ratio of any two eigenvalues since all eigenvalues scale identically with $|\Omega|$.

Problems of this type seem to have first been addressed by Payne, Pólya, and Weinberger [10, 11] in the mid-50's. In two dimensions with V = 0 (vibrating homogeneous membrane problem) they proved

$$\lambda_2/\lambda_1 \leq 3, \tag{2}$$

among other things. Also they conjectured that

$$\lambda_2/\lambda_1 \leq (\lambda_2/\lambda_1)|_{\Omega=\text{disk}} \approx 2.539. \tag{3}$$

Later, Brands [3] improved the bound (2) to $\lambda_2/\lambda_1 \leq 2.6861^+$ and subsequently de Vries [6] obtained the bound $\lambda_2/\lambda_1 < 2.6578$. A family of bounds containing that of Brands as a special case was later found by Hile and Protter [8].

Turning now to the general n-dimensional case, Thompson [13] proved the bound

$$\lambda_2/\lambda_1 \leq 1 + 4/n \tag{4}$$

for the operator $H_0 = -\Delta$ by a straightforward generalization of the technique of Payne-Pólya-Weinberger. He also gave the natural generalization of the P-P-W conjecture: he conjectured that

$$\lambda_2/\lambda_1 \leq (\lambda_2/\lambda_1)|_{\Omega=\text{n-dimensional ball}} \tag{5}$$

gives the optimal upper bound on λ_2/λ_1 in dimension n.

For purposes of comparison, Table 1 presents the bounds (4) and the conjectured best possible bounds (5) for low dimensions. These latter values are obtained as ratios of the squares of zeros of Bessel functions. In particular, they are $(j_{n/2,1}/j_{(n-2)/2,1})^2$ where $j_{\nu,m}$ denotes the mth zero of the Bessel function $J_\nu(x)$.

n	1+4/n	λ_2/λ_1 for a ball	$1 + 3/n = \lambda_2/\lambda_1$ for a cube
1	5	4	4
2	3	2.53874	2.5
3	$\frac{7}{3} \approx 2.33333$	2.04575	2
4	2	1.79639	1.75
5	1.8	1.64518	1.6
6	$\frac{5}{3} \approx 1.66667$	1.54340	1.5
7	$\frac{11}{7} \approx 1.57143$	1.47005	$\frac{10}{7} \approx 1.42857$

Table 1.

Extension of the investigations presented above to Schrödinger operators with nonnegative potentials was carried out only relatively recently by Harrell [7] and by Singer, Wong, Yau, and Yau [12]. These authors proved that the bound (4) applies also in the case of a Schrödinger operator with nonnegative potential and Dirichlet boundary conditions. In [2], Benguria gave a simple proof of the one-dimensional result, $\lambda_2/\lambda_1 \leq 5$, using a commutation formula [4, 5, 9]. We use a similar approach below in proving the optimal upper bound, $\lambda_2/\lambda_1 \leq 4$, in one dimension. More generally, we conjecture that the bound (5) holds for Schrödinger operators with nonnegative potentials and Dirichlet boundary conditions.

2. The Optimal Result in One Dimension

Theorem (Ashbaugh-Benguria [1]). Let $H_V = -d^2/dx^2 + V(x)$ be an operator on $L^2(\Omega)$ with Dirichlet boundary conditions where $\Omega = [a,b] \subset \mathbb{R}$, $V \in L^1(a,b)$, and $V \geq 0$. Then the first two eigenvalues of H_V obey the inequality $\lambda_2/\lambda_1 \leq 4$ and equality occurs if and only if $V = 0$.

First proof (by commutation [1]). Let u_1 be the groundstate wavefunction of H, i.e. $Hu_1 = \lambda_1 u_1$. By commutation (see [4, 5, 9]), it is known that the operator $\tilde{H} = H - 2(u_1'/u_1)'$ has eigenvalues $\{\lambda_2, \lambda_3, \lambda_4, \ldots\}$, i.e. the spectrum of $\tilde{H}$ is the same as that of H except for λ_1. Since λ_2 is the lowest eigenvalue of $\tilde{H}$ we may estimate it using the Rayleigh-Ritz inequality. Using u_1^2 as trial function we compute $\tilde{H}u_1^2 = 4\lambda_1 u_1^2 - 3Vu_1^2$ yielding

$$\lambda_2 \leq \frac{(u_1^2, \tilde{H}u_1^2)}{(u_1^2,\ u_1^2)} = 4\lambda_1 - \frac{3\int_a^b Vu_1^4 dx}{\int_a^b u_1^4 dx} \tag{6}$$

Thus $\lambda_2/\lambda_1 \leq 4$ with equality if and only if $V = 0$ a.e.

Second proof (comparison equations and integration by parts). Set $f = u_1^2$ and $g = (u_1'u_2 - u_1u_2')/u_1$ where u_2 is an eigenfunction of H corresponding to the eigenvalue λ_2. It is readily verified that

$$f'' = [2(u_1'/u_1)^2 + 2V - 2\lambda_1]f \equiv W_1 f \tag{7}$$

and that

$$g'' = [2(u_1'/u_1)^2 - V + 2\lambda_1 - \lambda_2]g \equiv W_2 g. \tag{8}$$

Now

$$W_2 - W_1 = 4\lambda_1 - \lambda_2 - 3V \tag{9}$$

and we can compute

$$\int_a^b (W_2 - W_1)u_1^4 dx = \int_a^b [g''/g - f''/f]u_1^4 dx$$

$$= \int_a^b (fg' - f'g)'\, f/g\, dx$$

$$= \int_a^b (fg' - f'g)^2/g^2\, dx$$

$$\geq 0 \tag{10}$$

where the third line follows by integration by parts. From equations (9) and (10) equation (6) follows and the proof concludes as before. To clear up some technicalities it should be observed that f and g are nonzero on (a,b) and that both approach 0 quadratically at a and b. The boundary terms which arise in the integration by parts therefore vanish. ∎

<u>Remarks</u>. 1. We used u_1^2 as trial and comparison functions because when $V = 0$ u_1^2 is exactly the groundstate wavefunction of $\tilde{H}_0$. Our second proof, in effect, uses commutation in disguise; the function g is precisely the groundstate wavefunction of $\tilde{H}_V$.

2. The ease with which we obtained the optimal bound by using u_1^2 as a trial function for estimating the lowest eigenvalue of $\tilde{H}$ suggests that a similar approach might work for estimating λ_2 using the Rayleigh-Ritz inequality for H directly. Taking $u = u_1 u_1'$ as trial function (observe that this is orthogonal to u_1 and reduces to u_2 when $V = 0$) one obtains

$$\lambda_2 \leq 4\lambda_1 + \frac{\int_a^b V(V - \lambda_1)u_1^4 dx}{\int_a^b u_1^2(u_1')^2 dx} \tag{11}$$

Unfortunately, although certain special cases can be dealt with in this way, we were unable to recover the optimal bound for arbitrary potentials $V \geq 0$. To complete the proof in this way, of course, one needs to prove the inequality

$$\int_a^b V(V - \lambda_1)u_1^4 dx \leq 0. \tag{12}$$

We leave this as an open problem.

3. The result $\lambda_2/\lambda_1 \leq 4$ also holds for problems on the line and half-line if $V(x) \to \infty$ as $x \to \pm\infty$. This follows by a norm resolvent convergence argument which was pointed out to us by Barry Simon at this conference.

4. The fact that $V = 0$ is the only optimizer in $L^1(a,b)$ actually obscures certain complications that enter into the problem of optimizing λ_2/λ_1. In fact, there are all sorts of maximizing sequences that do not approach $V = 0$ in any sense. These sequences go to potentials like

$$V(x) = \begin{cases} 0 \text{ on } [c,d] \\ \infty \text{ on } [a,b]\setminus[c,d] \end{cases} \tag{13}$$

or more complicated potentials (also outside L^1). This remark plays a larger role in higer dimensions, our next topic of discussion.

3. Optimal Upper Bounds on λ_2/λ_1 in Higher Dimensions

By Remark 4 above and the fact that $|\Omega|$ has no effect on λ_2/λ_1 one sees that in higher dimensions λ_2/λ_1 will be optimized for regions Ω of a particular optimal shape (possibly several shapes). This is because we can always use the $V \to \infty$ trick to "recut" Ω into a new shape; i.e. we send V to ∞ off the set we want to keep. Based on this argument and our knowledge of the one-dimensional case we conjecture

Generalized PPW Conjecture. For the Schrödinger operator $H = -\Delta + V(x)$ acting on $L^2(\Omega)$ in n dimensions with Dirichlet boundary conditions and nonnegative potential V, the ratio λ_2/λ_1 of the first two eigenvalues is less than or equal to the corresponding value for a ball in $\mathbb{R}^n$ with $V = 0$. Furthermore, the cases of equality are exactly those for which Ω is a ball and $V = 0$.

This result would follow from showing that λ_2/λ_1 for $-\Delta + V$ on a ball is maximized when $V = 0$.

4. Conjectures and (Non-optimal) Results Concerning Other Ratios in the One-Dimensional Case

For notational convenience we set $\mu_n = \lambda_n/n^2$ in this section. This corresponds to normalizing the eigenvalues with respect to those of the potential $V = 0$.

Conjecture 1. $\mu_n \le \mu_1$ for all n, i.e. $\lambda_n/\lambda_1 \le n^2$.

The best results known in this direction at present are

(i) $\lambda_3/\lambda_1 \le 9$ for V such that $0 \le V \le \lambda_1$ (Ashbaugh-Benguria) an

(ii) $\lambda_3/\lambda_1 < 14$ for $V \ge 0$ (Harrell [7] together with the bound $\lambda_2/\lambda_1 \le 4$).

Conjecture 2. $\mu_{n+1} \le \mu_n$ for all n when V is convex.

This conjecture could not be true for general $V \ge 0$ since a double well potential (see Figure 1) will have nearly degenerate pairs (λ_1, λ_2) and (λ_3, λ_4) and as the barrier is made higher the ratio

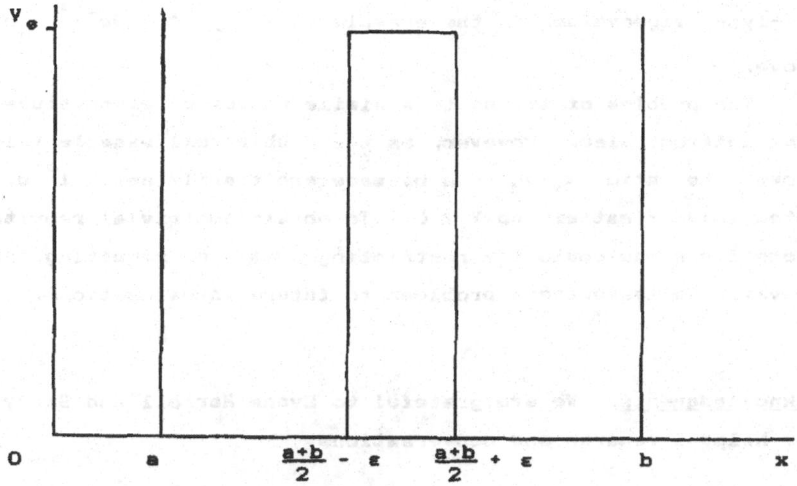

Figure 1. A double well potential.

λ_3/λ_2 will approach the value 4 (= λ_2/λ_1 for a single infinite square-well). Similar arguments apply to ratios other than λ_3/λ_2. This suggests the further conjecture

Conjecture 3. $\lambda_{n+1} \leq 4\lambda_n$ for all $V \geq 0$.

The best positive result of this nature is the bound $\lambda_3/\lambda_2 < 5$ for all $V \geq 0$ which follows from Harrell's work (and the bound $\lambda_2 > \lambda_1$).

Lastly, we give a conjectured inequality between sets of μ_i's taken four at a time.

Conjecture 4. For convex V, $\mu_i + \mu_{i+3} \geq \mu_{i+1} + \mu_{i+2}$ for $i = 1, 2, 3, \ldots$.

Again, this conjecture cannot hold for all $V \geq 0$ as is shown by considering the case where $i = 2$ for the potential $V(x) = \frac{1}{2}x^2 + 9e^{-x^2} - c$ where c is chosen so that the absolute minimum of $V(x)$ for $x \in \mathbb{R}$ is 0. Here we are taking $\Omega = \mathbb{R}$ which is relevant due to a convergence argument like that mentioned in Remark 3 of Section 2.

We have verified that all our conjectures hold as appropriate for the harmonic oscillator potential $V(x) = x^2$ as well as for low-lying eigenvalues in the example $V(x) = \frac{1}{2}x^2 + 9e^{-x^2} - c$ noted above.

The problem of trying to minimize ratios of eigenvalues is of some interest also. However, as the double well example (Figure 1) shows, the ratio λ_2/λ_1 can be made arbitrarily near 1 by potentials V satisfying $V \geq 0$. To obtain nontrivial results along these lines one could try restricting $\|V\|_p$ or requiring that V be convex. We leave these problems to future investigations.

Acknowledgments. We are grateful to Evans Harrell and Barry Simon for helpful remarks and conversations.

Note added (11/18/86). We have recently proved our Conjecture 3 above. In fact, one can prove the optimal bound $\lambda_{2n}/\lambda_n \leq 4$ and hence $\lambda_m/\lambda_n < 4$ for all m satisfying $n < m < 2n$ (and, in particular, for $m = n+1$ as in Conjecture 3.) That these bounds are best possible for $V \geq 0$ is shown by multiple-well examples.

REFERENCES

1. Ashbaugh, M.S., and R. Benguria, Best Constant for the Ratio of the First Two Eigenvalues of One-Dimensional Schrödinger Operators with Positive Potentials, to appear in Proc. Amer. Math. Soc.

2. Benguria, R., A note on the gap between the first two eigenvalues for the Schrödinger operator, J. Phys. A: Math. Gen. 19 (1986) 477-478.

3. Brands, J.J.A.M., Bounds for the Ratios of the First Three Membrane Eigenvalues, Arch. Rat. Mech. Anal. 16 (1964) 265-268.

4. Crum, M.M., Associated Sturm-Liouville systems, Quart. J. Math. Oxford (2) 6 (1955) 121-127.

5. Deift, P.A., Applications of a commutation formula, Duke J. Math. 45 (1978) 267-310.

6. DeVries, H.L., On the Upper Bound for the Ratio of the First Two Membrane Eigenvalues, Zeitschrift für Naturforschung 22A (1967) 152-153.

7. Harrell, E.M., unpublished (1982).

8. Hile, G.H., and M.H. Protter, Inequalities for Eigenvalues of the Laplacian, Indiana Math. J. 29 (1980) 523-538.

9. Marchenko, V.A., The construction of the potential energy from the phases of the scattered waves, Dokl. Akad. Nauk. SSSR 104 (1955) 695-698 [Math. Rev. 17 (1956) 740].

10. Payne, L.E., G. Pólya, and H.F. Weinberger, Sur le quotient de deux fréquences propres consécutives, Comptes Rendus Acad. Sci. Paris 241 (1955) 917-919.

11. Payne, L.E., G. Pólya, and H.F. Weinberger, On the Ratio of Consecutive Eigenvalues, J. Math. and Phys. 35 (1956) 289-298.

12. Singer, I.M., Wong, B., S.-T. Yau, and S.S.-T. Yau, An estimate of the gap of the first two eigenvalues in the Schrödinger operator, Ann. Scuola Norm. Sup. Pisa (series 4) 12 (1985) 319-333.

13. Thompson, C.J., On the Ratio of Consecutive Eigenvalues in N-Dimensions, Stud. Appl. Math. 48 (1969) 281-283.

INVERSE SCATTERING FOR SELF-ADJOINT N^{TH} ORDER DIFFERENTIAL OPERATORS ON THE LINE

R. Beals[1]
Mathematics Department
Yale University
New Haven, CT 06520

Percy Deift[2,3]
Mathematics Department
NYU-Courant
New York, NY 10012

Carlos Tomei[4]
Pontificia Univ. Catholica
Rio de Janiero (22453), Brazil

Recently the authors have developed a complete inverse scattering theory for generic, self-adjoint, n^{th} order differential operators on the line. The theory is closely related to the earlier work of Beals and Coifman [2] and Beals [1]. In this note we will describe the relation of the recent work to [2] and [1], and discuss some of the more salient features of the theory. Details will appear elsewhere.

The note is in 4 parts:

(A) The second order problem -- recast in a form that generalizes to n^{th} order operators;

(B) Inverse scattering and non-linear wave equations;

(C) Historical remarks on the inverse problem -- one man's view;

(D) The basic inverse theorem.

(A) The second order problem

Let $\mathbb{C}_\pm$ denote $\{\text{Im } z \gtrless 0\}$ respectively, and let $\mathbb{R}_\pm$ denote $\{z \gtrless 0\}$ respectively. For $q \in S(\mathbb{R})$, q real, and $z \in \mathbb{C}_+$, let $g(x,z)$, $f(x,z)$ be

---

[1]Research supported in part under NSF Grant MCS 8104234.

[2]Research supported in part under NSF Grant MCS 8301662 and ONR Grant N00014-76-C-0439.

[3]Talk given by Percy Deift at the International Conference on Differential Equations and Mathematical Physics at UAB, March, 1986.

[4]Research supported in part by CNPq, Brazil.

the Jost solutions

$$L_2 g \equiv -g'' + qg = z^2 g$$

$$g(x,z) \sim e^{-izx} \quad \text{as} \quad x \to -\infty$$

$$L_2 f \equiv -g'' + qf = z^2 f$$

$$f(x,z) \sim e^{izx} \quad \text{as} \quad x \to +\infty.$$

For fixed x, $g(x,z)$ and $f(x,z)$ are analytic functions of z in $\mathbb{C}_+$ and have continuous extensions to $\overline{\mathbb{C}}_+$.

The following results are standard (see e.g. [3]):

There exists a meromorphic function $T(z)$, the <u>transmission</u> coefficient, with simple poles at a finite set $\{z_1,\cdots,z_n\} \subset i\mathbb{R}_+$, such that

$$g(x) \sim \frac{1}{T(z)} e^{-izx} \quad \text{as} \quad x \to +\infty$$

and $$f(x) \sim \frac{1}{T(z)} e^{izx} \quad \text{as} \quad x \to -\infty$$

for $z \in \mathbb{C}_+\backslash\{z_1,\cdots,z_n\}$. The points $\lambda_j = -z_j^2 \in \mathbb{R}_-$ are the $L^2(\mathbb{R})$ eigenvalues of L_2.

The transmission coefficient $T(z)$ extends to a continuous function on $\overline{\mathbb{C}}_+\backslash\{z_1,\cdots,z_n\}$ and for $z \in \mathbb{R}$ there exist two functions $R_1(z), R_2(z) \in S(\mathbb{R})$, the <u>reflection</u> coefficients, with the properties

$$\text{(1)} \qquad \begin{aligned} T(z)f(x,z) &= R_2(z)g(x,z) + g(x,-z) \\ T(z)g(x,z) &= R_1(z)f(x,z) + f(x,-z) \end{aligned} \quad , \quad z \in \mathbb{R}.$$

The functions R_1 and T satisfy the relations

$$\text{(2)} \qquad \begin{aligned} T(z)T(-z) + R_2(z)R_2(-z) &= 1 \\ T(z)R_1(-z) + T(-z)R_2(z) &= 0 \end{aligned} \quad , \quad z \in \mathbb{R}.$$

Set $\psi(x,z) = \left[\psi_1(x,z), \psi_2(x,z)\right] \equiv \left[g(x,z), T(z)f(x,z)\right]$, $z \in \mathbb{C}_+$

$\psi(x,z) \equiv \psi(x,-z)$, $z \in \mathbb{C}_-$.

The function $\psi(x,z)$ has boundary values

$$\psi_\pm(x,z) \equiv \lim_{\epsilon \downarrow 0} \psi(x, z \pm i\epsilon) \quad , \quad z \in \mathbb{R}_+$$

$$\equiv \lim_{\epsilon \downarrow 0} \psi(x, z \mp i\epsilon) \quad , \quad z \in \mathbb{R}_-.$$

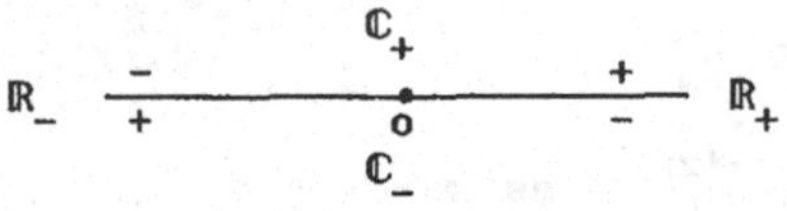

Figure (i)

Simple computations show that (1) and (2) may be re-expressed as

(3) $\psi_+(x,z)\pi = \psi_-(x,z)v(z), \quad z \in \mathbb{R}\backslash o$

where $v(z) = \begin{bmatrix} 1 - R_2(z)R_2(-z) & -R_2(-z) \\ R_2(z) & 1 \end{bmatrix}, \quad z \in \mathbb{R}_+$

$v(z) = v(-z) \quad , \quad z \in \mathbb{R}_-$

and $\pi = \begin{bmatrix} 0 & 1 \\ 1 & 0 \end{bmatrix}.$

For each x, the function $\psi(x,\cdot)$ is meromorphic in $\mathbb{C}^+$ with simple poles at the points of $Z \equiv \{z_1,\cdots,z_N,z_{N+1},\cdots,z_{2N}\}$, $z_{N+i} = -z_i$, $1 \leq i \leq N$, and at $z_i \in Z$, $g(x,z_i)$ and $f(x,z_j)$ are proportional. This may be re-expressed as

(4) $\operatorname{Res}_{z_j} \psi(x,\cdot) = \lim_{z\to z_j} \psi(x,z)v(z_i)$

where $v(z_i) = \begin{bmatrix} 0 & c(z_i) \\ 0 & 0 \end{bmatrix} = c(z_j)e_{12}$, for some constant $c(z_i)$.

The pair $S = (Z,v)$ where v maps $(\mathbb{R}\backslash 0) \cup Z \to M_2((\mathbb{C})$, is called the <u>scattering data</u> for L_2. To make the dependence on L_2 explicit we write $S = S(L_2) = \left[Z(L_2),v(L_2)\right]$. The map $F_2: L_2 \to S(L_2)$ is called the <u>scattering map</u>.

Let $\alpha_1(z)$, $\alpha_2(z)$ denote the square roots of unity, ± 1, ordered according to

$$\operatorname{Re}\left[i\alpha_1(z)z\right] > \operatorname{Re}\left[i\alpha_2(z)z\right] \quad , \quad z \in \mathbb{C}_+ \text{ and } z \in \mathbb{C}_-.$$

Thus $\left[\alpha_1(z),\alpha_2(z)\right] = (-1,1)$ if $\operatorname{Im} z > 0$,

$= (1,-1)$ if $\operatorname{Im} z < 0$.

Set $J(z) = \begin{bmatrix} \alpha_1(z) & 0 \\ 0 & \alpha_2(z) \end{bmatrix}$ for $z \in \mathbb{C}\backslash\mathbb{R}$, and let $J_\pm(z)$ be interpreted according to Figure (i) for $z \in \mathbb{R}\backslash 0$. Finally let $R = \begin{bmatrix} 0 & 1 \\ 1 & 0 \end{bmatrix} = \pi.$

<u>**Definition 1**</u>

$G_{sa}(2)$, the set of generic, self-adjoint, second order scattering data, consists of pairs

$$S = (Z, v)$$

where

(5) (a) Z is a finite subset of $i\mathbb{R}\backslash 0$, invariant under complex conjugation (and hence under multiplication by $\alpha = -1$).

(b) $v\colon (\mathbb{R}\backslash 0) \cup Z \to M_2(\mathbb{C})$ and the restrictions of v to $\mathbb{R}_\pm$ are in $I + \left[\mathcal{S}(\mathbb{R}_\pm)\right]^4$ respectively, and extend continuously with all their derivatives to $z = 0_\pm$.

(6)
$$v(-z_j) = -v(z_i), \quad z_j \in Z$$
$$v(-z) = v(z), \quad z \in \mathbb{R}\backslash 0,$$

(7)
$$v(\bar{z}) = -RJ(z_j)^* v(z_j)^* J(z_i)R, \qquad z_j \in Z$$
$$v(z) = J_-(z)^* v(z)^* J_-(z), \qquad z \in \mathbb{R}\backslash 0.$$

(8) For $z \in \mathbb{R}_\pm$, $v(z)$ is a 2×2 matrix of determinant 2 of form

$$\begin{bmatrix} 1 + a(z)b(z) & a(z) \\ b(z) & 1 \end{bmatrix}$$

where

(a) $1 + a(z)b(z) \neq 0 \qquad$ for $\qquad z \in \mathbb{R}_\pm$

(b) $$\begin{bmatrix} 1 + a(z)b(z) & a(z) \\ b(z) & 1 \end{bmatrix} = \begin{bmatrix} 0 & 1 \\ -1 & 1 \end{bmatrix} + \begin{bmatrix} \gamma_\pm z^2 + O(|z|^3) & O(|z|) \\ O(|z|) & 0 \end{bmatrix}$$

as $z \to 0$, $z \in \mathbb{R}_\pm$, where $\gamma_\pm \neq 0$.

(9) For each $z_i \in Z$ there exists a non-zero constant $c = c(z_i)$ such that

$$v(z_j) = c(z_j)e_{12}.$$

Moreover

$$i\left[\operatorname{sgn}(\operatorname{Im} z_j)\right]c(z_j) < 0.$$

(10)
$$\hat{v}_+\pi\hat{v}_-\pi = \hat{v}_-\pi\hat{v}_+\pi \equiv I$$

where $\hat{v}_\pm$ are the formal power series associated with v on $\mathbb{R}_\pm$ at $z = 0$, respectively.

Standard results show that for generic self-adjoint operators L_2 the pair $S(L_2) = \left[Z(L_2), v(L_2)\right] \in G_{sa}(2)$. For example the second identity in (7) is nothing more than the well known self-adjointness condition $\overline{R_2(z)} = R_2(-z)$.

$z \in \mathbb{R}$; similarly the asymptotic condition as $z \to 0$ in 8(b) is the familiar generic condition, $R_2(0) = -1$, $T(0) = 0$. Condition (10) is the preferred way of encoding the fact that $R_2(Z)$, defined a priori on $\mathbb{R}\backslash 0$, extends across $z = 0$ to a smooth function on $\mathbb{R}$.

The basic result of inverse scattering theory for generic, second order, self-adjoint operators on the line is

Theorem 1.

The map $F_2: L_2 \to S(L_2)$ is a bijection from the set of generic second order self-adjoint operators on $\mathbb{R}$ onto $G_{sa}(2)$.

Remark:

In the second order case there is no need to require genericity $\left[R_2(0) = -1, T(0) = 0\right]$; for the full result see, for example, [3].

(B) Inverse scattering and non-linear wave equations

The great discovery of Gardner, Greene, Kruskal and Miura [5] was that the non-linear Korteweg de Vries (kdV) equation of shallow water theory

$$q_t = 6qq_x - q_{xxx}, \quad -\infty < x < \infty, \quad t \geq 0$$
$$q(x,t)\big|_{t=0} = q_0(x)$$

is linearized via the scattering map F_2 of Section (A). More precisely, if we set $L_2(t) = -d^2/dx^2 + q(\cdot,t)$, then $S\left[L_2(t)\right]$ solves the linear equations

$$\frac{d}{dt} Z\left[L_2(t)\right] = 0$$

$$\frac{d}{dt} v\left[L_2(t), z\right] = 4[v, (izJ\_(z))^3], \qquad z \in \mathbb{R}\backslash 0$$

$$\frac{d}{dt} v\left[L_2(t), z_j\right] = 4[v, (iz_jJ\_(z_j))^3], \qquad z_j \in Z(L_2(t)).$$

These computations are most conveniently performed by writing KdV in Lax pair form [6] as an isospectral deformation

$$\frac{dL_2}{dt} = [B(L_2), L_2]$$

where $B(L_2) = -4\,d^3/dx^3 + 3\left[q(\cdot,t)\frac{d}{dx} + \frac{d}{dx}\,q(\cdot,t)\right] = -B(L_2)^*$. The linear equations integrate to

$$S\left[L_2(t)\right] = \begin{cases} Z\left[L_2(t)\right] = Z\left[L_2(t{=}0)\right] \\ v\left[L_2(t),z\right] = e^{-4(izJ_-(z))^3t}\,v\left[L_2(t{=}0),z\right]e^{4(izJ_-(z))^3t}, & z\in\mathbb{R}\setminus 0 \\ v\left[L_2(t),z_j\right] = e^{-4(iz_jJ_-(z_j))^3t}\,v\left[L_2(t{=}0),z_j\right]e^{4(iz_jJ_-(z_j))^3t}, & z\in\mathbb{R}\setminus 0. \end{cases}$$

In particular for $z > 0$ and $\operatorname{Im} z_j > 0$, say, we find

$$R_2(z,t) = R_2(z,t{=}0)e^{-8iz^3t}$$
$$c(z_i,t) = c(z_i,t{=}0)e^{-8iz_j^3t}.$$

<u>Most importantly, one easily verifies that if $S\left[L_2(t{=}0)\right] \in G_{sa}(2)$, then so does $S\left[(L_2(t)\right]$</u>. This means that the Cauchy problem can be solved by the following diagram and Theorem 1.

$$\begin{array}{ccc} q(x,t{=}0) & \xrightarrow{F_2} & S\left[L_2(t{=}0)\right] \\ & & \big\downarrow \\ q(x,t) & \xleftarrow{F_2^{-1}} & S\left[L_2(t)\right]. \end{array} \qquad \text{diagram (i)}$$

For each n, there is a natural class of flows (the Gel'fand Dikii flows) which give rise to isospectral deformations of linear n^{th} order differential operators

$$(11) \qquad L_n = D^n + \sum_{j=0}^{n-2} p_j(x)D^j, \qquad D = \frac{1}{i}\frac{d}{dx},$$

as in the case $n = 2$ above. For example, $n = 3$ gives rise to the Boussinesq equation, which also arises in shallow water theory,

$$q_{tt} = q_{xxxx} + (q^2)_{xx}$$
$$q(x,t{=}0) = q_0(x), \quad q_t(x,t{=}0) = q_0(x),$$

with associated linear operator

$$L_3 = D^3 + q(x)D + Dq(x) + p(x), \qquad q(x) \text{ and } p(x) \text{ real}.$$

The idea is to use inverse scattering theory for n^{th} order operators to solve these non-linear flows as indicated in the diagram for the case $n = 2$ above.

(C) Historical remarks

Theorem 1 is due to Faddeev and appears in his Ph.D. Thesis (1958) in a slightly different form.

The case $n = 3$ was analyzed in [4] in 1980. They obtained a partial solution to the inverse problem and used their results to prove a global stable/unstable manifold theorem for the Boussinesq equation in the ambient space of 3rd order self-adjoint operators L_3

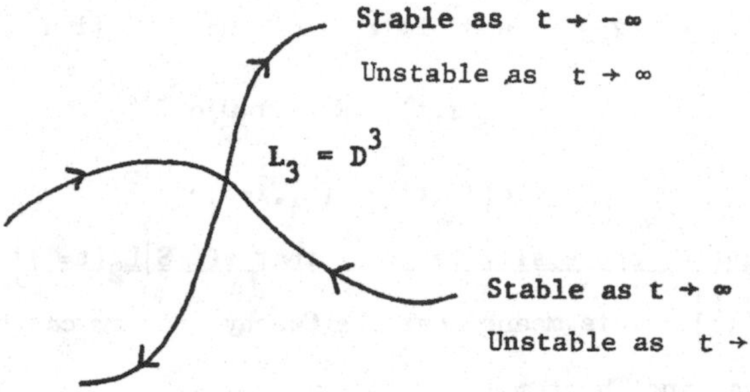

Figure (ii)

In 1982 Beals and Coifman [2] introduced a new approach to the inverse problem for first order $n \times n$ systems of the form

$$D\varphi = zJ\varphi + Q(x)\varphi \tag{12}$$

where J is a constant diagonal matrix. In 1982 Beals extended this approach to n^{th} order operators of the form (11) in [1].

In a similar way that the non-linear Gel'fand Dikki flows give rise to isospectral deformations of nth order differential operators L_n, other interesting non-linear flows such as the non-linear Schrödinger equation and the sine-Gordon equation, for example, give rise to "isospectral" deformations of systems of the form (12). In attempting to use the inverse theory of [2] and [1] to solve these flows in the manner of diagram (i), a curious problem arises: for technical reasons, the scattering map for general n^{th} order equations or $n \times n$ first order systems, F_n, is known to be invertible only on an (a priori) proper subset $\tilde{G}(n)$ of $G(n)$, the set of generic scattering data. Moreover $\tilde{G}(n)$ is not invariant under the associated non-linear flows for all time! This means that $F_n^{-1}\left[S(L_n(t))\right]$, say, need not exist and this means in turn that the inverse method as a general technique for solving non-linear equations breaks down (in special cases, however, for example when there is a priori control on the $L^2(dx)$ norm of the coefficients of $L_n(t)$, say, the method goes through).

As mentioned at the beginning of this note, the authors have developed a complete inverse theory for generic, self-adjoint n^{th} order differential operators. A principal consequence of this work is that F_n is a bijection from the set of generic n^{th} order self-adjoint operators onto $G_{sa}(n)$, the set of generic n^{th} order, self-adjoint scattering data: moreover, $G_{sa}(n)$ is invariant under the Gel'fand Dikii flows, $t \to S\left[L_n(t)\right]$. Thus the nth order analogue of diagram (1) can be completed, which solves the flows. At the technical level [2] and [1] pose the inverse problem as a Riemann-Hilbert factorization problem (see (19) below), which they then reduce to the solution of a linear Fredholm equation of index zero. To solve the inverse problem one must show that the kernel of the equation is empty. General techniques show this is so on a generic set $\tilde{G}(n) \subset G(n)$ (see above). In the self-adjoint case we show directly that the kernel of the equation is empty for <u>all</u> $S \in G_{sa}(n)$, by using a <u>vanishing theorem</u> of Liouville type. Precursors of this theorem were introduced in [3] and [4] for the cases $n = 2$ and $n = 3$ respectively.

(D) <u>The basic inverse theorem</u>

In this the final section we describe the basic inverse theorem and make some remarks on the hypotheses and the method of solution. For convenience, we only consider the odd order case, $n = 2\ell + 1$; the even order case $n = 2\ell$, $\ell \geq 2$, is similar but slightly more involved. The reader should refer to Section (A) to gain motivation for some of the definitions that follow.

Let $\alpha = e^{2\pi i/n}$, $\alpha^n = 1$.

$$\sum = \{z \in \mathbb{C}: \ \mathrm{Re}\left[i(\alpha^j - \alpha^k)z\right] = 0 \ \text{ for some } \ j \neq k\}$$

$$\sum\backslash 0 = \bigcup_{k=0}^{2n-1} \Sigma_k, \quad \text{where } \Sigma_0 = -i\mathbb{R}_+ \text{ and the } \Sigma_k\text{'s are numbered}$$

consecutively in the counterclockwise direction.

e.g. $n = 5$

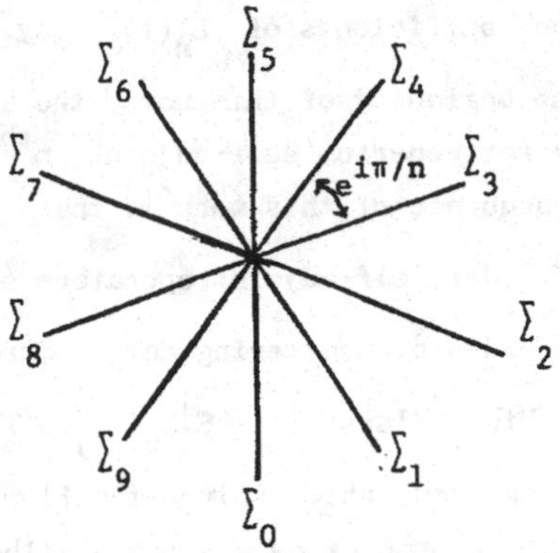

For $z \in \mathbb{C}\backslash\Sigma$ let α_i denote the ordering of the nth roots of unity given by

$$\mathrm{Re}(i\alpha_1 z) > \mathrm{Re}(i\alpha_2 z) > \cdots > \mathrm{Re}(i\alpha_n z).$$

For such z, set $J(z) \equiv \mathrm{diag}(\alpha_1,\cdots,\alpha_n)$ and for $z \in \Sigma$, let $J_\pm(z)$ denote the limit of $J(z')$ as z' approaches z in the clockwise (resp., anticlockwise) direction

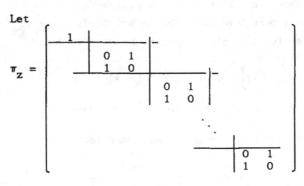

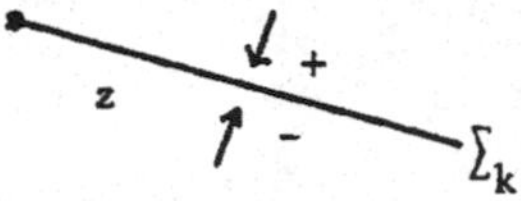

Figure (iii)

Let

$$\pi_z = \begin{pmatrix} 1 & & & & \\ & \begin{matrix} 0 & 1 \\ 1 & 0 \end{matrix} & & & \\ & & \begin{matrix} 0 & 1 \\ 1 & 0 \end{matrix} & & \\ & & & \ddots & \\ & & & & \begin{matrix} 0 & 1 \\ 1 & 0 \end{matrix} \end{pmatrix} \qquad \text{for } z \in \Sigma_{2k}$$

and

$$\pi_z = \begin{bmatrix} \begin{matrix} 0 & 1 \\ 1 & 0 \end{matrix} & & & & \\ & \begin{matrix} 0 & 1 \\ 1 & 0 \end{matrix} & & & \\ & & \ddots & & \\ & & & \begin{matrix} 0 & 1 \\ 1 & 0 \end{matrix} & \\ & & & & 1 \end{bmatrix} \qquad \text{for } z \in \Sigma_{2k+1}.$$

For $z \in \Sigma \backslash 0$, we say $j \sim j+1$ if j and $j+1$ belong to the same 2×2 block, e.g. $2 \sim 3$ in the case $z \in \Sigma_{2k}$ above.

Finally let $R = \begin{bmatrix} 0 & & & 1 \\ & & 1 & \\ & \cdot^{\cdot^{\cdot}} & & \\ 1 & & & 0 \end{bmatrix}$ and let Σ_B denote the bisectors of the 2n sectors $\mathbb{C} \backslash \Sigma$

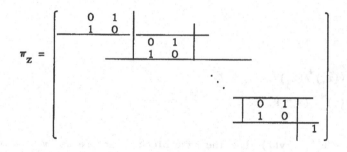

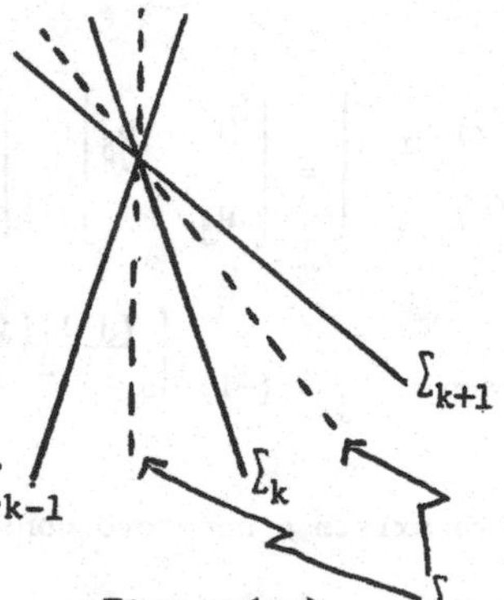

Figure (iv)

Definition 2

$G_{sa}(n)$, $n = 2\ell + 1$, the set of generic, odd order, self-adjoint scattering data consists of pairs

$$S = (Z, v)$$

where

(13)(a) Z is a finite subset of $\mathbb{C} \backslash (\Sigma \cup \Sigma_B)$, invariant under complex conjugation and multiplication by α.

(b) $v: (\Sigma \backslash 0) \cup Z \to M_n(\mathbb{C})$ and the restriction of v to Σ_k is in $I + \left[\mathcal{S}(\Sigma_k)\right]^{n^2}$ and extends continuously with all its derivatives to the origin for each k.

(14) $v(\alpha z_j) = \alpha v(z_j), \quad z_j \in Z$

$v(\alpha z) = v(z), \qquad z \in \Sigma\backslash o$

(15) $v(\overline{z}_j) = -R\ J(z_j)^* v(z_j)^* A(z_j) R, \quad z_j \in Z$

$v(\overline{z}) = R\Pi_z\ J_-^*(z) v(z)^* J_-(z) \Pi_z R, \quad z \in \Sigma\backslash 0.$

(16) For each k and $z \in \Sigma_k$, $v(z)$ has the same block structure as π_z. The one-by-one blocks consist of the number 1. The two-by-two blocks

$$\begin{bmatrix} v(z)_{j-1,j-1} & v(z)_{j-1,j} \\ v(z)_{j,j-1} & v(z)_{j,j} \end{bmatrix}, \quad z \in \Sigma_k$$

have the properties

(a) the determinant is 1

(b) $v(z)_{j,j} = 1$

(c) $v(z)_{j-1,j-1} \neq 0$

(d) $$\begin{bmatrix} v(z)_{j-1,j-1} & v(z)_{j-1,j} \\ v(z)_{j,j-1} & v(z)_{j,j} \end{bmatrix} = \begin{bmatrix} 0 & p_j \\ -p_j^{-1} & 1 \end{bmatrix} + \begin{bmatrix} \gamma_{kj} z^2 + O(|z|^3) & O(|z|) \\ O(|z|) & 0 \end{bmatrix}$$

as $z \to 0$, $\Sigma \in \Sigma_k$, where $p_j = (-1)^j \left[\alpha^{\frac{(j-1)(j-2)}{2}}\right]^{-1}$, and where $\gamma_{kj} \neq 0$

(17) For each $z_j \in Z$, there exists a non-zero constant $c = c(z_j)$ and an index $m = m(z_j)$, $1 \leq m \leq n-1$, such that

$$v(z_j) = ce_{m,m+1}.$$

Furthermore, if z_j lies in the hatched region

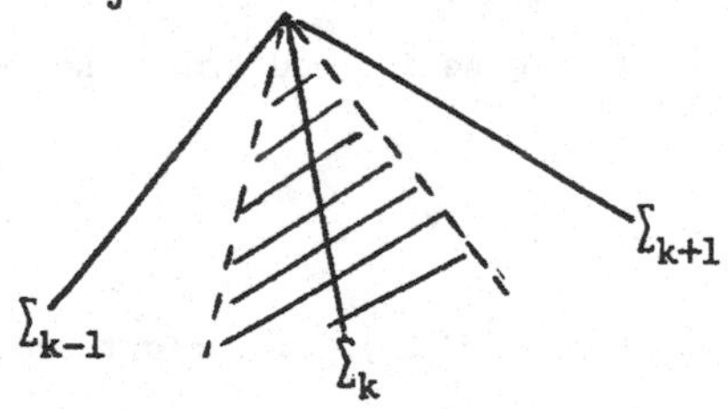

then $m \neq \ell + 1$ if $\ell \sim \ell + 1$ on Σ_k, and $m \neq \ell$ if $\ell + 1 \sim \ell + 2$ on Σ_k.

(18) For all j

$$\hat{v}_j \pi_j \hat{v}_{j+1} \pi_{j+1} \cdots \hat{v}_{j+2n-1} \pi_{j+2n-1} \equiv I$$

where $\hat{v}_j$ is the formal power series associated with v on Σ_k at $z = 0$, and $\pi_j = \pi_z$ for $z \in \Sigma_j$.

Self-adjoint operators L_n map into $G_{sa}(n)$ through their eigenfunctions in a similar way to the case $n = 2$ of Section (A). Let $\psi(x,z) = \left[\psi_1,(x,z),\cdots,\psi_n(x,z)\right]$, be the (unique) solutions of $L_n\psi = z^n\psi$ with

$$\psi e^{-ixzJ(z)} \to (1,1,\cdots,1) \quad \text{as} \quad x \to -\infty$$
$$\psi e^{-ixzJ(z)} \quad \text{bounded on } \mathbb{R}.$$

(The existence of these solutions and the analysis of their properties is far from trivial and occupies a large part of the theory. The construction of the solutions ψ_j that we use is new and considerably simpler than the constructions in [2] and [1].) Then for fixed x, $\psi(x,\cdot)$ is meromorphic in $\mathbb{C}\backslash\Sigma$ with simple poles at a finite set $Z = Z(L_n)$ and as in the case $n = 2$, there exists a map $v = v(L_n)$: $\Sigma \cup Z \to M_n(\mathbb{C})$ such that

$$(19) \quad \begin{cases} \psi_+(x,z)\pi_z = \psi_-(x,z)v(z) & z \in \Sigma\backslash o \\ \operatorname*{Res}_{z_j} \psi(x_i) = \lim\limits_{z \to z_j} \psi(x,z)v(z_i) & \end{cases}$$

Conversely (19), together with the asymptotic condition

$$\psi(x,z)e^{-ixzJ(z)} \to (1,1,\cdots) \quad \text{as} \quad z \to \infty,$$

poses a Riemann-Hilbert problem for $\psi(x,\cdot)$ (cf. Section C).

Set $F_n(L_n) = S(L_n) \equiv \left[Z(L_n),v(L_n)\right]$.

The basic result is

<u>Theorem 2</u>.

The map F_n: $L_n \to S(L_n)$ is a bijection from the set of generic n^{th} order, $n = 2\ell + 1$, self-=adjoint differential operators on $\mathbb{R}$ onto $G_{sa}(n)$. Moreover, $G_{sa}(n)$ is invariant under the Gel'Fand Dikii flows.

Theorem 2 says that $G_{sa}(n)$ codes all the information about generic, odd order self-adjoint operators $L = D^n + \sum_{j=0}^{n-2} p_j(x)D^j$. How is this accomplished?

(13)(a). The symmetry $z_j \to \alpha z_j$ indicates that the data arises from an nth order ode L_n, as opposed to a first order system. The symmetry $z_j \to \overline{z}_j$ indicates that $L_n = L_n^*$.

(13)(b) encodes $p_j(x) \in S(\mathbb{R})$.

(14) indicates that the data arises from an ode as opposed to a system.

(15) codes $L_n = L_n^*$.

(16)(d) indicates that the data arises from a generice ode as opposed to a generic system.

(17) The hatched regions reflect the fact that self-adjoint operators have only real L^2 eigenvalues.

(18) This condition, which is true for systems as well as ode's, plays a critical role in constructing a parametrix which reduces the Riemann-Hilbert problem for $\psi(x,z)$ to the solution of a Fredholm equation of index zero.

Finally we note that the single condition (18) replaces three conditions in [2] and [1] that were initially thought to be independent.

REFERENCES

[1] Beals, R., The inverse problem for ordinary differential operators on the line, Amer. J. of Math., 107, 281-366 (1985).

[2] Beals, R., and Coifman, R., Scattering and inverse scattering for first order systems, Comm. Pure Appl. Math., 37, 39-90 (1984).

[3] Deift, P., and Trubowitz, E., Inverse scattering on the line, Comm. Pure Appl. Math., 32, 121-251 (1979).

[4] Deift, P., Tomei, C., and Trubowitz, E., Inverse scattering and the Boussinesq equation, Comm. Pure Appl. Math., 35, 567-628 (1982).

[5] Gardner, C., Greene J., Kruskal, M., and Miura, R., Method for solving the Korteweg-de Vries equation, Phys. Rev. Lett, 1095-1097 (1967).

[6] Lax, P., Periodic solutions of the kdV equation, Comm. Pure Appl. Math., 28, 141-188 (1975).

ON THE DIRICHLET INDEX CONJECTURE

R.C. Brown
Mathematics Department
University of Alabama
Tuscaloosa, AL 35486

Don Hinton
Mathematics Department
University of Tennessee
Knoxville, TN 37996

1. Introduction

Let $p_0, p_1, \ldots, p_n$ be real valued functions on $I = [0, \infty)$ such that $p_i \geq 0$, $i = 0, \ldots, n-1$, $p_n \geq \varepsilon > 0$, and p_0^{-1}, p_i, $i = 1, \ldots, n$ are Lebesgue integrable on compact subintervals of I. Associated with the coefficients is the symmetric $2n^{th}$ order quasi-derivative expression

$$M[y] = \Sigma_{i=0}^{n} (-1)^{n-i} (p_i y^{(n-i)})^{(n-i)} . \tag{1.1}$$

Let T, T_0 signify respectively the densely defined closed maximal and minimal operators (see [10] for definitions) determined by M in the Hilbert space $\mathcal{L}^2(I)$ of square integrable complex valued functions. In this paper we are interested in obtaining information about the number of linearly independent solutions of $T[y] = 0$ such that the Dirichlet form

$$\underline{\underline{t}}[y] := \Sigma_{i=0}^{n} \int_I p_i |y^{(n-i)}|^2$$

is finite. We call this number the Dirichlet index (DI) of M. It is not difficult to show (see [4] or [7]) that $DI \geq n$; DI is then minimal if it is n. A natural conjecture is that under the conditions stated above the DI is always minimal. This conjecture under slightly less general condition is due to Kauffman [7].

It is now thought by many investigators that although almost always true that the conjecture is probably false. However, we will not settle this difficult problem here. Instead we will develop a new and rather simple interpretation of the index problem and apply it to obtain some incremental improvements in the theory and to simplify proofs of some known results.

Section 2 begins with a brief list of known results about the DI and concludes with a new proof of an upper bound of DI due to Niessen [11]. The core of the paper is section 3 which develops an equivalence (for $n = 2$) between $DI = n$ and a concept of limit-pointness for a certain vector-valued operator $\mathcal{K}$; $\mathcal{K}$ is associated with a boundary form which is simpler than the normal form corresponding to T. This theory is applied in section 4 to fourth order differential operators.

In what follows $N(L)$, $D(L)$, $R(L)$ denote the null space, domain, and range of an operator L. Inner products and norms of various Hilbert spaces are written $[\cdot, \cdot]$, $\|\cdot\|$ (reliance being placed on the context for precise definitions). AC_{loc}

signifies the class of locally absolutely continuous functions on I. For a positive locally Lebesgue integrable function w on I, $\mathcal{L}^2(w; I)$ denotes the Hilbert space of square integrable complex functions with weight w having inner product $[f, g] := \int_I w f\bar{g}$. If $y, g \in D(T)$ we define $D(y, g) := \Sigma_{i=0}^{n} y^{[i]}\bar{g}^{[2n-i-1]}$ where $y^{[i]}$ etc. stands for the ith quasi-derivative of y. Concerning T and T_0, it is known that $T_0^* = T$, $T^* = T_0$, and that both the Green's formula

$$[Ty, z] - [y, Tz] = \lim_{t\to\infty} \{D(y, z)(s) - D(z, y)(s)\}\Big|_{s=0}^{s=t} \tag{1.2}$$

and the Dirichlet formula

$$[Ty, y] = \lim_{s\to\infty} D(y, y)(s) - D(y, y)(0) + \underline{\underline{t}}[y]$$

hold. T is said to be limit-point (LP) if

$$\lim_{s\to\infty} \{D(f, g)(s) - D(g, f)(s)\} = 0 \tag{1.3}$$

for all $f, g \in D(T)$. T is said to be Dirichlet if $\underline{\underline{t}}[f] < \infty$ for all $f \in D(T)$. The deficiency index (DEF) of M is defined as half the dimension of the quotient space $D(T)/D(T_0)$. It is known that $n \leq \text{DEF} \leq 2n$ and that

$$\text{LP} \Leftrightarrow \text{DEF} = n \Leftrightarrow \text{dimension } \{N(M) \cap \mathcal{L}^2(I)\} = n . \tag{1.4}$$

Basic references for the above are [9], [10].

2. Dirichlet Index Properties and Proof of a Theorem of Niessen

The following is a short list of some known information about the DI.

(1) If T is LP, then $\text{DI} = n$ (this is an obvious consequence of (1.4)).

(2) In addition to $\text{DI} \geq n$, the DI is surprisingly $\leq 2n - 1$ (Niessen [11], Bennewitz [1]).

(3) There exists a 6th order operator of Euler type such that $\text{DI} = 3$ and $\text{DEF} > 3$ (Kauffman [8]).

(4) $\text{DI} \leq \text{DEF}$; $\text{DI} = \text{DEF} \Leftrightarrow T$ is Dirichlet (Brown [4]).

(5) $\text{DI} = n \Leftrightarrow \varliminf_{s\to\infty} D(y, z)(s) = 0$ for all $y \in D(\underline{\underline{t}})$ and $z \in D(\underline{\underline{t}}) \cap D(T)$ (Kauffman [7], Brown [4]).

(6) $\text{DI} = n \Leftrightarrow$ compact support functions are a core (cf. [6, p. 317]) of the restriction of $\underline{\underline{t}}$ to those y such that $y^{[i]}(0) = 0$, $i = 0,\ldots, n-1$, (Kauffman [7]).

(7) The DI is invariant with respect to a class of relatively bounded perturbations of $\underline{\underline{t}}$ (Brown [4]).

(8) If $\text{DI} = n$, then the greatest lower bound of $\underline{\underline{t}}$ can be characterized

as the least point of the spectrum of a certain self-adjoint extension of T_0 (Bradley, Hinton, and Kauffman [2]).

(9) DI = n if the coefficients p_i of M are of the form $a_i(t)\exp b_i(t)$ where a_i, b_i are polynomials (Kauffman [7]). For n = 2 or n = 3 more general coefficients are considered by Robinette [13].

(10) If n = 2 and $p_0(t) \geq m p_0(s)$ for some $m > 0$ and $t \geq s$, then DI = 2 (Read [12]). Note no other conditions on p_1, p_2 are required except the basic ones.

That DI $\leq 2n - 1$ emerges as a special case of a much more general theory developed by Niessen and Bennewitz. We end this section with a simple proof of a slightly more general result.

Theorem 2.1. Under the above stated assumptions on the p_i, $M[y] = 0$ has a solution which is not in $\mathcal{L}^2(p_n; I)$.

Proof. Suppose that $M[y] = 0$ has all solutions in $\mathcal{L}^2(p_n; I)$. Consider the maximal and minimal operator $\tilde{T}$, $\tilde{T}_0$ determined by $p_n^{-1}M[y]$ in $\mathcal{L}^2(p_n; I)$. Since $\tilde{T}_0$ is bounded below by one and $\tilde{T}$ is a finite dimensional extension of $\tilde{T}_0$, both have closed range. It follows [9, Theorem 4.2] that the deficiency indices of $\tilde{T}_0$ are (2n, 2n). Let K_0, K be the minimal and maximal operators on $\mathcal{L}^2(p_n; I)$ corresponding to $p_n^{-1}[(-1)^n(p_0 y^{(n)})^{(n)} + \ldots + (p_{n-1}y')']$. By a generalization of [10, Theorem 1, p. 192] to the weighted case, $\tilde{T}_0$ has the same deficiency indices as K_0. By [10, Theorem 4, p. 93], $K[y] = 0$ has all solutions in $\mathcal{L}^2(p_n; I)$. But $y_1 \equiv 1$ is a solution of $K[y] = 0$ and $y_1 \notin \mathcal{L}^2(p_n; I)$. This contradiction establishes the theorem.

3. The Operator $\mathcal{K}$ and the Boundary form $\{\cdot, \cdot\}$

A central difficulty in the DI problem is the complexity of the form $D(\cdot, \cdot)$. Here we define a new dual pair of operators "$\mathcal{K}$" and "$\mathcal{K}^+$", together with a boundary form $\{\cdot, \cdot\}$ which is simpler in structure than $D(\cdot, \cdot)$, and so that the "limit-pointness" of a minimal operator "$\mathcal{K}_0$" corresponds to DI = n. For technical simplicity we deal only with the case n = 2 and $p_1(t) > 0$, but the methods generalize to arbitrary n.

Definition 3.1. $\mathcal{K}: \mathcal{L}^2(p_2; I) \to H := \mathcal{L}^2(p_0; I) \times \mathcal{L}^2(p_1; I)$ is the operator given by $\mathcal{K}(y) = (y'', y')$ on

$$D := \{y \in \mathcal{L}^2(p_2; I) : y' \in AC_{loc};\ \mathcal{K}(y) \in H\}.$$

Definition 3.2. $\mathcal{K}^+ : H \to \mathcal{L}^2(p_2, I)$ is given by $\mathcal{K}^+(z_1, z_2) = p_2^{-1}[(p_0 z_1)' - p_1 z_2]'$ on

$$D^+ := \{(z_1, z_2) \in H :\ p_0 z,\ (p_0 z_1)' - p_1 z_2 \in AC_{loc};\ \mathcal{K}^+(z_1, z_2) \in \mathcal{L}^2(p_2; I)\}.$$

On H we construct the inner product

$$[(u_1, u_2), (v_1, v_2)]_H = \int_I p_0 u_1 \overline{v}_1 + \int_I p_1 u_2 \overline{v}_2 ;$$

with this inner product H is a Hilbert space.

Lemma 3.1. $[\mathcal{K}y, (z_1, z_2)]_H - [y, \mathcal{K}^+(z_1, z_2)] = \{y, (z_1, z_2)\}(\infty) - \{y, (z_1, z_2)\}(0)$ where $\{y, (z_1, z_2)\}(s) := [y' p_0 \overline{z}_1 - y(p_0\overline{z}_1)' + p_1 y \overline{z}_2](s)$.

Proof. Integrate by parts on $[0, s]$ and let $s \to \infty$.

Let $\mathcal{K}_0', \mathcal{K}_0^{+\prime}$ denote the restrictions of $\mathcal{K}$ and $\mathcal{K}^+$ to compact support functions. Let $\mathcal{K}_0, \mathcal{K}_0^+$ be the closures of $\mathcal{K}_0', \mathcal{K}_0^{+\prime}$. Routine but rather technical arguments similar to those of [3] establish that $\mathcal{K}_0, \mathcal{K}_0^+, \mathcal{K}, \mathcal{K}^+$ are densely defined operators such that

$$\mathcal{K}_0^* = \mathcal{K}^+ , \mathcal{K}^{+*} = \mathcal{K}_0 ; \mathcal{K}_0^{+*} = \mathcal{K} , \mathcal{K}^* = \mathcal{K}_0^+ .$$

Further $\mathcal{K}_0$ is the restriction of $\mathcal{K}$ to the domain

$$D_0 := \{y \in D: y(0) = y'(0) = 0 ; \{y, (z_1, z_2)\}(\infty) = 0 \quad \forall (z_1, z_2) \in D^+\} ,$$

and $\mathcal{K}_0^+$ is the restriction of $\mathcal{K}^+$ to the domain

$$D_0^+ := \{(z_1, z_2) \in D^+: z_1(0) = (p_0 z_1)'(0) = 0 ; \{y, (z_1, z_2)\}(\infty) = 0 \quad \forall y \in D\} .$$

Definition 3.3. $\mathcal{K}_0$ is "limit-point" if $\mathcal{K}$ is a 2-dimensional extension of $\mathcal{K}_0$.

Lemma 3.2. $\mathcal{K}_0$ is limit-point if and only if $\{y, (z_1, z_2)\}(\infty) = 0$ for all $(y, (z_1, z_2)) \in D \times D^+$.

The proof is formally the same as [9, Theorem 5.2, p. 19]. We are now in a position to relate the DI of M to the limit-pointness of $\mathcal{K}_0$.

Lemma 3.3. Let $n = 2$ and $p_1(t) > 0$. Then the DI of M is 2 if and only if $\mathcal{K}_0$ is limit-point.

Proof. The DI of M is 2 if and only if the functions of compact support are a core of the form $\underline{\underline{t}}[y]$ restricted to $y \in D$ such that $y(0) = y'(0) = 0$ (see (6) of section 2); for such y and $\delta > 0$ there exists a compact support function ϕ_δ such that $\underline{\underline{t}}[y - \phi_\delta] < \delta$ which implies $\|\mathcal{K}(y - \phi_\delta)\|_H^2 < \delta$ and $\|y - \phi_\delta\|^2 < \delta$ in $\mathcal{L}^2(p_2, I)$. Since δ is arbitrary (for fixed $y \in D$ and $(z_1, z_2) \in D^+$), this gives by Lemma 3.2 that $\mathcal{K}_0$ is limit-point since

$$|\{y, (z_1, z_2)\}(\infty)| = |\{y - \phi_\delta , (z_1, z_2)\}(\infty)|$$

$$= [\mathcal{K}(y - \phi_\delta), (z_1, z_2)]_H - [(y - \phi_\delta), \mathcal{K}^+(z_1, z_2)]| \leq \delta^{1/2}\|(z_1, z_2)\|_H + \delta^{1/2}\|\mathcal{K}^+(z_1, z_2)\| .$$

Conversely, if $\mathcal{H}_0$ is limit-point, then given $y \in D$ there exists $g \in D_0$ such that $y - g = \phi$ is a compact support function (cf. [9, Corollary 4.7, p. 18]). By definition of $\mathcal{H}_0$, given ε there exists θ_ε in $D(\mathcal{H}_0')$ such that $\|g - \theta_\varepsilon\| < \varepsilon/2$ in $\mathcal{L}^2(p_2; I)$ and $\|\mathcal{H}(g - \theta_\varepsilon\| < \varepsilon/2$. Set $\psi = \theta_\varepsilon - \phi$. Then $y - \psi = g - \theta_\varepsilon$ so $\|y - \psi\| < \varepsilon/2$ and $\|\mathcal{H}(y - \psi)\|_H < \varepsilon/2$ implying that $|\underline{t}(y - \psi)| < \varepsilon$.

4. Applications to Fourth-Order Operators

It follows from Lemmas 3.2-3.3 that we need to investigate $\{y, (z_1, z_2)\}$. Proofs of the following two lemmas are standard, e.g. see Lemma 5 of [5].

<u>Lemma 4.1.</u> $f \in \mathcal{L}[a, \infty) \Rightarrow \lim\limits_{X\to\infty} \frac{1}{X}\int_a^X tf(t)\,dt = 0$.

<u>Lemma 4.2.</u> $f \in \mathcal{L}^2[a, \infty) \Rightarrow \lim\limits_{X\to\infty} \frac{1}{X}\int_a^X t^{1/2} f(t)\,dt = 0$.

We require the following assumptions on p_0, p_2:

(A1) $\int_a^t p_2 \le Mtp_2(t)$ for some $M > 0$.

(A2) $p_0(t) = \Sigma_{i=1}^5 p_{i0}(t)$ where

(i) $|p_{10}(t)| \le a_1 t^2 p_1(t)$,

(ii) $p_{20}(t) = 0$, or $0 < \varepsilon \le p_0(t)$ and $|p_{20}(t)| \le a_2 t^2$,

(iii) $p_{30}(t) = 0$, or $0 < p_{30}(t) \le p_0(t)$ and $|p_{30}'(t)| \le a_3(p_{30}(t))^{3/4}$.

(iv) $p_{40}(t) = 0$, or $0 < \varepsilon \le p_{40}(t) \le p_0(t)$, $(p_{40}^{1/2}(t)'$ is bounded below, $(p_{40}^{1/2})''$ is continuous and for some $c < 1/4$, $2(p_{40}^{1/2}(t)'' + c\,p_{40}^{1/2}(t)/t^2 > 0$,

(v) $p_{50}(t) = 0$, or $0 < p_{50}(t) \le p_0(t)$, $\int_a^\infty 1/p_{50} < \infty$, and $\int_t^\infty 1/p_{50} \le a_4\, t/p_{50}(t)$.

<u>Theorem 4.1</u>. Suppose $n = 2$ and p_0, p_2 satisfy (A1) and (A2). Then the DI of M is 2.

<u>Proof</u>. By the perturbation theory of [4], we may without loss of generality assume $p_1(t) > 0$ so as to apply section 3. Also without loss of generality we take $a > 0$ in I and consider $y \in D$, $(z_1, z_2) \in D^+$ such that $y(a) = y'(a) = z_1(a) = z_2(a) = 0$. Set $f = (p_0 z_1)' - p_1 z_2$. Note that $\int_a^\infty p_2^{-1} |f'|^2 < \infty$ and that

$$|f(t)| = |\int_a^t f'| \le (\int_a^t p_2)^{1/2} (\int_a^t p_2^{-1} |f'|^2)^{1/2}. \tag{4.1}$$

If $DI \neq 2$, there exist $y \in D$ and $z = (z_1, z_2) \in D^+$ such that $\{y, z\}(\infty) = 1$ implying that

$$\lim_{t\to\infty} \frac{1}{t} \int_a^t \{y, z\} = 1 . \tag{4.2}$$

Now $\{y, z\} = -yf + y' p_0 z_1$. From (A1) and (4.1),

$$|y(t)f(t)| \leq M\, t^{1/2}\, p_2^{1/2}(t) \left(\int_a^\infty p_2^{-1} |f'|^2\right)^{1/2} |y(t)| .$$

Since $y \in \mathcal{L}^2(p_2; I)$, application of Lemma 4.2 yields that $t^{-1} \int_a^t |yf| \to 0$ as $t \to \infty$. Substitution of this into (4.1) gives

$$\lim_{t\to\infty} \frac{1}{t} \int_a^t (y'\, p_0 z_1) = 1 . \tag{4.3}$$

We claim that (A2) implies that

$$\lim_{t\to\infty} \frac{1}{t} \int_a^t (y'\, p_0 z_1) = 0 , \tag{4.4}$$

which is contrary to (4.3) and completes the proof. We begin by noting that $y' p_0 z_1 = (p_0^{1/2} y')(p_0^{1/2} z_1)$. Since $p_0^{1/2} z_1 \in \mathcal{L}^2(I)$, Lemma 4.1 completes the proof if $t^{-2} p_0 |y'|^2 \in \mathcal{L}(I)$, and Lemma 4.2 completes the proof if $t^{-1} p_0 |y'|^2$ is a bounded function. Thus we need only show one of these holds for each of $p_{io}|y'|^2$. Since $\mathcal{H}y \in H$, it is immediate that $t^{-2}|p_{10}||y'|^2 \in \mathcal{L}^2(I)$. For p_{20}, note that since $p_0 \geq \varepsilon > 0$, $y'' \in \mathcal{L}^2(I)$. This, together with $y \in \mathcal{L}^2(I)$ implies $y' \in \mathcal{L}^2(I)$. Thus $t^{-2}|p_{20}||y'|^2 \in \mathcal{L}(I)$. Cases (iii)-(iv) essentially follow the arguments of [13]. Consider now $p_{50}|y'|^2$. From (A2)-(v) and

$$|y'(t_2) - y'(t_1)| = \left|\int_{t_1}^{t_2} y''\right| \leq \left(\int_{t_1}^{t_2} p_{50}^{-1}\right)^{1/2} \left(\int_{t_1}^{t_2} p_{50}|y''|^2\right)^{1/2} \tag{4.5}$$

we conclude that $\lim y'(t)$ as $t \to \infty$ exists; $y \in \mathcal{L}^2(I)$ implies the limit is zero. Set then $t_2 = \infty$ in (4.5) and apply (A2)-(v) to get $|y'(t)| \leq M \cdot t^{1/2}/p_{50}^{1/2}(t)$; thus $t^{-1} p_{50}|y'|^2$ is bounded and the proof is complete.

The form $\{y, (z_1, z_2)\}$ may be expressed in different ways; for example,

$$\{y, (z_1, z_2)\} = -2y[(p_0 z_1)' - p_1 z_2] + [y p_0 z_1]' - y p_1 z_2 . \tag{4.6}$$

Each version of the form gives distinct criteria for minimality of DI. For example we may prove along the lines of the above proof the following.

Theorem 4.2. *Suppose* $p_0 = O(p_1)$ *and* $p_1 = O(p_2)$ *as* $t \to \infty$. *Then the* DI *of* M *is* 2.

Corollary 4.1. *If* $p_0 = p_1 = p_2$ *or if* p_0, p_1 *are bounded, then the* DI

<u>of</u> M <u>is</u> 2 .

References

1. C. Bennewitz, A generalization of Niessen's limit-circle criterion, <u>Proc. Roy. Soc. of Edinburgh</u>, 78A(1977), 81-90.

2. J. Bradley, D.B. Hinton, and R.M. Kauffman, On minimization of singular quadratic functionals, <u>Proc. Roy. Soc. of Edinburgh</u>, 87A(1981), 193-208.

3. R.C. Brown, A von Neumann factorization of some self adjoint extensions of positive symmetric differential operators and its application to inequalities, <u>Lecture Notes in Mathematics</u> 1032 (Springer Verlag, 1983).

4 . R.C. Brown, The Dirichlet Index under minimal conditions, <u>Proc. Roy. Soc. of Edinburgh</u> 96A(1984), 303-316.

5. D.B. Hinton, Limit-point criteria for positive definite fourth-order differential operators, <u>Quart. J. Math.</u> Oxford (2), 24(1973), 367-76.

6. T. Kato, Perturbation Theory for Linear Operators, (Springer Verlag, New York, 1966).

7. R.M. Kauffman, The number of Dirichlet solutions to a class of linear ordinary differential equations, <u>J. Differential Equations</u>, 31(1979), 117-129.

8. R.M. Kauffman, On the limit-n classification of ordinary differential operators with positive coefficients, <u>Proc. London Math. Soc.</u>, 35(3) (1977), 495-526.

9. R.M. Kauffman, T.T. Read, and A. Zettl, "The Deficiency Index Problem for Powers of Ordinary Differential Expressions," <u>Lecture Notes in Mathematics</u> 621 (Springer-Verlag, 1977).

10. M.A. Naimark, "Linear Differential Operators, Part II," (Ungar, New York, 1968).

11. H.D. Niessen, A necessary and sufficient limit-circle criteria for left definite eigenvalue problems, <u>Lecture Notes in Mathematics</u> 280(Springer-Verlag, 1974).

12. T.T. Read, The number of the Dirichlet solutions of a fourth order differential equation, <u>Proc. Roy. Soc. of Edinburgh</u> 92A(1982), 233-239.

13. J.B. Robinette, On the Dirichlet index of singular differential operators, M.S. thesis, University of Tennessee, 1979.

THREE-DIMENSIONAL INVERSE SCATTERING

Margaret Cheney
Department of Mathematics
Duke University
Durham, North Carolina 27706

James H. Rose
Center for NDE
Iowa State University
Ames, Iowa 50011

Brian DeFacio
Department of Physics and Astronomy
University of Missouri
Columbia, Missouri 65211

I. Introduction

In this paper we consider inverse scatttering for the equation on R^3

$$(-\nabla^2 + V - k^2)\psi = 0 , \tag{1}$$

where V is some linear operator that in some sense is small for large $|x|$ ($x \in R^3$). The following are examples of such linear operators.

a) $(V\psi)(x) = V(x)\psi(x)$, where V goes to zero at infinity. In this case, (1) is the Schrödinger equation and V is a potential energy.

b) $(V\psi)(x) = [k^2 - k^2 n^2(x)]\psi(x)$, where n is one for large x. In this case, (1) is the reduced wave equation and n is the index of refraction of the medium.

c) $(V\psi)(x) = [\rho^{-1}(x)\nabla\rho(x) \circ \nabla + k^2 - k^2 n^2(x)]\psi(x)$, where ρ is asymptotically constant and n is one for large x. In this case, (1) is the acoustic equation, and ρ is the density of the medium.

We define scattering solutions of (1) by a generalized Lippmann-Schwinger equation [1]:

$$\psi^{\pm}(k,e,x) = \exp(ike\cdot x) + \int G_0^{\pm}(k,x,y)(V\psi^{\pm})(k,e,y)dy . \tag{$2^{\pm}$}$$

where e is a unit vector in R^3 and

$$(3) \qquad G_0^{\pm}(k,x,y) = -(4\pi|x - y|)^{-1}\exp(\pm ik|x - y|) .$$

Equation (2) gives us two solutions of (1). Their physical interpretation can be understood by expanding (2) for large $|x|$:

$$(4) \qquad \psi^{\pm}(k,e,x) = \exp(ike\cdot x) + A(k,\hat{x},e)|x|^{-1}\exp(\pm ik|x|) + \ldots$$

where $\hat{x} = x/|x|$ and

$$(5) \qquad A(k,e,e') = \frac{-1}{4\pi}\int \exp(-ike\cdot y)(V\psi)(k,e',y)dy .$$

From (4) we see that ψ^+ corresponds to a plane wave plus an outgoing spherical wave, whereas ψ^- corresponds to a plane wave plus an incoming spherical wave. For problems a), b), and c), ψ^+ and ψ^- are related by

$$(6) \qquad \psi^+(k,e,x) = \psi^-(-k,-e,x) .$$

The function A is called the scattering amplitude; it contains the scattering data.

Our inverse problem is thus to recover the linear operator V from a knowledge of the scattering amplitude A. Our plan is to find an integral equation relating A and ψ. The equation we will obtain is

$$(7) \qquad \psi^+(k,e,x) = \psi^-(k,e,x) - \frac{ik}{2\pi}\int_{S^2} A(k,e',e)\psi^-(k,e',x)de' ,$$

where S^2 denotes the unit sphere in R^3.

Once we have equation (7), we can use it in several ways. If, given A, we are able to solve (7) to obtain ψ^+ or ψ^-, then we can solve the inverse problem, because there are a number of ways to obtain V from a solution ψ. On the other hand, if we are unable to solve (7), we can still obtain useful information from various approximations using (7). In the case of the three-dimensional Schrödinger equation (1a) with real-valued potential, (7) can be solved, and the resulting inverse scattering theory is in relatively good shape. Since (7) also holds for other wave equations, there is hope that it will play a similar role in inverse scattering theory for them.

Therefore we are interesting in deriving equation (7) for the general case of equation (1). For the Schrödinger equation, four different derivations of (7) are known. They are: 1) the spectral theory method, 2) the asymptotic method, 3) the Green's theorem method, and 4) a new method. This new method is contained in Theorems 1 and 2, which are the main results of this paper. This new method applies to examples a), b), and c). The four derivations of (7), together with their advantages and disadvantages, will be discussed in Section II.

II. Old and New Proofs of (7)

This method works for the Schrödinger equation (1a) with real-valued potential. First we recall that the scattering solutions defined by (2) can also be obtained by means of the "limiting absorption principle" [2]:

$$(8^{\pm})\qquad \psi^{\pm}(k,e,x) - \exp(ike\cdot x)$$

$$= (-\nabla^2 + V - [k^2 \pm i\epsilon])^{-1}[V(x)\exp(ike\cdot x)]$$

We then subtract equation (8-) from (8+), using the fact that the jump in the resolvent across the real axis is given by the spectral projection corresponding to the self-adjoint operator $-\nabla^2 + V$ [3]. This gives

$$(9)\qquad \psi^{+} - \psi^{-} = (\text{spectral projection at } k^2)[V\cdot\exp] .$$

However, the spectral projection can be written out explicitly in terms of the scattering solution ψ^- [2]. We use this eigenfunction expansion in (9) and interchange the order of integration to obtain (7).

This spectral theory method is mathematically appealing, but it has the disadvantage of being difficult to apply to equations besides Schrödinger's. It relies heavily on the fact that the Schrödinger equation can be written as an eigenvalue problem for a self-adjoint operator. The method also requires that a limiting absorption principle and an eigenfunction expansion both be known for the equation in question. Thus it would require some nontrivial work to apply this method to the reduced wave equation (1b), to the acoustic equation (1c), or even to the Schrödinger equation with a complex potential.

2). The asymptotic method.

This method has been worked out for both quantum scattering (1a) [4] and for the obstacle scattering problem [5]. In this method, one considers the difference φ between the two sides of (7), and then one shows that φ must be identically zero. For the quantum problem, Schmidt [4] shows that φ must satisfy equation (1) and moreover must decay at a certain rate as $|x|$ becomes infinite. For the case of obstacle scattering, Lax and Phillips [5] show that φ is a solution of the obstacle problem and decays rapidly at ∞. One then applies Kato's theorem [6] (or the Rellich uniqueness theorem) to conclude that φ must be identically zero.

This method has the disadvantage that it too is rather specialized, because it relies on the Kato or Rellich uniqueness theorem. It should, however, apply to both problems a) and b), because Kato's theorem applies to both.

3). The Green's Theorem Method. [7]

This method begins with the equation for $\psi^+ - \psi^-$

(10) $$(-\nabla^2 + V - k^2)(\psi^+ - \psi^-) = 0$$

and the equation that defines the Green's function

(11) $$(-\nabla^2 + V - k^2)G = \delta .$$

We multiply (10) by G, (11) by $(\psi^+ - \psi^-)$, and subtract the resulting equations. If V is a multiplication operator, this procedure gives

(12) $$G\nabla^2(\psi^+ - \psi^-) - (\psi^+ - \psi^-)\nabla^2 G = (\psi^+ - \psi^-)\delta .$$

We then integrate (12) over a large ball and use Green's theorem to convert the volume integral on the left side to a surface integral. Finally, this surface is allowed to expand to infinity, which allows us to use (4) and a similar asymptotic expansion for G. This results in (7). This method applies as well to 1b) as it does to 1a). However, the use of Green's Theorem in this method requires some smoothness of V, which is undesirable for inverse scattering applications. Moreover, tricks [8] are required to apply this method to 1c), and it is not clear whether other perturbations V containing derivatives could be treated this way.

4). <u>A new method</u>. [9]

Intuitively, this method can be understood by Fourier transforming G_0^+ and G_0^- into the time domain. The Fourier transformed G_0^+ and G_0^- can be interpreted as propagating waves "into the future" and "into the past", respectively. Thus (2+) can be interpreted as giving a solution ψ^+ "in the present" that looked like $\exp(ike\cdot x)$ "in the past", and similarly for (2-). We then do a short computation using (5) and (2+):

(13) $$\int_{S^2} A(k,e',e)\exp(ike'\cdot x)de' =$$

$$2\pi ik^{-1}[\psi^+ - \exp(ike\cdot x) - \int G_0^- V\psi^+] .$$

When we rearrange (13), we see that it can be interpreted as saying that "in the future", ψ^+ looks like

(14) $$\exp(ike\cdot x) - \frac{ik}{2\pi}\int_{S^2} A(k,e',e)\exp(ike'\cdot x)de' .$$

But by (2–) and the superposition principle, the solution that "in the future" looks like (14) must be precisely the right side of (7).

To state our result, we use the notation $L^{2,s} = \{u: (1 + |x|^2)^{s/2}u \in L^2\}$ and we denote by $H^{2,s}$ the Sobolev space of functions whose derivatives up to order 2 are in $L^{2,s}$.

<u>Theorem 1</u>. For $s = 3/2 + \epsilon$ with $0 < \epsilon < 1$, suppose that V maps $H^{2,-s}$ into $L^{2,s}$, and suppose that the operators $I - G_0^{\pm}V$ are invertible on $H^{2,-s}$. Then (7) holds.

<u>Sketch of Proof</u>. We make the above intuition into a rigorous argument by using operator notation. We write (2–) as

(15) $$(I - G_0^-V)\psi^- = \exp ;$$

by the superposition principle it follows that

(16) $$(I - G_0^-V)(\psi^- - \frac{ik}{2\pi}\int A\psi^-) = \exp - \frac{ik}{2\pi}\int A \exp .$$

We also rewrite (13) in operator notation:

(17) $$(I - G_0^-V)\psi^+ = \exp - \frac{ik}{2\pi}\int A \exp .$$

Comparison of (17) and (16) shows that if the operator $(I - G_0^-V)$ is invertible, then (7) must hold. This completes the outline of the proof.

We also need to know what linear operators V satisfy the hypotheses of Theorem 1.

<u>Theorem 2</u>. Each V in problems a), b), and c) satisfies the hypotheses of

Theorem 1 for almost every real k, provided that for each k,

i) the order zero terms [$V(x)$ in a), or $k^2 - k^2n^2(x)$ in b) and c)] are locally L^2 and are $O(|x|^{-3-\epsilon})$ at infinity.

ii) the order one terms [$\rho^{-1}(x)\nabla\rho(x)$ in c)] are locally bounded and are $O(|x|^{-3-\epsilon})$ at infinity.

Sketch of Proof. To prove Theorem 2, we verify that $V: H^{2,-s} \to L^{2,s}$ is compact [10]. We recall that $G_0^{\pm}: L^{2,s} \to H^{2,-s}$ is bounded [2]. Then, since the product $G_0^{\pm}V$ is compact, we can apply the analytic Fredholm theorem [3,11] to conclude that $I - G_0^{\pm}V$ is invertible for almost all real k. We need only check that $I - G_0^{\pm}V$ is invertible for <u>some</u> k. Unfortunately, this seems to be difficult in the general case, but for problems a), b), and c), we can use the fact that each is related to an eigenvalue problem for a self-adjoint operator [11]. This completes the sketch of the proof of Theorem 2.

Corollary. For problems a), b), and c) satisfying assumptions i) and ii) above, (7) holds for almost every real k.

We recall that Theorems 1 and 2 comprise the fourth method of deriving (7). This fourth method applies to all three of the problems a), b), and c), and moreover it requires no restrictions on the smoothness of the perturbation V.

We consider briefly the uses of (7) in inverse scattering. For the Schrödinger equation 1a) with a real potential and no bound states, (7) can be Fourier transformed in the k variable [12]. Equation (7) then becomes a Fredholm integral equation, which can always be solved [13]. If the original data (i.e., A) were good, then V can be obtained from the solution of (7) [14].

Unfortunately, this situation is more complicated in the case of the wave equation 1b). This is because the Fourier transform of (7) is no longer a Fredholm equation, and in fact is expected to have distribution solutions.

However, if a solution of (7) can be obtained by some means, then $n^2(x)$ can be recovered.

In the acoustic equation case 1c), equation (7) has not been studied, but undoubtedly the same difficulties arise as in the wave equation case.

Acknowledgements

The work of MC was partially supported by ONR contract number N00014-85-K-0224. JHR's work was supported by the NSF university/industry Center for NDE at Iowa State University.

References

[1] R. G. Newton, *Scattering Theory of Waves and Particles*, 2nd edition, Springer, New York, 1982.

[2] S. Agmon, "Spectral Properties of Schrödinger Operators and Scattering Theory," Annali della Scuola Norm. Sup. di Pisa, Classe di Science, Series IV, 2, 151-218 (1975).

[3] M. Reed and B. Simon, *Methods of Modern Mathematical Physics. I: Functional Analysis*. Academic Press, New York, 1972.

[4] E. G. Schmidt, "On the Representation of the Potential Scattering Operator in Quantum Mechanics," J. Diff. Eq. 7, 389-394 (1970).

[5] P. D. Lax and R. S. Phillips, *Scattering Theory*, Academic Press, New York, 1967.

[6] T. Kato, "Growth properties of solutions of the reduced wave equation with a variable coefficient," Comm. Pure and Appl. Math. 12, 403-425 (1959).

[7] J. H. Rose, M. Cheney, and B. DeFacio, "Three-dimensional inverse scattering: Plasma and variable velocity wave equations," J. Math. Phys. 26, 2803-2813 (1985).

[8] S. Coen, M. Cheney, and A. Weglein, "Velocity and density of a two-dimensional acoustic medium from point source surface data," J. Math. Phys. 25, 1857-1861 (1984).

[9] M. Cheney, J. H. Rose, and B. DeFacio, "On the direct relation of the wavefield to the scattering amplitude," in preparation.

[10] M. Schechter, *Spectra of Partial Differential Operators*, North-Holland, New York, 1971.

[11] M. Reed and B. Simon, *Methods of Modern Mathematical Physics. III: Scattering Theory*, Academic Press, New York, 1979.

[12] R. G. Newton, "Inverse Scattering. II. Three Dimensions," J. Math. Phys **21**, 1698-1715 (1980); **22**, 631 (1981); **23**, 693 (1982).

[13] R. G. Newton, "Variational principles for inverse problems," Inverse Problems, **1**, 371-380 (1985).

[14] R. G. Newton, "Inverse Scattering, III. Three Dimensions, continued," J Math. Phys. **22**, 2191-2200 (1981); **23**, 693 (1982).

SPECTRAL PROPERTIES OF SCHRÖDINGER OPERATORS WITH TRAPPING POTENTIALS IN THE SEMI-CLASSICAL LIMIT

J.M. Combes Ph. Briet P. Duclos

I. Introduction

Spectral properties of Schrödinger operators $H=-\hbar^2\Delta+V$ on $\mathcal{H} = L^2(\mathbb{R}^n)$ for small values of $\hbar$ have received considerable attention these last years. There exist now an impressive amount of results starting with the mere harmonic approximation for non-degenerate simple well ([CDS1], [HeSj1], [Si1]) and including now refined estimates on spectral shifts due to tunneling in multiple well problems (see e.g., [CDS2], [Ha1], [HeSj1], [Si2]). More recently the shape resonance problems, which differ from the previous one in that they involve non-selfadjoint Schrödinger operators, have also been considered ([AsHa], [CDS3], [CDKS], [HeSj3], [HiSig]). One common aspect of most approaches to tunneling in this Schrödinger operator framework is the essential role played by Agmon's metric [A] to control eigenfunctions decay and through it the effect of barriers on energy shifts. Another common aspect is the use in different forms of geometrical methods which consist roughly speaking in localizing the Schrödinger hamiltonian in those regions where states would be classically confined. In this talk I want to describe a joint work with Ph. Briet and P. Duclos [BCD3] where the above mentionned basic ingredients are synthetized in a method which appears as both simple and efficient. It uses some basic estimates of Agmon's type for Green's functions in some subsets of the classically forbidden regions. This leads to stability statements for the spectrum of localized hamiltonians when "tunneling is turned on". This method is very close in spirit to the interaction matrix formalism of Helffer and Sjöstrand ([HeSj1,2]). Although it leads to a general stability result it does not provide in some specific cases the refined predictions of the interaction matrix formalism. Applications include in particular exponentially small shifts in multiple wells, localization properties through arbitrarily small perturbations (the "flea on the elephant") and analysis of shape resonances spectrum. We emphasize that the small $\hbar$ limit for the spectrum of H could be replaced without changing the basic results by other types of limits like e.g., a large coupling constant limit for the potential or a large separation limit for the potential wells as occur for example in the analysis of electronic energy surfaces in molecular systems.

II. Basic estimates for tunneling

We present here a variant of Agmon's method to prove that resolvents of Schrödinger hamiltonians are exponentially small in the classically forbidden region in a sense to be specified below. The next theorem provides the basic estimates for the perturbative approach to tunneling introduced in the next section.

We consider $H=-\hbar^2\Delta + V$ on $L^2(W)$ where W is an open connected subset of $\mathbb{R}^n$ with Dirichlet boundary conditions on ∂W. For notational simplicity we omit indices $\hbar$, W to specify the dependance of H on these quantities. We assume that the potential V is in $C^1(W)$ and ∇V uniformly bounded on W (although this is only needed on some subsets of W as required in Theorem 2.1 below); V is assumed to be real. Given $E\in\mathbb{R}$ we define the classically forbidden region as the open subset of W:

$$G(E) = \{ x\in W,\ V(x) > E \}. \tag{2.1}$$

Agmon's distance $d^E(x,y)$ is given by [A] :

$$d^E(x,y) = \inf_{\gamma}\Big[\int_0^1 (V(\gamma(t))-E)_+{}^{1/2} |\gamma'(t)|\, dt \Big]$$

where γ runs through the set of all absolutely continuous paths in W with $\gamma(0) = x$ and $\gamma(1) = y$; here $(V(x)-E)_+ = \max(0, V(x)-E)$. Finally we define

(2.2) $$d_\Omega^E = d^E(\Omega, \partial G(E))$$

where $\partial G(E) = \{x \in W,\ V(x) = E\}$: Then one has :

Theorem 2.1. _Let_ $\Omega \subset W$ _and_ $z \in \mathbb{C}$ _satisfy for some_ $E > \mathrm{Re}\, z$:

i) $\Omega \subset G(E)$

ii) $\|(H-z)^{-1}\| \leq \exp -\hbar^{-1} d_\Omega^E$

iii) $|\nabla V(x)| \leq \alpha < \infty$ _for all_ $x \in G(\mathrm{Re}\, z)\backslash\overline{G(E)}$.

Then if X_Ω _is the characteristic function of_ Ω _one has_ :

(2.3) $$\|X_\Omega(H-z)^{-1}\| \leq 2(E-\mathrm{Re}\, z)^{-1}(1 + 2\hbar^2\alpha^2(E-\mathrm{Re}z)^{-1})$$

and

(2.4) $$\|X_\Omega\nabla(H-z)^{-1}\| \leq (2\hbar^2(E-\mathrm{Re}\, z))^{-1/2}(1 + 4\hbar^2\alpha^2(E-\mathrm{Re}z)^{-1}).$$

Sketch of the proof Define $\rho(x) = \hbar^{-1}d^E(x,\Omega)$; then (see e.g., [A]) ρ is a lipschitz function and $\hbar^2|\nabla\rho|^2 \leq (V-E)_+$. Let χ be equal to one on G(E), zero on $W\backslash G((E+\mathrm{Re}z)/2)$ and equal to $(2V-(E+\mathrm{Re}z))/(E-\mathrm{Re}z)$ in between. Then $\rho(x)=\exp-\hbar^{-1}d_\Omega^E$ on the support of $\nabla\chi$ and $\chi e^{-\rho} \geq \chi_\Omega$ Now by a straightforward estimate one has for all $u \in L^2(W)$, $\|u\|=1$, and $\hat{u} = (H-z)^{-1}u$:

$$\hbar^2\|\nabla(\chi e^{-\rho}\hat{u})\|^2 + (E-\mathrm{Re}z)/2\ \|\chi e^{-\rho}\hat{u}\|^2 \leq \mathrm{Re}\langle\chi e^{-2\rho}\hat{u}, (H-z)\chi\hat{u}\rangle$$
$$\leq \|\chi e^{-\rho}\hat{u}\|^2 + \hbar^2\|\nabla\chi\|_\infty^2.$$

The proof of (2.3) and (2.4) follows then from the estimate $\|\nabla\chi\|_\infty \leq 2\alpha/(E-\mathrm{Re}z)$.

Exponential decay properties of bound state-wave functions or wave-packets in the classically forbidden region follow easily from theorem 2.1. One has in particular :

Corollary 2.2. Let $H\varphi_0 = E_0\varphi_0$, $\|\varphi_0\|=1$ where E_0 is an isolated point in the spectrum of H. Then if $\Omega \subset G(E_0)$ one has $\forall\, E > E_0$ and $\hbar$ small enough :

$$\|\chi_\Omega \varphi_0\| < 2(E-E_0)^{-1} \exp -\hbar^{-1} d_\Omega^E \tag{2.5}$$

provided $|\nabla V|$ is uniformly bounded near $\partial G(E_0)$.

The proof follows from Theorem 2.1 by considering $z = E_0 + i\exp -\hbar^{-1} d_\Omega^E$ with $E > E_0$. Assuming $\Omega \subset G(E)$ without restricting generality all the conditions of Theorem 2.1 are met and (2.5) follows from :

$$\chi_\Omega (H-z)^{-1} P_0 (H-\bar{z})^{-1} \chi_\Omega = \exp(-2\hbar^{-1} d_\Omega^E)\, \chi_\Omega P_0 \chi_\Omega$$

where P_0 is the orthogonal projection on φ_0.

Remarks 2.3.

1) More elaborate forms of (2.3) concerning products like $\chi_{\Omega_1}(H-z)^{-1}\chi_{\Omega_2}$ will appear in [BCD3]. They allow in particular to improve (2.5) in case G(E) is not connected.

2) Exponential decay properties of eigenfunctions in x follow from the consideration of sets Ω of the type : $\Omega = \Omega_R = \{ x \in W,\ |x| > R\}$. However this will not be considered here since only the form (2.5) of these decay properties will be needed.

3) One can also use Theorem 2.1 to show decay properties of wave-packets having a given energy range [BCD3]. For example if $\Delta = (E_1, E_0)$ is a bounded

interval and $\Omega \subset G(E_0)$ then one can show using (2.3) that

$$\|\chi_\Omega P_\Delta \chi_\Omega\| < C_E \exp -2\hbar^{-1} d_\Omega^E$$

for all $E > E_0$ with a constant C_E depending only on $|E-E_0|$ and $|E_1-E_0|$ (here P_Δ is the spectral projection operator for H associated to Δ). This appears as very usefull in particular to discuss tunneling along the lines described below when there is continuous spectrum.

III. Spectral Stability Through Tunneling

We consider situations where the potential V has wells separated by barriers. It is not excluded that there exists an infinity of such wells, as would occur for example with periodic or hierarchical potentials, or that some of the wells extend to infinity and have their minimas there, as is the case in the analysis of shape resonances.

We first describe below a perturbative framework to relate the spectral properties of H to those of it's restrictions to the individual wells. To decouple the wells we introduce a C^2 partition of unity :

(3.1) $$1 = \sum_i J_i^2$$

with $\inf_i [\, |\mathrm{supp}\, \nabla J_i| \,] \geq \mathrm{cte} > 0$, so that the wells are supported in

(3.2) $$W_i = \mathrm{supp}\, J_i$$

and the barriers in

(3.3) $$\Omega = \cup_{i \neq j} (W_i \cap W_j)$$

The simple well hamiltonians are defined by :

(3.4) $$H_i = -\hbar^2\Delta + V$$

on $L^2(W_i)$ with Dirichlet boundary conditions on ∂W_i.

We now use the resolvent localization formula (see e.g., [BCD2]). Let

$$\mathcal{H}_0 = \bigoplus_i L^2(W_i) \quad \text{and} \quad H_0 = \bigoplus_i H_i$$

and the "Identification mapping" J from $\mathcal{H}_0$ to $\mathcal{H} = L^2(\mathbb{R}^n)$ be given by :

$$J(\oplus\varphi_i) = \sum_i J_i\varphi_i .$$

Notice that $J^*\varphi = \oplus J_i\varphi$ and $JJ^* = 1$. Also since each J_i is C^2 and vanishes as well as ∇J_i on ∂W_i:

$$J\,\mathfrak{D}(H_0) \subset \mathfrak{D}(H).$$

Then if $z \in \rho(H_0) \cap \rho(H)$, the intersection of the resolvent sets of H_0 and H, one has :

(3.5) $$(H-z)^{-1} = J(H_0-z)^{-1}J^* - (H-z)^{-1}M(H_0-z)^{-1}J^*$$

where the "interaction operator" $M := HJ - JH_0$ is given by $M(\oplus\ \varphi_i) = \sum_i M_i\ \varphi_i$ with

(3.6) $$M_i\varphi_i = -\hbar^2(\nabla.\nabla J_i + \nabla J_i.\nabla)\varphi_i.$$

Thus M_i is a first differential order operator with support on the barrier Ω

given by (3.3). Let

$$K(z) = M\,(H_0-z)^{-1}J^* = \sum_i M_i\,(H_i-z)^{-1}J_i\,. \tag{3.7}$$

Using familiar arguments from perturbation theory one shows that if $z \in \rho(H_0)$ and $\|K(z)\| < 1$ then $z \in \rho(H)$ and (3.5) has a solution:

$$(H-z)^{-1} = J(H_0-z)^{-1}J^*(1+K(z))^{-1}. \tag{3.8}$$

Assume now that I is an isolated part (non necessarily discrete) of $\sigma(H_0)$, the spectrum of H_0, and let $\Gamma \subset \rho(H_0)$ be some closed counterclockwise contour separating I from the other components of $\sigma(H_0)$. Let us write the spectral projection operator for H_0 associated to I as:

$$P_0 = -1/(2i\pi)\int_\Gamma (H_0-z)^{-1}dz.$$

Assume now that $\|K(z)\| < 1$ on Γ; then $\Gamma \subset \rho(H)$ and one can define

$$P = -1/(2i\pi)\int_\Gamma (H-z)^{-1}dz.$$

Then it follows from (3.8) that if $C = \sup_{z\in\Gamma}\|(1+K(z))^{-1}\|$:

$$\|P-JP_0J^*\| \leq C/(2\pi)\int_\Gamma \|(H_0-z)^{-1}J^*M(H_0-z)^{-1}\|dz\,. \tag{3.9}$$

Also if $\varphi \in P\mathcal{H}$, $\|\varphi\|=1$, and we define $E=\langle\varphi,H\varphi\rangle$, $E_0=\langle P_0J^*\varphi,H_0P_0J^*\varphi\rangle/\|P_0J^*\varphi\|^2$ one has:

$$|E-E_0| < C/(2\pi)\int_\Gamma |z-E_0|\,\|(H_0-z)^{-1}J^*M\,(H_0-z)^{-1}\|dz\,. \tag{3.10}$$

One obtains (3.9) and (3.10) from basic perturbation theory (see e.g., [K]). Together with the estimates of Section II, they provide the basic ingredients to the main spectral stability result presented below. The main idea is that since the interaction operator M has support on the barrier Ω one should be able to obtain, using (2.3) and (2.4) :

$$\|K(z)\| = O(\hbar) \tag{3.11}$$

uniformly on some contours $\Gamma \subset \rho(H_0)$, exponentially close to the spectrum of H_0, thus allowing to prove stability of components of $\sigma(H_0)$ separated from each others by distances which can be themselves exponentially small !

So let I be some isolated component of $\sigma(H_0)$, for example $I = \{E_0\}$ with E_0 an eigenvalue of H_0. It has to be understood that in general I depends on $\hbar$ since H_0 does. However in situations of interest, e.g., when the harmonic aproximation is valid in each of the wells, I will consist of eigenvalues varying slowly with respect to $\hbar$. We assume

i) $\Omega \subset G(I^+)$ where Ω is given by (3.3) and $I^+ = \sup I$.

(3.12) ii) $d_\Omega^{I^+} = \sup_{E > I^+} d_\Omega^E > 0.$

iii) There exists a contour Γ around I such that $\lim_{\hbar \downarrow 0} |\Gamma| = 0$ and $\operatorname{dist}(\Gamma, \sigma(H_0)) > \exp - a\hbar^{-1} d_\Omega^{I^+}$ for some a, $0 < a < 1$.

Assuming suitable smoothness properties for V in order to meet the conditions of Theorem 2.1 one has :

Lemma 3.1. Under conditions (3.12) i), ii), iii) one has for $\hbar$ small enough :

$$\sup_{z\in\Gamma} \| \chi_\Omega (H_0 - z)^{-1} \| < C$$

$$\sup_{z\in\Gamma} \| \chi_\Omega \nabla (H_0 - z)^{-1} \| < C\hbar^{-1}$$

for some constant C independent of $\hbar$.

This lemma together with (3.6) guarantees that (3.11) holds. Using estimates (3.9) and (3.10) immediately leads to :

Theorem 3.2. Let $I \subset \sigma(H_0)$ satisfy (3.12) i), ii), iii) and let P_0 be the corresponding spectral projection. Then there exists a projection operator P reducing H such that for $\hbar$ small enough :

For all $\varphi \in P\mathcal{H}$, $\|\varphi\| = 1$, there exists $\varphi_0 \in P_0 \mathcal{H}_0$ such that

$$\|\varphi - J\varphi_0\| < C\hbar\,|\Gamma| \tag{3.13}$$

and if $E = \langle \varphi, H\varphi \rangle$, $E_0 = \langle \varphi_0, H_0 \varphi_0 \rangle$:

$$|E - E_0| < C\hbar\,|\Gamma|^2$$

for some constant C independent of $\hbar$.

Roughly speaking this theorem says that a component I of the spectrum of H_0 separated from $\sigma(H_0)\backslash I$ by an exponentially small distance of the order of $\exp -\hbar^{-1} d_\Omega^{I^+}$ is stable under tunneling. In [BCD3] it is shown that the isolation distance can be taken in fact of the order $\exp -2a\hbar^{-1} d_\Omega^{I^+}$ for some $a<1$. In the next section we will use it to derive some of the most recent results on tunneling.

IV. Some examples

1) Symmetric multiple wells ([Ha1], [HeSj1], [JMSc1], [Si2]).

Assume that V has N wells W_i, $i = 1,\dots,N$ related to each others by euclidean transform so that the H_i's have the same spectrum. Then if E_0 is an eigenvalue of H_i with multiplicity p it is also an eigenvalue of H_0 with multiplicity Np. Assume V takes a finite absolute minimum in these wells; then the spacing of the lowest eigenvalues depend on the nature of this minimum ([CDS1], [HeSj1], [Si1]) (e.g., of the order $\hbar$ in the non-degenerate case, since then the harmonic approximation is valid, and of order $\hbar^2$ in the opposite extreme case of a completely degenerate (flat) minimum). If V is singular, i.e. it's minimum is $-\infty$, then the spacing is of negative order in $\hbar$, e.g. $\hbar^{-2}$ for a Coulomb like singularity. In all cases by choosing I in Theorem 3.2 to be one the lowest eigenvalues E_0 of H_0 one can choose Γ in (3.12) iii) to be a circle with center E_0 and radius $\exp -a\hbar^{-1} d_\Omega^{E_0^+}$ for any a, $0 < a < 1$. Then by Theorem 3.2 H has Np eigenvalues counting multiplicities in an neighbourhood of E_0 of diameter $O(\exp -2a\hbar^{-1} d_\Omega^{E_0^+})$

2) Periodic potentials ([Ha2] , [Si3] , [O])

A periodic potential corresponds to a situation of the previous type in which $N=\infty$, i.e., H_0 has infinitely degenerate spectrum. Let E_0 be one of the lowest eigenvalues of H_0, P_0 the corresponding infinite dimensional projection operator. Then by theorem 3.2 H has an invariant subspace $P\mathcal{H}$, consisting of periodic states (if one choose a "periodic" partition of unity) and the spectrum of H restricted to this infinite dimensional subspace has diameter smaller than $\exp -2a\hbar^{-1} d_\Omega^{E_0^+}$ for any a, $0 < a < 1$. This part of the spectrum (a "band") is separated from other bands by gaps which for non-degenerate minimum for example are of order $\hbar$.

3) Locally symmetric or slightly non-symmetric multiple wells

It can happen that only the bottom's of the wells arre related by euclidean transforms (i.e., some neighborhoods of the minimas). Then eigenvalues of H_0 naturally come into groups with the same order of separations as before but consisting now of eigenvalues exponentially close to each others. The same situation occurs if symmetric multiple wells are slightly perturbed outside a neighborhood of the minimas. One can again apply Theorem 3.2 with the result that the lowest part of $\sigma(H)$ also splits into groups with the same total multiplicity as those of $\sigma(H_0)$ and diameter of the same order as in 2). Let us also mention that a situation of this type occurs in hierachical potential [JMSc2] which appears as a suitable model for the analysis of Anderson's localization.

4) Small dissymmetries: The flea on the elephant ([HeSj2], [GGJ], [JMSc1], [Si4])

Consider for simplicity a symmetric double well (the elephant) with non-degenerate minimum; then the ground-state wave function for H is symmetric hence localized in both-wells. Assume now that the first well is slightly perturbed away from the minimum position. This produces an exponentially small shift of the ground-state energy for this well; this occurs for example for an arbitrarily small but with definite sign perturbation (the flea). So originally, in the symmetric case, H_0 had a degenerate ground-state energy E_0 as discussed in Example 1) above. The flea breaks this degeneracy and if it does it in such a way that the two lowest eigenvalues E_0 and $\tilde{E}_0$ are separated by a distance larger than $\exp -a\hbar^{-1} d_\Omega^{E_0^+}$ for some a, $0 < a < 1$, so that (3.12) iii) holds, then H will have according to

(3.13) it's ground state wave-function localized asymptotically in one well only. Thus the elephant has shifted his weight to one of the wells. This phenomenon might play a deep role in the explanation of some localization phenomena, e.g., in molecular physics, showing that fine structure effects might produce strong localization [Cl].

5)Shape resonances ([AHa], [BCD3], [BCDSig], [CDS3], [CDKS], [HeSj3], [HiSig])

We now consider situations where a finite number of wells, lying in some "interior " region W_1 and having finite minimum V_0, are surrounded by an infinite "exterior" region in which the potential is non-trapping for energies near these minima, the intersection $\Omega = W_1 \cap W_2$ supporting a finite barrier. In this case quantum particles having energies near V_0 can tunnel through barrier and escape to infinite whereas the corresponding classical particle would remain confined into the well.

The approch used before needs to be modified since $H_0 = H_1 \oplus H_2$ will have bound states (of H_1) embedded inthe continuum (of H_2). To deal with this type of situation it is convenient to use some type of complex scaling (see [BCDSig], [HiSig], for an elaborate general form of it) which implies that one has to deal with non-selfadjoint operators. Here we make the assumptions as in [CDKS]:

i) Let $V_0 = \inf\{ V(x);\ x \in W_1\}$; then $V_0 > \overline{\lim}_{|x| \to \infty, x \in W_2} V(x)$.

(4.1) ii) $\Omega = W_1 \cap W_2 \subset G(V_0 + \alpha_1)$ for some $\alpha_1 > 0$.

iii) There exits a sphere $K \subset W_2 \cap G(V_0 + \alpha_2)$ for some α_2, $0 < \alpha_2 < \alpha_1$.

This last condition is of purely technical nature (it does not appear in e.g. [BCD3], [HeSj3], [HiSig]) and is linked to the use of the so-called "exterior scaling" [Si5] in the approach presented below, defined by the transformation:

$$(U_\theta f)(r,\omega) := e^{\theta h(r)} f(r_\theta,\omega);\ \theta \in \mathbb{R}$$

where $\omega = x/|x|$, $r = |x|$, $h(r)$ is the characteristic function of the exterior of the sphere $K = \{x, |x| = r_0\}$ and $r_\theta = r_0 + e^{\theta h(r)}(r - r_0)$. We use the polar coordinate representation:

$$f \in L^2(\mathbb{R}^n) \to f(r,\omega) = r^{(n-1)/2} f(r\omega) \in L^2(\mathbb{R}^+ \times S^{n-1}).$$

It turns out that if V is dilation analytic for $r > r_0$ then $H_\theta = U_\theta H U_\theta^{-1}$ has an analytic extension to complex values of θ (see [CDKS] for a proof of this property). Eigenvalues of H_θ are known to be independant of θ as long as they remain isolated. Complex eigenvalues are the spectral resonances we are interesting in. To extend the perturbative analysis of section III with:

(4.2) $\quad H_{0,\theta} = H_1 \oplus H_{2,\theta}$

where $H_{2,\theta}$ is the analytic extension of $U_\theta H_2 U_\theta^{-1}$ to complex values of θ, and to complex eigenvalues, one has to control the resolvent and spectrum of $H_{2,\theta}$ in the inequalities (3.9), (3.10) and (3.11) using the following result of [BCD1] (generalised in [BCDSig]):

lemma 4.1. <u>Let</u> $1 = \chi^2 + \tilde{\chi}^2$ <u>be a</u> C^1 <u>partition of unity on</u> W_2 <u>with</u> $|\nabla\chi| < \infty$ <u>such that</u>: i) $\exists\, \alpha > 0$, supp $\chi \subset G(V_0 + \alpha)$

(4.3)

ii) $\exists\, S > 0$, $(r - r_0)/r\,(2(V - V_0) + r\partial V/\partial r) < -S$ on supp $\tilde{\chi}$.

<u>Then</u> $\exists\, \theta_0 > 0$ <u>and</u> $\nu := \{z \in \mathbb{C};\ |\mathrm{Re} z - V_0| < c_1,\ \mathrm{Im} z > -c_2 \mathrm{Im}\, \theta_0\}$ <u>for some positive constants</u> c_1 <u>and</u> c_2 <u>such that</u>

1) $\nu \subset \rho(H_{2,\theta_0})$

2) $\|(H_{2,\theta_0} - z)^{-1}\| = o(1)$ <u>as</u> $\hbar \searrow 0$ <u>uniformely in</u> $z \in \nu$.

Conditions (4.3) imply that the tail of the potential V in W_2 is non-trapping at energies close to V_0. Then the lemma asserts that in this case there is a neighbourhood of V_0 in the complex energy plane which is in the resolvent set of H_{2,θ_0}. This type of non-trapping assumption also plays a basic role in [HeSj3]. Then one gets the following generalization of a result of [CDKS] :

Theorem 4.2 (Existence of shape resonances). Let (4.1) and (4.3) hold and $I \subset \sigma(H_1)$ satisfy :

i) $\mathrm{dist}(I,V_0) = O(1)$ as $\hbar \searrow 0$

ii) $\mathrm{dist}(I, \sigma(H_1)\setminus I) > \exp -\hbar^{-1} d_\Omega^E$ for some $E > V_0$.

Let P_0 be the corresponding spectral projection for H_1; then there exists a projection operator P_{θ_0} reducing H_{θ_0} for some θ_0, $\mathrm{Im}\theta_0 > 0$, such that for $\hbar$ small enough :

1) $\dim P_{\theta_0} = \dim P_0$

2) For all $\varphi \in P_{\theta_0}\mathcal{H}$, $\|\varphi\|=1$ there exists $\varphi_0 \in P_0 L^2(W_1)$ such that

$$\|\varphi-\varphi_0\| \leqslant o(\exp -a\hbar^{-1} d_\Omega^{V_0^+})$$

for all a, $0 < a < 1$, and if $E = \langle\varphi, H_{\theta_0}\varphi\rangle$, $E_0 = \langle\varphi_0, H_0\varphi_0\rangle$:

$$|E-E_0| \leqslant o(\exp -2a\hbar^{-1} d_\Omega^{V_0^+})$$

where C is constant independent of $\hbar$.

6) Barrier top resonances [(BCD2] , [Sj]).

As at last example of the kind of ideas developed above we describe a result of [BCD2] concerning the resonance spectrum due to a potential V satisfying the following assumption :

i) V has an absolute maximum at $x = 0$.

ii) V is dilation analytic and $V_\theta(x) = V(e^\theta x)$ satisfies for $|\theta|$ small :

(4.4) $$V_\theta(x) = V_0 - e^{2\theta} x.Ax + O(x^3) \quad \text{with} \quad A \geqslant \alpha > 0$$

iii) V satisfies (4.3) ii) with $r_0 = 0$ and $\tilde{\chi}$ having support outside a neighbourhood of $x = 0$.

Let $K = -\Delta + x.Ax$ and $H_\theta = -e^{-2\theta}\hbar^2\Delta + V_\theta$; then one has :

Theorem 4.3. Let V satisfy (4.4); then for some θ_0, $\text{Im}\theta_0 > 0$, and for all $e \in \sigma(K)$ there exists a disk $B(\hbar)$ centered at $V_0 - i\hbar e$ with radius $O(\hbar^{3/2})$ such that for $\hbar$ small enough H_{θ_0} has only discrete spectrum inside $B(\hbar)$ with total algebraic multiplicity equal to the multiplicity of e.

This provides another example of a classical trapping situation (a particle with energy V_0 might take an infinite time to reach $x = 0$) leading to discrete spectrum for the quantum hamiltonian.

ACKNOWLEDGMENTS : One of us, [JMC] would like to thank P. Hislop and I. Sigal for many discussions about the material of this paper.

REFERENCES

[A] Agmon, S: Lectures on exponential decay of solutions of second order elliptic operators. Princeton University Press, (1982).

[AsHa] Asbaugh, M. Harrell, E. : Comm. Math. Phys. **83**, 151, (1982)

[BCD1] Briet, Ph., Combes, J.M., Duclos, P. : On the location of resonances in the semi-classical limit I : Preprint Marseille CPT86/P. 1829. To appear in Jour.Math.Anal.Appl.

[BCD2] Briet, P., Combes, J.M., Duclos, P. : On the location of resonances in the semi-classical limit II : Preprint Marseille CPT 86/P. 1884. To appear in Com. Part.Diff.Equ.

[BCD3] Briet, Ph., Combes, J.M., Duclos, P. : Spectral stability through tunneling. To appear.

[BCDSig] Briet, Ph., Combes, J.M., Duclos, P., Sigal, I. : Spectral deformations, non-trapping and resonances. To appear.

[CDS1] Combes, J.M., Duclos, P., Seiler, R. : J. Funct. Anal. **52**, 257, (1983).

[CDS2] Combes, J.M., Duclos,P., Seiler,R.: Comm. Math. Phys. **92**, 229, (1983).

[CDS3] Combes,J.M.,Duclos,P., Seiler, R. : Springer Lecture Notes in Physics, **211**, 64, (1984).

[CDKS] Combes, J.M., Duclos, P., Klein, M., Seiler, R. : The shape resonance, Marseille Preprint CPT85 / P. 1797. To appear in Com. Math. Phys.

[Cl] Claverie, P. , Jona-Lasinio, G. : Preprint LPTHE 84142.

[GGJ] Graffi, S., Grecchi, V., Jona-Lasinio, G. : J. Phys. **A17**, 2935, (1984).

[Ha1] Harrell, E. : Comm. Math. Phys. **119**, 351, (1979)

[Ha2] Harrell, E.: Ann.Phys.**75**,239,(1980).

[HeSj1] Helffer, B., Sjöstrand, J. : Comm. Part.Diff.Equ. **9**, 337, (1984).

[HeSj2] Helffer, B., Sjöstrand, J. : Ann. Inst.H.Poincaré **2**, 127, (1985).

[HeSj3] Helffer, B., Sjöstrand, J. : Resonances en limite semi-classique. To appear Bulletin SMF.

[HiSig] Hislop, P. Sigal, I. : Proceedings of this conference.

[Hu] W.Hunziker: Distortion Analyticity and molecular resonance Curves. To appear in Ann.Inst.H.Poincaré.

[JMSc1] Jona-Lasinio, G., Martinelli, F., Scoppola, E.: Comm. Math. PHys. **80**, 223, (1981).

[JMSc2] Jona-Lasinio, G., Martinelli, F., Scoppola, E.: Ann. Inst. H. Poincaré **43**, 2,(1985).

[K] Kato, T. : Perturbation theory for linear operators, Berlin, Heidelberg, New York, Springer 1966.

[O] Outassourt, A. : Comportement semi-classique pour l'opérateur de Schrödinger à potentiel périodique. Thèse de 3ème cycle, Nantes, (1985).

[Si1] Simon B. : Ann. Inst.H.Poincaré **38**, 295, (1983).

[Si2] Simon B. : Ann. Math. **120**, 89, (1984).

[Si3] Simon B. : Ann. Phys. **158**, 415, (1984).

[Si4] Simon B. : Jour. Funct. Anal. **63**, 123 (1985).

[Si5] Simon B. : Phys. Letters **71A**, 211, (1979).

[Sj] Sjöstrand,J. : Semi classical resonances generated by non-degenerate critical points. Preprint, Lund No.1986 : 1/ISSN0347-475.

J.M. Combes
Mathematics Department
University of California
Irvine, California 92717 USA

Ph. Briet*, P. Duclos*
Departement de Mathématiques
Université de Toulon, 83130 La Garde FRANCE

* Postal address: CPT, CNRS, Luminy Case 907, F13288 Marseille cedex 9 FRANCE

Discrete sets of coherent states and their use in signal analysis. *

Ingrid Daubechies **

Courant Institute of Mathematical Sciences
New York University
251 Mercer Street
New York NY 10012

Abstract. We discuss expansions of L^2-functions into $\{\phi_{mn} ; m,n \in Z\}$, where the ϕ_{mn} are generated from one function ϕ, either by translations in phase space, i.e. $\phi_{mn}(x) = e^{imp_0x}\phi(x-nq_0)$, $(p_0, q_0$ fixed), or by translations and dilations, i.e. $\phi_{mn}(x) = a_0^{-m/2}\phi(a_0^{-m}x-nb_0)$. These expansions can be used for phase space localization.

1. Introduction.

We present here some recent results concerning expansions of functions $f \in L^2(\mathbb{R})$ with respect to discrete sets of coherent states. We shall distinguish two cases, the Weyl-Heisenberg case, where

$$g_{mn}(x) = e^{imp_0x} g(x-nq_0) ,$$

and the affine case, where

$$h_{mn}(x) = a_0^{-m/2} h(a_0^{-m}x-nb_0) .$$

In both cases the parameters m,n range over all of Z, and we shall discuss the maps T_{WH}, T_{AFF} from $L^2(\mathbb{R})$ to $l^2(Z^2)$ defined by

$$(T_{WH}f)_{mn} = \langle g_{mn}, f\rangle \ , \ (T_{AFF}f)_{mn} = \langle h_{mn}, f\rangle .$$

These maps depend on the parameters $p_0,q_0>0$, $a_0>1$ and $b_0>0$, respectively, as well as on the functions $g, h \in L^2(\mathbb{R})$. The function h should satisfy the additional condition $\int |k|^{-1}|\hat{h}|^2 < \infty$, where $\hat{}$ denotes the Fourier transform.

The g_{mn} , h_{mn} are in fact coherent states associated with respectively the Weyl-Heisenberg group and the affine or $ax+b$-group, with labels restricted to discrete subsets of the parameter range. See [1] for more information concerning coherent states in general; a more extensive discussion of their connection to this paper can be found in the introduction of [2].

The maps T_{WH} , T_{AFF} and their properties are of interest for signal analysis. In engineering literature, the map T_{WH} is known as the "short-term Fourier transform". This is a procedure aimed at defining and computing a time-dependent frequency representation for a signal f. To do this,

* This paper is partially supported by NSF grant MCS 8301662.

** "Bevoegdverklaard Navorser" at the Belgian National Science Foundation; on leave from Vrije Universiteit Brussel, Belgium.

the signal f is multiplied by a "window function" g (often of compact support), and the Fourier coefficients of this product are computed. The process is repeated for different positions of the window g, leading to a frequency profile for f at different times. The resulting coefficients constitute exactly the series ($T_{WH} f$), with appropriately chosen p_0 , q_0 .

The use of T_{AFF} in signal analysis is not as wide-spread. It was first proposed by J.Morlet for the analysis of seismografic signals, where it seems to lead to better numerical results than T_{WH} [3]. The map T_{AFF} can probably be used for many other types of signals as well. Since the human ear analyses frequency in the same logarithmic way as T_{AFF} does (the number of m-levels needed to cover the region $\nu_1 < \nu < \nu_2$ is the same as for the region $2\nu_1 < \nu < 2\nu_2$), signal analysis based on T_{AFF} may be more efficient than the short-time Fourier transform for the analysis, filtering and reconstruction of speech or music.

It is a remarkable coincidence that the map T_{AFF} is also of interest to harmonic analysis, where techniques using dilations and translations have been extensively used for years (see e.g. [4]). Special choices for h_{mn} can be found in e.g. [5]; in [6] Y.Meyer constructs a function h, with $\hat{h} \in C_0^\infty$, such that the associated h_{mn} are an orthonormal base for $L^2(\mathbb{R})$ (this is generalized to more than one dimension in [7]); this base turns out to be an unconditional base for almost all useful function spaces [7][8].

We shall here discuss some mathematical properties of T_{WH} , T_{AFF} that are relevant for signal analysis. In remarks valid for both maps we shall drop the index WH or AFF, and we shall use the notation ϕ_{mn} in all formulas valid for both g_{mn} and h_{mn}.

It is clear that T should be injective if we want to be able to reconstruct f from Tf. In order to avoid instabilities in numerical computations, we impose the stronger condition that T should have a bounded inverse on its range. If we assume that T is defined on all of $L^2(\mathbb{R})$, this implies that T is bounded, by the closed graph theorem. All this means that we require, for some constants A, B , $0 < A \leq B < \infty$, that

$$A \leq T^*T \leq B \ , \tag{1}$$

or, equivalently, that

$$\forall f \in L^2(\mathbb{R}) \ : \ A \, ||f||^2 \leq \sum_{m,n} |<\phi_{mn}, f>|^2 \leq B \, ||f||^2 \ . \tag{2}$$

A set of vectors ϕ_{mn} satisfying (2) is called a *frame* (after Duffin and Schaeffer [9]). If the ϕ_{mn} are a frame, then, for all $f \in L^2(\mathbb{R})$,

$$f = \sum_{m,n} \psi_{mn} <\phi_{mn}, f> \ , \tag{3}$$

with $\psi_{mn} = (T^*T)^{-1}\phi_{mn}$, where the series converges in L^2-norm :

$$f = L^2\text{-}\lim_{K\to\infty} \sum_{\substack{m,n \\ |m|,|n| \leq K}} \psi_{mn} <\phi_{mn}, f>$$

Note that while (3) resembles the expansion with respect to biorthogonal bases, it definitely is not the same: the ϕ_{mn} need not be a base, and $<\psi_{mn}, \phi_{m'n'}> \neq \delta_{mm'}\,\delta_{nn'}$ in general.

We shall call the constants A, B in (1), (2) *frame bounds*. For some special frames, the largest lower frame bound and the smallest upper frame bound coincide, i.e.

$$T^*T = A\,\mathbf{1} \quad \text{or} \quad \forall f \in L^2(\mathbb{R}) : \sum_{m,n} |<\phi_{mn}, f>|^2 = A\,||f||^2 \ .$$

In this case we say that the ϕ_{mn} constitute a *tight frame*. For a tight frame the expansion formula (3) becomes

$$f = A^{-1} \sum_{m,n} \phi_{mn} <\phi_{mn}, f> \ . \tag{4}$$

Note again that the ϕ_{mn} in (4) need not be an orthogonal base.

This paper is organized as follows. Section 2 addresses frame questions. Basically we show two things, for the affine case as well as for the Weyl-Heisenberg case : 1) how to construct tight frames , 2) how to determine, for a *given* function g or h a range of parameters (p_0, q_0) or (a_0, b_0) such that the resulting g_{mn}, h_{mn} constitute a frame. The same methods can also be used to compute frame bounds, and we give a few numerical examples. In section 3 we discuss phase space localization. Let us illustrate what this means by an example in the Weyl-Heisenberg case. Take $g(x) = \pi^{-1/4} \exp(-x^2/2)$, $q_0 = p_0 = \pi^{1/2}$. In this case (see [10]) the g_{mn} do constitute a frame. Since $\int dx\, x\,|g(x)|^2 = 0 = \int dy\, y\,|\hat{g}(y)|^2$, it follows that g_{mn} is concentrated, in phase space, around the point (mp_0, nq_0) , i.e. $\int dx\, x\,|g_{mn}(x)|^2 = nq_0$, $\int dy\, y\,|(g_{mn})^\wedge(y)|^2 = mp_0$. Therefore the inner product $<g_{mn}, f>$ "measures" the phase space content of the signal f around the point (mp_0, nq_0) . If one knows a priori that f is mostly localized in phase space, i.e.

$$\int_{|t| \geq T} dt\ |f(t)|^2 \leq \delta \ , \quad \int_{|\omega| \geq \Omega} d\omega\ |\hat{f}(\omega)|^2 \leq \delta \ , \text{ for some } \delta \ll ||f||^2 \ ,$$

then it seems reasonable to expect that the partial reconstruction of f, using only those $<g_{mn}, f>$ for which $|m\,p_0| \leq \Omega$, $|n\,q_0| \leq T$, would be close to f. This is in fact correct; a precise statement is given in section 3. A similar result holds for h_{mn}-frames.

Part of the results in section 2 have been published before [2][10]; the rest of the material has not been published yet. An extended version of these results will appear elsewhere [11].

It is a great pleasure to thank Alex Grossmann for having drawn my attention to this subject, and for an always stimulating collaboration. I would also like to thank Percy Deift for many interesting discussions on this subject.

2. Frames and frame bounds.

2A. The Weyl-Heisenberg case

In this case there exists a critical value for the product p_0q_0. The following theorem states that if $p_0q_0 = 2\pi$, only functions g which are either not very smooth or don't decay very fast, give rise to a frame.

Theorem 1. *Choose $g \in L^2(\mathbb{R})$, $p_0q_0 = 2\pi$. If the associated g_{mn} constitute a frame, then either $xg \notin L^2$ or $g' \notin L^2$.*

This theorem was first formulated by R.Balian [12]; his proof contains a gap that was filled recently by R.Coifman and S.Semmes [13]. We give here a sketch of their proof.

Proof. The proof uses the Zak transform U_Z which maps $L^2(\mathbb{R})$ unitarily onto $L^2([0,1]^2)$,

$$(U_Zf)(t,s) = q_0^{1/2} \sum_{l \in Z} e^{2\pi i t l} f(q_0(s-l)) \ .$$

U_Zf can be defined for $(t,s) \in \mathbb{R}^2$ (instead of $[0,1]^2$) ; in that case it satisfies

$$(U_Zf)(t+1,s) = (U_Zf)(t,s) \ , \tag{5a}$$

$$(U_Zf)(t,s+1) = e^{2\pi i t} (U_Zf)(t,s) \ . \tag{5b}$$

Using $(U_Zg_{mn})(t,s) = e^{-2\pi i t n} e^{2\pi i m s} (U_Zg)(t,s)$, one finds that the g_{mn} constitute a frame, with frame bounds A, B , if and only if

$$A^{1/2} \le |(U_Zg)(t,s)| \le B^{1/2} \ . \tag{6}$$

If U_Zg were continuous, then (5), together with (6), would lead to a contradiction (see [14]; a very detailed proof can be found in [15]). Basically the argument is as follows. If U_Zg were continuous, then $\log U_Zg$ would be a continuous, one-valued function (because of (6)). But (5) allows one to compute two *different* values for $\log(U_Zg)(1,1)$, corresponding to the two paths $(0,0) \to (0,1) \to (1,1)$ and $(0,0) \to (1,0) \to (1,1)$. This is clearly a contradiction.
Unfortunately, U_Zg need not be continuous. One easily checks that if g' , $xg \in L^2$, then $\partial_t U_Zg$, $\partial_s U_Zg \in L^2_{loc}(\mathbb{R}^2)$. This does not imply that U_Zg is continuous (this is where Balian's proof failed : he implicitly assumes that $\nabla G \in L^2$ implies that G is continuous, which is true in 1 but not in 2 dimensions), but one can use this to show that there exists $r > 0$, with r small enough to ensure that the *continuous* function G_r ,

$$G_r(t,s) = r^{-2} \int_{|t'-t| \le r} dt' \int_{|s'-s| \le r} ds' \ (U_Zg)(t',s') \ ,$$

is close enough to U_Zg to satisfy

$$A^{1/2}/2 \le |G_r(t,s)| \le 2B^{1/2}$$

$$G_r(t+1,s) = G_r(t,s)$$

$$G_r(t,s+1) = e^{2\pi it}\, G_r(t,s) + R(t,s)$$

with $|R(t,s)| \le A^{1/2}\, \pi/16$. This then leads to a contradiction.

□

This theorem shows that it is impossible, starting from a reasonably smooth and decreasing function g, to construct a frame if $p_0q_0 = 2\pi$. An example of this phenomenon is the von Neumann lattice, also called the set of Gabor wave functions, which correspond to $p_0q_0 = 2\pi$, $g(x) = \pi^{-1/4}\exp(-x^2/2)$. The corresponding g_{mn} (which were, in fact, proposed by Gabor [16] for signal analysis purposes) lead to very bad numerical reconstruction results [17], which is due to the fact that the g_{mn} do not constitute a frame (see also [10]), even though T_{WH} is injective [18][19][14].

For $p_0q_0 > 2\pi$ it is known that T_{WH} is not even injective if $g(x) = \pi^{-1/4}\exp(-x^2/2)$ [18][19]. I believe this result to be generally true, i.e. T_{WH} should not be injective, if $p_0q_0 > 2\pi$, for *any* $g \in L^2$. For rational values of $p_0q_0/2\pi$ this can easily be shown, again by means of the Zak transform. The proof for irrational values of $p_0q_0/2\pi$ seems to be harder.

For any choice of $p_0q_0 < 2\pi$, it is possible to construct functions $g \in C_0^\infty(\mathbb{R})$ such that the associated g_{mn} constitute a tight frame. The following construction works if $\pi \le p_0q_0 \le 2\pi$. Similar constructions can be made if $p_0q_0 < \pi$ [2]. Let ν be an increasing C^∞-function from $\mathbb{R}$ to $\mathbb{R}$ such that $\nu(x) = 0$ if $x \le 0$, $\nu(x) = 1$ if $x \ge 1$. Define $g(x)$ by

$$g(x) = \begin{cases} \sin\left[\frac{\pi}{2}\,\nu\left(\frac{p_0x}{2\pi - p_0q_0}\right)\right] & \text{if } x \le \frac{2\pi}{p_0} - q_0 \\ 1 & \text{if } \frac{2\pi}{p_0} - q_0 \le x \le q_0 \\ \cos\left[\frac{\pi}{2}\,\nu\left(\frac{p_0(x-q_0)}{2\pi - p_0q_0}\right)\right] & \text{if } x \ge q_0 . \end{cases}$$

Since $\operatorname{supp} g = [0, \frac{2\pi}{p_0}]$, one finds, $\forall f \in L^2(\mathbb{R})$: $\sum_m |\langle g_{mn}, f\rangle|^2 = \frac{2\pi}{p_0}\int dx\, |g(x-nq_0)|^2\, |f(x)|^2$. Since, by construction, $\sum_n |g(x-nq_0)|^2 = 1$, this implies that $\sum_{m,n} |\langle g_{mn}, f\rangle|^2 = \frac{2\pi}{p_0}\, \|f\|^2$, i.e. the g_{mn} constitute a tight frame.

In the above construction, p_0 and q_0 are fixed, and an appropriate g is constructed. In situations where g is already fixed, the following theorem gives sufficient conditions on p_0, q_0 (and g) ensuring that the g_{mn} are a frame.

Theorem 2.

If 1. $m(q_0) = \inf_{x \in [0,q_0]} \sum_n |g(x-nq_0)|^2 > 0$

2. $\sup_{s \in \mathbb{R}} \left[(1+s^2)^{(1+\epsilon)/2} B(s,q_0) \right] = C_\epsilon < \infty$ *for some* $\epsilon > 0$,
where $B(s) = \sup_{x \in [0,q_0)} \sum_n |g(x-nq_0)|\,|g(x+s-nq_0)|$,

then there exists a $p_0^c > 0$ *such that*

$\forall p_0 < p_0^c$: *the* g_{mn} *are a frame*

$\forall p_0 > p_0^c$: $\exists\, p'_0 \in [p_0^c, p_0]$ *for which the* g_{mn} *are not a frame* .

Proof. Using the Poisson formula one finds

$$\sum_{m,n} |\langle g_{mn}, f\rangle|^2 = \frac{2\pi}{p_0} \sum_{n,k} \int dx\, g(x-nq_0)\, g(x-nq_0-\frac{2\pi}{p_0}k)^*\, f(x)^*\, f(x-\frac{2\pi}{p_0}k) \ . \tag{7}$$

Via the Cauchy-Schwarz inequality this leads to

$$\sum_{m,n} |\langle g_{mn}, f\rangle|^2 \geq \frac{2\pi}{p_0} \|f\|^2 \left\{ m(q_0) - 2C_\epsilon \sum_{k=1}^{\infty} \left[1 + \left(\frac{2\pi k}{p_0}\right)^2 \right]^{-(1+\epsilon)/2} \right\} .$$

Since this lower bound is positive for p_0 small enough, we find
$p_0^c = \inf\{\, p_0$; *the* g_{mn} *do not constitute a frame* $\} > 0$.

□

Remarks

1. If $xg,\, g' \in L^2$, then $p_0^c \leq 2\pi/q_0$, by theorem 1.
2. Equation (7) can also be used to show that T_{WH} is bounded for all p_0 .
3. Condition 2. is satisfied if e.g. $(1+x^2)^{(1+\epsilon)/2}\,(|g| + |g'|) \in L^2(\mathbb{R})$.
4. If $\sum_n |g(x-nq_0)|^2$ is continuous, then condition 1. is necessary.
5. One can use the argument in the proof to compute frame bounds. The following table lists some such frame bounds, for $g(x) = \pi^{-1/4} \exp(-x^2/2)$.

q_0	p_0	A	B
1.	π	.60	3.55
	$3\pi/2$	.03	3.55
2.	$\pi/2$	1.60	2.43
	$3\pi/4$	.58	2.09

2B. The affine case.

There is no analog of the Zak transform for the affine case, and therefore no analog of theorem 1. For any $a_0 > 1$, $b_0 > 0$ it is possible to construct h such that the associated h_{mn} constitute a frame. One such example is given by the following construction. We take the same function ν as in § 2A, and we define $l = 2\pi/[b_0(a_0^2-1)]$. Then $\hat{h}$ is defined by

$$\hat{h}(y) = \begin{cases} 0 & \text{if } y \le l \\ \sin\left[\frac{\pi}{2}\,\nu\left(\frac{y-l}{l(a_0-1)}\right)\right] & \text{if } l \le y \le a_0 l \\ \cos\left[\frac{\pi}{2}\,\nu\left(\frac{y-a_0 l}{a_0 l(a_0-1)}\right)\right] & \text{if } y \ge a_0 l \end{cases}$$

Clearly $\operatorname{supp} \hat{h} = [l, a_0^2 l]$. Define $h^+(y) = h(y)$, $h^-(y) = h(y)^*$. Then the $h^{\pm}_{mn}$ constitute a frame. Explicit calculation shows that

$$\sum_{+,-}\sum_{n} |\langle h^{\pm}_{mn}, f\rangle|^2 = \frac{2\pi}{b_0}\int_0^\infty dy\ |\hat{h}(a_0^m y)|^2 \left[\ |\hat{f}(y)|^2 + |\hat{f}(-y)|^2\ \right].$$

By construction $\sum_m |\hat{h}(a_0^m y)|^2 = \chi_{(0,\infty)}(y)$. Hence $\sum_{+,-}\sum_{m,n} |\langle h^{\pm}_{mn}, f\rangle|^2 = \frac{2\pi}{b_0}\|f\|^2$.

Remark

One can also choose $h^1 = \operatorname{Re} h$, $h^2 = \operatorname{Im} h$; the set $\{h^{\epsilon}_{mn}\,; m,n \in Z, \epsilon = 1 \text{ or } 2\}$ again constitutes a tight frame.

As in § 2A, one can again consider the situation in which h itself is fixed, and determine a_0, b_0 such that the associated h_{mn} constitute a frame. This allows, in particular, choices where h rather than $\hat{h}$ has compact support.

Theorem 3.

If 1. $m(a_0) = \inf_{1 \le |y| \le a_0} \sum_m |\hat{h}(a_0^m y)|^2 > 0$

2. $\sup_{s \in \mathbb{R}} \left[(1+s^2)^{(1+\epsilon)/2} B(s) \right] = C_\epsilon < \infty$ for *some* $\epsilon > 0$,
with $B(s) = \sup_{1 \le |y| \le a_0} \sum_m |\hat{h}(a_0^m y)|\ |\hat{h}(a_0^m y + s)|$,

then there exists $b_0^c > 0$ such that, for all $b_0 < b_0^c$, the associated h_{mn} constitute a frame.

Proof. The proof is entirely analogous to the WH-case : one uses $(h_{mn})^{\wedge}(y) = a_0^{m/2} \exp[inb_0 a_0^m y]\, \hat{h}(a_0^m y)$, and applies the Poisson formula.

□

Again this can be used to compute frame bounds for given h, a_0, b_0. The following two tables give the values of b_0^c for a few a_0-values, and give frame bounds for $b_0 = 1. < b_0^c$, for the two functions $h(x) = 2\, 3^{-1/2}\, \pi^{-1/4}\, (1-x^2) \exp(-x^2/2)$ and $h(x) = \sin x\ \chi_{|x| \le \pi}$.

$$h(x) = 2\,3^{-1/2}\,\pi^{-1/4}\,(1-x^2)\,\exp(-x^2/2)$$

a_0	b_0^c	A for $b_0=1.$	B for $b_0=1.$	B/A for $b_0=1.$
1.25	2.05	10.53	10.65	1.01
1.5	2.05	5.79	5.86	1.01
2.0	1.91	3.25	3.57	1.10

$$h(x) = \pi^{-1/2}\,\sin x\,\chi_{|x|\le\pi}$$

a_0	b_0^c	A for $b_0=1.$	B for $b_0=1.$	B/A for $b_0=1.$
1.25	2.95	18.51	20.81	1.12
1.5	2.91	10.09	11.55	1.14
2.0	2.80	5.70	6.98	1.22

3. Phase space localization.

We explained in the introduction the intuition behind the phase space localization results we present here. The precise statement is :

Theorem 4.
Let g be a function such that, for some $\delta > 0$, $\int dx\left[|g(x)|^2 + |g(x)|\,|g'(x)|\right](1+x^2)^{1+\delta} < \infty$, and $\int dy\left[|\hat{g}(y)|^2 + |\hat{g}(y)|\,|\hat{g}'(y)|\right](1+y^2)^{1+\delta} < \infty$. Let $p_0, q_0 > 0$. Suppose that the g_{mn} constitute a frame, with frame bounds A, B. Then, $\forall \epsilon > 0$, $\exists\ \alpha(\epsilon)$, $\beta(\epsilon)$ such that, for all $T, \Omega > 0$,

$$\Big\| f - \sum_{\substack{m,n\in Z\\ |mp_0|\le\Omega+\alpha(\epsilon)\\ |nq_0|\le T+\beta(\epsilon)}} \psi_{mn}\,\langle g_{mn}, f\rangle \Big\|$$

$$\le (B/A)^{1/2}\left\{\left[\int_{|t|\ge T} dt\ |f(t)|^2\right]^{1/2} + \left[\int_{|\omega|\ge\Omega} d\omega\ |\hat{f}(\omega)|^2\right]^{1/2} + \epsilon\,\|f\|\right\}$$

Proof. As in the introduction, $\psi_{mn} = (T^*T)^{-1} g_{mn}$. We denote $|mp_0| \le \Omega+\alpha$, $|nq_0| \le T+\beta$ as $(m,n) \in box$. Introducing $(P_Tf)(t) = f(t)\,\chi_{|t|\le T}$ and $(Q_\Omega f)^\wedge(\omega) = \hat{f}(\omega)\,\chi_{|\omega|\le\Omega}$, one finds, after some manipulations,

$$\| f - \sum_{m,n \in box} \psi_{mn} <g_{mn}, f> \| \leq \sup_{\|\phi\|=1} \sum_{m,n \notin box} |<\phi, \psi_{mn}>| \, |<g_{mn}, f>|$$

$$\leq (B/A)^{1/2} \left[\|(1-P_T)f\| + \|(1-Q_\Omega)f\| \right]$$

$$+ A^{-1/2} \left\{ \left(\sum_{\substack{m,n \\ |mp_0|>\Omega+\alpha}} |<g_{mn}, Q_\Omega f>|^2 \right)^{1/2} + \left(\sum_{\substack{m,n \\ |nq_0|>T+\beta}} |<g_{mn}, P_T f>|^2 \right)^{1/2} \right\} .$$

Using the Poisson formula and the Cauchy-Schwarz formula leads to

$$\sum_{\substack{m,n \\ |nq_0|>T+\beta}} |<g_{mn}, P_T f>|^2 \leq \frac{2\pi}{p_0} \left[\sum_k \left(1 + k^2 \frac{4\pi^2}{p_0^2} \right)^{-(1+\delta)/2} \right] \|P_T f\|^2$$

$$\times \int_{|x|>\beta} dx \left[|g(x)|^2 (1+x^2)^{1+\delta} + 2\, q_0\, |g(x)| \, (1+x^2)^\delta \left(|g'(x)| \, (1+x^2) + x\, |g(x)| \right) \right] .$$

The Q_Ω-term can be estimated similarly. The theorem then follows easily.

□

A similar theorem holds for the affine case.

The important fact about theorem 4 is that $\alpha(\epsilon)$, $\beta(\epsilon)$ *are independent* of T, Ω . Note that $\alpha(\epsilon)$, $\beta(\epsilon) \to \infty$ as $\epsilon \to 0$, i.e. infinite precision is only reached by taking the infinite collection $(<g_{mn}, f>)_{m,n \in Z}$. If both $g, \hat{g}$ have rapid decay at ∞ (e.g. g Gaussian), then α, β are still reasonable for fairly small ϵ (for g Gaussian, $q_0 = p_0 = \pi^{1/2}$, one has $\alpha(\epsilon) = \beta(\epsilon) \underset{\epsilon \to 0}{\sim} (C + |\ln \epsilon|)^{1/2}$).

References.

[1] J.R.Klauder and B.-S. Skagerstam, "Coherent States. Applications in Physics and Mathematical Physics". World Sci. Pub. (Singapore, 1985).

[2] I.Daubechies, A.Grossmann and Y.Meyer, "Painless non-orthogonal expansions." J. Math. Phys. 27 (1986) 1271-1283.

[3] J.Morlet, G.Arens, I.Fourgeau and D.Giard, "Wave propagation and sampling theory." Geophysics 47 (1982) 203-236.

[4] E.Stein, "Singular integrals and differentiability prperties of functions." Princeton University Press (1970).

[5] M.Frazier and B.Jawerth, "Decomposition of Besov spaces." To be published.

[6] Y.Meyer. "La transformation en ondelettes et les nouveaux paraproduits." To be published in Actes du Colloque d'Analyse non lineaire du Ceremade, Univ. de Paris-Dauphine.

[7] P.G.Lemarie and Y.Meyer, "Ondelettes et bases hilbertiennes." To be published in Revista Ibero-Americana.

[8] R.R. Coifman and Y.Meyer, "The discrete wavelet transform." To be published.
Y.Meyer, "Principe d'incertitude, bases hilbertiennes et algebres d'operateurs." Seminaire Bourbaki, 1985-1986, nr. 662.

[9] R.J.Duffin and A.C.Schaeffer, "A class of nonharmonic Fourier series." Trans. Am. Math. Soc. 72 (1952) 341-366.

[10] I.Daubechies and A.Grossmann, "Frames in the Bargmann space of entire functions."

Submitted for publication.

[11] I.Daubechies, "Frames of coherent states." In preparation.

[12] R.Balian, "Un principe d'incertitude fort en theorie du signal ou en mecanique quantique." C. R. Acad. Sci. Paris 292 (serie 2) (1981) 1357-1362.

[13] R.R.Coifman and S.Semmes, private communication.

[14] H.Bacry, A.Grossmann and J.Zak, "Proof of completeness of lattice states." Phys. Rev. B12 (1975) 1118-1120.

[15] A.J.E.M.Janssen, "Bargmann transform, Zak transform, and coherent states." J. Math. Phys. 23 (1982) 720-731.

[16] D.Gabor, "Theory of communication." J. Inst. Elec. Engrs. (London) 93 (1946) 429-457.

[17] A.J.E.M.Janssen, "Gabor representation and Wigner distribution of signals." Proc. IEEE Acoust., Speech and Signal Processing, April 1983, 41 B.2.1. - 41 B.2.4.

[18] V.Bargmann, P.Butero, L.Girardello and J.R.Klauder, "On the completeness of coherent states." Rep. Mod. Phys. 2 (1971) 221-228.

[19] A.M.Perelomov, "On the completeness of a system of coherent states." Theor. Math. Phys. 6 (1971) 156-164.

INFORMATION, UNCERTAINTY AND THE SINGULAR VALUE DECOMPOSITION OF THE FILTERED FOURIER TRANSFORMATION

B. DeFacio
Department of Physics & Astronomy University of Missouri-Columbia
Columbia, MO 65211 USA
and
O. Brander
Institute for Theoretical Physics Chalmers University of Technology
S 412 96 Göteborg, SWEDEN

1. Introduction

There are interesting and fundamental questions on the information which can, in principle, be extracted from a noisy signal. Many of these were raised by Wiener [1] in his monograph on time series. Much of communications theory [2-6], the weak scattering case of inverse scattering theory [7-9] and image restoration in optics [10-12] are based on the filtered Fourier transform. In d-dimensions; $\vec{x}, \vec{k} \in R^d$; the filtered Fourier transformation operator K is given by

$$(Kf)\ (\vec{k}) = \frac{1}{(2\pi)^d} \int P(\vec{k}) e^{-i\vec{k}.\vec{x}} Q(\vec{x}) f(\vec{x}), \qquad (1)$$

where P and Q are the wave number and direct space filter functions [9]. If P and Q are assumed to be (1) real, piecewise C^2 and (2) with (a) compact support or (b) to go to zero at infinity exponentially then

$$K : L^2(R^d) \to L^2(R^d)$$

is a compact operator [13]. There is no loss of generality in choosing P,Q real because Gori and Guattari [12] have shown that a continuously varying phase factor can be absorbed into the singular functions.

It is convenient to define the dimensionless variables

$$\vec{p} = \frac{\vec{k}}{k_o}, \quad \vec{q} = \frac{\vec{x}}{x_o}, \quad c = k_o x_o, \qquad (2)$$

where k_o is the wave-number bandwidth, x_o is the direct space bandwidth and c is the space-momentum bandwidth. The bandwidth product c is the most important parameter in the theory.

The scaled Fourier transform operator K_s and its adjoint $K_s^\dagger$ are given by

$$(K_s\psi)(\vec{p}) = \int P(\vec{p})e^{-ic\vec{p}\cdot\vec{q}}Q(\vec{q})\psi(\vec{q})d^d\vec{q}, \tag{3a}$$

and

$$(K_s^\dagger\chi)(\vec{q}) = \int Q(\vec{q})e^{ic\vec{p}\cdot\vec{q}}P(\vec{p})\chi(\vec{p})d^d\vec{p}, \tag{3b}$$

where P and Q denote scaled filter functions. The singular system $\{\alpha_n, u_n, v_n \mid n = 1,2,\ldots,\infty\}$ for a compact operator K satisfy

$$K u_n = \alpha_n v_n, \tag{4a}$$

$$K^\dagger v_n = \alpha_n u_n, \tag{4b}$$

$$\alpha_1 > \alpha_2 > \alpha_3 > \ldots > 0 \tag{5a}$$

and as $n \to \infty$

$$\alpha_n \to 0. \tag{5b}$$

This system is called the singular value decomposition and is abbreviated as SVD. Thus,

$$K^\dagger \cdot K = \left(\frac{c}{2\pi}\right)^d K_s^\dagger \cdot K_s \tag{6a}$$

and

$$\alpha_n^2 = \left(\frac{c}{2\pi}\right)^d \lambda_n \tag{6b}$$

where λ_n is the nth eigenvalue of $(K_s^\dagger \cdot K_s)$. All of the work by Slepian and collaborators [2-6] was with "windows" which have characteristic functions of symmetric intervals for the filter functions,

$$\chi_a = \begin{cases} 1, & |x| < a \\ 0, & |x| > a \end{cases} \tag{7}$$

and Grünbaum et al [14-16] have likewise restricted attention to this class. In order to generalize eq. (7), the authors [18] have defined Slepian symmetry between wave-number and direct space filters as:

(1) P and Q have the same functional form,

$$P(\vec{p}) = Q(\vec{p}), \tag{8}$$

$\vec{p} \in R^d$

(2) Even parity under coordinate reversal

$$P(-\vec{p}) = P(\vec{p}). \tag{9}$$

Property (1) of Slepian symmetry implies that $K_s^\dagger$ is the complex conjugate of K_s. Hence, they have common eigenfunctions, except for a complex conjugation, ie

$$K_s\psi_n = \beta_n\psi_n \Rightarrow K_s^\dagger\bar{\psi}_n = \bar{\beta}_n\bar{\psi}_n. \tag{10}$$

Then as proved by Slepian [5], property (2) leads to even and odd eigenfunctions with real even eigenvalues and pure imaginary odd eigenvalues. Upon choosing real eigenfunctions, eq. (10) yields

$$K_s^{\dagger} K_s \psi_n = |\beta_n|^2 \psi_n = \lambda_n \psi_n . \tag{11}$$

Slepian symmetry thus allows the singular values and singular functions to be solved from

$$\beta_n \psi_n(\vec{q}) = \int Q(\vec{q}) e^{-ic\vec{q}\cdot\vec{q}'} Q(\vec{q}')\psi_n(\vec{q}')d^d\vec{q}' , \tag{12}$$

instead of eq. (4). However, once eq. (12) has been solved the solutions to eq (4) can immediately be obtained from

$$u_n(\vec{x}) = \psi_n(\vec{x}/x_1) ,$$

$$\alpha_n = (\frac{c}{2\pi})^{d/2} |\beta_n| . \tag{13}$$

Slepian's wonderful insight was to find a differential operator which commutes with eq (12). Then the eigenvalues α_n, the eigenfunctions ψ_n and their asymptotics can be calculated from the differential equation. A particularly clear interpretation of the singular value decomposition was given by Bertero and De Mol [17]. Since eq (5b) in the presence of non-zero noise δ_N leads to some positive integer N (hopefully large) s.t. $\alpha_N > \delta_N$ with $\alpha_{N+1} < \delta_N$, N is called the <u>number of degrees of freedom</u> of the noisy filtered measurement. The spatial resolution is related to the spatial width of the oscillations of ψ_N, the Nth eigenfunction of the SVD. This depends upon the dependence of the α_n's on c and upon the noise δ_N.

2. Theorem:

Let D be the differential operator

$$D = -\vec{\nabla} \cdot (\alpha(q) \vec{\nabla}) + U(q), \tag{14}$$

where $\vec{\nabla}$ is the d-dimensional gradient operator, $\alpha(q)$ is a C^2 function of $q = |\vec{q}|$ and U(q) is continuous, except possibly for simple poles at the zeroes of α. If the filters P,Q are spherically symmetric and Slepian symmetric then the differential operator D must have α given by

$$\alpha(q) = a + bq^2 \tag{15}$$

with (a,b) arbitrary real constants and a $\neq$ 0. The filter is given by

$$O(q) = \frac{1}{\sqrt{\alpha(q)}} \exp \left\{ -\gamma \int_0^q \frac{q'dq'}{\alpha(q')} \right\} \tag{16}$$

with γ another constant s.t. $\gamma > |b|$. The "potential" U(q) is given by

$$U(q) = ac^2q^2 + \alpha(q) \frac{Q''(q)}{Q(q)} + [\alpha'(q) + \frac{(d-1)}{q} \alpha(q)] \frac{Q'(q)}{Q(q)} . \tag{17}$$

Proof: (Ref. 18)

A straightforward calculation, using partial integration, now gives an explicit expression for the commutator between K and D. The surface terms can be dropped either if taken over a hypersphere on which $\alpha(q)$ is zero or if taken at infinity, if $\alpha(q)O(q)$ and $\alpha(q)Q'(q)$ go to zero fast enough. The result of this calculation is

$$[(DK_s - K_sD)f(\vec{q}')](\vec{q}) = \int d\vec{q}'f(\vec{q}')O(q)\exp(-ic\vec{q} \cdot \vec{q}')O(q')F(\vec{q},\vec{q}') \tag{18}$$

with

$$F(\vec{q},\vec{q}')=c^2(q'^2\alpha(q)-q^2\alpha(q'))+ic\hat{q}\cdot\hat{q}'(q'G_1(q)-qG_1(q'))+G_2(q)-G_2(q') \tag{19}$$

where

$$G_1(q) = \alpha(q)\left(\frac{\alpha'(q)}{\alpha(q)} + 2\frac{O'(q)}{O(q)}\right) \tag{20}$$

and

$$G_2(q) = U(q) - \alpha(q) \frac{O''(q)}{O(q)} - \left(\alpha'(q) + \frac{d-1}{q} \alpha(q)\right)\frac{Q'(q)}{O(q)} . \tag{21}$$

Now, for the commutator to be zero, one must obviously have that

$$F(\vec{q},\vec{q}') = 0 \text{ (for all } \vec{q},\vec{q}' \in R^d). \tag{22}$$

This can only be true if the $\hat{q} \cdot \hat{q}'$ term is zero irrespective of the other terms, implying that

$$\frac{1}{q} G_1(q) = \text{constant} = -2\gamma , \tag{23}$$

with the solution

$$Q(q) = \frac{1}{\sqrt{\alpha(q)}} \exp\left(-\gamma \int_0^q \frac{q'dq'}{\alpha(q')}\right). \tag{24}$$

Moreover, whatever the form of the function G_2, the first term on the right-hand side of equation (19) must not contain mixed terms, e.g. of form q^2q', if equation (22) is to be satisfied. This puts a severe restriction on the functional form of $\alpha(q)$, with the only solution

$$\alpha(q) = a + bq^2 \tag{25}$$

where a and b are arbitrary constants, with a $\neq$ 0 to avoid problems at q = 0. We further choose $\gamma > |b|$ to avoid problems with the surface terms.

The corresponding potential U(q) is obtained from equation (22), using equations (18), (21), (23) and (24):

$$U(q) = ac^2q^2 + \alpha(q)\frac{Q''(q)}{Q(q)} + \left(\alpha'(q) + \frac{d-1}{q}\alpha(q)\right)\frac{Q'(q)}{Q(q)} . \tag{26}$$

This completes the construction of the commuting differential operator D of equation (14).

There is an interesting new direction suggested by this analysis. It is shown in ref. (18) that the choices

$$a = 1$$
$$b = 0 \tag{27a}$$

in eq. (14) then yield

$$Q(q) = e^{-\gamma q^2/2} \tag{27b}$$

and

$$U(q) = (c^2 + \gamma^2)q^2 \tag{27c}$$

This then generalizes earlier work by Starikov and Wolf [10] and Gori [11] to higher dimension d>1. Similarly, the choices

$$a = 1$$
$$b = -1 \tag{28a}$$

in eq. (14) were shown in ref. [18] to imply

$$Q(q) = (1-q^2)^{(\gamma-1)/2}\theta(1-q^2), \tag{28b}$$

and

$$U(q) = c^2q^2 + \frac{(\gamma-1)^2}{(1-q^2)} . \tag{28c}$$

If $\gamma = 1$, this reduces to Slepian's result [5] but for all other values it generalizes his equations. Using the differential operators of the theorem for the two cases in eqs (27), (28) the following asymptotics of the singular values are obtained. The Gaussian filter of eq. (27) has singular values in 3-d for large j and fixed c ref [20] shows go as

$$\alpha_j \sim \left(\frac{c}{\gamma + \sqrt{c^2+\gamma^2}}\right)^{j+\frac{3}{2}} \tag{29}$$

in contrast to the filter of eq. (28) whose singular values go as

$$\alpha_{\ell j} = \frac{\Gamma(j+1)\Gamma(j+\ell+3/2)}{\Gamma(2j+\ell+3/2)\Gamma(2j+\ell+5/2)}\left(\frac{c}{2}\right)^{2j+\ell+(3/2)} \times B_o(c), \tag{30}$$

$$B_o(c) \stackrel{\Delta}{=} [1 + \frac{(\ell + 1/2)^2 c^2}{4(2j+\ell+1/2)(2j+\ell+5/2)} + O(c^4)] ,$$

for small c and as

$$\alpha_{\ell j} = 1 - \frac{2\pi(4c)^{2j+\ell+1}\ e^{-2c}}{\Gamma(j+1)\Gamma(j+\ell+3/2)} [1 + O\ (1/c)]. \qquad (31)$$

for large c, keeping ℓ and j fixed. If c and j are both large

$$\alpha_{\ell j} = (1 + e^{\pi\delta})^{-1/2},$$

$$\delta = \frac{j\pi - c}{\ln[2\sqrt{c}]} . \qquad (32)$$

All of the results in eqs. (30)-(32) were obtained by Slepian and Pollack [2] in 1961 for the 1-d case.

3. Discussion

These equations show that the dependence of the $\alpha_{\ell j}$'s on (ℓ,j,c) is quite complicated, even for simple filters. There are two generalizations of the theorem which are most pressing.

(1) First one wishes to relax the condition on the functional forms of filters

$$P(\cdot) = Q(\cdot),$$

in Slepian symmetry. (The parity invariance remains for macroscopic filters.) Gori and Guattari [12] have solved a super-resolution problem which involves a window for Q and a Gaussian filter in P.

(2) The second is to relax the spherical symmetry condition. The differential operator becomes

$$D \stackrel{\Delta}{=} A(\partial_{q_i}, q_j) = \sum_{|\alpha|<m} a_\alpha(\vec{q})\ (-i\vec{\nabla}_q)^\alpha, \qquad (33)$$

and the idea needed is the uncertainty principle and microlocalization as discussed by Fefferman [19]. The operator D has been redefined as A so that Fefferman's SAK principle will not be the SAD principle here. The counting of eigenvalues $\alpha_{\ell k}(c)$ of A which are less than K_o is obtained by observing that the $u_n(\vec{q})$ is concentrated in a region of $\vec{q}$-space $|\vec{q}_o - \vec{q}| < \delta$ then $v_m(\vec{p})$ will be concentrated in a region of $\vec{p}$-space $|\vec{p}-\vec{p}_o| < 1/\delta$. The SAK_o principle [19] is that N, the number of eigenvalues of A less than K_o, is given by

$$N = \mathrm{Vol}(SAK_o) = \mathrm{Vol}\left\{(\vec{q},\vec{p})\,|\,A(\vec{p},\vec{q}) < K_o\right\}. \qquad (34)$$

This is just the number of degrees of freedom of Bertero and DeMol [18] discussed earlier. The boxes B_i which contain one eigenvalue of A are curved, even for the simple spherically-symmetric filters which led to eqs. (29)-(32). Their generalization to more general operators A ≡ D will be interesting and is in progress [20].

The phase space bound for $u \in L^2$,

$$c \max_{B_i} |p| \leq \max_{B_i} |q| + \text{error} \qquad (35)$$

of Fefferman should be related to the coherent state approach to filtering presented at this meeting by Daubechies [21]. Future work will address this question.

ACKNOWLEDGEMENT

This work was supported in part by STU under contract number 84-3865B and was performed at the University of Missouri-Columbia Department of Physics & Astronomy and at the RCP 264 at USTL in Montpellier Nov. 1985.

REFERENCES

1. N. Wiener, Time Series (MIT Press, Cambridge, MA, 1947).
2. D. Slepian and H.O. Pollak, Bell Syst. Tech. J. 40, 43-64, 1961.
3. H.J. Landau and H.O. Pollak, Ibid, 40, 65-85, 1961.
4. H.J. Landau and H.O. Pollak, Ibid, 41, 1295-1336, 1962.
5. D. Slepian, Ibid, 43, 3009-3058, 1964.
6. D. Slepian, SIAM Review 25, 379-393, 1983.
7. M. Bertero, C. De Mol and G.A. Viano, 1980, in "Inverse Scattering Problems in Optics", edited by H.P. Baltes, Topics in Current Physics Vol. 20 (Springer Verlag, Berlin, Heidelberg, New York) pp. 161-214.
8. M.Z. Nashed, IEEE AP 29, 220-231, 1981.
9. O. Brander and B. DeFacio, "The Role of Filters and the Singular value decomposion for the inverse Born approximation", University of Missouri-Columbia, Physics preprint, Nov. 1985. To be published in Inv. Probs.

10. A. Starikov and E. Wolf, J. Opt. Soc. Am. 72, 923-928, 1982.
11. F. Gori, Opt. Commun. 34, 301-305, 1980.
12. F. Gori and G. Guattari, Inv. Prob. 1, 67-85, 1985.
13. M. Reed and B. Simon, 1972, "Mathematical Physics, Volume I, Functional Analysis" (Academic Press, New York), p. 203.
14. F.A. Grünbaum, L. Longhi and M. Perlstadt, SIAM J. Appl. Math. 42, 941-955, 1982.
15. F.A. Grünbaum, SIAM J. Alg. Disc. Meth. 2, 136-141, 1981.
16. F.A. Grünbaum, J. Math. Anal. Appl. 88 (1982) 355-363 and 95 (1983) 491-500, 1982, 1983.
17. M. Bertero and C. De Mol, Atti della Fondazione Giorgio Ronchi Anno XXXVI, 619-624 (1981).
18. O. Brander and B. DeFacio, "A Generalisation of Slepian's solution for the singular value decomposition of filtered Fourier Transforms", Inv. Probs. 2, L9 (1986).
19. C.L. Fefferman, Bull. Am. Math. Soc. (new series) 9, 129-206 (1981).
20. O. Brander and B. DeFacio RCP 264 paper Cahiers Math Montpellier 1985 (in press) and Sherbrooke Workshop on Functional Integration (in press).
21. I. Daubechies, Paper presented at 1986 UAB meeting (these proceedings).

On Schrödinger Operators

With von Neumann-Wigner Type Potentials

Allen Devinatz
Department of Mathematics
Northwestern University
Evanston, Illinois 60201

Richard Moeckel
School of Mathematics
University of Minnesota
Minneapolis, Minnesota 55455

Peter Rejto
School of Mathematics
University of Minnesota
Minneapolis, Minnesota 55455

1. Introduction

In 1929 von-Neumann and Wigner [17],[13] introduced a potential such that the corresponding Schrodinger operator had an eigenvalue embedded in the continuous spectrum, and they called such an eigenvalue remarkable. Their potential was the sum of an oscillating and a short range one. Later Mochizuki and Uchiyama [12] introduced a class of oscillating potentials, which contains the von-Neumann Wigner potential. They also introduced a class of intervals and showed, among other things, that the part of the Schrodinger operator over their intervals is absolutely continuous. Recently, their class was extended by Saito [15],[16].

The problem of this paper is motivated by this one. We treat only short range perturbations of the original von-Neumann Wigner potential. However, the only restriction that we impose on our intervals is that they do not contain such a possible embedded eigenvalue. In [4] we have included the oscillating potential into the unperturbed operator and showed that resolvent estimates for such unperturbed operators, allow us to extend the Agmon short range perturbation theorem [1],[13] to them. In other words, we reduced this problem to the problem of proving resolvent estimates for such unperturbed operators. In [3] for a restricted class of intervals we proved such resolvent estimates. The reason for such restrictions was that we could prove Wronskian estimates only for such restricted intervals.

In Section 2 we formulate such estimates for the Wronskian. More specifically, we estimate from below the Wronskian of two normalized solutions of the basic equation where the spectral variable is in any compact interval not containing such a possible embedded eigenvalue. This is the statement of the main Theorem 2.1. We give evidence for the main Theorem 2.1 in the remaining three sections.

In Section 3, in Lemma 3.1, we observe that to estimate from below the Wronskian of two normalized solutions, it suffices to estimate from below, near infinity, the solution which is normalized near the left-endpoint.

In Section 4, in Theorem 4.1, we estimate a given solution of the basic equation in a given neighborhood of infinity. We prove Theorem 4.1 by adapting the notion of the approximate phase [9],[12],[15], to oscillating potentials. In Proposition 4.2 we show that if the basic equation admits such an approximate phase in a given neighborhood of infinity, then the estimate of Theorem 4.1 holds. We prove Proposition 4.2 with the help of Lemma 4.3, which is, essentially, a special case of Lemma 4.2 of [9]. This construction is similar to our previous construction of approximate potentials [3].

According to an informal communication of Professor Combes our potentials barely violate the trapping condition of Lavine, which is discussed in other parts of these Proceedings.

2. Formulation of the Result

Let the oscillating potential p_o be given by

$$p_o(\rho) = \frac{\sin b\rho}{\rho} \quad , \rho \in \mathbb{R}^+ \tag{2.1}$$

and let $-\Delta$ denote the Laplacian. Next let H_o be the closure of the operator

$$H_o f = (-\Delta + p_o)f \ , \ f \in \underline{C}_0^\infty (\mathbb{R}_3) \tag{2.2}$$

with respect to the $\underline{L}_2(\mathbb{R}_3)$-norm. As is well known H_o admits a complete family of reducing subspaces, on each of which it acts like an ordinary differential operator. More specifically, for each $j \in C^+$, define

$$p_o(j)(\rho) = j(j+1)\rho^{-2} + p_o(\rho) \ , \ \rho \in \mathbb{R}^+ \tag{2.3}$$

Then we know that on these reducing subspaces H_o is unitarily equivalent to $(2j+1)$ copies of the closure of the operator,

$$\begin{aligned} L(p_o(j)f(\rho) &= -f''(\rho) + p_o((j)(\rho)f(\rho) \ , \\ f &\in \underline{C}^\infty(\mathbb{R}^+) \ , \ f(0) = 0 \end{aligned} \tag{2.4}$$

The Weyl formula [13],[21] allows us to study the resolvent of this family of operators with the help of two solutions of the basic equation,

$$f''(\rho) + (\lambda - p_o(j,\rho))f(\rho) = 0 \ , \tag{2.5}$$

and their Wronskian. More specifically, for each j define

$$\nu = (j(j+1) + 1/4)^{1/2} \ , \tag{2.6}$$

and define $f_\ell(\lambda,j) = f_\ell(\lambda,j)(\rho) = f_\ell$ to be that solution for which

$$f_\ell(\rho) \sim (2\pi)(\lambda-\nu^2\rho^{-2})^{-1/4}\exp\Big(\int_{\nu\lambda^{-1/2}}^{\rho} -i \ (\lambda-\nu^2\sigma^{-2})^{1/2}d\sigma$$

$$\underline{\text{for}}\ \rho \to 0 \tag{2.7}$$

where

$$\operatorname{Im}(\lambda-\nu^2\rho^{-2})^{1/2} > 0 \ \underline{\text{for}}\ \ \lambda - \nu^2\rho^{-2} < 0 \tag{2.8}$$

Note that the branch of the square root function in the definition (2.8) has been so chosen that the absolute value of the exponential in the asymptotic formula (2.7) is at most 1 . Similarly, define $f_r = f_r(\lambda,j) = f_r(\lambda,j)(\rho)$ to be that solution for which

$$f_r(\rho) \sim \lambda^{-1/4}\exp(i\lambda^{1/2}\rho) \ \ \underline{\text{and}}\ \ f_r'(\rho) \sim i\lambda^{1/4}\exp(i\lambda^{1/2}\rho)$$

$$\underline{\text{for}}\ \ \rho \to \infty . \tag{2.9}$$

Then, with the usual definition of the Wronskian,

$$W(f_\ell,f_r) = f_\ell f_r' - f_r f_\ell' , \tag{2.10}$$

the resolvent kernel is given by,

$$(\lambda I - L(p_o(j))^{-1}(\xi,\eta) = \frac{1}{W(f_\ell,f_r)}\begin{cases} f_\ell(\eta)f_r(\xi) \ , \ \eta<\xi \\ f_r(\eta)f_\ell(\xi) \ , \ \eta>\xi \end{cases} .$$

THEOREM 2.1. <u>Let</u> $\mathcal{J}$ <u>be a compact subinterval of</u> $\mathbb{R}^+$ <u>such that</u>

$$0 \notin \mathcal{J} \ \ \underline{\text{and}}\ \ b^2/4 \notin \mathcal{J} \tag{2.11}$$

<u>and let</u> f_ℓ <u>and</u> f_r <u>be solutions to the basic equation</u> (2.5) <u>for which the asymptotic formulae (2.7) and (2.9) hold</u>. <u>Then</u>

$$\inf_{\lambda\in\mathcal{J}} \inf_{j\in\mathbb{Z}^+} |W(f_\ell(j,\lambda),f_r(j,\lambda))| \neq 0 \tag{2.12}$$

3. The First Part of the Proof of Theorem 2.1.

As a first part of the proof of Theorem 2.1 we factorize the Wronskian. To describe this factorization, define,

$$\vec{f}(\rho) = \begin{bmatrix} f(\rho) \\ f'(\rho) \end{bmatrix} \tag{3.1}$$

and

$$|\vec{f}(\rho)| = (|f(\rho)|^2 + |f'(\rho)|^2)^{1/2} . \tag{3.2}$$

If we choose f to be a non-trivial solution of the basic equation (2.5) then it follows from the uniqueness of the initial value problem that the positive function $|\vec{f}(\rho)|$ is strictly positive. Hence we may divide the Wronskian by it, and so

$$W(f_\ell,f_r) = |\vec{f}_\ell(\rho)||\vec{f}_r(\rho)| \quad |\vec{f}_\ell(\rho)|^{-1}|\vec{f}_r(\rho)|^{-1}W(f_\ell,f_r) . \tag{3.3}$$

Since the Wronskian of the solutions of the basic equation (2.5) is a constant and since the lim inf of a product is greater than the product of the lim infs, formula (3.3) yields,

$$|W(f_\ell,f_r)| \geq \liminf_{\rho\to\infty} |\vec{f}_r(\rho)| \cdot$$

$$\liminf_{\rho\to\infty} |\vec{f}_\ell(\rho)| \cdot \liminf_{\rho\to\infty} (|\vec{f}_\ell(\rho)|^{-1}|\vec{f}_r(\rho)|^{-1}W(f_\ell,f_r)|) . \tag{3.4}$$

Next we observe that the definitions (2.9) and (3.2) yield

$$\lim_{\rho\to\infty} |\vec{f}_r(\rho)| = (\lambda^{-1/2} + \lambda^{1/2})^{1/2} \tag{3.5}$$

and so, we see from assumption (2.11) that conclusion (2.12) holds for it. In the lemma that follows we show that this conclusion also holds for the third factor of formula (3.4).

LEMMA 3.1 Let the assumptions and notations of Theorem 2.1 hold. Then

$$\inf_{\lambda\in\mathcal{J}} \liminf_{\rho\to\infty} |\vec{f}_\ell(\rho)|^{-1}|\vec{f}_r(\rho)|^{-1}|W(f_\ell,f_r)| \neq 0 . \tag{3.6}$$

We illustrate the proof of this lemma, by showing that

$$W(f_\ell,f_r) \neq 0 \tag{3.7}$$

To see this, note that the asymptotic formula (2.7) shows that f_ℓ satisfies a real boundary conditions at $\rho = 0$. This fact, and the fact that the coefficients of the basic equation (2.5) are real, together imply that

$$\mathrm{Im}\, f_\ell'(\rho)f_\ell(\rho)^{-1} \neq 0 \;\; \underline{\text{or}} \;\; \mathrm{Im} f_\ell(\rho)f_\ell'(\rho)^{-1} \neq 0 \, , \, \rho \in \mathbb{R}^+ \tag{3.8}$$

Similarly we see from the asymptotic formula (2.9) that,

$$\lim_{\rho\to\infty} \mathrm{Im} f_r'(\rho)f_r(\rho)^{-1} = i\lambda^{1/2} \tag{3.9}$$

Relations (3.8) and (3.9) show that for large enough ρ these two vectors are not parallel. Since this Wronskian is independent of ρ , this proves relation (3.7), and completes the illustration of the proof of Lemma 3.1.

4. The Second Part of the Proof of Theorem 2.1.

Our next aim is to give evidence that the conclusion of the main Theorem 2.1 holds for the second factor of estimate (3.4);

$$\inf_{\lambda \in \mathcal{J}} \inf_{j \in \mathbb{Z}^+} \liminf_{\rho \to \infty} |\vec{f}_\ell(\rho)| \neq 0 .$$

As a second part of the proof of Theorem 2.1 we show that this key estimate holds under the additional assumption that

$$\inf_{\lambda \in \mathcal{J}} \inf_{j \in \mathbb{Z}^+} |\vec{f}_\ell(\tau\delta)| \neq 0 , \tag{*}$$

where

$$\tau = \nu\lambda^{-1/2} . \tag{4.1}$$

This is implied by the theorem that follows.

THEOREM 4.1. <u>Let the interval</u> $\mathcal{J}$ <u>satisfy assumption</u> (2.10), <u>let</u> f <u>be a given solution of the basic equation</u> (2.5). <u>Then, to each constant</u> $\delta > 1$ <u>there is another constant</u> $\gamma > 0$ <u>such that for each</u> $\lambda \in \mathcal{J}$,

$$\inf_{\rho > \tau\delta} |\vec{f}(\rho)| \geq \gamma |\vec{f}(\tau\delta)| . \tag{4.2}$$

The proof of Theorem 4.1 is based on an adaptation of the notion of the approximate phase [9],[12],[15] to our problem. We say that $\theta = \theta(\lambda,j)(\rho)$ is an approximate phase of the basic equation (2.5) over the given interval $\mathcal{J}$, if the error potential

$$e(\lambda,j) = e(\lambda) = e = -i\theta' + \theta^2 - (\lambda - p_o(j)) \tag{4.3}$$

is such that

$$\sup_{j \in \mathbb{Z}^+} \sup_{\lambda \in \mathcal{J}} \int_{\mathcal{J}} |e(\sigma)| d\sigma < \infty . \tag{4.4}$$

The real part of θ is such that

$$\inf_{j \in \mathbb{Z}^+} \inf_{\lambda \in \mathcal{J}} \inf_{\rho \in \mathcal{J}} \operatorname{Re} \theta(\rho) > 0 \tag{4.5}$$

and its imaginary part is such that

$$\sup_{j \in \mathbb{Z}^+} \sup_{\lambda \in \mathcal{J}} \sup_{\rho \in \mathcal{J}} |\operatorname{Im}\theta(\rho) \operatorname{Re}\theta(\rho)^{-1}| < 1 \tag{4.6}$$

and

$$\sup_{j \in \mathbb{Z}^+} \sup_{\lambda \in \mathcal{J}} \sup_{\rho_1,\rho_2 \in \mathcal{J}} \left| \int_{\rho_1}^{\rho_2} \operatorname{Im}\theta(\sigma) d\sigma \right| < \infty . \tag{4.7}$$

Incidentally, note that for a given approximate phase θ the approximate solution

$$y(\rho) = \exp(i \int^{\rho} \theta(\sigma) d\sigma) \tag{4.8}$$

satisfies the differential equation

$$y'' - (i\theta' - \theta^2)y = 0 \ ,$$

and so, we shall refer to the potential

$$q = i\theta' - \theta^2 + \lambda \qquad (4.9)$$

as the corresponding approximate potential.

We continue the proof of Theorem 4.1 by showing that if the basic equation (2.5), admits such an approximate phase over the interval $\tau(\delta,\infty)$ then conclusion (4.2) holds. This is implied by the proposition that follows.

PROPOSITION 4.2. Let f be a given solution of the basic equation (2.5), and for each $\lambda \in \mathcal{J}$ let this basic equation admit an approximate phase $\theta = \theta(j,\lambda)(\rho)$ for which assumptions (4.4),(4.5),(4.6) and (4.7) hold over the given interval $\mathcal{J}$. Then, there is a constant $\gamma > 0$ such that for each $\lambda \in \mathcal{J}$

$$\inf_{\rho \in \mathcal{J}} |\vec{f}(\rho)| \geq \gamma |\vec{f}(\inf \mathcal{J})| \ . \qquad (4.10)$$

The proof of Proposition 4.2 is based on the following lemma, which is, essentially, a special case of Lemma 4.2 of [9]. In it, with the help of the approximate phase θ, for each function f define

$$F(f) = (D + \mathrm{Im}\theta)f \cdot \overline{(D + \mathrm{Im}\theta)f} + \mathrm{Re}\theta f \cdot \overline{\mathrm{Re}\theta f} \ . \qquad (4.11)$$

LEMMA 4.3. Let f be a solution of the basic equation (2.5) and for the given approximate phase θ let the error potential e be given by the definition (4.3). Then the function of the definition (4.11) satisfies the differential equation

$$F(f)' = 2\mathrm{Im}\theta \cdot F(f) + 2\mathrm{Re}(\mathrm{Re}(e)f \cdot \overline{Df})$$
$$- 2\mathrm{Re}(\mathrm{Im}(e)\mathrm{Re}\theta f \cdot \bar{f}) \ . \qquad (4.12)$$

Thus to prove Theorem 4.1 it suffices to construct an approximate phase. In [3] we adapted the JWKB-approximation method [6],[19],[18] to construct approximate potentials. Now we observe that the same construction yields an approximate phase.

5. The Third Part of the Proof of Theorem 2.1.

The third and final part of the proof of Theorem 2.1 is to verify the additional assumptions (*) of Section 4. Now this is a technical point and for this verification we refer to the report [20].

References

1. Agmon, S.: Spectral properties of Schrodinger operators and scattering theory, Ann. Scuola Norm. Sup. Pisa Sci. Mat. (1975), 151-218.

2. Benz-Artzi, M. and Devinatz, A.: Spectral and scattering theory for the adiabatic oscillator and related potentials, J. Math. Phys. 111(1979), 594-607.

3. Devinatz, A. and Rejto, P.: A limiting absorption principle for Schrodinger operators with oscillating potentials. I.J. Diff. Equations. 49, (1983), 85-104.

4. Devinatz, A. and Rejto, P.: A limiting absorption Principle for Schrodinger operators with oscillating potentials. II. J. Diff. Equations. 49, (1983), 85-104.

5. Erdelyi, A.: Asymptotic Expansions. Dover, New York, 1956.

6. Pauli Lectures on Physics, Vol. 5, Wave Mechanics, Enz, C., P., (Ed.), MIT Press, Cambridge, Mass. 1977, See Sect. 27, The WKB-method.

7. Jager, W. and Rejto, P.: On the absolute continuity of the spectrum of Schrodinger operators with long range potentials. Oberwolfach Tagungsberichte, 17.7.-23.7. 1977.

8. Jager, W. and Rejto, P.: Limiting absorption principle for some Schrodinger operators with exploding potentials I. J. Math. Anal. Appl. 91(1983), 192-228.

9. Jager, W. and Rejto, P.: Limiting absorption principle for some Schrodinger operators with exploding potentials II. J. Math. Anal. Appl. 95 (1983), 169-194.

10. Kato, T.: Perturbation Theory for Linear Operators, Springer-Verlag, 1973.

11. Langer, R.E.: The asymptotic solutions of ordinary linear differential equations for the second order with special reference to a turning point. Trans. Amer. Math. Soc. 67 (1949), 461-490.

12. Mochizuki, M., and Uchiyama, J.: Radiation conditions and spectral theory for 2-body Schrodinger operators with "oscillating" long range potentials I. J. Math. Kyoto Univ. 18(2) (1978), 377-408.

13. Reed. M., and Simon, B.: Analysis of operators, Methods of Modern Mathematical Physics. Vol. IV, Academic Press. 1978, Section XIII.13, Example 1.

14. Saito, Y.: On the asymptotic behavior of the solutions of the Schrodinger equation. Osaka. J. Math. 14(1977), 11-35. See the Ricatti equation (3.11).

15. Saito, Y.: Schrodinger Operators with a nonspherical radiation condition,to appear.

16. Saito, Y.: These Proceedings.

17. von Neumann, J. and Wigner, E.: Über merkwurdige diskrete Eigenwerte. Phys. Zschr 30(1929), 465-467.

18. Olver, F., T.,W.: Asymptotics and Special Functions. Academic Press. 1974.

19. Wasov, W.: Asymptotic Expansions for Ordinary Differential Equations. Wiley-Interscinec, 1965.

20. Devinatz, A., Moeckel, R., and Rejto, P.: A Wronskian Estimate For Schrodinger Operators With von Neumann-Wigner Type Potentials. Univ. Minn. Math. Report 1986.

21. Atkinson, F.: Discrete and Continuous Boundary Problems, Academic Press 1964, See Theorem 88.1.

22. Atkinson, F.: The asymptotic solutions of second order differential equations, Ann. Mat. Pura Appl. 37 (1954), 347-378.

23. Isozaki, Hiroshi: On the generalized Fourier transform associated with Schrodinger operators with long-range perturbations. J. reine angewandte Math. 337 (1982), 18-67.

24. Reed, M. and Simon, B.: Scattering Theory. Methods of Modern Mathematical Physics Vol. III, Academic Press 1979. See Section XI.8 and the notes to it where references to the original works of Natveev-Skriganov, Combescure-Ginibre and Schechter are also given.

25. Harris, W.A. and Lutz, D.A. A unified theory of asymptotic integration. J. Math. Anal. Appl. 57 (1977), 571-586.

NONLINEAR CONSERVATIVE SYSTEMS

Ronald J. DiPerna
Department of Mathematics
University of California
Berkeley, CA 94720

We shall discuss singularities and oscillations in certain non-linear conservative systems arising in fluid dynamics. In the setting of compressible flows we shall consider hyperbolic systems of conservation laws in one space dimension,

$$\partial_t u + \partial_x f(u) = 0, \quad u \in R^n. \tag{1}$$

Here f is a smooth nonlinear map whose Jacobian has n real and distinct eigenvalues. In the setting of incompressible flows we shall discuss the Euler equations in two space dimensions,

$$\partial_t u + \operatorname{div} u \otimes u + \nabla p = 0$$

$$\operatorname{div} u = 0. \tag{2}$$

§1. Singularities

In the geometric theory of conservation laws a central problem deals with the structure and stability of singularities. It is well known that, even if the initial data are smooth, singularities develop in finite time due to the nonlinear structure of the eigenvalues. Such singularities represent shock waves. For hyperbolic systems of conservation laws in one space dimension, the spatial maximum norm and the total variation norm serve as natural metrics in which to measure the solution amplitude and gradient. The corresponding spaces L^∞ and BV provide a framework to analyze singularities. The relevance of these spaces for hyperbolic systems of conservation laws in one space dimension was established by a constructive existence theorem of Glimm [4] dealing with the Cauchy problem with small data.

<u>Theorem</u>. If TVu_0 is sufficiently small there exists a globally defined distributional solution u of (1) that takes on the data u_0 at $t = 0$ and satisfies the following bounds,

$$|u(\cdot,t)|_\infty \le c|u_0|_\infty \tag{3}$$

$$TVu(\cdot,t) \le c\ TVu_0. \tag{4}$$

The solution u is constructed by the random choice method. At any time t its amplitude is dominated by the amplitude of the data (3). The total amount of wave magnitude at time t is bounded by the total amount of wave magnitude in the data, as measured by the total variation norm TV. In general the constant c is greater than one. The stability estimates (3) and (4) provide a foundation for the regularity theory of solutions.

In the regularity theory for hyperbolic systems in one space dimension one of the main problems concerns the structure of the solution u as a function of x and t. In this connection the notion of function of bounded variation of several variables is useful. A locally integrable function w of m variables is said to lie in the space $BV(R^m)$ if each of its first order partial derivatives is represented by a Borel measure σ_j with finite total mass:

$$\int w\partial_j\varphi dy = \int \varphi d\sigma_j$$

for all smooth functions φ where $\partial_j = \partial/\partial y_j$.

The prototypical example of a BV function of several variables is provided by a piecewise smooth function that experiences singularities across isolated piecewise smooth manifolds of codimension 1. In the special case of functions of two variables the sets of jump discontinuity are curves. In the setting of mechanics they represent shock waves.

The subject of geometric measure theory has dealt with several topics including the classification and structure of singularities of BV functions. It has been shown that, for a general function w in the $BV(R^m)$, most of the points of R^m are regular points for w in the sense of Lebesgue: controlled limiting behavior on the average. Furthermore the regular points corresponding to jump discontinuities (shock waves) are contained in sets with a rectifiable structure: limiting normals on the average [12,13].

Geometric measure theory has provided a basis for the regularity theory of solutions to conservation laws in one space dimension. It is a straightforward consequence of the sharp-time estimates (3) and (4) that the solution generated by the random choice method is a BV function of x and t. Thus, the problem arises of analyzing the structure of BV solutions to hyperbolic systems of conservation laws in one space dimension.

In recent years a fairly complete regularity theory has been developed for solutions constructed by the random choice method [1,8]. Some of the basic questions are the following. Can all of the singularities of a general BV function be found in a BV solution? Do there exist mechanisms which convert controlled averaged behavior into controlled pointwise behavior? It turns out that a BV solution is substantially more regular than a general BV function as a consequence of the entropy condition which incorporates a slight amount of dissipation into the solution. We refer the reader to [1] for regularity results in the context of systems with nondegenerate eigenvalues such as isentropic gas dynamics and to [8] in the context of systems with degenereate eigenvalues such as nonisentropic gas dynamics.

The aforementioned work deals with solutions constructed by the random choice method. It remains an open problem to develop an a priori regularity theory for conservation laws. The question of uniqueness in the space BV is also open. A partial result establishing uniqueness of the fundamental solution within BV is given in [2].

One is led to consider analogous questions in the setting of incompressible fluids. At the level of the two dimensional Euler equations, global existence of solutions to the Cauchy problem is an open problem if the initial vorticity, $w_0 = \text{curl}\, u_0$ is a bounded measure and if the initial u_0 has finite total kinetic energy:

$$\int_{R^2} |u_0(x)|^2 \, dx < \infty.$$

This case arises naturally in the study of flows with vortex sheets: a jump in the velocity field across a line yields a singular contribution to the vorticity field in the form of a measure concentrated along a line, namely one-dimensional Hausdorff measure.

In contrast, if the data has finite kinetic energy and if w_0 lies in L^p with $p > 1$ then global existence for 2 - D Euler can be established in a variety of ways. Furthermore the solution u can be realized as the strong L^2 limit of Navier-Stokes solutions u_ε,

$$\partial_t u_\varepsilon + \text{div}\, u_\varepsilon \otimes u_\varepsilon + \nabla p_\varepsilon = \varepsilon \Delta u_\varepsilon$$
$$\text{div}\, u_\varepsilon = 0$$
$$u_\varepsilon(x,0) = u_0(x),$$

as the diffusion coefficient ϵ tends to zero. This fact is a consequence of standard elliptic regularity theory, and is established in the following theorem.

<u>Theorem</u>. If $u_0 \in L^2(R^2)$ and if $\operatorname{curl} u_0 \in L^p(R^2)$ with $p > 2$ then a subsequence u_{ϵ_k} of 2 - D Navier-Stokes solutions converges in the strong topology of L^2_{loc} to a globally defined distributional solution of the 2 - D Euler equations.

<u>Proof</u>. The vorticity-stream formulation, see (5), indicates that at each fixed time t the velocity u is obtained from the vorticity w by inverting the Laplacian and taking one derivative:

$$\psi = c \int_{R^2} \log|x - y| w(y)\, dy$$

$$u = \nabla\psi^{\perp} = c \int \frac{(x - y)^{\perp}}{|x - y|} w(y)\, dy.$$

It follows from L^p elliptic regularity theory with $p > 1$ that ψ lies locally in $W^{2,p}(R^2)$ and hence that u lies locally in $W^{1,p}(R^2)$. The Sobolev imbedding theorem implies that u lies in a compact subset of L^q where $q = np/(n - p)$. In the special case $n = 2$ and $p > 1$ one finds that $q > 2$. In particular u lies in a compact set of $L^2(R^2)$.

Uniform control in time follows from the behavior of the distribution function of the vorticity w. Indeed, w is constant along particle paths modulo diffusion:

$$\partial_t w + (u\cdot\nabla)w = \epsilon\Delta w.$$

Since u is divergence-free we have

$$\partial_t\eta(w) + \operatorname{div}\left[\eta(w)u\right] \leq \epsilon\Delta\eta(w),$$

for all smooth convex functions η. Integrating with respect to x shows that

$$\frac{d}{dt} \int \eta\left[w(x,t)\right] dx \leq 0$$

and in particular that

$$\int |w(x,t)|^p\,dx \le \int |w(x,0)|^p\,dx$$

if we choose $\eta(z) = |z|^p$. The special choice $p > 1$ produces a temporally uniform L^p-bound on the vorticity $w(\cdot,t)$ which induces a temporally uniform compactness statement for the corresponding velocity fields.

Since the estimates are independent of the diffusion parameter ϵ it follows that $U(x,t) = U(x,t,\epsilon)$ lies in a compact set of $L^2(R^2)$ locally for all t. A compactness argument based on the Aubin-Lions lemma then leads to compactness of u in space-time. The application of this lemma relies on the fact that $\partial_t u$ lies in a bounded set of a negative Sobolev space,

$$\partial_t u = -\mathrm{div}(u \otimes u)\ \nabla p + \epsilon \Delta u,$$

uniformly with respect to ϵ.

Remark. A similar argument applies to computational vortex methods. If the initial vorticity lies in L^p with $p > 1$ then a subsequence converges strongly in L^2. Strong convergence in L^2 is sufficient to pass to the limit since the nonlinear term is quadratic.

The situation changes substantially if the initial vorticity lies in L^1 or in the space of Borel measures with finite total mass. In the latter case it is an open problem to determine whether or not the solutions u_ϵ of the 2 - D Navier-Stokes equations converge strongly in L^2. If the answer is yes, the limiting field is necessarily a distributional solution of the 2 - D Euler equations.

In connection with singularities it is natural to ask if the space BV is sufficiently large to encompass all 2 - D incompressible flows. The initial data for an isolated vortex sheet corresponds to a field in $L^2 \cap BV$. Does the solution leave BV in finite time due to Helmhotz instability? Certain evidence tends to favor the conjecture for finite time blow up in the space BV due to vortex roll-up. Such a result, however, would not necessarily interfere with global existence in a weaker space.

The vorticity-stream formulation of (2), namely

$$\partial_t w + \mathrm{div}(wu) = 0$$
$$\Delta\psi = w \qquad (5)$$

indicates that the total mass of the vorticity w is (formally) conserved with time:

$$\text{total mass } w(\cdot,t) = \text{total mass } w(\cdot,0).$$

Indeed, equation (5) asserts that the vorticity is constant along particle paths. Vorticity may rearrange itself in a complex fashion but the total amount is invariant. Thus the velocity field has certain partial derivates represented by measures with finite total mass:

$$\begin{aligned} \operatorname{div} u &= 0 \\ \operatorname{curl} u &= w. \end{aligned} \tag{6}$$

In general one cannot conclude from the elliptic system (6) that u is in BV if w has finite total mass. It would be interesting to have a precise regularity statement for solutions to (6) in order to contrast the structure with the BV situation.

§2. Oscillations

For the purpose of constructing solutions to systems (1) and (2) one may consider the associated zero diffusion limit, e.g.

$$\partial_t u + \partial_x f(u) = \epsilon\, D\partial_x^2 u, \tag{7}$$

where D is a constant nonnegative matrix. Motivated by the stability estimates (3) and (4) it is natural to ask if the L^∞ norm and the total variation norm of the solution u_ϵ of the parabolic system (7) remain bounded as the diffusion coefficient ϵ tends to zero. Uniform control on both the amplitude and derivatives of a family of vector fields guarantees the existence of a sequence that converges in the strong topology. Strong convergence permits passage to the limit in the nonlinear flux function $f(u)$. The right hand side of (7) vanishes in the sense of distributions. In short, classical compactness arguments allow one to convert uniform bounds of the form

$$|u_\epsilon(\cdot,t)|_\infty \le C|u_0|_\infty \tag{8}$$

$$TVu_\epsilon(\cdot,t) \le C\, TVu_0 \tag{9}$$

into an existence theorm for the unperturbed system (1). The pair of estimates (8), (9) is sufficient but not necessary for compactness in the strong topology.

One of the main goals in the theory of singular perturbations for hyperbolic conservation laws is to establish estimates on the maximum norm and total variation norm which are independent of the diffusion coefficient in the setting of the Cauchy problem. For certain special systems of two equations, pointwise maximum principles lead to the desired L^∞ bound. The problem of uniform total variation bounds is open even in the case of small data.

Recent work in the theory of oscillations has lead to convergence results for the viscosity method (7) applied to strictly hyperbolic systems of two equations with nondegenerate eigenvalues [1,14]. No a priori derivative estimates are required. The analysis makes use of the compensated compactness theory of Tartar and Murat [9,10,11] and the Young measure. We refer the reader to [10,11] for a general introduction and for a discussion of the viscosity method for a scalar conservation law. A typical result for nondegenerate systems is the following [1].

<u>Theorem</u>. If f is strictly hyperbolic and genuinely nonlinear in the sense of Lax and if the viscosity solutions u_ϵ of the parabolic system (7) are bounded in L^∞ uniformly with respect to ϵ then there exists a sequence u_{ϵ_k} that converges in the strong topology of L^1_{loc} to a solution u of the underlying system (1).

<u>Outline of proof</u>.

We consider the Young measure $\nu_{(x,t)}$ associated with the sequence u^ϵ. It describes all weak limits of u^ϵ in the sense that

$$\lim g(u^\epsilon) = \langle \nu_{(x,t)}, g(\lambda)\rangle$$

for all continuous functions g where the bracket indicates average values.

It is known that strong convergence of u^ϵ is equivalent to the reduction of the Young measure to a Dirac mass. Reduction is achieved using the divergence-curl lemma of Tartar and Murat which states that if the divergence of a sequence of fields z^ϵ is H^{-1}-compact and if the curl of a sequence of fields w^ϵ is H^{-1}-compact then the inner product is weakly continuous, i.e., the weak limit of the inner product coincides with the inner product of the weak limits:

$$\lim\langle z^\epsilon, w^\epsilon\rangle = \langle \lim z^\epsilon, \lim w^\epsilon\rangle.$$

Applying this lemma to the entropy fields

$$\partial_t \eta(u^\epsilon) + \partial_x q(u^\epsilon)$$

which can be shown to be H^{-1}-compact leads to a functional equation for the Young measure at each point (x,t), namely

$$\langle \nu, \eta_1 q_2 - \eta_2 q_1 \rangle = \langle \nu, \eta_1 \rangle \langle \nu, q_2 \rangle - \langle \nu, \eta_2 \rangle \langle \nu, q_1 \rangle$$

for all entropy pairs (η_j, q_j). An appropriate choice of entropy pairs given by the Lax geometrical optics construction

$$\eta_k = e^{k\varphi} \sum_{n=1}^{\infty} V_n / k^n$$

$$q_k = e^{k\varphi} \sum_{n=1}^{\infty} H_n / k^n$$

indicates that the functional equation for ν connects terms of different order with respect to k as k tends to infinity with the only exception that $\nu = \delta$. In short an asymptotic analysis of the functional equation using a variant of the Laplace transform leads to the desired statement that the Young measure reduces to a Dirac mass.

Remark 1.. Uniform control on the L^∞ norm guarantees the existence of a sequence that converges in the weak topology, i.e., in the sense of average values or distributions. However, nonlinear maps are not continuous in the weak topology in general. For this reason additional arguments are required in order to pass to the limit.

Remark 2. The source of the compactness lies at the hyperbolic level. The focusing of acoustic waves into shocks produces a loss of information and a gain in entropy. This is recorded algebraically in the Lax shock condition and captured functional analytically with compensated compactness.

§3. Concentrations

A general problem is to analyze weak limits of solutions to conservative systems associated with both compressible and incompressible flow. Weak limits may manifest themselves through the presence of both oscillations and concentrations.

In the setting of $1-D$ compressible flows, uniform amplitude bounds preclude the development of concentrations. The principle mechanism for the loss of strong compactness is the development of oscillations. Fortunately, in one space dimension the entropy condition annihilates substantial oscillations and leads to theorems asserting strong compactness without a priori control on the derivatives.

In the setting of $2-D$ incompressible flows, the natural amplitude bound is associated with L^2. Here one may lose strong compactness due to the presence of both oscillations and concentrations. Thus, the problem arises of describing the defects associated with weakly convergent $2-D$ Euler sequences. An ongoing program with A. Majda contains several results in this direction.

Explicit examples of $2-D$ Euler sequences can be constructed which are not compact in L^2 despite the fact that both the total kinetic energy and total vorticity are uniformly bounded. The examples are based on Rankine and Kirchoff vorticies and they deposit a finite amount of kinetic energy on a set with zero two-dimensional Lebesgue measure, namely a point.

General problem is to estimate the size of the set in space and in space-time on which L^2 defects due to concentrations may condense. In the case of solution sequences with uniformly bounded energy and vorticity the author and A. Majda have shown that the exceptional sets associated with L^2 defects have Hausdorff dimension zero in space and one in space-time. One of the main results is given by the following theorem which provides for the uniformization of arbitrary familes of measures.

<u>Theorem</u>. Suppose w_k is a sequence of nonnegative measures on R^n with uniformly bounded total mass. Fix parameters δ and γ so that $\gamma > n\delta > 0$. Then there exists a subsequence, which we still label as w_k with the following property. For each $r > 0$ the set of local algebraic decay,

$$\{x: w_k(s,x) \le Ks^\delta, \quad 0 \le s \le 1, \quad k \ge 1/r\},$$

where $w(s,x) = w\{B_r(x)\}$ denotes the weight of a ball of radius r centered at x, contains a closed set F_r whose γ-order Hausdorff premeasure at level r satisfies

$$H_r^\gamma(F_r^c) \le c(\delta,\gamma).$$

Here $K = c + cr^{-\delta}$ and c is a universal constant.

Proof. We establish uniform pointwise bounds on the associated Riesz potentials

$$\varphi_k(x) = \int_{R^n} \frac{1}{|x - y|^\delta} dw_k(y).$$

Specifically, we show that there exists large closed sets F_r such that

$$\varphi_k(x) \leq K(r) \quad \text{if} \quad k \geq 1/r$$

and if $x \in F_r$. It then easily follows that

$$s^{-\delta} w_k(s,x) \leq \int_{B_s(x)} \frac{1}{|x - y|^\delta} dw_k \leq \varphi_k(x) \leq K$$

if $x \in F_r$.

The idea to prove the uniform pointwise bound on the Riesz potentials is to show that they lie in a compact subset of $W^{1,p}$ if $p < n/(\delta + 1)$. This fact is a straightforward consequence of the structure of the Fourier transform of the kernel. We then show that strong convergence in Sobolev space implies uniform convergence on the complement of an exceptional set with finite Hausdorff premeasure. This result is an extension of the classical result in L^p which states that L^p convergence is uniform on the complement of an exceptional set with small Lebesgue measure. By replacing L^p by $W^{1,p}$ and by replacing Lebesgue measure by Hausdorff premeasure one obtains an analogous statement.

Remark. In the application to the Euler equations we consider the measures associated with the vorticity. The uniformization theorem says roughly that the vorticity looks like an L^p functional locally on the complement of a small set.

This program is motivated in part by the concentration compactness results of P. L. Lions [6,7] that deal with the quantification of losses of compactness in classical function inequalities such as the Sobolev inequality.

REFERENCES

1. DiPerna, R. J. "Singularities of solutions of nonlinear hyperbolic systems of conservation laws," Arch. Rational Mech. Anal. 60(1975), 75-100.
2. ______, "Uniqueness of solutions to hyperbolic conservation laws," Indiana Math. J. 28(1979), 137-188.
3. ______, "Convergence of approximate solutions to conservation laws," Arch. Rational Mech. Anal. 82(1983), 27-70.
4. Glimm, J. "Solutions in the large for nonlinear hyperbolic systems of equations," Comm. Pure Appl. Math. 18(1965), 697-715.
5. Lax, P. D. "Shock waves and entropy," in: Contributions to Non-linear Functional Analysis, E. A. Zarantonello, ed., Academic Press, 1971.

6. Lions, P. L. "The concentration-compactness principle, the locally compact case," parts I and II, Ann. Inst. Henri Poincare, 1(1984), 109-145.
7. ______, "The concentration-compactness principle in the calculus of variations, the limit case," parts I and II. Riv. Mat. Ibero-americana, 1(1984), 145-201 and 1(1985), 45-121.
8. Liu, T. P. Admissible Solutions to Systems of Conservation Laws, Amer. Math. Soc. Memoirs, 1982.
9. Murat, Compacite par compensation, Ann. Scuola Norm. Sup. Pisa 5(1978), 69-102.
10. Tartar, L. Compensated compactness and applications to p.d.e., Pitman Research Notes in Mathematics, vol. 39, 1979.
11. ______, The compensated compactness method applied to systems of conservation laws, in Systems of Nonlinear P. D. E., NATO ASI Series, J. M. Ball ed., Reidel Pub. 1983.
12. Federer, H. Geometric Measure Theory, Springer 1969.
13. Vol'pert, A. I. "The spaces BV and quasilinear equations," Math. USSR Sb. 2(1967), 256-267.
14. DiPerna, R. J. "Convergence of the viscosity method for isentropic gas dynamics," Comm. Math. Phys. 91(1983), 1-30.

THE INITIAL VALUE PROBLEM FOR THE NONLINEAR EQUATIONS FOR ZERO MACH NUMBER COMBUSTION

Pedro Embid
Department of Mathematics and Statistics
University of New Mexico
Albuquerque, NM 87131

1. Introduction

Here we consider the short time existence of classical solutions for the equations for zero Mach number combustion in a bounded domain, in the absence of viscous, heat conductive, and chemical species diffusion effects.

The equations studied here were formulated by Majda in (5) under the assumptions of small Mach number for the flow, the nearly constant initial pressure and the approximate chemical-fluid balance for the initial data. This last condition introduced in (5) to guarantee the formal validity of the limiting equations for zero Mach number combustion is also needed in producing the solution of the initial value problem.

In the absence of viscous, heat conductive and species diffusion effects, the equations for zero Mach number combustion for a mixture of M chemical species in a bounded domain Ω become the following system in nondimensional form.

1.1 Nonlinear ODE for the mean pressure

$$\frac{dp}{dt} = H(t) = \frac{\int_\Omega \gamma^{-1} G dx}{\int_\Omega \gamma^{-1} dx} \tag{a}$$

Elliptic equation for the potential part of the velocity field

$$\Delta\phi = (\gamma p)^{-1}(G - H(t)) \tag{b}$$

$$\left.\frac{\partial \phi}{\partial n}\right|_{\partial\Omega} = 0$$

Nonhomogeneous incompressible Euler equations for the solenoidal part of the velocity field

$$\rho\frac{Dw}{Dt} + \nabla\pi = -\rho\frac{D\nabla\phi}{Dt} \tag{c}$$

$$\text{div } w = 0$$

$$w.n|_{\partial\Omega} = 0$$

Convective equations for the temperature and the mass fractions

$$\rho c_p \frac{DT}{Dt} = \frac{\Gamma-1}{\Gamma} H(t) - \sum_{i=1}^{M-1} (h_i - h_M)\, \Phi_i \tag{d}$$

$$\rho \frac{DY_i}{Dt} = \Phi_i \qquad i = 1,\ldots,M-1 . \tag{e}$$

Here $\frac{D}{Dt} = \frac{\partial}{\partial t} + v.\nabla$. Initially, $\theta = (P;T,Y_1,\ldots,Y_{M-1})$ and v are given by

$$p(0) = p_0 \tag{1.2a}$$

$$T(x,0) = T_0(x) \tag{b}$$

$$Y_i(x,0) = Y_{i0}(x), \qquad i = 1,\ldots,M-1 \tag{c}$$

$$v(x,0) = v_0(x) = w_0(x) + \nabla\phi_0(x) \tag{d}$$

where $w_0(x)$ is the solenoidal component of v_0

$$\operatorname{div} w_0 = 0 \tag{e}$$

$$w_0.n\Big|_{\partial\Omega} = 0$$

and the potential component $\nabla\phi_0(x)$ of $v_0(x)$ satisfies the chemical-fluid balance condition

$$\Delta\phi_0(x) = (\rho_0\gamma)^{-1}(\theta_0)\left(G(\theta_0) - \frac{\int_\Omega (\gamma^{-1}G)(\theta_0)\,dx}{\int_\Omega \gamma^{-1}(\theta_0)\,dx} \right). \tag{f}$$

$$\frac{\partial\phi_0}{\partial n}\Big|_{\partial\Omega} = 0$$

Here p is the pressure, T is the temperature, and $Y_i (i = 1,\ldots,M)$ is the mass fraction for the i-th chemical species.

We assume

$$p > 0 \;,\; T > 0 \tag{1.3a}$$

$$Y_i \geq 0 \quad \text{for} \quad i = 1,\ldots,M \tag{b}$$

$$\sum_{i=1}^{M} Y_i = 1 \tag{c}$$

Therefore, Y_M is determined from the constraint (1.3) (c). We also define $Y = (Y_1,\ldots,Y_{M-1})$.

The average velocity v of the mixture has been written in terms of its solenoidal and potential components (12) w and $\nabla\phi$ satisfying

$$v = w + \nabla\phi \text{ with} \tag{1.3d}$$

$$\operatorname{div} w = 0 \ , \ w.n|_{\partial\Omega} = 0 \text{ and}$$

$$\Delta\phi = \operatorname{div} v \ , \ \left.\frac{\partial\phi}{\partial n}\right|_{\partial\Omega} = v.n|_{\partial\Omega}$$

Hence $v.n|_{\partial\Omega} = 0$.

ρ is the density and h_i is the enthalpy of the i-th chemical species. $\gamma = c_p/c_v$ where c_p, c_v are the heat capacity at constant pressure and volume, respectively. $\Gamma > 1$ is a nondimensional constant. All these quantities are functions of P, T, and Y_i. The formulas in the case of a mixture of ideal gases with molecular weight W_i and γ-gas constant γ_i can be found in (13), (1). In particular

$$\rho = PT^{-1}\left(\sum_{i=1}^{M} \frac{Y_i}{W_i}\right)^{-1} \ . \tag{1.4}$$

We assume that the source terms Φ_i satisfy

$$\sum_{i=1}^{M} \Phi_i = 0 \tag{1.5a}$$

$$\Phi_i \geq 0 \quad \text{on} \quad Y_i = 0 \ . \tag{b}$$

In particular (1.5) is satisfied when Φ_i is given by the law of mass action ((13), (1)). Finally G is given by

$$G(p,T,Y) = \sum_{i=1}^{M}\left(\gamma T\left(\frac{1}{W_i} - \frac{1}{W_M}\right) - \frac{\gamma-1}{\Gamma-1}\,\Gamma(h_i - h_M)\right)\Phi_i \ . \tag{1.6}$$

The chemical-fluid balance condition (1.2) (f) for the initial data requires that equation (1.1) (b) be satisfied initially. This condition guarantees the formal validty of (1.1) and it is also needed to show that the initial value problem is well-posed. Recently Schochet (9) has shown rigorously the validity of (1.1) for the case of a binary mixture of gases with some γ-gas constant undergoing a one-step irreversible reaction.

The equations for zero Mach number combustion have also been studied numerically under the infinitely thin flame approximation in unconfined channels (2) and confined chambers (7).

For the formulation of the zero Mach number combustion in bounded or unbounded regions as well as other approximations like the infinitely thin flame structure limit mentioned before, the reader is referred to (5).

Different formulations given for the equations for zero Mach number combustion (10) keep the conservation of mass as one of the basic equations. However, it is straightforward to verify that (1.1), (1.4) imply that $1/\rho \, D\rho/Dt + \operatorname{div} v = 0$.

Finally we mention that for classical solutions of (1.1), if (1.3) (b), (c) are satisfied initially, then they remain satisfied for as long as the solution is defined. This can be verified easily by writing (1.1) (e) in Lagrangian coordinates and using (1.5).

2. The Initial Value Problem for the Nondiffusive Zero Mach Number Combustion Equations

Next we consider the initial-boundary value problem (1.1), (1.2). From the previous considerations we require from the initial data $\theta_0 = (p_0, T_0, Y_{10}, \dots, Y_{M0})$ and v_0 that

$$\theta_0(x) \quad G_0 \text{ where } G_0 \text{ has compact closure in the set} \tag{2.1a}$$

$$\Theta = \{\theta \quad R^{M+1} \,|\, p > 0, T > 0, Y_i > 0 \quad i = 1, \dots, M-1,$$

$$\text{and } \sum_{i=1}^{M-1} Y_i \leq 1\}$$

v_0 satisfies (1.2) (d), (e) and (f). (b)

Based upon the energy estimates for hyperbolic equations like in (1.1) we define $X([0,T],H^s) = C([0,T],H^0) \cap L^\infty([0,T],H^s)$ with norm $|||f|||_{s,T} = \operatorname*{ess\,sup}_{0 \leq t \leq T} ||f(t)||_s$. Here H^s is the Sobolev space of order s on Ω.

The initial-boundary value problem (1.1), (1.2) is well-posed:

<u>Theorem 2.1.</u> Assume $\theta_0, v_0 \quad H^s(\Omega)$, $s > \frac{N}{2} + 1$, and satisfy (2.1). Then there is a bounded open set G_1 with $\overline{G}_0 \quad G_1$ and there is $T > 0$ depending only on s, Ω, G_1, $||\theta_0||_s$ and $||v_0||_s$, so that the initial-boundary value problem (1.1), (1.2) has a unique classical solution

θ, v, $\nabla\pi$ with

$$\theta,\ v,\ \nabla\pi \in X([0,T],H^s) \text{ and}$$

$$\frac{\partial\theta}{\partial t},\ \frac{\partial v}{\partial t} \in X([0,T],H^{s-1})$$

This is a short time existence result. s is an integer larger than $\frac{N}{2}+1$ to guarantee that the solution constructed is classical.

The technical details of the proof can be found in (1). Here we only discuss the main ideas involved.

The proof of Theorem 2.1 is based on the method of successive approximations. We found it convenient to use the following iteration scheme (1):

(2.2) Set $\theta^0(x,t) = \theta_0(x)$ and $v^0(x,t) = v_0(x)$, and for k = 0,1,2,..., define $\theta^{k+1}(x,t)$ and $v^{k+1}(x,t)$ inductively in three steps.

<u>Step 1.</u> Construct θ^{k+1} from the previous iterates θ^k and v^k by solving

$$\frac{dp^{k+1}}{dt} = H^k(t) := \frac{\int_\Omega (\gamma^{-1}G)(\theta^k)dx}{\int_\Omega \gamma^{-1}(\theta^k)dx} \qquad (2.3a)$$

$$p^{k+1}(0) = p_0$$

$$\frac{D^k}{Dt} T^{k+1} = (\rho c_p)^{-1}(\theta^k)H^k(t) - \sum_{i=1}^{M-1} (\rho c_p)^{-1}(h_i - h_M)\Phi_i(\theta^k) \qquad (b)$$

$$T^{k+1}(x,0) = T_0(x)$$

$$\frac{D^k}{Dt} Y_i^{k+1} = (\rho^{-1}\Phi_i)(\theta^k) \qquad (c)$$

$$Y_i^{k+1}(x,0) = Y_{i,0}(x)\ ,\ i = 1,\dots M-1\ ,$$

where $\frac{D^k}{Dt} = \frac{\partial}{\partial t} + v^k.\nabla$.

<u>Step 2.</u> Construct $\nabla\phi^{k+1}$ from the iterate θ^{k+1} computed in Step 1 by solving

$$\Delta\phi^{k+1} = (\gamma p)^{-1}(\theta^{k+1})\ G(\theta^{k+1}) - H^{k+1}(t)\ , \qquad (d)$$

$$\left.\frac{\partial \phi^{k+1}}{\partial n}\right|_{\partial\Omega} = 0 ,$$

$$\Delta\phi^{k+1}(x,0) = (\gamma p)^{-1}(\theta_0)\left(G(\theta_0) - \frac{\int_\Omega (\gamma^{-1}G)(\theta_0)dx}{\int_\Omega \gamma^{-1}(\theta_0)dx} \right).$$

Step 3. Construct w^{k+1} and $\nabla\phi^{k+1}$ computed in Step 2 and the previous iterates θ^k and v^k by solving

$$\rho(\theta^k)\,\frac{D^k}{Dt}\,w^{k+1} + \nabla\pi^{k+1} = -\rho(\theta^k)\,\frac{D^k}{Dt}\,\nabla\phi^{k+1} , \tag{e}$$

$$\operatorname{div} w^{k+1} = 0 ,$$

$$\left. w^{k+1}.n\right|_{\partial\Omega} = 0 ,$$

$$w^{k+1}(x,0) = w_0(x) .$$

Convergence is proved in two steps. In the first step stability is established by showing the uniform boundedness of the approximating sequence in $X([0,T],H^s)$ with the strong norm $|||\circ|||_{s,T}$. This step is crucial for the success of the method. The second step is simpler and consists in showing contraction of the sequence in a weak norm involving only a few L^2 derivatives.

Next, we comment on the derivation of the stability estimates for the iterates in (2.3). We use elliptic regularity estimates and energy estimates for hyperbolic equations.

If u is a regular enough solution of

$$\frac{Du}{Dt} = f \tag{2.4a}$$

$$u(0) = u_0$$

$$\text{where } \frac{D}{Dt} = \frac{\partial}{\partial t} + v.\nabla \text{ and } \left. v.n\right|_{\partial\Omega} = 0 , \tag{b}$$

then for any $r \geq 0$, u satisfies the energy estimate

$$||u(t)||_r \leq e^{\alpha(t)}\,(||u_0||_r + \int_0^t e^{-\alpha(\tau)}\,C||f(\tau)||_r d\tau) \tag{2.5a}$$

where $\alpha(t)$ is given by

$$\alpha(t) = C \int_0^t ||Dv(\tau)||_{r_1} d\tau, \text{ with } r_1 = \max(r-1, s_0) \quad \text{(b)}$$

and C depends on r and Ω .

Since the components of θ^{k+1} satisfy equations of the form (2.4), we can use (2.5) and derive the estimate

$$|||\theta^{k+1}|||_{s,T} \leq e^{CT}(||\theta_0||_s + CT), \text{ where C depends on} \quad (2.6)$$

$|||\theta^k|||_{s,T}$ and $|||v^k|||_{s,T}$.

Next, we consider $\nabla\pi$. In (11) Temam studed the Euler equations for a homogeneous fluid and derived an elliptic equation for the artificial pressure term $\nabla\pi$.

We can adapt his proof to the variable density case in (2.3) (e) and derive for $\nabla\pi^{k+1}$ the elliptic equation

$$\operatorname{div}(\rho^{-1}(\theta^k)\nabla\pi^{k+1}) = -(\nabla v^k)^T : (\nabla v^{k+1}) - \frac{D^k}{Dt}\nabla\phi^{k+1} \quad (2.7a)$$

$$\left.\frac{\partial\pi^{k+1}}{\partial n}\right|_{\partial\Omega} = \rho(\theta^k)(v^k.\nabla n).v^{k+1}\Big|_{\partial\Omega} \quad \text{(b)}$$

Using elliptic regularity for (2.7) we estimate $||\nabla\pi^{k+1}|||_s$ and $||\rho(\theta^k)\, D^k/Dt\, \Delta\phi^{k+1}||_{s-1}$. To estimate this last term we evaluate $D^k/Dt\, \Delta\phi^{k+1}$ from (2.3) (d) and (2.3) (a), (b), (c), and use the estimate (2.6). Here it is important to exploit the fact that the right side of (2.3) (d) only depends on θ. The resulting estimate is

$$||\nabla\pi^{k+1}||_s \leq C(1 + ||v^{k+1}||_s), \text{ where C depends on}$$

$$|||\theta^{k+1}|||_{s,T}, \ |||\theta^k|||_{s,T} \text{ and } |||v^k|||_{s,T} . \quad (2.8)$$

Finally, we consider v^{k+1}. Rather than estimating $\nabla\phi^{k+1}$ and w^{k+1} separately, we rewrite (2.3) (e) as

$$\rho(\theta^k)\frac{D^k v^{k+1}}{Dt} + \nabla\pi^{k+1} = 0 \quad (2.9)$$

Using (2.5) and (2.8) provides the estimate

$|||v^{k+1}|||_{s,T} \leq e^{CT}(||v_0||_s + CT)$ where C depends on

$|||\theta^{k+1}|||_{s,T}$, $|||\theta^k|||_{s,T}$ and $|||v^k|||_{s,T}$. (2.10)

Therefore, if T is sufficiently small, (2.6), (2.6), (2.8), and (2.10) provide uniform stability estimates for the approximating sequence.

Finally, we make few comments on the linearized versions of the equations in (1.1) that were used in 1 to build up the iteration scheme.

For the linearized version of equations (1.1) (d), (e) we utilized (2.2) with v(x,t) and f(x,t) given. The solution of (2.2) is then constructed by using Galerkin's method with a special basis. For special basis we chose the eigenfunctions ϕ_i of the elliptic operator $L\phi = \Sigma_{j=0}^{s} (-\Delta)^j \phi$ with Neumann boundary conditions. It is known that these eigenfunctions are infinitely differentiable 8 when $\partial\Omega$ is smooth.

For the linearized version of the inhomogeneous Euler equation (1.1) (c) we worked with the equivalent elliptic-hyperbolic system:

$$\frac{Dw}{Dt} = Q(v.\nabla Pw) - P(\rho^{-1}\nabla\pi) - P(\rho^{-1}f) \qquad (2.6a)$$

$$\operatorname{div}(\rho^{-1}\nabla\pi) = -(\nabla v)^T:\nabla(Pw) - \operatorname{div}(\rho^{-1}f) \qquad (b)$$

$$\left.\frac{\partial\pi}{\partial n}\right|_{\partial\Omega} = \rho(v.\nabla n).w - f.n|_{\partial\Omega} \,. \qquad (c)$$

$$w(0) = w_0 \text{ with } \operatorname{div} w = 0 \,,\ w.n|_{\partial\Omega} = 0 \qquad (d)$$

Here P is the orthogonal projection of L^2 onto the solenoidal vector fields and Q = I-P is the orthogonal projection of L^2 onto the gradient fields.

The solution of (2.6) is then obtained by the method of successive approximations and showing contraction of the sequence in the high norm.

The artifice of using the equivalent system (2.6) was inspired by earlier work of Lai (4) and Kato and Lai (3) where they studied the Euler equations for a homogeneous fluid. Here we had further complications because the density is variable.

To conclude we also remark that in (1) it was also studied the initial value problem for the zero Mach number combustion equations when all the diffusive effects are present and under periodic boundary conditions.

Acknowledgements

This paper contains partial results of my Ph.D. thesis, done under the guidance of Professor Andrew Majda. I want to take this opportunity to thank him for his generous help. I am also grateful for the support from grants ARO No. 483964-25530 and from CONICIT, which helped me at different stages of this research.

References

(1) Embid, P. "Well-posedness of the Nonlinear Equations for Zero Mach Number Combustion," Ph.D. Thesis, Univ. Calif. Berkely, 1984.

(2) Ghoniem, A.F., A.J. Chorin, and A.K. Oppenheim, "Numerical modelling of turbulent flow in a combustion tunnel," Philos. Trans. Roy. Soc. London Series A (1981), 1103-1119.

(3) Kato, T., and C.Y. Lai, "Nonlinear evolution equations and the Euler flow," J. Functional Anal. 56 (1984), 15-28.

(4) Lai, C.Y., "Studies on the Euler and the Navier-Stokes equations," Ph.D. thesis, Univ. Calif., Berkely, 1975.

(5) Majda, A., "Equations for low Mach number combustion," Center for Pure and Appl. Math., Univ. Calif., Berkely, Rep. #112, Nov. 1982.

(6) Majda, A., Compressible Fluid Flow and Systems of Conservation Laws in Several Space Variables, Appl. Math. Sci. Series, Springer-Verlag, New York, 1984.

(7) Majda, A., and J. Sethian, "The derivation and numerical solution of the equations for zero Mach number combustion," Center for Pure and Appl. Math., Univ. Calif., Berkeley, Rep. #197, Jan. 1984 (to appear in Comb. Sci. and Tech.).

(8) Niremberg, L., "Remarks on strongly elliptic partial differential equations," Comm. Pure Appl. Math. 8 (1955), 649-675.

(9) Schochet, S., "Singular limits in bounded domains for quasi-linear symmetric hyperbolic systems having a vorticity equations" (preprint).

(10) Sivashinsky, G.I., "Hydrodynamic Theory of Flame Propagation in an Enclosed Volume," Acta Astronautics 6 (1979), 631-634.

(11) Temam, R., "On the Euler equations of incompressible perfect fluids," J. Functional Anal. 20 (1975), 32-43.

(12) Temam, R., Navier-Stokes Equations, North Holland-Elsevier, Amsterdam, New York, 1976.

(13) Williams, F.A., Combustion Theory, Addison-Wesley, Reading, MA, 1964.

LONGTIME SOLUTIONS FOR A CLASS OF CONVECTION DIFFUSION SYSTEMS

W. E. Fitzgibbon
Department of Mathematics, University of Houston
Houston, Texas 77004

1. Introduction

We shall be concerned with convection diffusion systems of the form:

$$U_t = DU_{xx} + f(U)_x , \qquad x \in \mathbb{R} , \qquad t > 0 \tag{1.1a}$$

$$U(x,o) = U_0(x) \tag{1.1b}$$

Here $U = (u_1,\ldots, u_n)^T$ is a vector in R^n, D is a constant diagonal $n \times n$ matrix with positive entries along the diagonal and $f(\ ) : \mathbb{R}^n \to \mathbb{R}^n$ is a quadratic function having componentwise definition,

$$f_i(U) = \sum_{j,k=1}^{n} a_{ijk} u_j u_k \qquad i \le i \le n. \tag{1.2}$$

In what follows we establish the global existence of solutions to (1.1a-b) under the assumption of a priori L^2 bounds. Such bounds can be shown to exist if the system satisfies an energy inequality. In the manner of [8] a priori bounds can be obtained by demonstrating the existence of an entropy-entropy flux pair. Variational structure can be exploited [14] to obtain global existence for parabolic systems. Our approach involves a mixture of energy estimates and semigroup methods which has been applied to Burgers' systems [2], [3], [5].

2. The Abstract Setting and Local Theory

We shall make use of the following which states basic inequalities; a proof appears in [4].

Lemma 2.1 If $u \in H^1(R)$ and v, $w \in L^2(R)$ then the following are true:

(i) $\|u\|_\infty \le \sqrt{2}\, \|u\|^{1/2} \|u'\|^{1/2}$

(ii) $\int_{-\infty}^{\infty} |u(x)v(x)wx)|dx \le \sqrt{2}\, \|u\|^{1/2} \|u'\|\ \|v\|\ \|w\|$

We shall be working with a variety of spaces $L^2(R)$, $\oplus_{i=1}^{n} L^2(R)$, $H^1(R)$, $\oplus_{i=1}^{n} H^1(R)$, $L^\infty(\mathbb{R})$. Unless otherwise noted $\|\ \|$ shall denote the L^2 norm. We hope that our use of $\|\ \|$ to denote the norm in $L^2(\mathbb{R})$ and $\oplus_{i=1}^{n} L^2(\mathbb{R})$ does not lead to confusion.

We now discuss local solutions to (1.1a-b). We let $K(x, t)$ be the vector Green's function associated with the vector differential operator $(\partial/\partial t - d_i\partial^2/\partial x^2)_i$, $i = 1, \ldots, n$. Equivalently $K(x, t)$ is the n-component vector whose components are given by

$$K_i(x,t) = \frac{1}{\sqrt{4\pi d_i t}} \exp\,[-x^2/4d_i t] \tag{2.2}$$

The solution (1.1a-b) has the integral representation

$$U(t) = K(t) * U_0 - \int_0^t K_x(t-s)\ f(U(s))ds \tag{2.3}$$

where $*$ denotes convolution in space taken componentwise. The integral representation may be used to obtain a local existence result via application of the contraction mapping theorem. The following lemma summarizes results which appear in [8].

<u>Lemma 2.4</u> If $U_0 = (u_{01}, \ldots, u_{0n}) \in \bigoplus_{i=1}^{n} L^2(\mathbb{R}) \cap L^\infty(\mathbb{R})$ and $\|U_0\|_\infty = S$ then there exists an unique solution U of (1.1a-b) defined on a strip $[0, T(U_0, S)]$. Moreover, U_t, U_x, U_{xx} are continuous in t; $U_t(t)$, $U_x(t)$, $U_{xx}(t)$ belong to $\bigoplus_{i=1}^{n} H^2(\mathbb{R})$ for $t \geq \tau > 0$.

By virtue of this local result, solutions to (1.1a-b) belong to $\bigoplus_{i=1}^{n} H^2(\mathbb{R})$. We use the theory of analytic semigroups to represent solutions for $t > 0$. We work in the space $\overline{X} = \bigoplus_{i=1}^{n} L^2(\mathbb{R})$. We define the operator $A : \overline{X} \to \overline{X}$ componentwise in the following manner:

$$(AU)(x) = d_i u_i''(x) \qquad x \in \mathbb{R}, \quad i = 1, \ldots, n \tag{2.5}$$

with

$$D(A_i) = H^2(\mathbb{R}).$$

It is well known that each component of A is the infinitesional generator of an analytic semigroup in $L^2(\mathbb{R})$ and it is trivial to observe that A with $D(A) = \bigoplus_{i=1}^{n} L^2(\mathbb{R})$ generates an analytic semigroup $\{T(t) \mid t \geq 0\}$ on $\overline{X}$. A nonlinear and unbounded operator F on $\overline{X}$ is defined,

$$(F_i(U))(x) = \Big[\sum_{j,k=1}^{n} a_{ijk}\ u_j(x)\ u_k(x) \Big]_x . \tag{2.6}$$

The solution to (1.1a-b) has the semigroup variation of parameters formula,

$$U(t) = T(t)U_0 + \int_0^t T(t-s)\ F(U(s))ds . \tag{2.7}$$

If $t > 0$, $U(t)$ is continuously differentiable and satisfies

$$\dot{U}(t) = AU(t) + F(U(t)). \tag{2.8}$$

3. Global Existence

We begin by stating our assumption concerning the existence of a priori bounds

B There exists a continuous function $C : \mathbb{R}^+ \to \mathbb{R}^+$ so that if $U(\ )$ is a solution to $U_t = DU_{xx} + f(U)_x$ for $t \in [0, T)$ then $\|U(t)\| \le C(\|U_0\|)$.

The following results which adapts ideas appearing in [1] provides a uniqueness results and serves to underpin of our global theory.

Theorem 3.1 Assume that Condition B holds if U and V are solutions to (1.1a-b) on $[0, T)$ with initial data $U_0, V_0 \in \bigoplus_{i=1}^{n} L^2(\mathbb{R}) \cap L^\infty(\mathbb{R})$ then there exists an $L > 0$ depending only upon $C(\max\{\|U_0\|, \|V_0\|\})$ where $C(\ )$ is a nondecreasing function so that

$$\begin{aligned} \|U(t) - V(t)\| &= \Big\{ \sum_{i=1}^{n} \|u_i(t) - v_i(t)\| \Big\}^{1/2} \\ &\le L \Big\{ \sum_{i=1}^{n} \|u_i(0) - v_i(0)\|^2 \Big\}^{1/2} \qquad (3.2) \\ &= L\, \|U_0 - V_0\| \end{aligned}$$

Proof We set $W = U - V$ and observe that

$$\partial w_i/\partial t = d_i \partial^2 w_i/\partial x^2 + \partial/\partial x(f_i(U)) - \partial/\partial x(f_i(V)) \qquad (3.3)$$

Consequently, we multiply each side by w_i, compute the L^2 inner product and observe that,

$$\begin{aligned} &\langle \partial w_i/\partial t, w_i\rangle + d_i\langle \partial w_i/\partial x, \partial w_i/\partial x\rangle = \\ &\qquad \int_{-\infty}^{\infty} \Big\{ \sum_{j,k=1}^{n} a_{ijk}(u_j u_k)_x w_i - \sum_{j,k=1}^{n} a_{ijk}(v_j v_k)_x w_i \Big\} \qquad (3.4) \\ &= \sum_{j,k=1}^{n} a_{ijk} \int_{-\infty}^{\infty} (w_j)_x u_k w_i + \sum_{j,k=1}^{n} a_{ijk} \int_{-\infty}^{\infty} (v_j)_x w_k w_i \\ &+ \sum_{j,k=1}^{n} a_{ijk} \int_{-\infty}^{\infty} w_j (u_k)_x w_i + \sum_{j,k=1}^{n} a_{ijk} \int_{-\infty}^{\infty} v_j (w_k)_x w_i . \end{aligned}$$

We wish to estimate the rightmost terms of (3.4). Using integration by parts we have,

$$\begin{aligned} a_{ijk} \int_{-\infty}^{\infty} w_j (u_k)_x w_i &= -a_{ijk} \int_{-\infty}^{\infty} (w_j)_x u_j w_i \\ &\quad - a_{ijk} \int_{-\infty}^{\infty} w_j u_k (w_i)_x . \end{aligned} \qquad (3.5)$$

These expressions may in turn be estimated,

$$\left|\int_{-\infty}^{\infty} w_j\, u_k (w_i)_x\right| \le \sqrt{2}\, \|(w_j)_x\|^{1/2}\, \|w_j\|^{1/2}\, \|(w_i)_x\|\, \|u_k\|$$

$$\le \sqrt{2}\left[\frac{C_1}{2}\|(w_i)_x\|^2\, \|u_k\|^2 + \frac{1}{2C}\|(w_j)_x\|\, \|w_j\|\right] \tag{3.6}$$

$$\le \sqrt{2}\left[\frac{C_1}{2}\|(w_i)_x\|\, \|u_k\|^2 + \frac{C_2}{4C_1}\|(w_j)_x\| + \frac{1}{4C_1C_2}\|w_j\|^2\right].$$

By virtue of Condition B $\|u_i\|$ and $\|w_i\|$ satisfy a priori bounds we can choose C_i's so that

$$\Sigma d_i - \Sigma \sup[\text{coefficients of } \|(w_i)_x\|^2] > 0\,. \tag{3.7}$$

We add the inequalities resulting from (3.4)-(3.7) and obtain

$$\frac{1}{2}\frac{d}{dt}\sum_{i=1}^{n}\|u_i - v_i\|^2 \le L\sum_{i=1}^{n}\|u_i - v_i\|^2\,. \tag{3.8}$$

Our desired result thus follows by integration.

We now are in a position to state our global result.

__Theorem 3.9__ Assume that Condition B is satisfied and that A, F and $\{T(t)/t \ge 0\}$ are defined via (2.5) and (2.6). If $U_0 = (u_{01}, \ldots, u_{0n})^T \in \bigoplus_{i=1}^{n} L^2(\mathbb{R}) \cap L^\infty(\mathbb{R})$ and $T > 0$ there exists an unique function $U(t) : (0, T] \to \bigoplus_{i=1}^{n} H^2(\mathbb{R})$ which satisfies

$$U(r) = T(t)U_0 + \int_0^t T(t-s)\, F(U(s))ds\,. \tag{3.10}$$

The function is continuously differentiable for $t > 0$ and satisfies

$$\dot{U}(t) = AU(t) + F(U(t))\,. \tag{3.11}$$

__Proof__ A local solution is guaranteed by Lemma 2.4. To obtain solutions on $[0, T]$ for any $T > 0$ we follow classical ordinary differential equations techniques. We assume that a solution exists on $[0, T_0)$ and continue it past T_0.

We first argue that $\|U_t\|$ and $\|U_x\|$ are uniformly bounded on $[0, T_0)$. From Theorem 3.1 we see that if $0 < \tau < t < t+h < T_0$

$$\|U(t+h) - U(t)\| \le L\|U(\tau+h) - U(\tau)\|$$

Dividing by h and letting $h \downarrow 0$ we thus bound $\|\dot{U}(t)\|$ in terms of $\|\dot{U}(\tau)\|$. To see that $\|U_x\|$ is bounded we take the L^2 inner product of (1.1a) with U. Componentwise we have

$$\langle \partial u_i/\partial t,\ u_i\rangle + d_i\|(u_i)_x\|^2 = \sum_{j,k=1}^{n} a_{ijk}\int_{-\infty}^{\infty}(u_j)_x\, u_k\, u_i + \sum_{j,k=1}^{n} a_{ijk}\int_{-\infty}^{\infty} u_j (u_k)_x\, u_i \tag{3.12}$$

We estimate the terms on the right of 3.13 by observing

$$\left|\int_{-\infty}^{\infty}(u_i)_x\, u_j\, u_k\right| \le \sqrt{2}\left[\frac{C_1}{2}\|(u_1)_x\|^2\ \|u_j\| + \frac{C_2}{4C_1}\|(u_j)_x\|^2 + \frac{1}{4C_1C_2}\|u_k\|^2\right].$$

Because $\|u_i\|$ satisfies an a priori bound we can choose the C_i's so that

$$\Sigma d_i - \Sigma\sup\,[\text{coefficients of } \|(u_i)_x\|^2] \ge \frac{1}{2}\Sigma d_i .$$

Because $\|\dot{U}(t)\|$ and $\|U(t)\|$ are bounded on $[0,T_0]$ we can therefore sum the components (3.13) to produce bounds for $\Sigma d_i\|(u_i)_x\|^2$ and establish our claim.

Lemma 2.1 can be applied to bound $\|u_i\|_\infty$ uniformly on $[\tau, T_0)$. The foregoing observations allow us to conclude that $F(U(t))$ is uniformly bounded on $[\tau, T_0)$ and we can use our variation of parameters formula to compute

$$U(T_0) = \lim_{t\to T_0} U(t) = T(T_0) + \int_0^{T_0} T(T_0-s)F(U(s))ds .$$

We now argue that $U(T_0) \in D(A)$. Because $U(t)$, $U_t(t)$ and $F(U(t))$ are uniformly bounded on $[\tau, T_0)$, $AU(t)$ is uniformly bounded. We are working in a Hilbert space and we can extract a weakly convergent sequence $\{AU(t_n)\}$ as $t_n \to T_0$. The closedness of the linear operator A now quarantees $AU(T_0) = w - \lim AU(t_n)$ and $U(T_0) \in D(A) = \bigoplus_{i=1}^{n} H^2(\mathbb{R})$. We observe

$$\|u_i(\ ,T_0)\|_\infty \le \sqrt{2}\left\|\frac{d^2}{dx^2}u_i(\ ,T_0)\right\|^{1/2}\left\|\frac{d}{dx}u_i(\ ,T_0)\right\|^{1/2}.$$

Because $U(T_0) \in \bigoplus_{i=1}^{n} L^2(\mathbb{R}) \cap L^\infty(\mathbb{R})$ we can apply the local theory to continue our solution past T_0 and thereby obtain our desired result. The time regularity is a consequence of the local regularity theory and the regularity for abstract Cauchy initial value problems, cf [10].

As a simple example to illustrate our theory we consider the following system of 3 convection diffusion equations:

$$\begin{aligned}\partial u/\partial t &= d_1\partial^2 u/\partial x^2 - (u^2-v^2)_x + (uw)_x\\ \partial v/\partial t &= d_2\partial^2 v/\partial x^2 + (2uv)_x - 2(vw)_x\\ \partial w/\partial t &= d_3\partial^2 w/\partial x + (w^2-v^2)_x + UU_x\end{aligned} \tag{3.13a}$$

$$u(x,o) = u_0(x) \qquad v(x,o) = v_0(x) \qquad W(x,o) = W_0(x) \tag{3.13b}$$

Let $[u, v, w]^T$ be a solution of (3.13a) on some interval $[0, T]$. Computing the L^2 inner product of (3.13a) with $[u, v, w]^T$ we obtain the energy equality

$$\frac{1}{2}\frac{d}{dt}[\|u\|^2 + \|v\|^2 + \|w\|^2] + d_1\|u_x\|^2 + d_2\|v_x\|^2 + d_3\|w_x\|^2 \le 0$$

and thereby produce the a priori bounds sufficient to guarantee the global existence of solutions. This example immediately generalizes to the following corollary.

Corollary 3.14 Assume that for all $U \in \bigoplus_{i=1}^{n} H^1(\mathbb{R})$, $\langle F(U)_x, U \rangle \le 0$. If $U_0 = [u_{01}, \ldots, u_{0n}]^T \in \bigoplus_{i=1}^{n} L^2(\mathbb{R}) \cap L^\infty(\mathbb{R})$ and $T > 0$ then there exists an unique $U(t) : [0, T] \to \bigoplus_{i=1}^{n} L^2(\mathbb{R})$ which satisfies (3.11) and (3.12).

We remark that our techniques will provide global existence for more general systems. For example we could have considered systems of the form

$$\partial u_i/\partial t = \partial/\partial x(a_i(x)\partial u_i/\partial x) + \sum_{i=1}^{n} g_{ij}(U)(u_j)_x$$

where the convection terms satisfy and inequality of the form

$$|g_{ij}(U)| \le K \sum_{i=1}^{n} |u_k| .$$

BIBLIOGRAPHY

[1] T. Dlotko, "The two dimensional Burgers' turbulence model," J. Kyoto Univ., 21 (1981), 809-823.

[2] W. Fitzgibbon, "A semigroup approach to Burgers' systems," Differential Equations, Elsevier Science Publishers, New York, 1984, 213-217.

[3] ____, "A two dimensional model for turbulence," Proceedings International Conference on Nonlinear Phenomena, Marcel Decker, New York, 1984, 177-184.

[4] ____, "Global existence for a particular convection diffusion system," Nonlinear Analysis TMA (to appear).

[5] ____, "A variation of parameters formula for Burgers' system," Infinite-Dimensional Systems, Lecture Notes in Mathematics 1076, Springer-Verlag, Berlin, 1984, 78-85.

[6] A. Friedman, Partial Differential Equations, Holt, Rhinehart, and Winston, New York, 1969.

[7] J. Goldstein, "Semigroups of operators and abstract Cauchy problems," Lecture Notes, Tulane University, 1970.

[8] D. Hoff and J. Smoller, "Solutions in the large for certain nonlinear parabolic systems" (to appear).

[9] D. Henry, Geometric Theory of Semilinear Parabolic Equations, Lecture Notes in Mathematics 840, Springer-Verlag, Berlin, 1981.

[10] A. Pazy, Semigroups of Linear Operators and Applications to Partial Differential Equations, Springer-Verlag, Berlin, 1983.

[11] J. Smoller, Shock Waves and Reaction-Diffusion Equations, Springer-Verlag, Berlin, 1983.

[12] P. E. Sobolevskii, "On equations of parabolic type in a Banach space," Trudy Moscov. Mat. Orsc, 10 (1961), 297-350.

[13] G. Webb, "Exponential representation of solutions to an abstract semilinear differential equation" Pac. J. Math., 70 (1977), 268-279.

[14] W. Wieser, "On parabolic systems with variational structure," Manuscript Mathematica 54 (1985), 53-82.

A CLOSED FORM FOR THE SYMBOL OF THE RESOLVENT PARAMETRIX OF AN ELLIPTIC OPERATOR

S. A. Fulling* and G. Kennedy
Mathematics Department, Texas A & M University
College Station, Texas, 77843

This exposition is addressed primarily to people who do not know much about pseudodifferential operators but have some background in quantum physics. In fact, however, every physicist is already familiar with pseudodifferential operators, if not by that name. Apart from some technical conditions (which will be ignored in this informal presentation), a pseudodifferential operator is what in quantum mechanics is called *a function of both position and momentum.*

The Fourier transformation,

$$\hat{\psi}(\mathbf{p}) = \int d^n x\, e^{-i\mathbf{p}\cdot\mathbf{x}} \psi(\mathbf{x}),$$

translates differentiation ($-i\nabla$) into multiplication by $\mathbf{p}$, and hence identifies every (linear) differential operator with a polynomial in $\mathbf{p}$ (whose coefficients are, in general, $\mathbf{x}$-dependent). The function of $\mathbf{x}$ and $\mathbf{p}$ corresponding to an operator A is called the *symbol* of A. For example, we can write

$$A = -\nabla^2 + m^2 = \mathrm{Op}(a)$$

if

$$a(\mathbf{x}, \mathbf{p}) = \mathbf{p}^2 + m^2 = \mathrm{Sy}(A).$$

(The standard notations in the mathematics literature are "ξ" for "p" and "σ" for "Sy".) The inverse of this operator also has a symbol, which is not a polynomial:

$$\mathrm{Sy}(A^{-1}) = \frac{1}{\mathbf{p}^2 + m^2}.$$

This simply means that this function appears in the Fourier inversion formula,

$$(A^{-1}\psi)(\mathbf{x}) = (2\pi)^{-n} \int d^n p\, e^{i\mathbf{p}\cdot\mathbf{x}} \frac{1}{\mathbf{p}^2 + m^2} \hat{\psi}(\mathbf{p}),$$

in the same way that the polynomial symbol would if A^{-1} were differential.

Now consider a differential operator with variable coefficients,

$$A = -\nabla^2 + V(\mathbf{x}).$$

---

*speaker

What is $\mathrm{Sy}(A^{-1})$ this time? That is, can one find a function $b(\mathbf{x},\mathbf{p})$ such that

$$(A^{-1}\psi)(\mathbf{x}) = (2\pi)^{-n}\int d^n p\, e^{i\mathbf{p}\cdot\mathbf{x}}\, b(\mathbf{x},\mathbf{p})\,\hat{\psi}(\mathbf{p})\,?$$

One needs to have $A(A^{-1}\psi) = \psi$; because the differential operator acts on the $\mathbf{x}$-dependence of b as well as on the exponential, the naive answer does not work:

$$b \neq \frac{1}{\mathbf{p}^2 + V(\mathbf{x})}\,.$$

(In other words, as is well known, variable-coefficient differential equations are not solvable by the method of Fourier transforms.)

Nevertheless, b can be constructed, as an asymptotic series, through the *symbolic calculus* of pseudodifferential operators. Its foundation is the *product rule*: Suppose that $A = \mathrm{Op}(a)$ and $B = \mathrm{Op}(b)$ are two pseudodifferential operators. Then their product is one also, and its symbol is equal to the product of the symbols a and b, plus "quantum corrections":

$$\mathrm{Sy}(BA) \sim \sum_{u=0}^{\infty} \frac{i^{-u}}{u!}(D^u b)(\partial^u a) \equiv \sum_{u=0}^{\infty} P_u(b,a).$$

Here we use the notation

$$D \equiv \nabla_{\mathbf{p}}, \qquad \partial \equiv \nabla_{\mathbf{x}},$$

and tensor contractions are implied:

$$(D^3 b)(\partial^3 a) = \sum \frac{\partial^3 b}{\partial p_\alpha \partial p_\beta \partial p_\gamma}\,\frac{\partial^3 a}{\partial x^\alpha \partial x^\beta \partial x^\gamma}\,,$$

for instance. Notice that if b and a are polynomials then this formula is just the familiar Leibnitz rule; u is the number of ∂'s in B that fall on A instead of on ψ when the products in $A\psi$ are differentiated.

Now we shall let $B = A^{-1}$ and try to find b by a kind of perturbation theory. Assume that

$$b \sim b_0 + b_1 + \cdots.$$

Substitute into the product formula and group terms of the same order together, treating $D^n b_m$ as of the same order as b_{n+m}:

$$\underbrace{1}_{0} \sim \underbrace{b_0 a}_{0} + \underbrace{b_1 a - iDb_0\,\partial a}_{1} + \cdots.$$

(The left-hand side is the symbol of the identity operator.) The zeroth-order equation gives

$$b_0 = \frac{1}{a} = \frac{1}{\mathbf{p}^2 + V}\,,$$

as one might expect. The first-order equation is $0 = b_1 a - iDb_0\partial a$, yielding

$$b_1 = \frac{i}{a}\frac{\partial b_0}{\partial p_\alpha}\frac{\partial a}{\partial x^\alpha} = \frac{-2i}{(\mathbf{p}^2+V)^3}\,\mathbf{p}\cdot\nabla V.$$

We could continue, finding b_2, etc. Usually this expansion is combined with an expansion with respect to $V/\mathbf{p}^2$, so that the perturbation parameter is effectively $1/|\mathbf{p}|$.

Let G be the integral kernel of A^{-1}:

$$(A^{-1}\psi)(\mathbf{x}) = \int d^n y\, G(\mathbf{x},\mathbf{y})\,\psi(\mathbf{y}).$$

Since

$$G(\mathbf{x},\mathbf{y}) = (2\pi)^{-n}\int d^n p\, e^{i\mathbf{p}\cdot(\mathbf{x}-\mathbf{y})} b(\mathbf{x},\mathbf{p}),$$

our series for b yields a series for G, each successive term of which is one degree smoother in the limit $\mathbf{y}\to\mathbf{x}$. This series does not converge to an exact solution, or even give a numerically accurate approximation, in general. However, it does correctly reproduce the singular behavior of $G(\mathbf{x},\mathbf{y})$ as $\mathbf{y}$ approaches $\mathbf{x}$ — information of crucial importance in many applications.

This calculation serves as a model for a more serious one. Consider the general second-order linear differential operator,

$$A = A^{\mu\nu}\partial_\mu\partial_\nu + B^\mu\partial_\mu + C = \mathrm{Op}(a).$$

Here $A^{\mu\nu}(\mathbf{x})$, $B^\mu(\mathbf{x})$, and $C(\mathbf{x})$ may be matrices (for each μ and ν). More precisely, the argument of A is a section of a vector bundle, E, over a manifold, M,

$$\psi\in\Gamma^\infty(E),$$

and thus the coefficient functions in A are tensor-valued sections of the corresponding endomorphism bundle:

$$B\in\Gamma^\infty(E\otimes E^*\otimes T(M)),$$

for example. In fact, we can just as easily consider an operator of arbitrary order l, writing the symbol in the notation

$$a = \sum_{r=0}^{l} a_r, \qquad \text{where} \quad a_r\propto|\mathbf{p}|^{l-r}.$$

We do need to assume that A is elliptic and that the coefficient tensors are C^∞ in their dependence on $\mathbf{x}$.

We shall construct an asymptotic series for the symbol of the inverse of $A-\lambda$, where $\lambda\in\mathbf{C}$. More precisely, we do this for a *resolvent parametrix*, B_λ: an operator such that $B_\lambda(A-\lambda)$ differs from the identity at most by an operator with a C^∞ integral kernel. (Nothing has been said about the boundary conditions or other global information needed to specify the function space which serves as the domain of A. Until that is done, $(A-\lambda)^{-1}$ is not well-defined; after it is done, $(A-\lambda)^{-1}$ may or may not exist for a given λ. The existence and usefulness of the (nonunique) parametrix is independent of such global issues; it answers purely local questions about the differential operator A.)

Much as before, try

$$b\sim\sum_{s=0}^{\infty} b_s, \qquad \text{where} \quad b_s\sim|\mathbf{p}|^{-(l+s)}.$$

The product formula gives

$$1 \sim -\lambda \sum_{s=0}^{\infty} b_s + \sum_{u,s=0}^{\infty} \sum_{r=0}^{l} P_u(b_s, a_r).$$

Looking successively at the terms of each order, one sees that

$$b_0 = (a_0 - \lambda)^{-1}$$

and, for $s > 0$,

$$b_s = - \sum_{\substack{r,u,q \geq 0 \\ r+u \geq 1}}^{r+u+q=s} P_u(b_q, a_r) b_0 \equiv - \sum_{\substack{r,u,q \geq 0 \\ r+u \geq 1}}^{r+u+q=s} \frac{i^{-u}}{u!} D^u b_q \, \partial^u a_r \, b_0. \tag{$*$}$$

All that has been said so far is well known (and, of course, presented in much greater rigor in treatises on the subject [1]). The new result which we announce here is that *the recursion relation* (∗) *is solvable!* In fact, it is easy to show that

$$b_s = \sum_{J=1}^{s} (-1)^J \sum_{r_j + u_j \geq 1}^{|r+u|=s} P_{u_1}\Big(P_{u_2}(\;\cdots\; P_{u_J}(b_0, a_{r_J}) b_0 \;\;\cdots\;, a_{r_2}) b_0, a_{r_1}\Big) b_0.$$

Here

$$|r+u| \equiv \sum_{j=1}^{J} (r_j + u_j).$$

However, this formula is not very useful until we evaluate the derivatives contained in the operations P_{u_j}. This is where most of the hard work lies. Note first that

$$a_r = i^{l-r} A_r(\mathbf{x}) \left(\otimes^{(l-r)} \mathbf{p} \right),$$

where A_r is the coefficient tensor of rank $l - r$. For example, if $l - r = 2$, then

$$a_r = -A_r^{\alpha\beta} p_\alpha p_\beta$$

in more classical notation. After a lengthy calculation (involving elementary calculus and substantial combinatorics), we arrive at a formula of the structure

$$b_s = \sum_{\substack{\text{many} \\ \text{indices}}} i^{\text{something}} \frac{\text{many factorials}}{\text{more factorials}} \times$$

$$b_0 A_0 b_0 \cdots b_0 \partial^{\overline{n}_j} A_{r_j} b_0 A_0 b_0 \cdots b_0 \left(\otimes^{\text{something}} \mathbf{p} \right),$$

where all the covariant tensor indices on the ∂'s are contracted with some of the contravariant indices of the various A's standing farther to the left. Some of the many summation indices serve to describe the precise pattern of this tensor index contraction for each term. (The theorem is stated in full at the end of this article.) For instance, when $l = 2$ a typical term in b_2 is

$$-2b_0 A_0^{\alpha\beta} b_0 A_1^{\gamma} b_0 \frac{\partial A_0^{\delta\epsilon}}{\partial x^\alpha} b_0 \, p_\beta p_\gamma p_\delta p_\epsilon .$$

There are 13 more terms where that one came from, and the number increases rapidly with the order, s.

The calculation we have just described is not the one which we actually did. Rather, we performed the analogous calculation for what is called the *intrinsic* symbolic calculus [2], in which ip corresponds, not to the conventional partial derivative, but to a *covariant* derivative. In the most general case, to define a covariant derivative requires both a connection on the [tangent bundle of the] manifold (represented in coordinates by Christoffel symbols) and a connection on the vector bundle E (such as an electromagnetic or Yang–Mills potential). (Thus, if $u(\mathbf{x})$ is both vector-valued and E-valued, then its covariant derivative is given by a formula

$$\nabla_\mu u^\nu = \partial_\mu u^\nu + \Gamma^\nu_{\mu\sigma} u^\sigma + i A_\mu u^\nu ,$$

where A_μ (for each μ) is a matrix acting on the suppressed E-index of u.) A connection defines a notion of parallel transport, and this makes it possible to define a "manifestly covariant" Fourier transformation in the neighborhood of any point of M. Therefore, pseudodifferential operators and their symbols can be introduced in a geometrically covariant, or intrinsic, way.

The product formula for the intrinsic calculus is considerably more complicated than that for the conventional calculus; in each order there are new terms involving the curvature tensors of the two connections. The result is that b_s contains many more terms (and many more indices), involving as factors not only the coefficient tensors of A and their covariant derivatives, but also the curvature of the manifold ($R^\alpha{}_{\beta\gamma\delta}$), the curvature of the bundle ($F_{\alpha\beta}$, alias the gauge field strength), and their covariant derivatives. (For many operators of interest the coefficient tensors will themselves have geometrical interpretations; for example, $A_0^{\alpha\beta}$ might be a metric tensor on M.)

Why bother? The great advantage of the intrinsic calculus is that it yields answers directly in terms of invariantly defined quantities with direct physical meaning. The corresponding conventional pseudodifferential calculation would yield an intractable morass of partial derivatives of components of the metric tensor, etc. The task of grouping such quantities into geometrical objects in each concrete calculation is avoided by the intrinsic approach; this is the return on the investment of developing a more complex general theory.

In any case, the results we have obtained are still rather formidable as the basis for hand calculations. The point, however, is that they lend themselves to computerization. Shrewdly designed computer programs ought to make possible calculations at a level of detail formerly regarded as infeasible. (Work along this line is under way.) This will allow further advances in the many areas of mathematics and physics where the local asymptotics of differential operators (i.e., of their spectral resolutions and Green functions, such as the heat kernel) has applications: index theory, inverse problems, renormalization in quantum field theory, effective Lagrangians, partition functions, Wigner distribution functions, inhomogeneous elastic media, and quantum gravity. The last two of these are particularly significant, since in them are encountered types of operators whose asymptotics apparently cannot be investigated concretely without pseudodifferential methods [3]. These are fourth-order operators, and second-order operators of the structure

$$(A\psi)^i = \left[A^{\mu\nu i}{}_j \nabla_\mu \nabla_\nu + B^{\mu i}{}_j \nabla_\mu + C^i{}_j\right] \psi^j,$$

where $A^{\mu\nu i}{}_j \mathbf{p}_\mu \mathbf{p}_\nu$ is not a multiple of the identity matrix.

Remark: The curvatures enter the formula for the b_s of the intrinsic calculus through certain functionals which have nothing to do with pseudodifferential operators in particular; they are covariant derivatives of the inverse exponential mapping and the parallel-transport mapping. (Some physicists know them as DeWitt's $[-\sigma_{;\alpha'\beta\gamma\ldots}]$ and $[I_{;\alpha\beta\ldots}]$.) Unfortunately, the recursion relations determining these quantities have *not* been solved in closed form. Several methods for calculating them recursively to high order are under investigation.

To close, we quote from our manuscript in preparation [4] the complete theorem on the formula for b_s in the conventional symbolic calculus. (The formula appropriate to the intrinsic calculus will be announced elsewhere [5].) A multi-index notation is used, where

$$q_j = (q_{j1}, \ldots, q_{jT_j}) \in \mathbf{Z}_+^{T_j} \quad \text{for all } 1 \le j \le J,$$

$$q = (q_{jt}) \quad \text{where } 1 \le j \le J \text{ and } 1 \le t \le T_j,$$

$$|q| = \sum_{j=1}^{J} |q_j| = \sum_{j=1}^{J} \sum_{t=1}^{T_j} q_{jt},$$

$$q! = \prod_{j=1}^{J} (q_j!) = \prod_{j=1}^{J} \prod_{t=1}^{T_j} (q_{jt}!),$$

with similar but simpler conventions for the singly-subscripted indices. (One of the multi-indices is $\overline{p}$, which can be distinguished by context from the momentum variable $\mathbf{p}$.)

Theorem: Let $A \in L^l(M, E, E)$ be a differential operator of order $l > 0$ which is elliptic with respect to a ray Γ in $\mathbf{C}$, let $\lambda \in \Gamma$, and let $B_\lambda \in L^{-l}(M, E, E)$ be a resolvent parametrix of A. If the expression of A relative to a choice of local coordinate chart for an open set $U \subset M$ and frame for $E|U$ is $A^\psi = \sum_{r=0}^{l} A_r^\psi \partial^{l-r}$, where $A_r^\psi \in C^\infty(V, GL(\dim E, \mathbf{K}) \otimes S(\otimes^{l-r}\mathbf{R}^n))$, V is the image of U under the coordinate diffeomorphism, and $\mathbf{K} = \mathbf{R}$ or $\mathbf{C}$ according as E is real or complex, then the conventional symbol of B_λ with respect to this same choice of chart and frame has asymptotic expansion $b \sim \sum_{s \ge 0} b_s$, where $b_0 = \left(A_0^\psi(\otimes^l(i\mathbf{p})) - \lambda\right)^{-1}$ and, for all $s \ge 1$,

$$b_s = \sum_{J=1}^{s} \sum_{\substack{T_j \geq 0 \\ |T| \geq \frac{s}{l} - J}}^{\sum_{i=1}^{j} T_i \leq s-J+j} \sum_{\substack{r_j \geq 0, \overline{n}_j \geq 0 \\ r_j + \overline{n}_j \geq 1 \\ \sum_{i=1}^{j} \overline{n}_i \geq \sum_{i=1}^{j} T_i}}^{|r+\overline{n}|=s, r_j \leq l} \sum_{\substack{\overline{p}_j \geq 0, q_{jt} \geq 1 \\ \overline{p}_1 = 0}}^{\substack{|\overline{p}|+|q|=|\overline{n}| \\ \sum_{i=1}^{j}(\overline{p}_{i+1}+|q_i|) \leq \sum_{i=1}^{j} \overline{n}_i \\ \overline{p}_j \leq l-r_j, q_{jt} \leq l}}$$

$$i^{(l+2)(J+|T|)-s}\,(l!)^{|T|}\,\overline{N}!\,(l-r)!\,\left(\overline{n}!\,\overline{p}!\,q!\,(\overline{N}-\overline{n})!\,(l-q)!\,(l-r-\overline{p})!\right)^{-1}$$

$$\Big[\bigotimes_{j=1}^{J}{}_R \Big[\Big(\bigotimes_{t=1}^{T_j} b_0 A_0^{\psi}\Big) b_0(\partial^{\overline{n}_j} A_{r_j}^{\psi})\Big]\Big] b_0(\otimes^{l(J+|T|)-s}\mathbf{p}).$$

($\otimes_R$ denotes a tensor product in reversed order.) The index contractions are described inductively as follows: For all $1 \leq j \leq J-1$ the $\overline{n}_j$ derivatives of the form $\partial^{\overline{n}_j}$ are combined with the $\sum_{i=1}^{j-1}(\overline{n}_i - \overline{p}_{i+1} - |q_i|)$ as yet uncontracted derivatives of the form $\partial^{\overline{n}_i}$ for all $1 \leq i \leq j-1$, and the total of $\overline{N}_j = \overline{n}_j + \sum_{i=1}^{j-1}(\overline{n}_i - \overline{p}_{i+1} - |q_i|)$ derivatives is symmetrized. Of these, q_{jt} are contracted with q_{jt} of the l indices of the tth factor of A_0^{ψ} in $\bigotimes_{t=1}^{T_j} b_0 A_0^{\psi}$ for all $1 \leq t \leq T_j$ and $\overline{p}_{j+1}$ are contracted with $\overline{p}_{j+1}$ of the $l - r_{j+1}$ contravariant indices of $\partial^{\overline{n}_{j+1}} A_{r_{j+1}}^{\psi}$. The remaining $\overline{N}_j - \overline{p}_{j+1} - |q_j| = \sum_{i=1}^{j}(\overline{n}_i - \overline{p}_{i+1} - |q_i|)$ are combined with the $\overline{n}_{j+1}$ derivatives of the form $\partial^{\overline{n}_{j+1}}$ and the process is repeated. After the final such operation (when $j = J-1$), the remaining $\overline{N}_{J-1} - \overline{p}_J - |q_{J-1}| = \sum_{i=1}^{J-1}(\overline{n}_i - \overline{p}_{i+1} - |q_i|)$ derivatives are combined with the $\overline{n}_J$ derivatives of the form $\partial^{\overline{n}_J}$ and the total of $\overline{N}_J = \overline{n}_J + \sum_{i=1}^{J-1}(\overline{n}_i - \overline{p}_{i+1} - |q_i|) = |q_J|$ derivatives is symmetrized. Of these, q_{Jt} are contracted with q_{Jt} of the l indices of the tth factor of A_0^{ψ} in $\bigotimes_{t=1}^{T_J} b_0 A_0^{\psi}$ for all $1 \leq t \leq T_J$. Finally, $\otimes^{l(J+|T|)-s}\mathbf{p}$ is contracted with the remaining $l(J+|T|)-s$ uncontracted indices.

Outline of proof: An operation P_u has been defined in (*), and a formula for b_s in terms of nested P_u operations has been given below that equation. The nesting has been arranged to occur in the argument of P_u which experiences differentiations with respect to the "fiber" variable, $\mathbf{p}$. One's task is to evaluate these nested fiber derivatives from the outside in. At each step one encounters a derivative of a product of matrix-valued functions; it is evaluated by the Leibnitz formula in the form

$$D^N\Big(\bigotimes_{i=1}^{I} h_i\Big) = \sum_{q_i \geq 0}^{|q|=N} \frac{N!}{q!} S_N\Big(\bigotimes_{i=1}^{I} D^{q_i} h_i\Big),$$

where $q = (q_1, \ldots, q_I) \in \mathbf{Z}_+^I$ and S_N indicates symmetrization in the N contravariant indices of the $\{D^{q_i}\}$. One of the factors to be differentiated is b_0; the formula for its derivative is easily seen to be

$$D^N b_0 = \sum_{T=1}^{N} (-1)^T \sum_{q_t \geq 1}^{|q|=N} \frac{N!}{q!} S_N\Big(\bigotimes_{t=1}^{T} b_0 D^{q_t} a_0\Big) b_0.$$

The other factors are coefficient tensors of the differential operator; these are homogeneous functions of **p**, so their differentiation is straightforward. Fitting all the ingredients together, with due attention to many notational and combinatorial details, one eventually arrives at the relatively compact (!) formula stated in the theorem.

REFERENCES

1. F. Treves, *Introduction to Pseudodifferential and Fourier Integral Operators*, Vol. 1, Plenum, New York, 1980; M. E. Taylor, *Pseudodifferential Operators*, Princeton Univ. Press, Princeton, 1981; H. Kumano-Go, *Pseudo-Differential Operators*, M.I.T. Press, Cambridge, Mass., 1982; B. E. Petersen, *Introduction to the Fourier Transform and Pseudo-Differential Operators*, Pitman, Boston, 1983; P. B. Gilkey, *Invariance Theory, the Heat Equation, and the Atiyah-Singer Index Theorem*, Publish or Perish, Wilmington, 1984.
2. H. Widom, *A complete symbolic calculus for pseudodifferential operators*, Bull. Sci. Math. 104, 19-63 (1980); L. Drager, *On the Intrinsic Symbol Calculus for Pseudo-Differential Operators on Manifolds*, Ph.D. Dissertation, Brandeis University, 1978; J. Bokobza-Haggiag, *Opérateurs pseudo-différentiels sur une variété différentiable*, Ann. Inst. Fourier (Grenoble) 19, 125-177 (1969).
3. N. H. Barth and S. M. Christensen, *Quantizing fourth order gravity theories: The functional integral*, Phys. Rev. D **28**, 1876-1893 (1983).
4. S. A. Fulling and G. Kennedy, *The resolvent parametrix of the general elliptic linear differential operator: A closed form for the intrinsic symbol*, in preparation.
5. S. A. Fulling and G. Kennedy, *A closed form for the intrinsic symbol of the resolvent parametrix of an elliptic operator*, in the proceedings of the First International Conference on the Physics of Phase Space (College Park, 1986), Springer Lecture Notes in Physics, to appear.

EXISTENCE AND FINITE-DIMENSIONALITY OF ATTRACTORS FOR THE LANDAU-LIFSCHITZ EQUATIONS

Tepper L. Gill*
Department of Physics
Virginia Polytechnic Institute
and State University
Blacksburg, Virginia 24061

W.W. Zachary
Naval Research Laboratory
Washington, DC 20375

ABSTRACT

The Landau-Lifschitz equations describe the time-evolution of magnetization in classical ferro- and antiferromagnets and are of fundamental importance for the understanding of nonequilibrium magnetism. We sketch a proof that, under quite general conditions, dissipative forms of these equations have attracting sets which are finite-dimensional in a suitable sense. In particular, upper bounds are obtained for the Hausdorff and fractal dimensions of these sets.

1. INTRODUCTION

The Landau-Lifschitz equations

$$\frac{d}{dt}\mathbf{M} = \gamma\,\mathbf{M}\times\mathbf{H}_e - \lambda\,\mathbf{M}\times(\mathbf{M}\times\mathbf{H}_e),\ \gamma,\lambda > 0, \tag{1}$$

for the time-evolution of the magnetization $\mathbf{M}$ in a ferromagnet form a basic system of equations for the classical (i.e., non-quantum) description of magnetism [1]. A general form for the effective magnetic field, $\mathbf{H}_e$, is

$$\mathbf{H}_e = \mathbf{H}_{ext}(t) + \frac{C}{|\mathbf{M}|}\,\Delta\,\mathbf{M} - \frac{2A}{|\mathbf{M}|}\,(\mathbf{M}\cdot\mathbf{n})\mathbf{n} + \mathbf{H}_d\,, \tag{2}$$

(C, A = constant) where $\mathbf{H}_{ext}$ denotes the external fields, the second and fourth terms represent the respective contributions from exchange interactions (specialized to the case of cubic crystals) and demagnetization effects. In the latter case we use the magnetostatic approximation, curl $\mathbf{H}_d = 0 = div\,(\mathbf{H}_d + 4\pi\mathbf{M})$, together with the usual boundary conditions on the surface of the magnetic material.

It is easily shown that $\mathbf{H}_d$ can be written in the form

$$\mathbf{H}_d(\mathbf{x},t) = \nabla\int_\Omega \frac{div\,\mathbf{M}(\mathbf{x}',t)}{|\mathbf{x}-\mathbf{x}'|}\,dx' - \nabla\int_\Gamma \frac{M_n(\mathbf{x}',t)}{|\mathbf{x}-\mathbf{x}'|}\,d\sigma(x'), \tag{3}$$

where Ω represents the volume of 3-space occupied by the ferromagnet. For the purposes of the present discussion, the boundary of Ω, Γ, will be assumed to be a C^∞ surface. This assumption can be significantly weakened. The third term in (2) represents the contribution from anisotropy effects in uniaxial crystals with the unit vector $\mathbf{n}$ pointing along the magnetization easy-axis. The term in (1) containing λ is a phenomenological damping term which represents various interactions.

It is only in recent years that nonperturbative solutions of nonlinear versions of (1) have begun to be investigated. These studies have usually dealt with special cases in which the exchange term in $\mathbf{H}_e$ is replaced by a one-dimensional second derivative (but $\mathbf{M}$ still considered a 3-vector) and special choices (usually zero) made for the other terms in (2) [2]. In these situations those authors found that (1) is equivalent to a nonlinear Schrödinger equation, i.e., it is a completely integrable system.

In the three-dimensional case, although soliton solutions are known when suitable integrals of the motion exist [3], it is expected that with the general form (2) for $\mathbf{H}_e$, Eq. (1) possesses "chaotic solutions." There is evidence for such behavior from both experimental results [4] and numerical investigations of finite-dimensional models [5]. The only study of (1)-(3) without drastic simplifying assumptions known to the authors is the work of one of us [6], where the existence and uniqueness of periodic solutions were discussed. In the present work, we continue this program by proving existence and finite-dimensionality of attractors for the system (1)-(3).

2. TRANSFORMATION OF THE LANDAU-LIFSCHITZ EQUATIONS

The system of nonlinear integro-(partial) differential equations (1)-(3) is very complicated and very little is known about it. Therefore, it is convenient to transform to a form more amenable to analysis. To do this, we note that it follows from (1) that $|\mathbf{M}(\mathbf{x},t)|^2 =$ const. so that we can, after suitable normalization, consider the magnetization on the unit two-sphere, S^2. We then use the well-known result that the stereographic projection maps S^2 minus one point homeomorphically onto $\mathbb{R}^2$. Then, writing this map in terms of a complex quantity ψ,

$$\psi = \frac{m_x + i m_y}{1 + m_z}\,, \quad \mathbf{m} = \frac{\mathbf{M}}{|\mathbf{M}|}\,, \tag{4}$$

where we have chosen the excluded point on S^2 as the "south pole" $\mathbf{m} = (0,0,-1)$, we use the procedure described in [7] to write (1)-(3) in the form

$$\frac{d}{dt}\begin{pmatrix}\psi\\ \overline{\psi}\end{pmatrix} + \begin{pmatrix}B & 0\\ 0 & B^*\end{pmatrix}\begin{pmatrix}\psi\\ \overline{\psi}\end{pmatrix} = \begin{pmatrix}f\\ \overline{f}\end{pmatrix} \tag{5}$$

where $B = -i\,(1 - i\,\lambda)\gamma\, C\,\Delta\,, \gamma, C > 0$, and

$$\begin{aligned} f\,(\psi,\overline{\psi},t) = {} & i\,(1-i\,\lambda)\,\gamma\,\Big[-2\,C\,\overline{\psi}(\nabla\psi)^2\,(1 + |\psi|^2)^{-1} \\ & + \frac{h_+}{2} - \frac{h_-}{2}\,\psi^2 - h_z\,\psi - \\ & - 2A\,(\psi n_- + \overline{\psi}\, n_+ + (1 - |\psi|^2)\, n_z)\;(1 + |\psi|^2)^{-1}\left[\frac{n_+}{2} - \frac{n_-}{2}\,\psi^2 - n_z\,\psi\right]\Big]. \end{aligned} \tag{6}$$

We have grouped the external and demagnetization fields together in the quantity $\mathbf{h} = \mathbf{H}_{ext} + \mathbf{H}_d$ and have used the combinations $h_\pm = h_x \pm i\,h_y$, $n_\pm = n_x \pm i n_y$. We impose zero Neumann boundary conditions on Γ. This is the simplest choice of boundary condition consistent with the comparison of the two well-known forms of the exchange energy of a ferromagnet [1] combined with the fact that zero Dirichlet conditions are not possible since $\mathbf{m} \in S^2$. The self-adjointness problem for the Neumann Laplacian has a well-known solution on a maximal domain in $X = L^2(\Omega)$. Moreover, since Γ is smooth, $-\Delta$ has compact resolvent on X and so has a nonnegative discrete spectrum. Consequently, the spectrum $\sigma(B)$ of B is discrete with $Re\,\sigma(B) \geq 0$ so that B is a sectorial operator [8,9].

Thus, (5) looks formally like a system of semilinear parabolic differential equations in a Hilbert space. Quite a lot is known about such systems and this is the advantage of (5) compared to the original Landau-Lifschitz equations (1)-(3). We investigate the properties of attractors for (5) and then relate these results to (1)-(3) via the stereographic projection (4).

3. EXISTENCE OF MAXIMAL ATTRACTORS

The operator B has zero as its lowest eigenvalue, so it is useful to define $B_\epsilon = B + \epsilon I, \epsilon > 0$, where I denotes the identity operator, in order to have operators simply related to B with $\mathrm{Re}\,\sigma(B_\epsilon) > 0$. We can then define the fractional operators $B_\epsilon^\alpha, \alpha \in (0,1)$, and fractional Banach spaces X^α in the usual manner [8]. The latter have the following useful properties when $\alpha > \beta \geq 0 : X^\alpha$ is dense in X^β and, since B_ϵ has compact resolvent in $X = X^0$, the inclusions $X^\alpha \subset X^\beta$ are compact.

We make the following assumptions.

I. Ω is a convex connected bounded open subset of $\mathbb{R}^3$ with C^∞ boundary Γ.

II. The initial data of (5), $\psi_0, \bar{\psi}_0$, are assumed small in the sense that $\|\psi_0\|_\alpha = \|\bar{\psi}_0\|_\alpha < 1, \frac{7}{8} < \alpha < 1$.

III. The external fields can be decomposed in the form $\mathbf{H}_{ext}(\mathbf{x},t) = \mathbf{H}_s(\mathbf{x}) + H_{rf}(\mathbf{x},t)$ into static, $\mathbf{H}_s$, and time-dependent, $\mathbf{H}_{rf}$, parts, where the second derivatives of the former are uniformly bounded on $\bar{\Omega}$ and the latter are locally Hölder continuous and 2π-periodic in the temporal variable, and the second spatial derivatives are uniformly bounded on $\bar{\Omega}$.

Our first task is to prove local existence and uniqueness of solutions of (5). We first obtain some useful properties of f:

Lemma 1. *Assume* I *and* III *and suppose that* $\alpha > \frac{7}{8}$. *Then* $f(\psi,\bar{\psi},t)$ maps $X^\alpha \times X^\alpha \times [0,T]$ *into* $L^2(\Omega \times [0,T])$ *for all* $T>0$. *Furthermore, f is locally Hölder continuous in t and locally Lipschitz in* ψ and $\bar{\psi}$ *on* $X^\alpha \times [0,T]$, $\|f(\psi,\bar{\psi},t) - f(\chi,\bar{\chi},t)\|_X \leq L\|\psi(\cdot,t) - \chi(\cdot,t)\|_\alpha$, *where L depends upon the parameters in (6), a number of imbedding constants, and the α-norms of ψ and χ.*

We combine this result with hypothesis II, the additional condition $\alpha<1$, and the Picard iteration procedure in the Banach spaces X^α, $\frac{7}{8} < \alpha < 1$, to prove that solutions of (5) exist on suitable time intervals $[0,T_{max}]$, and that these solutions are uniquely determined by the initial data at $t = 0$. We then use techniques of the theory of semilinear parabolic differential equations [8], classical Schauder estimates [10], and estimates for Bf analogous to those for f given in Lemma 1, to show that these solutions are classical, i.e., that $T(t)\,\psi_0 \in C^0([0,T_{max}], C^{2+\nu}(\bar{\Omega}))$ with $0 \leq \nu < 2\alpha - \frac{3}{2}$ where, in addition to assumptions I-III, we require that $\psi_0, \bar{\psi}_0 \in X^\eta$ for some $\eta > 1+\alpha$.

A priori, it appears that we should use a complete atlas description of S^2 to transform (1)-(3) to equations of the type (5). However, the fact that we obtain classical solutions of (5) justifies the transformation of (1)-(3) to (5) by the use of (4) alone, since the uniform boundedness of ψ guarantees that the magnetization will never pass through the south pole of S^2.

The proof of Lemma 1 involves a (large) number of standard estimates which use imbeddings of the fractional spaces X^α into Sobolev spaces. For example, the most restrictive of these is $X^\alpha \subset W_4^1(\Omega), \alpha > \frac{7}{8}$. This imbedding theorem, as well as the others used in the proof of Lemma 1, can be proved in the present context by using a similar argument as in ([8], p. 75) by making use of the simple relation between B and Δ and the fact that we use zero Neumann conditions on Γ. The contributions arising from the first term in (3) are estimated by coupling an inequality involving Riesz potentials with some well-known L^p inequalities. The corresponding surface integral contributions are estimated by following a procedure discussed in [6].

We now use the solutions of (5) to define a map

$$T(t): \begin{bmatrix} \psi_0 \\ \overline{\psi}_0 \end{bmatrix} \to \begin{bmatrix} \psi(t) \\ \overline{\psi}(t) \end{bmatrix}.$$

In order for this to make sense it is essential that we be able to extend $\psi(t)$ to a global solution for all positive times. This is done by using the standard result that, if this is not possible, i.e., if $T_{\max} < \infty$, there exists a sequence of times $\{t_n\}$, $t_n \uparrow T_{\max}^-$, such that $||T(t_n)\psi_0||_\alpha \uparrow +\infty$.

We prove that $T_{\max} = \infty$ by showing that, under appropriate conditions, this limit must be finite. To do this, we again impose I and II but we modify III by requiring that the fourth derivatives of $\mathbf{H}_s$ and $\mathbf{H}_{rf}$ be uniformly bounded on $\overline{\Omega}$ and that $\psi_0 \in X^\eta$ for some $\eta > 2 + \alpha$. We then use estimates for $B^2 f$ analogous to those obtained for f and Bf previously, and similar techniques to those used to prove that $T(t)\psi_0$ exist and are classical, to show that $T(t)\psi_0 \in C^{4+\nu}(\overline{\Omega})$, where ν satisfies the same inequalities mentioned previously. The proof is then completed by the use of techniques closely related to weighted energy methods [11] to show that $||T(t)\psi_0||_\alpha$ is a decreasing function of $t \in [0, T_{\max}]$. The proof of this requires an additional assumption. Namely,

1) if either $H_x^{\text{ext}} \neq 0$ or $H_y^{\text{ext}} \neq 0$, then $\max_{i=x,y} ||H_i^{\text{ext}}||_{C^2(\overline{\Omega})}$ (which we recall is time-independent) is sufficiently small, and $||\psi_0||_\alpha$ is sufficiently small, or

2) if $H_x^{\text{ext}} = 0 = H_y^{\text{ext}}$, then it is enough to assume that $||\psi_0||_\alpha$ is sufficiently small.

The existence of a maximal attractor, i.e., a compact maximal invariant set which attracts all bounded sets, follows from the following properties of the semigroup $T(t)$:

Lemma 2. Assume the hypotheses just discussed for the proof of the global property of $T(t)\psi_0$. Then:

a) *$T(t)$ is uniformly bounded on X^α.*

b) $T(t)$ is compact on X^α, $t > 0$.

c) $T(t)$ is strongly continuous for $t \geq 0$.

d) *There exists a bounded absorbing set* Y_0 in X^α, i.e., for all bounded sets $Y \subset X^\alpha$, there exists $t_0 > 0$ such that $T(t)Y \subset Y_0$ for $t \geq t_0$.

Proof. (Sketch) (a) is proved by using the integral equation form of (5), standard properties of the semigroups generated by B and B^*, and properties of f established in Lemma 1. The proof of (b) then follows from (a), the fact that B has compact resolvent in X, and Lemma 2.3 of [9]. To prove (d), let Y denote a bounded set in X^α and take $||u||_\alpha \leqslant R$ *for* $u \in Y$. We can take $R \geqslant 1$ without loss in generality. By estimates of the type in Theorem 3.3.3 of [8] we have for $\psi_0 \in Y$,

$$|| T(t) \psi_0 ||_\alpha \leqslant C_\alpha \left[t^{-\alpha} R + \frac{\delta M}{2(M + L\delta)} \right], \tag{7}$$

where $|| \psi - \psi_0 ||_\alpha \leqslant \delta$ and

$$M = ||f||_X \leqslant \gamma_0 + \gamma_1 R + \gamma_2 R^2 + \gamma_3 R^3, \tag{8}$$

$$L \leqslant \gamma_4 + \gamma_5 R + \gamma_6 R^2 + \gamma_7 R^3,$$

(8) having been established in Lemma 1, where the γ_i are independent of ψ and $\bar{\psi}$. For large R we have $\delta = 0(R)$, so that the second term in (7) is 0 (1). Then, since $t^{-\alpha}$ is a decreasing function of t, we can choose t_0 such that for $t \geqslant t_0(R)$,

$$|| T(t) \psi_0 ||_\alpha \leqslant \frac{C_\alpha \delta M}{M + L\delta} < \delta'$$

with δ' independent of R. Thus, we can take $Y_0 = \{ u \in X^\alpha : ||u||_\alpha < \delta'(\alpha) \}$ as our desired absorbing set.

Now we have

Theorem 1. Assume the hypotheses of Lemma 2. Then $T(t)$ *has a compact maximal attractor*

$$A = \bigcap_{t \geqslant 0} \overline{\bigcup_{s \geqslant t} T(s) Y_0}$$

in X^α *if* $\frac{7}{8} < \alpha < 1$.

Proof. Follows from Lemma 2 and some general results of Babin and Vishik [12].

It can be proved that the corresponding maximal attractor for the original system (1)-(3) is $\tilde{A} = \underline{P}A$, with $\underline{P}$ the transformation inverse to (4). Moreover, if A has finite topological dimension, then so does $\tilde{A}$ and these dimensions are equal.

4. FRÉCHET DERIVATIVE OF T(t)

Now that we have existence of the maximal attractor A, we establish its finite-dimensionality by proving that $T'(t)$, the Fréchet derivative of $T(t)$, is a compact linear operator on the fractional spaces X^α, $\frac{7}{8} < \alpha < 1$. The finiteness of the topological and Hausdorff, $d_H(A)$, dimensions and of the fractal dimension, $d_F(A)$, (upper capacity) of A then follows from results of Mallet-Paret [13] and Mañé [14], respectively.

To prove that $T'(t)(t > 0)$ is compact, we first prove boundedness by using Taylor expansions with remainder for maps of Banach spaces, properties of the integral equation form of (5), and estimates, analogous to those in the proof of Lemma 1, obtained by means of a number of imbedding theorems; and, in addition, estimates similar to those used to prove the global existence of $T(t)$.

5. UPPER BOUNDS FOR $d_H(A)$ AND $d_F(A)$

We see from the preceding discussion that A has finite Hausdorff and fractal dimensions, but bounds on these dimensions are not easily obtained by the methods used to prove those results. In the present section we obtain upper bounds for these dimensions by investigating the evolution of finite-dimensional volume elements following the work of Constantin et al [15] on the Navier-Stokes equations.

Consider two solutions ψ, χ of (5). From our analysis of $T'(t)$ we obtain the system of equations

$$\frac{d}{dt} W(t) = -\tilde{B}(t) W(t) \tag{9}$$

where

$$W(t) = \begin{bmatrix} \psi(t) - \chi(t) \\ \bar{\psi}(t) - \bar{\chi}(t) \end{bmatrix},$$

and $\psi(t) = T(t)\psi_0$ with a corresponding expression for $\chi(t)$. $\tilde{B}(t)$ is defined by

$$\tilde{B}(t) = \begin{bmatrix} B & 0 \\ 0 & B^* \end{bmatrix} - i \begin{bmatrix} F & G \\ -G^* & -F^* \end{bmatrix}, \tag{10}$$

where F and G are complicated integro-differential operators with coefficients depending upon $\psi(t)$ in a nonlinear manner. We obtain from (9) the following system of equations in $\Lambda^m X$, $m \in Z_+$:

$$\frac{d}{dt}(W_1 \Lambda \ldots \Lambda W_m) = -\tilde{B}(t)^{\Lambda} W_1 \Lambda \ldots \Lambda W_m,$$

with

$$\tilde{B}(t)^{\Lambda} = \tilde{B}(t) \Lambda I \Lambda \ldots \Lambda I + I \Lambda \tilde{B}(t) \Lambda I \Lambda \ldots \Lambda I$$
$$+ \ldots + I \Lambda \ldots \Lambda I \Lambda \tilde{B}(t).$$

From an analysis similar to that in [15], we conclude that $\| W_1 \Lambda \ldots \Lambda W_m \|_\alpha$ decays exponentially in time if the quantity

$$Tr\left[(B(t)^* + B(t))\, Q_m\right]$$

has suitable properties. Here, Q_m denotes the orthogonal projection in the Hilbert space $X^\alpha \times X^\alpha$ onto the subspace spanned by $W_1, \ldots, W_m$. This idea can be formulated quantitatively by defining, for $m \geqslant 1$:

$$-q_m = \limsup_{t\to+\infty}\left\{\inf_{W_0\in X^a\times X^a}\frac{1}{t}\int_0^t \inf_{\operatorname{rank}Q=m} Tr\left[(B(s)^* + B(s))\,Q\right] ds\right\}.$$

We obtain the following upper bounds for the Hausdorff and fractal dimensions of A:

Theorem 2. Assume the hypotheses of Lemma 2. Then

$$d_H(A) \leqslant m \text{ and } d_F(A) \leqslant m + \frac{(m-1)(\phi-1)}{m(m^{2/3}-\phi)}, \tag{11}$$

where m is a uniquely determined positive integer $\geqslant 2$, *and m and* $\phi \in (1, m^{2/3})$ *depend on the parameters in* (2) *as well as on a number of imbedding constants.*

Proof. (Sketch) From [15] we have

$$d_H(A) \leqslant m \text{ and } d_F(A) \leqslant m\left[1 + \max_{1\leqslant j\leqslant m-1}\frac{-q_j}{q_m}\right],$$

where m denotes the least positive integer such that $q_m > 0$. For the first term in (10), $\tilde{B}_0$, we use some results of Lieb [16] concerning eigenvalues of the Dirichlet Laplacian. Then, using an analogous argument to one in [15], we obtain the estimate

$$Tr\left[(\tilde{B}_0^* + \tilde{B}_0)\,Q_m\right] \geqslant 4\lambda\gamma C a m^{5/3}, \quad a = \text{const} > 0.$$

For the second term in (10) we find

$$\left| Tr\left[(\tilde{B}_1(t)^* + \tilde{B}_1(t))\,Q_m\right]\right| \leqslant m\gamma K,$$

where K is independent of m. In obtaining this estimate we use some of the estimates obtained in the proof of existence of global solutions to prove that $W_j(t) \in C^3(\overline{\Omega})$, $j = 1, 2, \ldots, m$.

We obtain (11) with $\phi = \dfrac{K}{4Ca\lambda}$.

Remark. The bound on $d_F(A)$ in (11) depends on the volume of Ω through the various imbedding constants.

ACKNOWLEDGMENTS

The second author would like to thank I. Knowles and Y. Saitō for the invitation to present these results at their conference, and to thank C. Foiaş, H. Amann, S. Newhouse, J. A. Yorke, S. Antman, and R. Cawley for valuable remarks. We would also like to thank J. K. Hale and I. Herron for providing copies of Refs. 9 and 11, respectively, prior to their publication.

REFERENCES

*On leave from the Department of Mathematics, Howard University, Washington, D.C. 20059.

1. Akhiezer, A.I., V.G. Bar'yakhtar, and S.V. Peletminskii, *Spin Waves*, North Holland, Amsterdam, 1968.

2. Nakamura, K., and T. Sasada, Solutions and Wave Trains in Ferromagnets, Phys. Lett. **48A** (1974), 321-322; Lakshmanan, M., Continuum spin system as an exactly solvable dynamical system, Ibid. **61A** (1977), 53-54.

3. Kosevich, A.M., B.A. Ivanov, and A.S. Kovalev, Magnetic Solitons: A New Type of Collective Excitation in Magnetically Ordered Crystals, Sov. Sci. Rev. A. Phys. **6** (1985), 161-260.

4. Gibson, G. and C. Jeffries, Observation of period doubling and chaos in spin-wave instabilities in yttrium iron garnet, Phys. Rev. **A29** (1984) 811-818; Yamazaki, H., Oscillations and Period-Doubling of Magnon Amplitude under Parallel Pumping in anti-ferromagnetic $Cu\,Cl_2\,2H_20$, J. Phys. Soc. Japan **53** (1984), 1155-1159; de Aguiar, F.M., and S.M. Rezende, Observation of Subharmonic Routes to Chaos in Parallel-Pumped Spin Waves in Yttrium Iron Garnet, Phys. Rev. Lett. **56** (1986), 1070-1073.

5. Nakamura, K., S. Ohta, and K. Kawasaki, Chaotic states of ferromagnets in strong parallel pumping fields, J. Phys. **C15** (1982), L143-L148; Ohta, S., and K. Nakamura, Power spectra of chaotic states in driven magnets, Ibid. **C16** (1983), L605-L602; Waldner, F., D.R. Barberis, and H. Yamazaki, Route to chaos by irregular periods: Simulations of parallel pumping in ferromagnets, Phys. Rev. **A31** (1985), 420-431; Zhang, X.Y., and H. Suhl, Spin-wave-related period doublings and chaos under transverse pumping, Ibid. **A32** (1985), 2530-2533.

6. Zachary, W.W., Existence and Uniqueness of Periodic Solutions of the Landau-Lifschitz Equations with Time-Periodic External Fields, submitted to Lett. Math. Phys.; Some Approaches to the Study of Realistic Forms of the Landau-Lifschitz Equations, in *14th International Colloquium on Group Theoretical Methods in Physics* (ed. by Y.M. Cho), World Scientific, Singapore, 1986, pp. 417-420.

7. Lakshmanan, M., and K. Nakamura, Landau-Lifschitz Equation of Ferromagnetism: Exact Treatment of the Gilbert Damping, Phys. Rev. Lett. **53** (1984), 2497-2499.

8. Henry, D., *Geometric Theory of Semilinear Parabolic Equations*, Lecture Notes in Mathematics, Vol. 840, Springer-Verlag, Berlin-Heidelberg-New York, 1981.

9. Hale, J.K., Asymptotic Behavior and Dynamics in Infinite Dimensions, Research Notes in Math., Pitman, Boston-London-Melbourne, 1985.

10. Ladyženskaja, O.A., V.A. Solonnikov, and N.N . Ural'ceva, *Linear and Quasi-Linear Equations of Parabolic Type*, Amer. Math. Society, Providence, Rhode Island, 1968.

11. Galdi, G.P., and S. Rionero, *Weighted Energy Methods in Fluid Dynamics and Elasticity*, Lecture Notes in Mathematics, Vol. 1134, Springer-Verlag, Berlin-Heidelberg-New York, 1985.

12. Babin, A.V., and M.I. Vishik, Attractors of partial differential evolution equations and estimates of their dimension, Uspekhi Mat. Nauk **38** (1983), 133-187; English translation: Russian Math. Surveys **38** (1983), 151-213.

13. Mallet-Paret, J., Negatively Invariant Sets of Compact Maps and an Extension of a Theorem of Cartwright, J. Diff. Eqns. **22** (1976), 331-348.

14. Mañé, R., On the dimension of the compact invariant sets of certain non-linear maps, in *Dynamical Systems and Turbulence, Warwick 1980*, (ed. by D.A. Rand and L.-S. Young), Lecture Notes in Mathematics, Vol. 898, Springer-Verlag, Berlin-Heidelberg-New York, 1981.

15. Constantin, P., C. Foiaş, and R. Temam, Attractors representing turbulent flows, Memoirs Amer. Math. Soc., no. 314, 1985.

16. Lieb, E.H., The Number of Bound States of One-Body Schrödinger Operators and the Weyl Problem, Proc. Symposia in Pure Math, 36 (1980), 241-252.

THE COULOMB POTENTIAL IN HIGHER DIMENSIONS

Jerome A. Goldstein and Gisèle Ruiz Rieder
Department of Mathematics and Quantum Theory Group
Tulane University, New Orleans, LA. 70118, USA

1. INTRODUCTION

Of concern is the quantum mechanical Hamiltonian

(1) $$H = -\sum_{j=1}^{N} \Delta_j + \sum_{j=1}^{N} V(x_j) + \frac{1}{2}\sum_{j \neq k} |x_j - x_k|^{-\alpha} .$$

Here $x_j \in \mathbb{R}^d$ $(d \geq 3)$ represents the position of the j-th electron, Δ_j is the d-dimensional Laplacian in the variable x_j, and V is a real locally integrable function on $\mathbb{R}^d$. The operator H acts on $L^2_a(\mathbb{R}^{dN})$, the space of functions ϕ in $L^2(\mathbb{R}^{dN})$ satisfying the antisymmetry property

$$\phi(x_{\pi 1},\ldots,x_{\pi N}) = (\operatorname{sgn} \pi)\phi(x_1,\ldots,x_N)$$

for each $x_j \in \mathbb{R}^d$ and each permutation π of N objects. The most important case of V is the (generalized) molecular Coulomb potential

(2) $$V(x) = -\sum_{k=1}^{M} Z_k |x - R_k|^{-\beta} .$$

When $d = 3$, $\alpha = \beta = 1$ (see (1) and (2)); that is, both the electron-electron repulsion potential and the electron-nuclear attraction potential are given by multiples of the same function, $1/r$.

A *wave function* is a unit vector ψ in $L^2_a(\mathbb{R}^{dN})$; it gives rise to an *electron density* ρ , $0 \leq \rho \in L^1(\mathbb{R}^d)$, defined by the formula

(3) $$\rho(x) = N \int_{\mathbb{R}^d} \cdots \int_{\mathbb{R}^d} |\psi(x,x_2,\ldots,x_N)|^2 \, dx_2 \ldots dx_N .$$

Here $\int_{\mathbb{R}^d} \rho(x)dx = N$. The *ground state wave function* is a wave function satisfying

$$\langle H\psi,\psi\rangle = \inf\{\langle H\phi,\phi\rangle : \|\phi\| = 1\} .$$

The corresponding electron density is called a ground state density. In systems of bulk matter, N is huge, being of the order of magnitude of 10^{26}, so ground state wave function computations become unwieldy because of the large number of variables involved. Thus one wishes to construct a functional $E(\rho)$ of densities ρ on $\mathbb{R}^d$ which approximately equals $\langle H\psi,\psi\rangle$ when ρ comes from ψ via equation (3), and then to minimize $E(\rho)$. More precisely, let

$$E : \mathrm{Dom}(E) \subset \{\rho \in L^1(\mathbb{R}^d) : \rho \geq 0\} \longrightarrow \mathbb{R}$$

and let $D_N = \{\rho \in \mathrm{Dom}(E) : \int_{\mathbb{R}^d} \rho(x)dx = N\}$. Given $N > 0$ we want to find a $\rho_0 \in D_N$ such that

$$E(\rho_0) = \inf\{E(\rho) : \rho \in D_N\} .$$

The first developers of such a theory were L. Thomas [13] and E. Fermi [5], working independently about sixty years ago. Rigorous mathematical results in Thomas-Fermi theory were obtained in the 1970s by E. Lieb and B. Simon [10], [11] using the direct methods of the calculus of variations. By reducing the (convex) minimization problem to its Euler-Lagrange equation, Ph. Bénilan and H. Brezis [2], [3], [4] extended the Lieb-Simon results to a much wider class of energy functionals E. We have obtained some extensions of the Bénilan-Brezis results by working in d dimensions rather than 3 dimensions and by introducing a weight function into the kinetic energy term appearing in E (see below).

After stating a relevant existence theorem (and providing a very brief sketch of the proof) we shall deal with the question: What is the Coulomb potential in d dimensions? Our contention is that there are two different Coulomb potentials when $d \geq 4$, that is, $\alpha > \beta$ when when $d \geq 4$. (See (1), (2), and recall $\alpha = \beta = 1$ when $d = 3$.)

2. THE EXISTENCE THEOREM

Our energy functional of the electron density is

$$(4)\quad E(\rho) = \int_{\mathbb{R}^d}\rho(x)^p w(x)dx + \int_{\mathbb{R}^d}V(x)\rho(x)dx + c_1\int_{\mathbb{R}^d}\int_{\mathbb{R}^d}\frac{\rho(x)\rho(y)}{|x-y|^\alpha}\,dx\,dy;$$

the three terms in (4) correspond (in order) to the three terms in $\langle H\psi,\psi\rangle$ with H given by (1). Below will be made assumptions on the weight function w, the potential V, and the exponent p. But first we indicate how a scaling argument gives rise to the Thomas-Fermi exponent $p = 1 + 2/d$, which reduces to $p = 5/3$ when $d = 3$. For ψ a wave function in $L_a^2(\mathbb{R}^{dN})$ and $\lambda > 0$, let $\psi_\lambda(x) = \lambda^{dN/2}\psi(\lambda x)$ be the wave function "ψ scaled by λ". Then $\langle -\Delta\psi_\lambda, \psi_\lambda\rangle = \lambda^2\langle -\Delta\psi, \psi\rangle$. For ρ a density of mass N (i.e. $\rho \geq 0$, $\int_{\mathbb{R}^d} \rho(x)dx = N$), $\rho_\lambda(x) = \lambda^d \rho(\lambda x)$ is the density "ρ scaled by λ", also of mass N. If w is a positive constant,

$$\int_{\mathbb{R}^d} \rho_\lambda(x)^p w \, dx = \lambda^{dp-d} \int_{\mathbb{R}^d} \rho(x)^p \, w \, dx ,$$

which reduces to $\lambda^2 \int_{\mathbb{R}^d} \rho(x)^p w \, dx$ when $p = 1 + 2/d$.

The most common choice of c_1 is $c_1 = 1/2$, but sometimes the Fermi-Amaldi choice $c_1 = (N-1)/2N$ is used; this reduces to zero when there is only one electron $(N = 1)$ and is approximately $1/2$ for large N.

Minimizing $E(\rho)$ subject to $\rho \in D_N$ is a convex minimization problem. Formally this is equivalent to solving the Euler-Lagrange problem, which takes the following precise form.

EULER-LAGRANGE PROBLEM. *Find* $\rho_0 \geq 0$ *such that* $\int_{\mathbb{R}^d} \rho_0(x)\,dx = N$ *and find* $\lambda \in \mathbb{R}$ *such that*

$$p w \rho_0^{p-1} + V + c_2 B\rho_0 + \lambda = 0 \quad \text{a.e. in } [\rho_0(x) > 0] ,$$

$$p w \rho_0^{p-1} + V + c_2 B\rho_0 + \lambda \geq 0 \quad \text{a.e. in } [\rho_0(x) = 0] .$$

Here $B\rho = C_d|x|^{2-d} * \rho$, so that $B = (-\Delta)^{-1}$, and λ is the Lagrange multiplier corresponding to the constraint $\int\rho = N$. The inequality arises from the constraint $\rho \geq 0$.

A basic hypothesis on V and w is as follows.

(H1) w *is measurable,* $V \in L^1_{loc}(\mathbb{R}^d)$, *and* $\{x : V(x) < 0\}$ *has positive Lebesgue measure. There are points* $\{X_1,\dots,X_Z,Y_1,\dots,Y_P\}$ *in* $\mathbb{R}^d$ $(Z, P \in \{0,1,2,\dots\})$ *and* $\varepsilon > 0$, $\delta > 0$ *such that outside of a* δ*-neighborhood of* $\{X_1,\dots,Y_P\}$, $\varepsilon \leq w(x) \leq \varepsilon^{-1}$, *and* $w(x) \to 0$ *[resp.* $w(x) \to \infty$*] as* $x \to X_j$ *[resp.* $x \to Y_k$*] at the rate* $|x - X_j|^{\alpha_j}$ *[resp.* $|x - Y_k|^{-\beta_k}$*].*

In the following conditions, q is the dual exponent of p $(p^{-1} + q^{-1} = 1)$, and $r_+ = \max(r,0)$.

(H2) *For some real number* M,

$$w[(M - V)_+ w^{-1}]^q \in L^1(\mathbb{R}^d) .$$

(H3) *Let one of the following three conditions hold.*

(a) $\Delta V \in L^1(\mathbb{R}^d)$ *and* $w^{-1/(p-1)} \in L^1_{loc}(\mathbb{R}^d)$.

(b) ΔV *is a bounded signed measure on* $\mathbb{R}^d$, $w^{-1/(p-1)} \in L^1_{loc}(\mathbb{R}^d)$, *and* $(|x|w)^{-1/(p-1)} \in L^1$ *(neighborhood of* 0).

(c) $[(M_1 - V)_+ w^{-1}]^{1/(p-1)} \in L^1(\mathbb{R}^d)$ *for some* $M_1 \in \mathbb{R}$.

EXISTENCE THEOREM. *Let* (H1) *hold and let* $N > 0$.

(i) *Let* ρ_0 *be a solution of the minimization problem for* E. *Then there is a unique* $\lambda \in \mathbb{R}$ *such that* (ρ_0, λ) *is a solution of the Euler-Lagrange problem.*

(ii) *Conversely, let* (ρ_0, λ) *solve the Euler-Lagrange problem. If* (H2) *holds, then* E *is bounded below and* ρ_0 *is a solution of the minimization problem.*

(iii) *Let* (H3) *hold. There exists* $N_0 > 0$ *such that the Euler-Lagrange problem has a unique solution if and only if* $0 < N \leq N_0$ *and no solution for* $N > N_0$. *If* $V(x) \to 0$ *as* $|x| \to \infty$ *and* $0 < N < N_0$, *then the solution density* ρ_0 *has compact support in* $\mathbb{R}^d$. *Moreover, if* (a) *or* (b) *of* (H3) *holds, then*

$$N_0 \leq (2c_1)^{-1} \int_{\mathbb{R}^d} (\Delta V)_+ .$$

If furthermore $\int_{|x|>1} (|x|^{2-d} w^{-1})^{1/(p-1)} dx = \infty$, *then*

$$(2c_1)^{-1} \int_{\mathbb{R}^d} (\Delta V) \leq N_0 \leq (2c_1)^{-1} \int_{\mathbb{R}^d} (\Delta V)_+ .$$

When $\Delta V > 0$ and $\int_{\mathbb{R}^d} (\Delta V) = Z$, the last expression reduces to $N_0 = Z$ or $Z + 1$, according as $c_1 = 1/2$ or $c_1 = (N-1)/2N$.

The proofs of parts (i), (ii) are patterned after lectures of Brezis [3], [4]. Finding the solution (ρ_0, λ) of the Euler-Lagrange problem reduces to solving an elliptic partial differential equation equation for $u = -V - c_2 B\rho_0$. More precisely we can find a function u and a constant λ such that $u + V \in \text{Weak } L^{d/(d-2)}(\mathbb{R}^d)$,

$$(5) \qquad -\Delta u + c_2 k((u-\lambda)_+ w^{-1})^{1/(p-1)} = \Delta V ,$$

$$\int_{\mathbb{R}^d} k[(u-\lambda)_+ w^{-1}]^{1/(p-1)} \, dx = N$$

where $k = p^{-1/(p-1)}$. For each $\lambda \geq 0$, one can solve (5) with the aid of results of Gallouët and Morel [6], [7]. Let

$$N(\lambda) = \int_{\mathbb{R}^d} k[(u_\lambda - \lambda)_+ w^{-1}]^{1/(p-1)} \, dx$$

where u_λ is the solution just obtained. Then $(N_0 =)$ $N(0) > 0$, $\lim_{\lambda\to\infty} N(\lambda) = 0$, and N is continuous and strictly decreasing where it is positive. (The arguments are based on repeated applications of maximum principles.) Much of the theorem follows from this statement. For a fuller explanation of the idea see Brezis [3], [4], and for full details and a more general version of the existence theorem see Rieder's dissertation [12].

3. WHAT IS THE COULOMB POTENTIAL IN d DIMENSIONS?

To simplify the exposition we consider an atom with nucleus at the origin, so that

$$(6) \qquad V(x) = -Z|x|^{-\beta}$$

(see (2)). At issue is the appropriate value of β. We take α in (1) to be $d - 2$, because then convolution with $C_d|x|^{-\alpha}$ equals $(-\Delta)^{-1}$, and this facilitates the reduction of the energy minimization problem to a nonlinear elliptic equation involving the Laplacian. This seems to provide a substantial mathematical motivation for the choice of $\alpha = d - 2$.

Earlier we worked with $\beta = d - 2$ (see [8]), because it seemed natural in some respects to have $\alpha = \beta$. With this choice of β and $w \equiv$ constant, the existence theorem implies the following result. The Euler-Lagrange problem has a solution iff $p > 1 + 1/d$, while the minimization problem has a solution iff $p > d/2$. In fact it can be shown [12] that $\inf E(\rho) = -\infty$ if $p \leq d/2$. Thus the Euler-Lagrange problem, which can be written formally as $E'(\rho_0) = 0$, admits solutions in Dom(E'), which is a bigger set than Dom(E). Thus solutions ρ_0 outside of Dom(E) can exist. This phenomenon occurs because solving the Euler-Lagrange problem reduces to solving an equation of the form $-\Delta u + F(x,u) = G$, and both the Laplacian and the composition operator $F(x,u)$ can be defined on domains larger than Dom(E). In particular, for the Thomas-Fermi exponent $p = 1 + 2/d$, we can solve the Euler-Lagrange problem in all dimensions (≥ 3), but for $p = 1 + 2/d$ we can solve the minimization problem only in dimension 3.

One way to remedy this situation is by allowing the weight function to have a pole at the origin (or at each nucleus if we consider the molecular Coulomb potential (2)). Precise results will be stated below.

Let us now turn to the wave function version of the problem for a one electron atom (or ion). Of concern is the Hamiltonian $H = -\Delta - Z|x|^{-\beta}$ on $L^2(\mathbb{R}^d)$, $d \geq 4$. If $\beta > 2$ then (any self-adjoint realization of) H satisfies $\inf \sigma(H) = -\infty$, so the notion of a ground state wave function fails to make sense. If $\beta = 2$, then $\inf \sigma(H) = -\infty$ if $Z > (d-2)^2/4$, while for $0 < Z \leq (d-2)^2/4$, $\sigma(H) = [0,\infty)$ consists of only continuous spectrum; there is no eigenfunction to play the role of a ground state wave function. (Cf. Baras-Goldstein [1].) If we want $\inf \sigma(H)$ to be a negative eigenvalue, we must necessarily have $\beta < 2$.

To return to the density theory let (6) hold and let $0 < \varepsilon \leq w(x) \leq \varepsilon^{-1}$ outside of a δ-neighborhood of the origin but with $w(x) \sim c|x|^\gamma$ near $x = 0$; here $\gamma > 0$, $\gamma < 0$, or $\gamma = 0$ according as the origin a zero, a pole, or a "regular point" for w, respectively. (In the latter case the theory differs only inessentially from the case of $w \equiv$ constant.) Then the Euler-Lagrange problem has a solution for $1 \leq \beta < d-2$ if $p > 1 + (\beta + \gamma)_+/d$, or $p > 1 + 1/d$ if $\beta = d - 2$. (Note that $\inf p$ is not a continuous function of β. See [12].) Moreover, the minimization problem has a solution iff $p > (d+\gamma)/(d-\beta)$. Thus to solve the Thomas-Fermi minimization problem ($p = 1 + 2/d$) with exponent $\beta = d - 2$ we need $\gamma < 2 - d + 4/d$. In particular, γ must be negative when $d \geq 4$. For $\beta = 1$, $\gamma = 0$, the cutoffs become $p > 1 + 1/(d-1)$ for the minimization problem and $p > 1 + 1/d$ for the Euler-Lagrange problem. Thus $p = 1 + 2/d$ fits comfortably into this range in all dimensions $d \geq 3$. In order that the minimization problem is solvable when β is a parameter but when $\gamma = 0$, we need $1 + 2/d > 1 + \beta(d-\beta)^{-1}$, which is equivalent to $\beta < 2d/(d+2)$. This confirms our earlier requirement that $\beta < 2$ and gives an upper bound less than 2 for β, viz. $\beta < 2 - 4(d+2)^{-1}$. We view this as evidence that $\beta = 1$ is the "natural" d-dimensional generalization of the exponent in the electron-nuclear Coulomb attraction term.

Throughout our work on Thomas-Fermi theory we have benefitted from helpful conversations and correspondence with Haim Brezis and Elliott Lieb. We gratefully acknowledge their help. This work was partially supported by an NSF grant.

REFERENCES

[1] Baras, P. and J. A. Goldstein, Remarks on the inverse square potential in quantum mechanics, in Differential Equations (ed. by I. W. Knowles and R. T. Lewis), North-Holland, Amsterdam (1984), 31-35.

[2] Bénilan, Ph. and H. Brezis, The Thomas-Fermi problem, to appear.

[3] Brezis, H., Nonlinear problems related to the Thomas-Fermi equation, in Contemporary Developments in Continuum Mechanics and Partial Differential Equations (ed. by G. M. de la Penha and L. A. Medieros), North-Holland, Amsterdam (1978), 81-89.

[4] Brezis, H., Some variational problems of Thomas-Fermi type, in Variational Inequalities and Complementary Problems: Theory and Applications (ed. by R. W. Cottle, F. Giannessi, and J.-L. Lions), Wiley, New York (1980), 53-73.

[5] Fermi, E., Un metodo statistico per la determinazione di alcune prioretà dell'atome, Rend. Acad. Naz. Lincei 6 (1927), 602-607.

[6] Gallouët, Th. and J.-M. Morel, Resolution of a semilinear equation in L^1, Proc. Roy. Soc. Edinburgh 96A (1984), 275-288 and 99A (1985), 399.

[7] Gallouët, Th. and J.-M. Morel, On some semilinear problems in L^1, Boll. Un. Mat. Ital. 4A (1985), 123-131.

[8] Goldstein, J. A. and G. R. Rieder, Some extensions of Thomas-Fermi theory, in Proceedings of the Conference on Nonlinear Partial Differential Equations (ed. by A. Favini), Springer Lecture Notes in Mathematics, to appear in 1986.

[9] Lieb, E. H., Thomas-Fermi theory and related theories of atoms and molecules, Rev. Mod. Phys. 53 (1981), 603-641.

[10] Lieb, E. H. and B. Simon, Thomas-Fermi theory revisited, Phys. Rev. Lett. 33 (1973), 681-683.

[11] Lieb, E. H. and B. Simon, The Thomas-Fermi theory of atoms, molecules and solids, Adv. Math. 23 (1977), 22-116.

[12] Rieder, G. R., Mathematical Contributions to Thomas-Fermi Theory, Ph.D. Thesis, Tulane University, New Orleans, 1986.

[13] Thomas, L. H., The calculation of atomic fields, Proc. Camb. Phil. Soc. 23 (1927), 542-548.

A REGULARITY THEOREM FOR DIFFERENTIAL EQUATIONS OF CONSTANT STRENGTH

Gudrun Gudmundsdottir
Department of Mathematics, University of Uppsala
Thunbergsvägen 3, 752 38 Uppsala, Sweden

1. Introduction. Let $P(x,D)$, $D=-i\partial/\partial x$, be a linear partial differential operator defined in an open set in $\mathbb{R}^n$. The problem of determining which open sets can occur as the singular support of u if $u\in\mathcal{D}'$ and $P(x,D)u\in C^\infty$ is central in the theory of partial differential equations. It is closely related to the question of existence of solutions for the adjoint equation. In Hörmander [4] there was given a necessary and sufficient condition on a linear subspace V of $\mathbb{R}^n$ for the existence of $u\in\mathcal{D}'(\mathbb{R}^n)$ such that $\emptyset \neq \text{sing supp } u\subset V$ and $P(D)u\in C^\infty(\mathbb{R}^n)$, when $P(D)$ has constant coefficients. This paper concerns the generalization of this result to operators with variable coefficients.

The necessary and sufficient condition of [4] depends on a number $\sigma_P(W)$ which is associated to a linear subspace W of $\mathbb{R}^n$ and a differential operator $P(D)$ with constant coefficients. It gives some measure of how much P differentiates in directions not orthogonal to W. The precise definition is as follows:

After a linear change of coordinates we have

$$W = \{\xi;\ \xi''=0\},$$

where $\xi=(\xi',\xi'')$, $\xi'=(\xi_1',\dots,\xi_k')$, $\xi''=(\xi_1'',\dots,\xi_{n-k}'')$ is a splitting of the coordinates in two groups. Let

$$\tilde{P}(\xi,t)=(\sum|P^{(\alpha)}(\xi)|^2t^{2|\alpha|})^{1/2},\quad \tilde{P}_W(\xi,t)=(\sum_{\alpha''=0}|P^{(\alpha)}(\xi)|^2t^{2|\alpha|})^{1/2}$$

for $\xi\in\mathbb{R}^n$, $t\in\mathbb{R}^+$. Then

$$\sigma_P(W) = \inf_{t>1}\ \liminf_{\xi\to\infty} \tilde{P}_W(\xi,t)/\tilde{P}(\xi,t).$$

The condition $\sigma_P(W)=0$ is satisfied if and only if

$$\liminf_{\xi\to\infty} \tilde{P}_W(\xi,\lambda\log|\xi|)/\tilde{P}(\xi,\lambda\log|\xi|)=0 \qquad (1.1)$$

for some, or equivalently for all, $\lambda>0$. For the proof of this fact and examples illustrating the meaning of $\sigma_P(W)$ we refer to [4]. We formulate the result of [4] mentioned above for reference in

Theorem 1.1. Let V be a linear subspace of $\mathbb{R}^n$ and V' its orthogonal space. There exists $u\in\mathcal{D}'(\mathbb{R}^n)$ such that $P(D)u\in C^\infty(\mathbb{R}^n)$ and $\emptyset\neq$ sing supp $u\subset V$ if and only if $\sigma_P(V')=0$.

Consider now a differential operator $P(x,D)$ with variable C^∞ coefficients defined near a point $x_o \in \mathbb{R}^n$. If x is a point near x_o and W is a linear subspace we set

$$\sigma_P(W,x) = \sigma_{P_x}(W),$$

where $P_x(D)$ is the operator with constant coefficients obtained by freezing the coefficients of P at a point x near x_o. In general $\sigma_P(W,x)$ may be positive for some points x and zero for others. However, it was shown in [4] that if P has constant strength then $\sigma_P(W,x)=0$ for all x if and only if this is true when $x=x_o$. We recall that the operator P is said to have constant strength if for every x there are constants c_x and C_x such that

$$0 < c_x \leq \tilde{P}_x(\xi)/\tilde{P}_{x_o}(\xi) \leq C_x, \quad \xi \in \mathbb{R}^n.$$

Here as usual $\tilde{P}(\xi)=\tilde{P}(\xi,1)$ (see [3], section 13.1).

Many local results which are valid for operators with constant coefficients can be extended to operators of constant strength. Since the validity of the condition $\sigma_P(W,x)=0$ does not depend on x one could therefore expect that the statement of Theorem 1.1 holds for such an operator locally. In [2] it was proved that if

$$P(x,D) = P_o(D) + \sum_{k=1}^{r} c_k(x)P_k(D)$$

where $\tilde{P}_k(\xi)/\tilde{P}_o(\xi) \to 0$ when $\xi \to \infty$ and c_k are real analytic functions then

$$P(x,D)u \in C^\infty \text{ near } x_o, \text{ sing supp } u \subset V+\{x_o\} \Rightarrow u \in C^\infty \text{ near } x_o,$$

if $\sigma_P(V') \neq 0$. Here we will prove that the same conclusion holds for any P with constant strength if u is a L^1_{loc} function in $V+\{x_o\}$ considered as a measure in $\mathbb{R}^n$.

Results on the existence of distributions u, defined near x_o, such that $P(x,D)u \in C^\infty$ and sing supp $u = V+\{x_o\}$, when $\sigma_P(V')=0$ and P has constant strength, are given in [1].

2. Localizations at infinity.

The method used to prove regularity theorems in [2] and [4] was to construct a parametrix for the operator. Here we will employ a different technique which involves so called localizations at infinity of an operator with constant coefficients.

Let $P(D)$ be a partial differential operator with constant coefficients and denote its characteristic polynomial also by P. Consider a sequence $\xi_j \to \infty$ in $\mathbb{R}^n$ and set $t_j = \log|\xi_j|$. All derivatives at 0 of the polynomials

$$P_j(\xi) = P(\xi_j + t_j\xi)/\tilde{P}(\xi_j, t_j) \qquad (2.1)$$

have then absolute value less than or equal to 1. Thus there is a subsequence j_k such that all the coefficients of P_{j_k} have limits when $k\to\infty$. Then there is a polynomial $Q(\xi)$ such that

$$P_{j_k}(\xi) \to Q(\xi) \text{ as } k \to \infty \tag{2.2}$$

in the sense that all coefficients of P_{j_k} tend to corresponding coefficients of Q. We shall call such a limit Q a <u>localization of P at infinity</u> and denote the set of all these by L(P).

Note that if $P_j\to Q$ as in (2.2) then $P_j(D)u \to Q(D)u$ in $\mathcal{D}'(\mathbb{R}^n)$ for any $u \in \mathcal{D}'(\mathbb{R}^n)$. Moreover if $u_j\to u$ in $\mathcal{D}'(\mathbb{R}^n)$ then $P_j(D)u_j \to Q(D)u$ in $\mathcal{D}'(\mathbb{R}^n)$.

The following lemma shows how the condition $\sigma_P(W)>0$ is related to the operators of L(P).

<u>Lemma 2.1.</u> Let W be a linear subspace of $\mathbb{R}^n$. If P is a differential operator with constant coefficients and $\sigma_P(W)>0$ then every $Q\in L(P)$ has some non-characteristic direction in W. On the other hand if $\sigma_P(W)=0$ then there is some $Q \in L(P)$ which is identically zero in W.

<u>Proof.</u> Assume first that $\sigma_P(W)>0$ and let $q\in L(P)$ be the limit of (2.1) for some sequence $\xi_j\to\infty$ in $\mathbb{R}^n$, $t_j=\log|\xi_j|$. For any $\xi\in\mathbb{R}^n$, $s\in\mathbb{R}$ we have

$$\frac{\tilde{Q}_W(\eta,s)}{\tilde{Q}(\eta,s)} = \lim_{j\to\infty} \frac{\tilde{P}_W(\xi_j+t_j\eta,st_j)}{\tilde{P}(\xi_j+t_j\eta,st_j)}. \tag{2.3}$$

It is easy to see that there is a constant a such that for every we have

$$a^{-1}\le st_j/(s(\log|\xi_j+t_j\eta|)) \le a,$$

when j is large. Then (1.1) with $\lambda=s$ gives that the expression (2.3) is greater than a constant $C > 0$, that is

$$\tilde{Q}_W(\eta,s) \ge C\tilde{Q}(\eta,s). \tag{2.4}$$

Denote the principal part of Q by q and let m be the order of Q. Replace η and s by $t\eta$ and ts in (2.4). After division by t^m it follows then that

$$\tilde{q}_W(\eta,s) \ge C\tilde{q}(\eta,s)$$

when $t\to\infty$. In particular q cannot be identically 0 in W, for then $\tilde{q}_W$ would be identically 0 in W and $\tilde{q}(\eta,s)$ is greater than a positive constant for all η and s.

Conversely if $\sigma_P(W)=0$ then (1.1) gives that

$$\tilde{P}_W(\xi_j,t_j)/\tilde{P}(\xi_j,t_j) \to 0$$

as $j\to\infty$ for some sequence $\xi_j\to\infty$, $t_j=\log|\xi_j|$. This means that

$$\tilde{Q}_W(0,1) = 0$$

if $Q \in L(P)$ is defined by a subsequence of ξ_j. Then q must be identically zero in W. This finishes the proof of the lemma.

Next we will define a localization at infinity of an operator $P(x,D)$ with constant strength defined near a point $x_o \in \mathbb{R}^n$. We can write

$$P(x,D) = P_o(D) + \sum_{k=1}^{r} c_k(x)P_k(D),$$

where $c_k \in C^\infty$, $c_k(x_o)=0$ and

$$\tilde{P}_k(\xi)/\tilde{P}_o(\xi) \leq C \text{ for all } \xi \in \mathbb{R}^n,\ k=1,\dots,r. \tag{2.5}$$

(See [3], Chapter 13). The condition (2.5) is equivalent to

$$\tilde{P}_k(\xi,t)/\tilde{P}_o(\xi,t) \leq C,\ \xi \in \mathbb{R}^n,\ t \geq 1,\ k=1,\dots,r. \tag{2.6}$$

(Theorem 10.4.3 of [3]). The coefficients of the polynomials

$$P_k(\xi_j+t_j\xi)/\tilde{P}_o(\xi_j,t_j),\ k=1,\dots,r,$$

are therefore bounded functions of j if ξ_j is a sequence defining a polynomial $Q(\xi) \in L(P_o)$. Let

$$P_j(x,\xi) = P(x_o+(x-x_o)/t_j, \xi_j+t_j\xi)/\tilde{P}_o(\xi_j,t_j). \tag{2.7}$$

Then $P_j(x,\xi) \to Q(\xi)$ as $j\to\infty$ in the sense that all coefficients of P_j tend to the corresponding constant coefficients of Q in $C^\infty(\Omega)$ if Ω is a neighborhood of x_o. Thus $P_j(x,D)u_j \to Q(D)u_o$ in $\mathcal{D}'(\Omega)$ if $u_j \to u_o$ in $\mathcal{D}'(\Omega)$. We will call an operator Q obtained in this way a localization of P at infinity. Note that Q has constant coefficients and $Q \in L(P_o)$.

Now let V be a linear subspace of $\mathbb{R}^n$ such that $\sigma_P(V')>0$ and let Ω be a neighborhood of x_o. Our problem is to show that if $U \in \mathcal{D}'(\Omega)$, $P(x,D)U \in C^\infty(\Omega)$ and sing supp $U \subset \{x_o\}+V$ then $U \in C^\infty$ near x_o. Here we specialize to measures with support in $\{x_o\}+V$, which are defined by a function $u \in L^1_{loc}$. If u is an L^1_{loc} -function in $V+\{x_o\}$ then u defines a measure U in $\mathbb{R}^n$ by

$$U(\phi) = \int_V u\phi,\quad \phi \in C_o^\infty(\mathbb{R}^n).$$

Then the singular support of U is $(V+\{x_o\} \cap \text{supp } u)$ and $P(x,D)U \in C^\infty \Leftrightarrow P(x,D)U = 0$. We will prove that if $P(x,D)U = 0$ then u is zero almost everywhere.

Define

$$U_j = U(x_o+(x-x_o)/t_j).$$

The exponential function

$$e_j(x) = \exp(-i\langle \xi_j, (x-x_o)/t_j\rangle)$$

is equal to 1 in $\{x_o\}+V$. Since U_j is a measure with support in $\{x_o\}+V$ we have

$$e_jU_j = U_j.$$

The definition (2.7) of P_j gives that

$$P_j(x,D)U_j = P_j(x,D)(e_jU_j) = a_je_j(x)(PU)(x_o+(x-x_o)/t_j)/\tilde{P}_o(\xi_j,t_j).$$

This vanishes near x_o since PU=0 there. After possibly passing to a subsequence we have $P_j(x,D) \to Q(D) \in L(P_o)$. The measures U_j tend to U_o where

$$U_o(\phi) = u(x_o)\int_V \phi,$$

if x_o is a Lebesgue point for u. We can assume that this is the case since the set of non-Lebesgue points is of measure zero and x_o can be exchanged with another point in V nearby. Moreover, x_o can be chosen as a Lebesgue point where $u(x_o) \neq 0$, unless u is zero almost everywhere. Then

$$Q(D)U_o = 0.$$

By Lemma 2.1 there is some non-characteristic direction η for Q in V'. The support of U_o is contained in $\{x_o\}+V$ and hence in the half space where $\langle x-x_o,\eta\rangle \geq 0$. Holmgren's uniqueness theorem gives that U_o is zero near x_o, which is a contradiction.

Thus we have proved the following theorem:

Theorem 2.2. Let P(x,D) be a differential operator of constant strength defined near a point $x_o \in \mathbb{R}^n$. If V is a linear subspace of $\mathbb{R}^n$ such that $\sigma_P(V') > 0$ then

$$Pu \in C^\infty \text{ near } x_o,\ \text{sing supp } u \subset \{x_o\}+V \Rightarrow u \in C^\infty \text{ near } x_o,$$

if u is a measure defined by an L^1_{loc}- function in $\{x_o\}+V$.

References:

1. G.Gudmundsdottir, Global properties of differential operators of constant strength. Ark.Mat, Vol 15 (1977) No. 1, p. 169-198.
2. G.Gudmundsdottir, On continuation of regularity for differential equations of constant strength. Report, University of Lund, 1977.
3. L.Hörmander, The Analysis of linear partial differential operators, Springer Verlag, 1983.
4. L.Hörmander, On the singularities of solutions of partial differential equations with constant coefficients, Israel J. Math. Vol 13 (1972) No. 1-2, p. 82-105.

INTERMITTENT BIFURCATION OF VORTEX FLOWS

Karl Gustafson

University of Colorado
Boulder, Colorado USA

We describe newly found vortex bifurcation sequences in full Navier-Stokes flow. These are initiated by a bursting effect near a separation point and possess intermittent behavior thereafter. Some appear to be transient while others are not. The relationship and implications to the standard bifurcation theory and diagrams will be discussed.

1. INTRODUCTION

Recent studies (see Benjamin and Mullin [1], Cliff and Mullin [2], Bolstad and Keller [3], and the references therein) have been concerned with questions of flow multiplicity higher than previously expected in the Taylor Problem of flow between rotating cylinders. For the most part these studies consider steady cellular flows at Reynolds numbers reasonably near those at which the Taylor vortices appear. Quoting

> "[1, p. 219] A prime contention of the previous discussions has been that although the realistic hydrodynamic problem modelling the Taylor experiment is yet unsolved in closed form, it must have a high multiplicity of isolated solutions when R lies well above the quasi-critical range wherein Taylor cells are first easily demonstrable by standard flow-visualization techniques.
>
> "[2, p. 256] A striking feature of anomalous modes, particularly those with a larger number of cells, is the distortion of the cell boundary adjacent to the anomalous cell."
>
> "[3, p. 16] A new phenomenon is ... the splitting of the extra vortices into two smaller vortices."

Re [1], while admitting that I have only recently become aware of these recent new higher multiplicities found for the Taylor problem, nontheless I would first like to advance here the hypothesis that in some cases the end effects in the

Taylor Problem imply even higher multiplicities that just haven't been found yet, in some situations infinite multiplicities. All of this depends on the exact experimental or numerical model employed, but when a corner with no slip conditions prevailing is encountered, or a corner with slip conditions on one side only, e.g., an intersurface separation interface, and when the angle is not too large, one should expect an infinite set of smaller vortices descending into it. An example of a sequence of ten of these that we have found in a corner will be given below.

A second thought I advance here is that the existence of higher multiplicities in real flow depends more on certain "parity rules" established by the fluid during its actual dynamic evolution, than on the bifurcation parameter homotopy arguments followed in [1,2,3]. The latter "homotopy model" is a valuable technique in connection with the numerical continuation methods used in [1,2,3] to enable the tracking of the "full" bifurcation diagram as, say, the Reynolds number Re or the aspect ratio A is varied. But in the end it would appear to be limited to the analyses of the steady flow equations and can therefore generate mathematically valid but physically spurious solutions. I will illustrate below the development of such a "parity rule" structure governing a full Navier-Stokes flow. Moreover, as will be seen, the parity rules explain the cell boundary distortion referred to in [2].

Finally, I will illustrate the mechanisms of the splitting of vortices into smaller vortices. This can occur [6] as a function of the varying of the key parameters (e.g., Re, A) of the problem in a steady flow as in [3] but more interestingly is found to occur dynamically [6,7] in unsteady flow, with both splitting and coalescence sequences found.

2. END EFFECTS AND CORNER VORTEX SEQUENCES

As pointed out in [2, p. 257], the anomalous modes are not surprising, should be expected, and are due to the end effects on the Taylor annulus.

In [4,5] we concentrated on finding similar "anomalous modes" for corner flow in a driven unit cavity, and thus far have succeeded in finding twenty of them. There are (mathematically) an infinite number there, although (computationally) they will, depending on the precision carried, drop into the noise level because their intensities fall off $O(10^{-4})$, and (physically) experimentally only three or four at most have been seen. For full details about this interesting problem see [8,9]. Here are the first 10 corner modes reported in [4], measured both by stream function intensity ψ_i and in terms of the zeros z_i between them on the 45° diagonal angle bisector extending out from the lower left corner of the cavity. There are such vortex sequences in other cavity corners but we will omit discussion of those here. The sign changes on the ψ_i intensities are in accordance with the parity rules I will discuss next.

Local Maximum Stream Function Intensity	Stream Function Zero Measured Along Diagonal
1.0006×10^{-1}	6.97×10^{-2}
-2.232×10^{-1}	4.205×10^{-3}
6.165×10^{-11}	2.534×10^{-4}
-1.703×10^{-15}	1.536×10^{-5}
4.71×10^{-20}	9.247×10^{-7}
-1.30×10^{-24}	5.602×10^{-8}
3.59×10^{-29}	3.370×10^{-9}
-9.93×10^{-34}	2.040×10^{-10}
2.75×10^{-38}	1.236×10^{-11}
-7.59×10^{-43}	7.421×10^{-13}

3. PARITY RULES AND PROXIMITY LIMITATIONS

As pointed out in [3, p. 4], the demonstration of additional "hidden" vortices remove all difficulties with "wrong" odd numbers of vortices found in previous experiments.

Such "hidden" very weak vortices are known in the aerodynamic literature as "ornamentation" vortices. I preferred the term "intermediating" vortices in [8] to indicate that they are not ornamental in any sense of the word but are in fact topological necessities to the flow. Aerodynamics is characterized by open regions and often the smaller vortices do indeed flow away, but in a closed flow such as the Taylor geometry of [1,2,3] or in the cavity geometry of [6,7] they do not disappear once they have managed to enter the flow. Whether or not they can enter appears to depend not only on their parity but also on the proximity of their potential development region to ends, corners, walls, and even to intersurface separation lines. Here are some details of their evolution as reported in [6,7]. Note that the deformed cell contours occur very naturally in terms of the parity signs plotting along their boundaries.

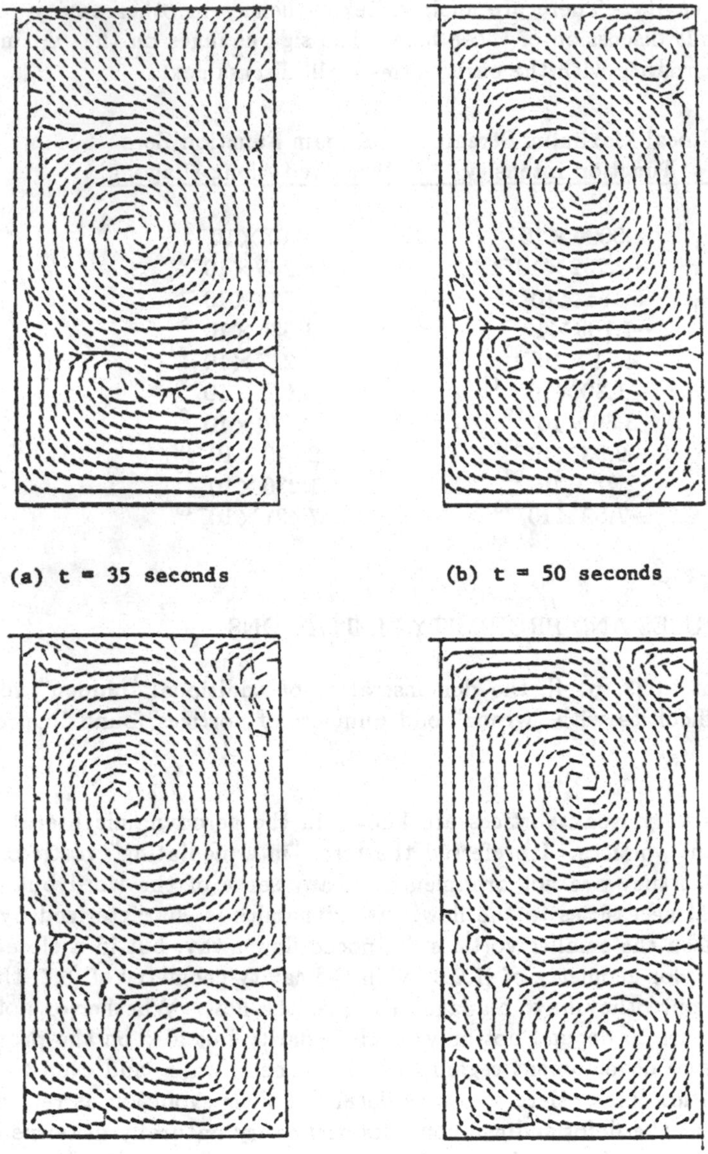

(a) t = 35 seconds

(b) t = 50 seconds

(c) t = 60 seconds

(d) t = 75 seconds

Figure 1. Separation Region "Lubrication" Dynamics

(a) Fission into 3, almost 4, tertiary eddies for "self-lubrication".
(b) Right "corner lubrication" begins, to continue the energy cascade.
(c) The last two eddies report in, causing "temporary mass confusion".
(d) "Final Resting Place", as the basic final flow topology is determined.

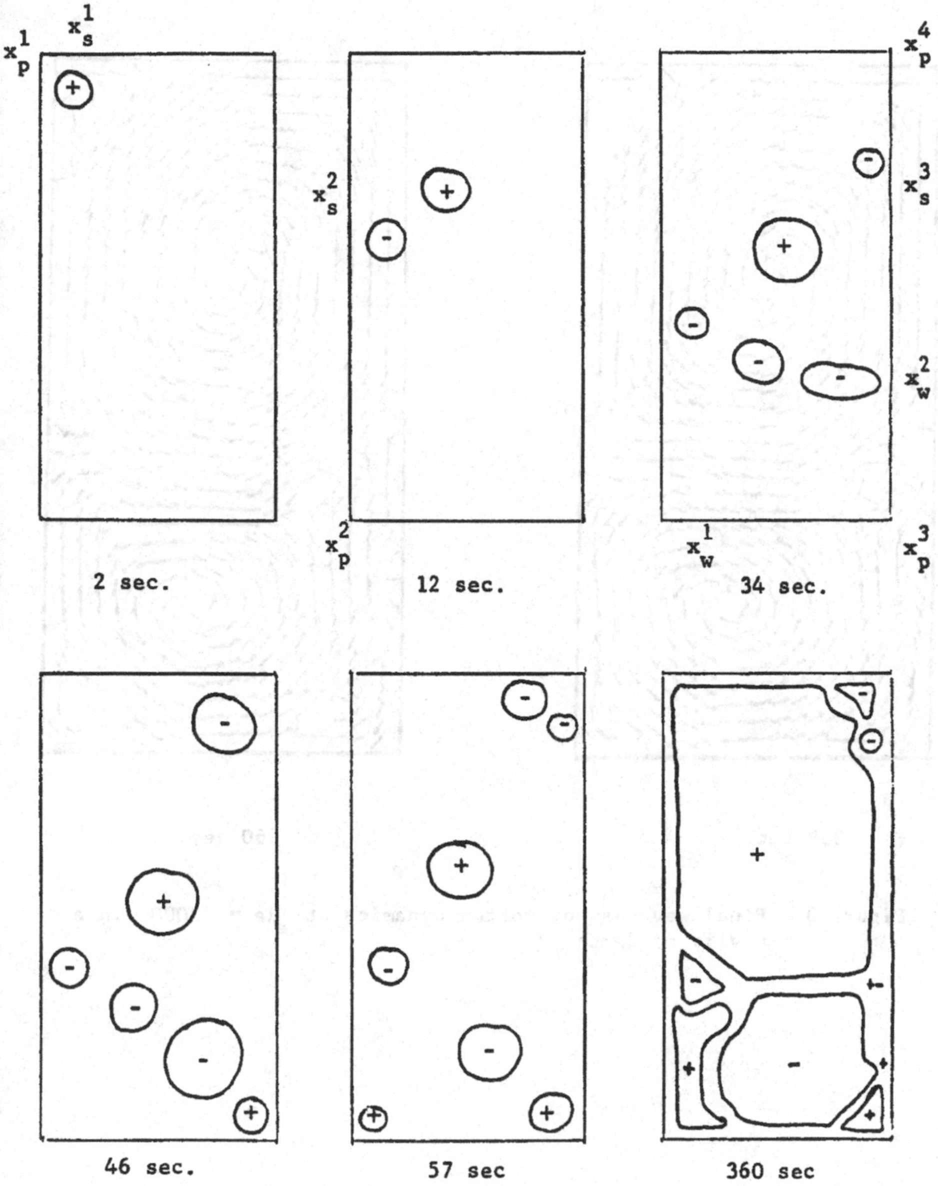

Figure 2. Wall, corner, provocation, separating, and intermediating effects in cavity flow dynamics.

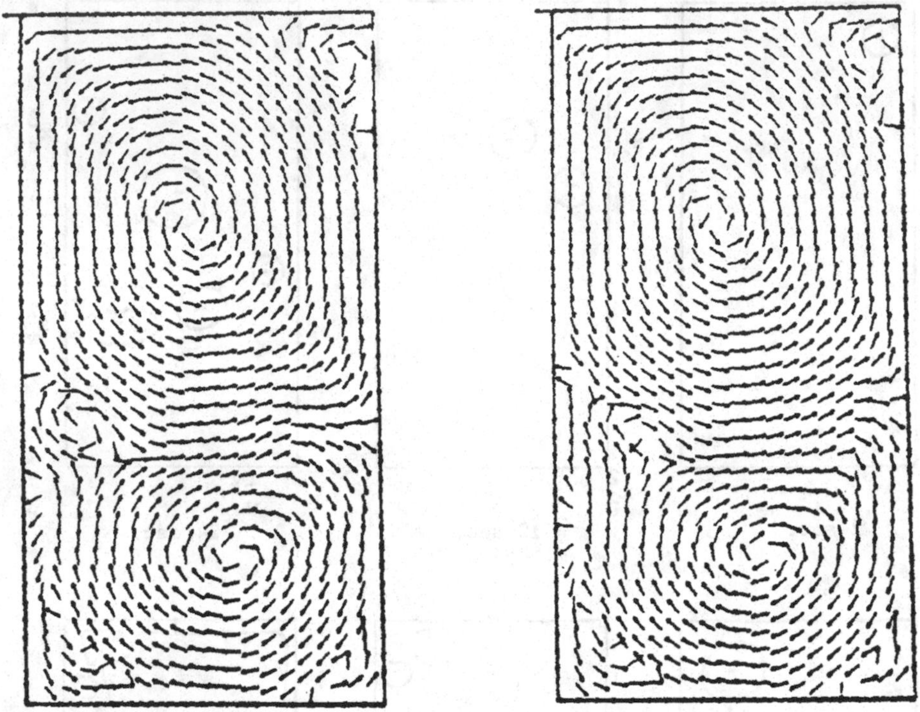

359 sec. 360 sec.

Figure 3. Final wavyness of vortex dynamics at Re = 10000 in a cavity of depth 2.

4. INTERMITTENT BIFURCATION

As can be seen from the above, there are many intermittent bifurcations (e.g., wall bursting, the splitting-coalescence sequences) that appear during the dynamic development of the flow. These must be taken into account in any attempt at a physically correct understanding of a final steady flow bifurcation diagram. Some of these sequences are transient, thus "ornamental" in one sense, yet their temporary existence is absolutely essential in the "mediating" of the development of a final flow pattern. Others, such as the final wavy flow states shown above, must be distinguished as possible Hopf bifurcations, rather than, say, multiple candidates for a final steady state.

REFERENCES

1. T. Brooke Benjamin and T. Mullin, Notes on the multiplicity of flows in the Taylor experiment, *J. Fluid Mech.* 121 (1982), 219-230.

2. K. Cliffe and T. Mullin, A numerical and experimental study of anomalous modes in the Taylor experiment, *J. Fluid Mech.* 153 (1985), 243-258.

3. J. Bolstad and H. Keller, Computation of anomalous modes in the Taylor experiment, *J. Computational Physics,* to appear.

4. K. Gustafson and R. Leben, Multigrid computation of subvortices, *Applied Math. and Computation* (1986), to appear.

5. K. Gustafson and R. Leben, to appear.

6. K. Gustafson and K. Halasi, Vortex dynamics of cavity flows, *J. Computational Physics* (1986), to appear.

7. K. Gustafson and K. Halasi, Cavity flow dynamics at higher Reynolds number and higher aspect ratio, to appear.

8. K. Gustafson, Vortex separation and fine structure dynamics, *Applied Numerical Mathematics* (1986), to appear.

9. K. Gustafson, *Partial Differential Equations,* 2nd Edition, Wiley, New York (1986), to appear.

Remarks on the Scattering Problem for Nonlinear Schrödinger Equations

Nakao Hayashi
Department of Applied Physics
Waseda University
Tokyo 160, Japan

Yoshio Tsutsumi
Faculty of Integrated Arts and Sciences
Hiroshima University
Hiroshima 730, Japan

§1. Introduction and main results.

We consider the asymptotic behavior as $t \to \pm\infty$ of solutions and the scattering theory for the following nonlinear Schrödinger equation with power interaction:

$$i\frac{\partial u}{\partial t} = -\Delta u + |u|^{p-1}u, \quad t \in \mathbb{R}, \quad x \in \mathbb{R}^n, \tag{1.1}$$

$$u(0,x) = u_0(x), \quad x \in \mathbb{R}^n. \tag{1.2}$$

Let $U(t)$ be the evolution operator associated with the free Schrödinger equation. For $1 \leqq q < \infty$, L^q and $\|\cdot\|_q$ denote the standard q-integrable function space on $\mathbb{R}^n$ and its norm, respectively. $(\cdot\,,\,\cdot)$ denotes the scalar product in L^2. Let Σ denote the Hilbert space

$$\Sigma = \{ v \in L^2 \ ; \ \nabla v \in L^2 \text{ and } xv \in L^2 \}$$

with the norm $\|v\|_\Sigma^2 = \|v\|_2^2 + \|\nabla v\|_2^2 + \|xv\|_2^2$. We put

$$\alpha(n) = \begin{cases} \infty, & n = 1, 2, \\ \dfrac{n+2}{n-2}, & n \geq 3, \end{cases}$$

$$\gamma(n) = \frac{n + 2 + \sqrt{n^2 + 12n + 4}}{2n}.$$

We note that $\gamma(n) < 1 + \frac{4}{n} < \alpha(n)$. We put $f(z) = |z|^{p-1}z$ for $z \in \mathbb{C}$. For $z \in \mathbb{C}$ we denote the complex conjugate of z by $\bar{z}$.

There are many papers concerning the asymptotic behavior as $t \to \pm\infty$ of solutions for (1.1)-(1.2) (see, e.g., [1]-[9], [12]-[15] and [17]-[19]). Recently Y. Tsutsumi has shown the following two theorems in [17, 18] (see also Ginibre and Velo [4]).

<u>Theorem A.</u> Assume that $\gamma(n) < p < \alpha(n)$.

(i) For any $u_+ \in \Sigma$ there exists a unique $u_0 \in \Sigma$ such that

$$\|u_+ - U(-t)u(t)\|_\Sigma \to 0 \quad (t \to +\infty), \tag{1.3}$$

where $u(t)$ is a solution in $C(\mathbb{R}\,;\Sigma)$ of (1.1) with $u(0) = u_0$.

(ii) For any $u_- \in \Sigma$ there exists a unique $u_0 \in \Sigma$ such that

$$\|u_- - U(-t)u(t)\|_\Sigma \to 0 \quad (t \to -\infty), \tag{1.4}$$

where $u(t)$ is a solution in $C(\mathbb{R};\Sigma)$ of (1.1) with $u(0) = u_0$.

Theorem B. Assume that $\gamma(n) < p < \alpha(n)$. For any $u_0 \in \Sigma$ there exist unique $u_\pm \in \Sigma$ such that the solution $u(t)$ in $C(\mathbb{R};\Sigma)$ of (1.1) with $u(0) = u_0$ satisfies

$$\|u_\pm - U(-t)u(t)\|_\Sigma \to 0 \quad (t \to \pm\infty). \tag{1.5}$$

Remark 1.1. (i) If $1 < p < \alpha(n)$, then for any $u_0 \in \Sigma$ there exists a unique weak solution in $C(\mathbb{R};\Sigma)$ of (1.1)-(1.2) (see [3, Theorem 3.1] and [4, Proposition 3.5]).

(ii) Theorem A implies that if $\gamma(n) < p < \alpha(n)$, the wave operators $W_\pm: u_\pm \to u_0$ are well defined on Σ as a mapping from Σ to Σ. Theorem B implies that Range (W_+) = Range $(W_-) = \Sigma$ and that $W_\pm$ are one to one. Therefore, we can construct the scattering operator $S = W_+^{-1}W_-: u_- \to u_+$ as a mapping from Σ to Σ, if $\gamma(n) < p < \alpha(n)$.

In [4] and [6] Ginibre and Velo proved the fine results concerning the construction of the wave operators and their asymptotic completeness for $1 + \frac{4}{n} < p < \alpha(n)$. But, in [14] and [15] Strauss conjectured that the construction of the wave operators and their asymptotic completeness could be brought down from $1 + \frac{4}{n}$ to $\gamma(n)$. Theorems A and B answer Strauss' conjecture. However, the proof in [18] is rather complicated and is based on the following transform:

$$u(t,x) = (it)^{-\frac{n}{2}} e^{\frac{ix^2}{4t}} \overline{v(\frac{1}{t}, \frac{x}{t})} \tag{1.6}$$

(for the details of (1.6), see [5] and [17-20]). In this paper we give a simpler proof of Theorems A and B without the transform (1.6). In addition, we prove that if $\gamma(n) < p < \alpha(n)$, the wave operators and the scattering operator are continuous from Σ to Σ. The continuity in the natural topology of the wave operators and the scattering operator is proved in [4] and [6] for $1 + \frac{4}{n} < p < \alpha(n)$, but it has not been proved yet for $\gamma(n) < p < 1 + \frac{4}{n}$. We put

$$J_j u = e^{ix^2/4t}(2it)\frac{\partial}{\partial x_j}(e^{-ix^2/4t}u) = U(t)\, x_j\, U(-t)u = (x_j + 2it\frac{\partial}{\partial x_j})u, \tag{1.7}$$

$$1 \leq j \leq n.$$

Our proof is based on the fact that J_j and $i\frac{\partial}{\partial t} + \Delta$ are commutative (see, e.g., [9]-[11] and [20]) and the Strichartz space-time estimate (see [6] and [16]).

Our main theorem in this paper is the following.

Theorem 1.1. Assume that $\gamma(n) < p < \alpha(n)$. Then, the assertions of Theorems A and B hold. In addition, the wave operators and the scattering operator given by Theorems A and B are homeomorphisms from Σ onto Σ.

In the course of calculations below various constants will be simply denoted by C. C(*,$\cdots$,*) denotes a constant depending only on the quantities appearing in parentheses.

§2. Proof of Theorem 1.1.

In this section we give a sketch of the proof of Theorem 1.1. We first summarize three lemmas needed for the proof of Theorem 1.1.

Lemma 2.1. (i) Let q and r be positive numbers such that $1/q + 1/r = 1$ and $2 \leq q \leq \infty$. Then,

$$\|U(t)v\|_q \leq (4\pi|t|)^{\frac{n}{q}-\frac{n}{2}} \|v\|_r, \quad v \in L^r, \quad t \neq 0. \tag{2.1}$$

(ii) Let q and r be positive numbers such that $1 \leq q - 1 < \alpha(n)$ and $(\frac{n}{2} - \frac{n}{q})r = 2$. Then,

$$\|U(\cdot)v\|_{L^r(\mathbb{R};L^q)} \leq C \|v\|_2, \quad v \in L^2, \tag{2.2}$$

where $C = C(n, q)$.

For Lemma 2.1, see, e.g., Ginibre and Velo [6, 7] and Strichartz [16].

Lemma 2.2. For the nonlinear term $f(z)$, we have

$$\left|\frac{\partial}{\partial z} f(z_1) - \frac{\partial}{\partial z} f(z_2)\right| \leq C(|z_1| + |z_2|)^\sigma |z_1 - z_2|^\beta, \tag{2.3}$$

$$\left|\frac{\partial}{\partial \bar{z}} f(z_1) - \frac{\partial}{\partial \bar{z}} f(z_2)\right| \leq C(|z_1| + |z_2|)^\sigma |z_1 - z_2|^\beta, \tag{2.4}$$

for $z_1, z_2 \in \mathbb{C}$, where $\sigma = \max(0, p-2)$, $\beta = \min(p-1, 1)$ and $C = C(p)$.

Lemma 2.2 is clear.

Lemma 2.3. Assume that $1 < p < \alpha(n)$. Then, the solution $u(t)$ of (1.1)-(1.2) with $u_0 \in \Sigma$ satisfies

$$\|u(t)\|_2 = \|u_0\|_2, \tag{2.5}$$

$$\|\nabla u(t)\|_2^2 + \frac{2}{p+1}\|u(t)\|_{p+1}^{p+1} = \|\nabla u_0\|_2^2 + \frac{2}{p+1}\|u_0\|_{p+1}^{p+1}, \tag{2.6}$$

$$\|u(t)\|_{p+1} \leq C(1 + |t|)^{-\theta}, \tag{2.7}$$

$$\|xU(-t)u(t)\|_2 \leq C\,(1+|t|)^{a(p)} \tag{2.8}$$

for $t \in \mathbb{R}$, where $C = C(n, p, \|u_0\|_\Sigma)$, $\theta = \frac{n(p-1)}{2(p+1)}$ and $a(p) = \max\,(0\ ,\ 1 - \frac{n}{4}(p-1))$.

For Lemma 2.3, see, e.g., Ginibre and Velo [3, Theorem 3.1], [4, Proposition 3.5], Barab [1, Lemma 3] and Tsutsumi and Yajima [19].

Let $r = \frac{4(p+1)}{n(p-1)}$ and $\theta = \frac{n(p-1)}{2(p+1)}$, unless specified otherwise. We next prove the following proposition.

Proposition 2.4. Assume that $\gamma(n) < p < \alpha(n)$. Then, the solution $u(t)$ of (1.1)-(1.2) with $u_0 \in \Sigma$ satisfies

$$\|J_j u\|_{L^r(\mathbb{R};L^{p+1})} \leq C\ , \quad 1 \leq j \leq n, \tag{2.9}$$

$$\|\nabla u\|_{L^r(\mathbb{R};L^{p+1})} \leq C\ , \tag{2.10}$$

where $r = \frac{4(p+1)}{n(p-1)}$ and $C = C(n, p, \|u_0\|_\Sigma)$.

Proof. We give a formal calculation, but it can be justified by the regularizing technique of Ginibre and Velo [3, 4].

We first prove (2.9). We put $v_j(t) = (J_j u)(t)$. Since J_j and $i\frac{\partial}{\partial t} + \Delta$ are commutative, we have

$$v_j(t) = U(t)x_j u_0 - i\int_0^t U(t-s)J_j f(u(s))\,ds, \quad t \in \mathbb{R}, \quad 1 \leq j \leq n. \tag{2.11}$$

For any $R > 0$, we take the $L^r(-R,R\,;L^{p+1})$ norm of (2.11) and use Lemmas 2.1, 2.3, the generalized Young inequality and Hölder's inequality to obtain

$$\|v_j\|_{L^r(-R,R\,;\,L^{p+1})} \leq C\,\|x_j u_0\|_2 \tag{2.12}$$
$$+ C\left(\int_{\mathbb{R}} (1+|t|)^{-(p-1)(1-\eta)\theta q + a(p)\varepsilon q}\,dt\right)^{1/q} \|v_j\|^{1-\varepsilon}_{L^r(-R,R\,;\,L^{p+1})}\ ,$$
$$R > 0, \quad 1 \leq j \leq n,$$

where $q = \frac{4(p+1)}{2\{(n+2)-(n-2)p\}+n(p-1)\varepsilon}$, $\eta = \frac{2n\varepsilon}{n+2-(n-2)p}$, ε is a sufficiently small positive constant such that $0 < \eta < 1$ and $(p-1)(1-\eta)\theta q - a(p)\varepsilon q > 1$, and $C = C(n, p, \|u_0\|_\Sigma)$. Here we have used the following inequality.

$$\|J_j f(u(s))\|_{1+1/p} \leq C\,\|u(s)\|^{(p-1)(1-\eta)}_{p+1}\,\|\nabla u(s)\|_2^{(p-1)\eta} \tag{2.13}$$
$$\times\ \|v_j(s)\|_2^{\varepsilon}\,\|v_j(s)\|^{1-\varepsilon}_{p+1}\ , \quad 1 \leq j \leq n,$$

where $C = C(n, p)$. Since $\frac{n(p-1)^2}{n+2-(n-2)p} > 1$ for $\gamma(n) < p < \alpha(n)$, we can choose $\varepsilon > 0$ so small that $(p-1)(1-\eta)\theta q - a(p)\varepsilon q > 1$. Therefore, (2.12) implies

$$\|v_j\|_{L^r(-R,R\,;\,L^{p+1})} \leq C\,, \quad 1 \leq j \leq n, \tag{2.14}$$

for any $R > 0$, where $C = C(n,\ p,\ \|u_0\|_\Sigma\,)$. This shows (2.9).

In the same way we can prove (2.10).

(Q. E. D.)

We are now in a position to prove Theorem 1.1.

Proof of Theorem 1.1. The calculations below are rather formal, but they can be justified by the regularizing thechnique (see, e.g., [3] and [4]).

We first prove Theorem B. We consider only the case of $t \to +\infty$, since we can treat the case of $t \to -\infty$ in the same way. We have only to prove

$$\|U(-t)u(t) - U(-s)u(s)\|_\Sigma \to 0 \quad (t,\ s \to +\infty). \tag{2.15}$$

Since J_j and $i\,\frac{\partial}{\partial t} + \Delta$ are commutative, we have by (1.7)

$$x_jU(-t)u(t) - x_jU(-s)u(s) = -\,i\int_s^t U(-\tau)J_jf(u(\tau))\,d\tau,\quad t,\ s \in \mathbb{R}, \quad 1 \leq j \leq n. \tag{2.16}$$

We take the $L^2(\mathbb{R}^n)$ norm of (2.16) and use Lemma 2.1(i) to obtain

$$\begin{aligned}\|x_jU(-t)u(t) - x_jU(-s)u(s)\|_2^2 &= \Big\|\int_s^t U(-\tau)J_jf(u(\tau))\,d\tau\Big\|_2^2 \\ &= \Big(\int_s^t U(-\tau)J_jf(u(\tau))\,d\tau\,,\ \int_s^t U(-\tilde{\tau})J_jf(u(\tilde{\tau}))\,d\tilde{\tau}\Big) \\ &\leq C\int_s^t \|u(\tau)\|_{p+1}^{p-1}\,\|J_ju(\tau)\|_{p+1} \\ &\quad\times\int_s^t |\tau-\tilde{\tau}|^{-\theta}\,\|u(\tilde{\tau})\|_{p+1}^{p-1}\,\|J_ju(\tilde{\tau})\|_{p+1}\,d\tilde{\tau}\,d\tau,\quad t,\ s \in \mathbb{R}\,,\ 1 \leq j \leq n,\end{aligned} \tag{2.17}$$

where $C = C(n,\ p)$. Lemma 2.3, Proposition 2.4 and the assumption that $\gamma(n) < p < \alpha(n)$ imply that the integrand at the right hand side of (2.17) is integrable on $\mathbb{R} \times \mathbb{R}$. Therefore, we obtain

$$\|x_jU(-t)u(t) - x_jU(-s)u(s)\|_2 \to 0 \quad (t,\ s \to +\infty), \quad 1 \leq j \leq n. \tag{2.18}$$

In the same way we have

$$\|\nabla U(-t)u(t) - \nabla U(-s)u(s)\|_2 \to 0 \quad (t,\ s \to +\infty), \tag{2.19}$$

$$\|U(-t)u(t) - U(-s)u(s)\|_2 \to 0 \quad (t,\ s \to +\infty). \tag{2.20}$$

(2.18), (2.19) and (2.20) show (2.15).

In order to prove Theorem A, we consider the following initial value problems with the initial data $u_\pm \in \Sigma$ given at $t = \pm\infty$

$$u(t) = U(t)u_\pm - i\int_{\pm\infty}^{t} U(t-s)f(u(s))\,ds\ , \quad t \in \mathbb{R}. \tag{2.21}$$

We obtain Theorem A by evaluating (2.21) in the same way as the proof of Theorem B.

Finally we consider the continuity in Σ of the wave operators and the scattering operator. We first show the continuity in Σ of the inverse of W_+. Let $u(t)$ and $u_k(t)$, $k = 1,2,\cdots$, be the solutions of (1.1)-(1.2) such that $u(0) = u_0 \in \Sigma$, $u_k(0) = u_{0k} \in \Sigma$, $k = 1,2,\cdots$, and $u_{0k} \to u_0$ in Σ $(k \to \infty)$. We can easily prove that

$$U(-t)u_k(t) \to U(-t)u(t) \quad \text{in } \Sigma \quad (k \to \infty), \quad t \in \mathbb{R}, \tag{2.22}$$

$$\sup_{t\in\mathbb{R}} (1+|t|)^\theta \, \|u_k(t) - u(t)\|_{p+1} \to 0 \quad (k \to \infty) \tag{2.23}$$

(for the proofs of (2.22-23), see, e.g., [4, §5]). Noting Lemma 2.2, we obtain by (2.22-23) in the same way as Proposition 2.4

$$\|J_j u_k - J_j u\|_{L^r(T,\infty\,;\,L^{p+1})} \to 0 \quad (k \to \infty), \quad 1 \leq j \leq n, \tag{2.24}$$

$$\|\nabla u_k - \nabla u\|_{L^r(T,\infty\,;\,L^{p+1})} \to 0 \quad (k \to \infty) \tag{2.25}$$

for sufficiently large $T > 0$. Let $u_+ = \lim_{t\to+\infty} U(-t)u(t)$ and $u_{+k} = \lim_{t\to+\infty} U(-t)u_k(t)$. Then, we obtain by Lemma 2.2 and (2.22-25) in the same way as (2.18)

$$\|u_{+k} - u_+\|_\Sigma \to 0 \quad (k \to \infty). \tag{2.26}$$

This shows the continuity in Σ of the inverse of W_+. By using the same argument as above we can show the continuity in Σ of W_+. Therefore, W_+ is a homeomorphism from Σ onto Σ. We can similarly prove that W_- is a homeomorphism from Σ onto Σ and so that $S = W_+^{-1}W_-$ is a homeomorphism from Σ onto Σ.

(Q. E. D.)

ACKNOWLEDGEMENT. One of the authors, (Y.T.), would like to thank Professor T. Kato for his helpful suggestion.

REFERENCES

[1] J. E. Barab, Nonexistence of asymptotic free solutions for a nonlinear Schrödinger equation, J. Math. Phys., 25 (1984), 3270-3273.

[2] G. C. Dong and S. Li, On the initial value problem for a nonlinear Schrödinger equation, J. Diff. Eqs., 42 (1981), 353-365.

[3] J. Ginibre and G. Velo, On a class of nonlinear Schrödinger equations. I: The Cauchy problem, J. Funct. Anal., 32 (1979), 1-32.

[4] J. Ginibre and G. Velo, On a class of nonlinear Schrödinger equations. II: Scattering theory, J. Funct. Anal., 32 (1979), 33-71.

[5] J. Ginibre and G. Velo, Sur une équation de Schrödinger non linéaire avec interaction non locale, in "Nonlinear parial differential equations and their applications", College de France Seminair, Vol. II, Pitman, Boston, 1981.

[6] J. Ginibre and G. Velo, Scattering theory in the energy space for a class of nonlinear Schrödinger equations, J. Math. Pur. Appl., 64 (1985), 363-401.

[7] J. Ginibre and G. Velo, Time decay of finite energy solutions of the non linear Klein-Gordon and Schrödinger equations, Ann. I. H. P. (Phys. Theor.), 43 (1985), 399-442.

[8] N. Hayashi and M. Tsutsumi, $L^\infty(\mathbb{R}^n)$-decay of classical solutions for nonlinear Schrödinger equations, to appear.

[9] N. Hayashi, K. Nakamitsu and M. Tsutsumi, On solutions of the initial value problem for the nonlinear Schrödinger equations, to appear in J. Funct. Anal.

[10] W. Hunziker, On the space-time behavior of Schrödinger wavefunctions, J. Math. Phys., 7 (1966), 300-304.

[11] A. Jensen, Commutator methods and a smooth property of the Schrödinger evolution group, Math. Z., 191 (1986), 53-59.

[12] J. E. Lin and W. A. Strauss, Decay and scattering of solutions of a nonlinear Schrödinger equation, J. Funct. Anal., 30 (1978), 245-263.

[13] M. Reed. Abstract nonlinear wave equations, Lecture Notes in Math., 507, Springer-Verlag, Berlin-Heidelberg-New York, 1976.

[14] W. A. Strauss, Everywhere defined wave operators, in "Nonlinear Evolution Equations", pp. 85-102, Academic Press, New York, 1978.

[15] W. A. Strauss, Nonlinear scattering theory at low energy, J. Funct. Anal., 41 (1981), 110-133.

[16] R. S. Strichartz, Restrictions of Fourier transforms to quadratic surfaces and decay of solutions of wave equations, Duke Math. J., 44 (1977), 705-714.

[17] Y. Tsutsumi, Global existence and asymptotic behavior of solutions for nonlinear Schrödinger equations, Doctor thesis, University of Tokyo, 1985.

[18] Y. Tsutsumi, Scattering problem for nonlinear Schrödinger equations, Ann. I. H. P. (Phys. Theor.), 43 (1985), 321-347.

[19] Y. Tsutsumi and K. Yajima, The asymptotic behavior of nonlinear Schrödinger equations, Bull. (New Series) Amer. Math. Soc., 11 (1984), 186-188.

[20] K. Yajima, The surfboard Schrödinger equations, Comm. Math. Phys., 96 (1984), 349-360.

Asymptotics of Solutions and Spectra of Perturbed Periodic Hamiltonian Systems

D. B. Hinton
Mathematics Department
University of Tennessee
Knoxville, TN 37996

J. K. Shaw
Mathematics Department
Virginia Tech
Blacksburg, VA 24061

1. Introduction

We will be concerned with a 2x2 linear Hamiltonian system

(1.1) $J\vec{y}\,' = [\lambda R(x)+Q(x)]\vec{y}, \ -\infty<x<\infty, \ \vec{y} = \begin{pmatrix} y \\ \hat{y} \end{pmatrix}, \ J = \begin{pmatrix} 0 & -1 \\ 1 & 0 \end{pmatrix},$

where $R(x)$ and $Q(x)$ are real, symmetric, piecewise continuous and periodic of period 1, and an auxiliary perturbed equation

(1.2) $J\vec{y}\,' = [\lambda R(x)+Q(x)+P(x)]\vec{y}, \ -\infty<x<\infty,$

with a suitably small symmetric perturbation term $P(x)$. We take it that $R(x)$ is either positive definite or that $R(x) = \begin{pmatrix} r_1(x) & 0 \\ 0 & 0 \end{pmatrix}$, with $r_1(x)>0$. In this way (1.1) and (1.2) include both Dirac systems ([3],[6]) and scalar, ordinary differential equations ([5]).

We are going to examine conditions under which the essential spectra of the operators associated with (1.1) and (1.2) are the same; i.e., the essential spectrum of (1.1) is invariant under the perturbation $P(x)$. The usual approach in this kind of study is through some variation of the Weyl theorem on relatively compact perturbations ([9],[8],[7],[1]). In this paper, however, we present some classes of perturbations which exhibit the invariance property but to which none of the standard techniques of perturbation theory apply. Instead, we rely on asymptotics of solutions of (1.2) and the Titchmarsh-Weyl $M(\lambda)$ coefficient. Separate sets of hypotheses on $P(x)$ and the corresponding asymptotic forms of solutions will be given in §2 and §3.

The unperturbed equation (1.1), in the case where $R(x)>0$, has been studied by Harris ([3]) who showed that the spectrum of the operator $T\vec{y} = R^{-1}(J\vec{y}\,' - Q\vec{y})$, acting in the Hilbert space L_R^2 of measurable functions $\vec{f}(x)$ such that $\int_{-\infty}^{\infty} \vec{f}^*R\vec{f}dx<\infty$, consists of the stability intervals associated with the Floquet theory of (1.1). We extended this result to the case $R = \begin{pmatrix} r_1 & 0 \\ 0 & 0 \end{pmatrix}$ in [6]. In both instances the stability intervals are defined as follows. Let $Y_0(x,\lambda)$ be the fundamental matrix solution of (1.1) satisfying $Y(0,\lambda) = I$ for all complex λ. Partition Y_0 into scalar components by writing $Y_0(x,\lambda) = \begin{pmatrix} \theta_0(x,\lambda) & \varphi_0(x,\lambda) \\ \hat{\theta}_0(x,\lambda) & \hat{\varphi}_0(x,\lambda) \end{pmatrix} = [\vec{\theta}_0(x,\lambda), \vec{\varphi}_0(x,\lambda)]$ and set

$D(\lambda) = \Theta_0(1,\lambda)+\hat{\varphi}_0(1,\lambda)$. The stability intervals are those open λ-intervals such that $-2<D(\lambda)<2$. The specturm of (1.1) is then the union of the closures of the stability intervals.

2. Absolutely Integrable Case

We proved in [6] that if $P\in L^1(-\infty,\infty)$ (all entries absolutely integrable) then the continuous spectra of (1.1) and (1.2) have equal interiors. Thus the perturbation can introduce at most eigenvalues into the gaps between stability intervals. We want to briefly discuss this case now in order to introduce the notation and prepare to weaken the assumption $P\in L^1(-\infty,\infty)$. We give some remarks on how even the absolutely integrable case falls outside the range of the usual tools of perturbation theory.

Let I be an open interval whose closure is contained in the interior of one of the stability intervals of (1.1). By [6] (also compare [5]) there is a region $\Omega_I = \{\lambda=\lambda_1+i\lambda_2 \mid \lambda_1\in I, -\delta<\lambda_2<\delta, \delta>0\}$ such that λ_2 Im $D(\lambda_1+i\lambda_2)$ is of constant sign for $\lambda\in\Omega_I$. Furthermore, there is an analytic function $\alpha(\lambda)=u(\lambda)+iv(\lambda)$, $\lambda\in\Omega_I$, such that $u(\lambda)>0$ for $\lambda_2>0$, $u(\lambda)=0$ for $\lambda_2=0$ and $u(\lambda)<0$ for $\lambda_2<0$, and which yields the Floquet solutions $\vec{\Phi}_1(x,\lambda)=e^{\alpha(\lambda)x}\vec{p}_1(x,\lambda)$ and $\vec{\Phi}_2(x,\lambda)=e^{-\alpha(\lambda)x}\vec{p}_2(x,\lambda)$. The $\vec{p}_k$ are periodic in x and analytic in λ; $[\vec{\Phi}_1,\vec{\Phi}_2]$ is a fundamental matrix solution of (1.1). Not having these rather precise ways of expressing the solutions of (1.2), we use instead their asymptotics. Let $Y(x,\lambda)$ be the fundamental matrix solution of (1.2) which satisfies $Y(0,\lambda) = I$, and partition Y, as we did Y_0, by writing $Y(x,\lambda) = \begin{pmatrix}\Theta(x,\lambda) & \varphi(x,\lambda)\\ \hat{\Theta}(x,\lambda) & \hat{\varphi}(x,\lambda)\end{pmatrix} = [\vec{\Theta}(x,\lambda),\vec{\varphi}(x,\lambda)]$. By variation of parameters, and suppressing the λ-dependence,

$$Y(x) = Y_0(x) + \int_0^x K(x,t)P(t)Y(t)dt, \tag{2.1}$$
$$K(x,t) = Y_0(x)Y_0^{-1}(t)J^{-1},$$

and we clearly have $\|Y_0(x)\|\leqslant M_1(\lambda)e^{u(\lambda)x}$ and $\|K(x,t)\|\leqslant M_2(\lambda)e^{u(\lambda)(x-t)}$, where M_1 and M_2 are constants, $0\leqslant t\leqslant x$, $\lambda_2\geqslant 0$ and $\|\cdot\|$ is the matrix norm gotten by summing moduli of entries. Feeding these into (2.1) and using Gronwall's inequality yields a bound of the form $\|Y(x,\lambda)\|\leqslant K(\lambda)e^{u(\lambda)x}$, $0\leqslant x<\infty$, $\lambda_2\geqslant 0$, where $K(\lambda)$ is a constant. Based on these estimates we go on to prove in [6] that there exist row vector functions $C_1^{(+)}(\lambda)$, $C_2^{(+)}(\lambda_1)$, $C_1^{(-)}(\lambda_1)$ and $C_2^{(-)}(\lambda)$, $\lambda=\lambda_1+i\lambda_2\in\Omega_I$, $\lambda_2\geqslant 0$ such that

$$Y(x,\lambda_1)=e^{\alpha(\lambda_1)x}\vec{p}_1(x,\lambda_1)C_1^{(\pm)}(\lambda_1)+e^{-\alpha(\lambda_1)x}\vec{p}_2(x,\lambda)C_2^{(\pm)}(\lambda_1)+o(1),\ x\to\pm\infty, \tag{2.2}$$

(2.3) $Y(x,\lambda) = e^{\alpha(\lambda)x}[\vec{p}_1(x,\lambda)C_1^{(+)}(\lambda)+o(1)],\ x\to\infty,\ \lambda_2>0,$

(2.4) $Y(x,\lambda) = e^{-\alpha(\lambda)x}[\vec{p}_2(x,\lambda)C_2^{(-)}(\lambda)+o(1)],\ x\to-\infty,\ \lambda_2>0.$

These asymptotic formulas hold under the hypotheses $P\in L^1(-\infty,\infty)$. We will derive them in §3 for an alternative hypothesis.

The functions $C_1^{(+)}(\lambda) = [C_{11}^{(+)}(\lambda),C_{12}^{(+)}(\lambda)]$ and $C_2^{(-)}(\lambda) = [C_{21}^{(-)}(\lambda),C_{22}^{(-)}(\lambda)]$ do not vanish identically and are analytic in Ω_1. From the relations (2.2)-(2.4) follow the expressions $m^{(+)}(\lambda)=-C_{11}^{(+)}(\lambda)/C_{12}^{(+)}(\lambda)$ and $m^{(-)}(\lambda)=-C_{21}^{(-)}(\lambda)/C_{22}^{(-)}(\lambda)$ for the Titchmarsh-Weyl coefficients of (1.2) associated with the endpoints $\pm\infty$. These representations completely determine the spectrum of (1.2). Based on the analyticity properties of $m^{(\pm)}$ we were able to conclude in [6] that the perturbed and unperturbed spectra have equal interiors; in particular, the essential spectrum of (1.1) is invariant under the perturbation P(x).

This result is not a consequence of Weyl's classical theorm on relatively compact perturbations, or its extension to relatively form compact perturbations ([9],[8],[7],[1]). Let $Ly=-y''$, $-\infty<x<\infty$, with domain $D(L)=\{f\in L^2(-\infty,\infty)\,|\,f'$ is locally absolutely continuous and $f''\in L^2(-\infty,\infty)\}$, and put $y_0(x)=|x|^{-3/4}$, $|x|\geq 2$, and extend y_0 smoothly to $(-2,2)$ so that $y_0\in D(L)$. Define a potential term V(x) on $(-\infty,\infty)$ to be 0 except on intervals $(n-\delta_n,n)$, $n=2,3,4,\ldots$, where $V(x) = a_n$. We choose $a_n = 2^n n^{1/2}$, $\delta_n = 1/(2^{2n}n^{1/2})$ and consider the perturbed operator L+V. Let $\varphi_n(x)$ be any "characteristic" function of $[-n,n]$ which vanishes smoothly near the endpoints, and set $y_n=y_0\varphi_n$. Then $\|y_n\|_2+\|LY_n\|_2$ is uniformly bounded but $\|Vy_n\|_2\to\infty$ as $n\to\infty$. Therefore L+V is not a relatively compact perturbation. However, $V\in L^1(-\infty,\infty)$ and so the main results of [6] outlined above assert that the essential spectrum is unchanged.

3. Conditionally Integrable Perturbation

We now remove the assumption of absolute integrability of P(x) and replace it with conditions (3.1)-(3.3) below, which basically allow large-amplitude, oscillating perturbations. Our aim is to establish the asymptotic formulas (2.2)-(2.4). Once these are in hand the representations for $m^{(\pm)}(\lambda)$ follow in the same way as before, with coefficients $C_K^{(\pm)}$ which are different but have the same properties. From this point, the results from [6] on the spectra of (1.1) and (1.2) carry over because they are based completely on the representations of $m^{(\pm)}(\lambda)$. Therefore we claim invariance of the essential spectrum under the new assumptions (3.1)-(3.3).

To further simplify the presentation, we will discuss only asymptotics on the ray $0\leq x<\infty$, the dual formulas for $x\to-\infty$ being direct analogues.

Let us now suppose that P(x) satisfies

(3.1) $P(x)$ is conditionally integrable on $(0,\infty)$,

(3.2) $W(x) = \int_x^\infty P(t)dt$ is absolutely integrable on $(0,\infty)$,

(3.3) $W(x)JP(x)$ is absolutely integrable on $(0,\infty)$.

These conditions are neither linearly weaker nor stronger than $P(x)\in L^1(0,\infty)$. They are suggested by analogous conditions for scalar equations in the paper [10] of Wintner, and by a technique used in [4].

Observe that the potential $V(x)$ in the example of §2 satisfies (3.1)-(3.3). Therefore potentials satisfying (3.1)-(3.3) still do not fall under the Weyl theorem on relatively compact or relatively form compact perturbations.

We now proceed with the derivations of (2.1) and (2.2). Begin by substituting the relation $W'=-P$ from (3.2) into (2.1), and then integrating by parts, to obtain

$$Y(x)=Y_0(x)-[K(x,\cdot)WY]_0^x + \int_0^x [K_t(x,\cdot)WY + K(x,\cdot)WY']dt.$$

Since $[K(x,\cdot)WY]_0^x = J^{-1}WY - Y_0J^{-1}W(0)$ and $K_t(x,t) = -Y_0(x)Y_0^{-1}(t)J^{-1}(\lambda R(t)+Q(t))J^{-1}$ then the previous equation yields after rearranging and multiplying by e^{-ux} that

$$Y_1(x)=e^{-ux}Y_0(x)[I+J^{-1}w(0)]+\int_0^x e^{-u(x-t)}K(x,t)W_1(t)Y_1(t)dt,$$

$Y_1(x)=[I+J^{-1}W(x)]e^{-ux}Y(x)$, $W_1=[WJ^{-1}(\lambda R+Q+P)-(\lambda R+Q)J^{-1}W](I+J^{-1}W)^{-1}$. The term $I+J^{-1}W(x)$ is invertible for x sufficiently large because $W(x)\to 0$. Using the bounds given on $\|Y_0\|$ and $\|K\|$ after (2.1) we get $\|Y_1(x)\|\leqslant M_1C+\int_0^x M_2\|W_1\|\|Y_1\|dt$, where $\|I+J^{-1}W(0)\|\leqslant C$, and so by Gronwall's inequality $\|Y_1(x)\| \leqslant M_1C\exp(M_2\int_0^x\|w_1\|dt)$. By (3.2) and (3.3), $\|Y_1(x)\|$ is bounded, and this means

(3.4) $\|Y(x,\lambda)\|\leqslant M_3(\lambda)e^{u(\lambda)x}$, $\lambda=\lambda_1+i\lambda_2\in\Omega_1, \lambda_2\geqslant 0, x\geqslant 0$,

where M_3 is a constant.

Let $\Phi(x,\lambda)=[\vec{\phi}_1(x,\lambda),\vec{\phi}_2(x,\lambda)]$ be the fundamental matrix of Floquet solutions. By comparing initial values with $Y_0(x,\lambda)$, one finds that these fundamental matrices are related by

$$\Delta(\lambda)Y_0(x,\lambda)=\Phi(x,\lambda)\begin{pmatrix}\hat{p}_2(0,\lambda) & -p_2(0,\lambda)\\ -\hat{p}_1(0,\lambda) & p_1(0,\lambda)\end{pmatrix}, \tag{3.5}$$

where $\Delta(\lambda)=p_1(0,\lambda)\hat{p}_2(0,\lambda)-\hat{p}_1(0,\lambda)p_2(0,\lambda)$ (see [6]). The basis for (2.2)-(2.3) is then obtained by placing (3.5) into (2.1), which yields

$$\Delta(\lambda)Y(x,\lambda)=\Phi(x,\lambda)\begin{pmatrix}\hat{p}_2(0,\lambda) & -p_2(0,\lambda)\\ -\hat{p}_1(0,\lambda) & p_1(0,\lambda)\end{pmatrix} \cdot [I+\int_0^x Y_0^{-1}(t,\lambda)J^{-1}P(t)Y(t,\lambda)dt]. \tag{3.6}$$

The proof of (2.2) is accomplished by setting $\lambda=\lambda_1$ in (3.6) and showing $\int_0^\infty Y_0^{-1}J^{-1}PYdt$ converges.

We proceed, as in the derivation of (3.4), to integrate by parts and obtain $\int_0^x Y_0^{-1}J^{-1}PYdt = J^{-1}W(0)-Y_0^{-1}J^{-1}WY+\int_0^x[Y_0^{-1}J^{-1}WJ^{-1}(\lambda R+Q+P)Y-Y_0^{-1}J^{-1}(\lambda R+Q)J^{-1}WY]$.

The matrix Y_0^{-1} can be computed from $Y_0^*JY_0=J$ (see [2]) so that it and also $Y(x,\lambda_1)$, by (3.4) with $\lambda_2=0$, are bounded. Hence absolute integrability of $Y_0^{-1}J^{-1}PY$ follows from (3.2) and (3.3). The bracketed term in (3.6) can be written $I+\int_0^\infty Y_0^{-1}J^{-1}PY-\int_x^\infty Y_0^{-1}J^{-1}PY=B(\lambda_1)-\int_x^\infty Y_0^{-1}J^{-1}PY$ where $\lambda=\lambda_1$. Therefore (2.2) can be read off from (3.6) with appropriate choices of $C_1^{(+)}$ and $C_2^{(+)}$.

Specifically, $C_1^{(+)}(\lambda_1) = \Delta^{-1}(\lambda_1)[\hat{p}_2(0,\lambda_1,-p_2(0,\lambda_1)]B(\lambda_1)$, $C_2^{(+)}(\lambda_1) = \Delta(\lambda_1)^{-1}[-\hat{p}_1(0,\lambda_1),p_1(0,\lambda_1)]B(\lambda_1)$, and the o(1) in (2.2) comes from $\int_x^\infty Y_0^{-1}J^{-1}PY$. Invertibility of $\Delta(\lambda_1)$ for $\lambda_1\in I$ is a fact from the Floquet theory of (1.1); see [3] or [6].

There remains only to prove (2.3). By (3.5) we have $\vec{\psi}_2(x,\lambda)=Y_0(x,\lambda)\vec{p}_2(0,\lambda)$. This can be used in the expression defining $B(\lambda_1)$ to derive

(3.7) $\quad \Delta(\lambda_1)C_1^{(+)}(\lambda_1)=[\hat{p}_2(0,\lambda_1),-p_2(0,\lambda_1)]+\int_0^\infty \vec{\psi}^T{}_2(t,\lambda_1)P(t)Y(t)dt,$

where T denotes the ordinary transpose. We will now show that the right side of (3.7) continues to exist for $\lambda=\lambda_1+i\lambda_2\in\Omega_I,\lambda_2>0$, and then defining $C_1^{(+)}(\lambda)$ by the expanded expression we will derive (2.3). Integrating by parts as before,

$$\int_0^x\vec{\psi}_2^TPYdt=-[\vec{\psi}_2^TWY]_0^x+\int_0^x\vec{\psi}_2^T[(\lambda R+Q)^TJW+WJ^{-1}(\lambda R+Q+P)]Ydt,$$

and since $\|\vec{\psi}_2^T(x,\lambda)\|\leqslant K(\lambda)e^{-u(\lambda)x}$ for a constant $K(\lambda)>0$, (3.2) and (3.3) imply that $\int_0^\infty\vec{\psi}_2^T PYdt$ converges absolutely for $\lambda_2\geqslant 0$.

Placing (3.5) into the variation of parameters formula (2.1), and using the fact that $\int_0^\infty\vec{\psi}_2^TPY$ converges we have (suppressing the λ-dependence)

(3.8) $\quad \Delta Y(x)=e^{\alpha x}\vec{p}_1(x)\{[\hat{p}_2(0),-p_2(0)]+\int_0^\infty\vec{\psi}_2^TPY-\int_x^\infty e^{-\alpha t}\vec{p}_2^TPY\}$

$\quad -e^{-\alpha x}\vec{p}_2(x)[\hat{p}_1(0),-p_1(0)]-\int_0^x e^{-\alpha(x-t)}\vec{p}_2(x)\vec{p}_1^T(t)P(t)Y(t)dt.$

The term $e^{-\alpha x}\vec{p}_2(x)[\hat{p}_1(0),-p(0)]=e^{\alpha x}H(x)$, where $H(x)\to 0$, and so this expression can go into the o(1) term of (2.3). The same is true for the integral term $\int_x^\infty(*)$ of (3.8). Comparing (3.7) and (2.3), we will be finished if we show that the integral term $\int_0^x(*)$ in (3.8) is a o(1) term in (2.3). Proceeding as before, this term equals

(3.9) $\quad -e^{-\alpha x}[\vec{\psi}_1^TWY]_0^x+e^{-\alpha x}\int_0^x[(J^{-1}(\lambda R+Q)\vec{\psi}_1)^TWY-\vec{\psi}_1^TWJ^{-1}(\lambda R+Q+P)Y]dt.$

By (3.4) and (3.2), the first term is $o(e^{\alpha x})$, $x\to\infty$. The second term in (3.9), by (3.4) and the definition of $\vec{\psi}_1$, has order of magnitude $\int_0^x e^{u(2t-x)}\|\vec{p}_1\|(\|\lambda R+Q\|\|W\|+\|WJ^{-1}(\lambda R+Q+P)\|)$. Break the integral up into ranges $0\leqslant t\leqslant(x/2)$ and $(x/2)\leqslant t\leqslant x$, so that the preceeding expression has order of magnitude

$$\int_0^{x/2}\|W\|+\int_0^{x/2}(\|W\|+\|WJP\|)+e^{ux}\int_{x/2}^{x}\|W\|+e^{ux}\int_{x/2}^{x}(\|W\|+\|WJP\|)=o(e^{\alpha x}),x\to\infty.$$

REFERENCES

1. N.I. Akhiezer and I.M. Glazman, "Theory of Linear Operators in Hilbert Space", Vol. II, Pitman, Boston, 1980.

2. F.V. Atkinson, "Discrete and Continuous Boundary Problems", Academic Press, New York, 1964.

3. B.J. Harris, On the spectra and stability of periodic differential equations, Proc. London Math. Soc. (3), 41 (1980), 161-192.

4. D.B. Hinton and J.K. Shaw, Absolutely continuous spectra of Dirac systems with long range, short range and oscillating potentials, Quart. J. Math. Oxford (2), 36(1985), 183-213.

5. D.B. Hinton and J.K. Shaw, On the absolutely continuous spectrum of the perturbed Hill's equation", Proc. London Math. Soc. (3), 50(1985), 175-192.

6. D.B. Hinton and J.K. Shaw, Absolutely continuous spectra of perturbed periodic Hamiltoniam systems, to appear in Rocky Mtn. J. Math.

7. E. Müller-Pfeiffer, "Spectral Theory of Ordinary Differential Equations", Ellis Horwood Limited, Chichester, 1981.

8. M. Reed and B. Simon, "Methods of Modern Mathematical Physics", Vol. IV, Academic Press, New York, 1978.

9. J. Weidmann, "Linear Operators in Hilbert Spaces", Springer-Verlag, New York, 1980.

10. A. Wintner, Addenda to the paper on Bôcher's Theorem, Amer. J. Math., 78(1956), 895-897.

BEHAVIOR OF EIGENFUNCTIONS AND THE SPECTRUM OF SCHRÖDINGER OPERATORS

Andreas M. Hinz
Mathematisches Institut
Universität München
D-8000 München 2

0. Introduction

The eigenfunction asymptotics and the spectrum of a Schrödinger operator $-\Delta+q$ are inter-related. Some of these relations will be discussed here.

Eigenfunction asymptotics: Let the potential q satisfy the following condition

(a) $q\in L_{2,loc}(\mathbb{R}^n)$ $(n\geq 2)$ real-valued, and for any compact $U\subset\mathbb{R}^n$ there are positive numbers ν and c such that

$$\forall x\in\mathbb{R}^n: \quad \int_{B(x,1)\cap U} \frac{|q(y)|^2}{|x-y|^{n+\nu-4}}\, dy \leq c .$$

(In particular, this is fulfilled if q is locally bounded.)

If for a $\lambda\in\mathbb{R}$ we have a distributional solution $v\in L_{2,loc}(\mathbb{R}^n)$ of the equation $-\Delta v+qv=\lambda v$, then we know by Lemma 1 in [2] that v is continuous, and we can consider the pointwise asymptotic behavior of v near ∞.

Spectrum: If we assume in addition

(b) $\exists R_0,\beta>0\ \forall|x|\geq R_0: q(x)\geq-\beta|x|^2$,

then $-\Delta+q$ defined on $C_0^\infty(\mathbb{R}^n)$ is essentially self-adjoint in $L_2(\mathbb{R}^n)$ (Satz 7 in [5]), and we may ask for the position of λ with respect to the different parts of the spectrum of $\overline{-\Delta+q}$.

An immediate prototype of a connection between the asymptotics of solutions of $-\Delta v+qv=\lambda v$ and the spectrum of $\overline{-\Delta+q}$ is

$$\exists \text{ non-trivial solution } v(x)=O(|x|^{-(n+\mu)/2}) \text{ for a } \mu>0 \Rightarrow \lambda\in\sigma_p(\overline{-\Delta+q})\subset\sigma(\overline{-\Delta+q}) \quad (1).$$

(The order of growth or decay is always meant near ∞, not at 0!)

In what follows, we will assume q to satisfy conditions (a) and (b).

1. Bounds on eigenfunctions

The first interesting question is a kind of inversion of (1): Do L_2-eigenfunctions decay? The answer is "yes", if q is bounded from below (as proved by Shnol' [4]), or if q_-, the negative part of q, lies in K_n (Simon [6]); then any L_2-eigenfunction v tends to 0 pointwise for $|x|\to\infty$. This is, in general, no longer true if q is unbounded, as can be seen by the following example: Define a positive function v having ever increasing peaks centered at the points $x_k := (k,0,\ldots,0)$, $k\in\mathbb{N}$, but such that v is still in $L_2(\mathbb{R}^n)$.

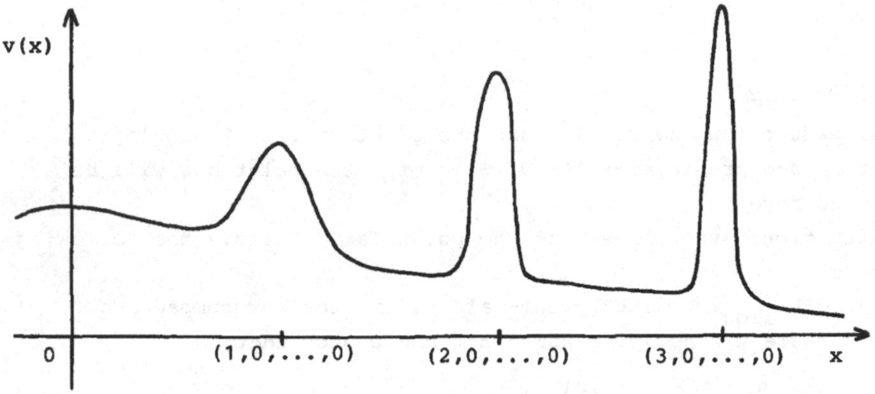

It can be arranged that $v(x_k) \geq |x_k|^{(n\gamma-\varepsilon)/2}$ and $q(x) := \frac{\Delta v(x)}{v(x)} \geq -\beta|x|^{2\gamma}$ ($0<\gamma\leq 1$, $\varepsilon>0$ arbitrarily small; see [2], Example 1, for details). Here $0\in\sigma_p(\overline{-\Delta+q})$, but v is not even bounded.

However, we have

<u>Theorem 1</u>. If $\lambda\in\sigma_p(\overline{-\Delta+q})$, then for all solutions $v\in L_2(\mathbb{R}^n)$ of $-\Delta v+qv=\lambda v$, $v(x)=o(|x|^{n/2})$.

<u>Proof</u>: By Kato's inequality and our assumption (b), we have outside a compact set:

$$\Delta|v| \geq \operatorname{sign} v\, \Delta v = (q-\lambda)|v| \geq -(\beta+|\lambda|)\,|\cdot|^2\,|v| \ .$$

This allows us to deduce a mean value inequality (Lemma 2 in [2]) for $|v|$, which reads (for some $R_1>0$ and $c>0$):

$$\forall |x|\geq R_1: \ |v(x)|^2 \leq c|x|^n \int_{B(x,1/2|x|)} |v(y)|^2\, dy \ .$$

The integral on the right-hand side tends to 0 as $|x|\to\infty$, because $v\in L_2$ ■

The above example (put $\gamma=1$) shows that the estimate of Theorem 1 is the best we can expect.

2. Exponential decay of eigenfunctions

If $\lambda\in\sigma_d(\overline{-\Delta+q})$, we expect the L_2-eigenfunctions to decay exponentially. The extensive literature on this subject is summarized, at least partially, in [2]. For potentials q which are bounded from below, Shnol' proved in [4] for every L_2-eigenfunction : $|v(x)|\le ce^{-a|x|}$ with an a depending on $\operatorname{dist}(\lambda,\sigma_e)$. The same conclusion holds for $q_-\in K_n$ (Simon [6]).

Let us define

$$v(x)\to 0 \text{ exponentially } :\Leftrightarrow \forall\delta>0:\ v(x)=O(|x|^{-\delta}) \ .$$

Then we have

<u>Theorem 2</u>. Let $\lambda\in\sigma_d(\overline{-\Delta+q})$.

i) If $q(x)\ge -o(|x|^2)$ or $\sigma(\overline{-\Delta+q})=\sigma_d(\overline{-\Delta+q})$, then for all solutions $v\in L_2(\mathbb{R}^n)$ of $-\Delta v+qv=\lambda v$, $v(x)\to 0$ exponentially;

ii) otherwise for any solution $v\in L_2(\mathbb{R}^n)$ of $-\Delta v+qv=\lambda v$ there is a $\mu>0$ such that
$$v(x)=O(|x|^{(n-\mu)/2}) \ .$$

<u>Proof</u>: Let $\beta>0$ such that $q(x)\ge-\beta|x|^2$ outside a ball, and assume $(1+|x|)^{(\mu-n)/2}v(x)$ to be unbounded for a $\mu>0$. Then, by a mean value argument as above, $r^\mu \int_{r\le|y|} |v(y)|^2 dy$ is also unbounded.

Cutting off v around the origin by a smooth function $\Theta_R(x):=\Theta(\frac{|x|}{R}-1)$ $(R>0)$, $\Theta(r)$ being 0 for $r\le 0$ and 1 for $r\ge 1$, we arrive at a sequence $v_R:=\Theta_R v$ with

$$\| (\overline{-\Delta+q}-\lambda)v_R \|^2 \le 2\,\|v\Delta\Theta_R\|^2 + 8\,\|\nabla v\cdot\nabla\Theta_R\|^2 \ .$$

We can cope with the last term, if we apply the distributional identity $\Delta(v^2)=2\Delta vv+2|\nabla v|^2$ to the testing function $|\nabla\Theta_R|^2$.

This leads to (R large enough):

$$\| (\overline{-\Delta+q}-\lambda)v_R \|^2 / \|v_R\|^2 \le C\beta\, \frac{\int_{R\le|y|\le 2R} |v(y)|^2\, dy}{\int_{2R\le|y|} |v(y)|^2\, dy} \ ,$$

where C depends on the shape of Θ only.

By construction of our sequence v_R, an upper bound for the $\liminf_{R\to\infty}$ of the right-hand side is also an upper bound for the square of the distance of λ to the essential spectrum of $\overline{-\Delta+q}$ (see Lemma 1 in [3]). To get such an estimate, we define for R fixed

$$b := \inf_{r\geq R} \frac{\int_{r\leq|y|\leq 2r} |v(y)|^2\,dy}{\int_{2r\leq|y|} |v(y)|^2\,dy} + 1 .$$

An iteration process yields

$$\forall k\in\mathbb{N}: \quad \int_{2^kR\leq|y|} |v(y)|^2\,dy \leq b^{-k} \int_{R\leq|y|} |v(y)|^2\,dy .$$

With $r_k:=2^kR$, we get $\forall k\in\mathbb{N}\ \forall r_k\leq r\leq r_{k+1}$:

$$r^\mu \int_{r\leq|y|} |v(y)|^2\,dy \leq 2^\mu r_k^{\ \mu} \int_{r_k\leq|y|} |v(y)|^2\,dy \leq (2R)^\mu \left(\frac{2^\mu}{b}\right)^k \int_{R\leq|y|} |v(y)|^2\,dy .$$

As the left side is unbounded for $k\to\infty$, we must have $b<2^\mu$.
So we arrive at

$$\operatorname{dist}(\lambda,\sigma_e(\overline{-\Delta+q}))^2 \leq C\beta(2^\mu-1) ,$$

and the different cases of Theorem 2 can be treated immediately ∎

For an application of Theorem 2, we go back to the example in 1. If we take for instance $\gamma=\frac{1}{2}$, we see that $0\in\sigma_p(\overline{-\Delta+q})\cap\sigma_e(\overline{-\Delta+q})$. More quantitative results can be found in [3].

3. Bounded solutions and the spectrum

It is an old question, whether the existence of a bounded non-trivial solution v of $-\Delta v+qv=\lambda v$ implies $\lambda\in\sigma(\overline{-\Delta+q})$. Again the affirmative answer has been given by Shnol' [4] for q bounded from below and by Simon [6] for $q_-\in K_n$. But we also get

<u>Theorem 3</u>. Let $q(x)\geq -o(|x|^2)$.
If for a $\lambda\in\mathbb{R}$ there is a (polynomially) bounded solution $v\neq 0$ of $-\Delta v+qv=\lambda v$, then $\lambda\in\sigma(\overline{-\Delta+q})$.

<u>Proof</u>: Assuming without loss that $v\notin L_2(\mathbb{R}^n)$, we construct as in the proof of Theorem 2, but using $1-\Theta_R$ as cut-off function, upper bounds on $\operatorname{dist}(\lambda,\sigma_e)$, this time in terms of

$$\frac{\int_{R\leq|y|\leq 2R} |v(y)|^2\,dy}{\int_{|y|\leq R} |v(y)|^2\,dy} .$$

If for a $\mu>0$ we suppose $(1+|\cdot|)^{-\mu/2}v\in L_2(\mathbb{R}^n)$, then again

$$\operatorname{dist}(\lambda,\sigma_e(\overline{-\Delta+q}))^2 \leq C\beta(2^\mu-1) \ ∎$$

Using the same method in the general case, we can only state

Theorem 4. If for a $\lambda\in\mathbb{R}$ there is a (polynomially) bounded solution $v\notin L_2(\mathbb{R}^n)$ of $-\Delta v+qv=\lambda v$, then $\sigma_e(\overline{-\Delta+q})\neq\emptyset$.

Furthermore:

$$\exists \text{ non-trivial solution } v(x)=O(|x|^{-n/2}) \rightarrow \lambda\in\sigma(\overline{-\Delta+q}) \quad (2).$$

Although (2) seems to be a very weak statement, especially if compared with (1), an example of Halvorsen [1] for one dimension indicates that for a certain q there may be a bounded solution $v\neq 0$ of $-\Delta v+qv=\lambda v$, but with the λ not in the spectrum of $\overline{-\Delta+q}$!

References

1. Halvorsen, S.G.: Counterexamples in the spectral theory of singular Sturm-Liouville operators. In: Ordinary and Partial Differential Equations, Lecture Notes in Mathematics 415, pp. 373-382. Berlin: Springer 1974.
2. Hinz, A.M.: Pointwise bounds for solutions of the equation $-\Delta v+pv=0$. J. Reine Angew. Math. 370, 83-100(1986).
3. Hinz, A.M.: Asymptotic behavior of solutions of $-\Delta v+qv=\lambda v$ and the distance of λ to the essential spectrum. Math. Z., to appear.
4. Shnol', Ė.Ė.: O povedenii sobstvennykh funktsiĭ uravneniya Shredingera. Mat. Sb. 42, 273-286(1957); erratum 46, 259(1958).
5. Simader, C.G.: Bemerkungen über Schrödinger-Operatoren mit stark singulären Potentialen. Math. Z. 138, 53-70(1974).
6. Simon, B.: Schrödinger Semigroups. Bull. Amer. Math. Soc. 7, 447-526(1982).

Shape Resonances in Quantum Mechanics

by

P.D. Hislop[1] and I.M. Sigal[*1]
Mathematics Department
University of California
Irvine, Calfornia 92717 USA

ABSTRACT

We prove the existence of shape resonances for Schrödinger operators of the form $H(\lambda) = -\Delta + \lambda^2 V + U$, $\lambda = \hbar^{-1}$, in the semiclassical limit in any number of dimensions. The potential V is non-negative, vanishing at infinity as $O(|x|^{-\alpha}), \alpha > 0$, and forms a barrier about a compact region in which V has finitely many zeros. $U \in L^2_{loc}$ is any real function which is bounded above and continuous except at a finite number of points. In addition, V and U are assumed to be dilation analytic in a neighborhood of infinity. The resonances shown to exist correspond as $\lambda \to \infty$ to the eigenvalues of a particle confined to the region containing the zeros of V. The width of a resonance near one of these eigenvalues λE is proved to be bounded above by $c \exp(-2\beta(\lambda)(\rho_E - \varepsilon))$, for any $\varepsilon > 0$ and where $c > 0$ is a constant. $\beta(\lambda)$ depends upon α and is given by λ for $\alpha > 2$, $\lambda \ln \lambda$ for $\alpha = 2$, and $\lambda^{1/\alpha + 1/2}$ for $0 < \alpha < 2$. The factor ρ_E satisfies $\lim_{\lambda \to \infty} \rho_E(\lambda) < \infty$, and $\beta(\lambda)\rho_E$ is the leading asymptotic to the geodesic distance in the Agmon metric $ds^2 = (\lambda^2 V + U - \lambda E)_+ dx^2$ between the turning surfaces given by $\{x | \lambda^2 V + U = \lambda E\}$.

---

*supported in part by US NSF Grant No. DMS-8507040.
[1]present address: Mathematics Department, University of Toronto, Toronto, Canada M5S1A1.
For the Proceedings of the International Conference on Differential Equations and Mathematical Physics, University of Alabama, Birmingham, March, 1986.

1. Introduction:

We describe new results on the existence and location of spectral resonances of the Schrödinger operator

$$H(\lambda) = -\Delta + \lambda^2 V + U \quad \text{on} \quad L^2(\mathbb{R}^n)$$

in the semi-classical limit $\lambda \to \infty$. The potential V satisfies the following conditions:

(V1) $V \geq 0$, $\lim_{|x|\to\infty} V(x) = 0$, and $V \in C^3(\mathbb{R}^n)$;

(V2) V has finitely many, non-degenerate zeros located at x_k ; non-degeneracy means that the Hessian of V satisfies:

$$A_{ij} = (\partial^2 V/\partial x_i \partial x_j)|_{x=x_k} > 0 ;$$

(V3) $\exists\, R >> 0$ such that V is dilation analytic in $\mathbb{R}^n \backslash B_R(0)$; specifically we assume $\exists$ bounded functions $f_i : \mathbb{R}^n \to \mathbb{R}_+$, homogeneous of degree $-\alpha_i$, $\alpha_i > 0$, such that

$$V(x) = \sum_{i=1}^{k} f_i(x) \quad \text{for} \quad |x| \geq R$$

and $0 < \delta_0 \leq \sup(|x|^{\alpha_i} f_i(x)) < \delta_1$, for constants $\delta_0, \delta_1 > 0$.

The potential U is a real-valued function which is bounded above and satisfies:

(U1) $U \in L^2_{loc}(\mathbb{R}^n)$ with a finite number of singularities located at y_i ;

(U2) $U \in C(\mathbb{R}^n \backslash \{y_i\})$ and $\lim_{|x|\to\infty} U(x) = 0$;

(U3) U is U_θ^λ- analytic for all λ suitably large where U_θ^λ is defined in Section 3B below.

The form of V in (V3) is stronger than necessary but convenient for the calculations. We let $\alpha = \min \alpha_i$ and we will distinguish the case of a short-range tail: $\alpha > 2$, and a long-range tail: $\alpha \leq 2$. For simplicity, we assume that V has a single non-degenerate zero at $x_0 = 0$.

The potential V has a barrier which separates an attractive potential well about $x_0 = 0$ from the outside region (see Figure 2.1). Any state with energy below the top of the barrier and localized in the well will eventually decay due to quantum mechanical tunneling through the barrier. Such states correspond to spectral resonances of $H(\lambda)$ which are called shape resonances. Models of this type were introduced in quantum mechanics in 1928 to describe alpha decay of nuclei [1].

The spectral resonances of $H(\lambda)$ are complex eigenvalues of an analytic family of non-self-adjoint operators $H(\lambda,\theta)$ constructed from $H(\lambda)$ by analytically continuing $U_\theta H(\lambda) U_\theta^{-1}$ where $U_\theta, \theta \in \mathbf{R}$, is a unitary group; see [2] - [3] for a discussion of this approach. It is a standard result [2] - [6] that the spectral resonance are independent of $\theta \in \mathbf{C}$ provided they remain isolated from $\sigma_{ess}(H(\lambda,\theta))$. Moreover, they are independent of the group used provided the groups have a common dense domain of analytic vectors and a certain spectral condition is satisfied. Consequently, the unitary group may be chosen according to the problem at hand.

To connect these resonances with decay phenomena, it is known [2] - [6] that they coincide with the poles of the meromorphic continuation of the resolvent of $H(\lambda)$ across its continuous spectrum. Furthermore, whenever the S-matrix for $H(\lambda)$ exists, it has been shown [7], [8] that these resonances coincide with the poles of its meromorphic continuation.

Previous work on shape resonances has been restricted to the one-dimensional case or the three-dimensional case with spherically symmetric potentials [9]. We mention recent independent work of Helffer and Sjöstrand [10] and of Combes, et al, [11] in which the shape resonances are studied. These authors show the existence of the shape resonances in the non-threshold case which would correspond to $V(x_i) > 0$ in our condition (V2).

2. The Results.

Let $H(\lambda) = -\Delta + \lambda^2 V + U$ with V satisfying (V1) - (V3) and U satisfying (U1) - (U3). Note that for all λ sufficiently large, $\sigma(H(\lambda)) \subset [0,\infty)$ (except possibly for $n = 1,2$). Let K be a comparison harmonic oscillator Hamiltonian:

$$K = -\Delta + \sum_{ij} A_{ij} x_i x_j$$

where A_{ij} is defined in (V2). The eigenvalues of K are denoted by e_n. For λ large, a particle localized in the well should have energy levels approximately given by λe_n and hence the real parts of the resonance energies should be near λe_n. Fix n such that $B_R(0)$ is contained in the set bounded by $\{x | V(x) + U(x) - e_n = 0\}$.

Result 1: Existence of Resonances

<u>For all</u> λ <u>suitably large,</u> $H(\lambda)$ <u>has spectral resonances with energy</u> $z_{n(i)}(\lambda)$ <u>satisfying</u>

$$\lim_{\lambda \to \infty} \lambda^{-1} \mathrm{Re}\, z_{n(i)}(\lambda) = e_n$$

<u>and the total multiplicity of the</u> $z_{n(i)}(\lambda)$ <u>is greater than or equal to that of</u> e_n.

Remark 2.1. In certain situations, it can be shown that the multiplicity of the resonance eigenvalues equals the multiplicity of e_n.

To describe the second main result, we must introduce some geometric quantities associated with the potential $\lambda^2 V + U$ (see Figure 2.1). Let $S_n(\lambda)$ be the classical turning surface for potential $\lambda^2 V + U$ at energy λe_n:

$$S_n(\lambda) = \{x | \lambda^2 V(x) + U(x) = \lambda e_n\} .$$

We assume that $S_n(\lambda)$ is the union of two disjoint regular surfaces $S_n^{\pm}(\lambda)$. Let $F_n(\lambda)$ be the region bounded by $S_n^-(\lambda)$ and $S_n^+(\lambda)$; $\lambda^2 V(x) + U(x) - \lambda e_n > 0$ for $x \in F_n(\lambda)$. We will occassionally omit writing the explicit dependence of F_n , etc., on λ .

Figure 2.1 Profile of a typical potential $\lambda^2 V + U$ with a partition of unity

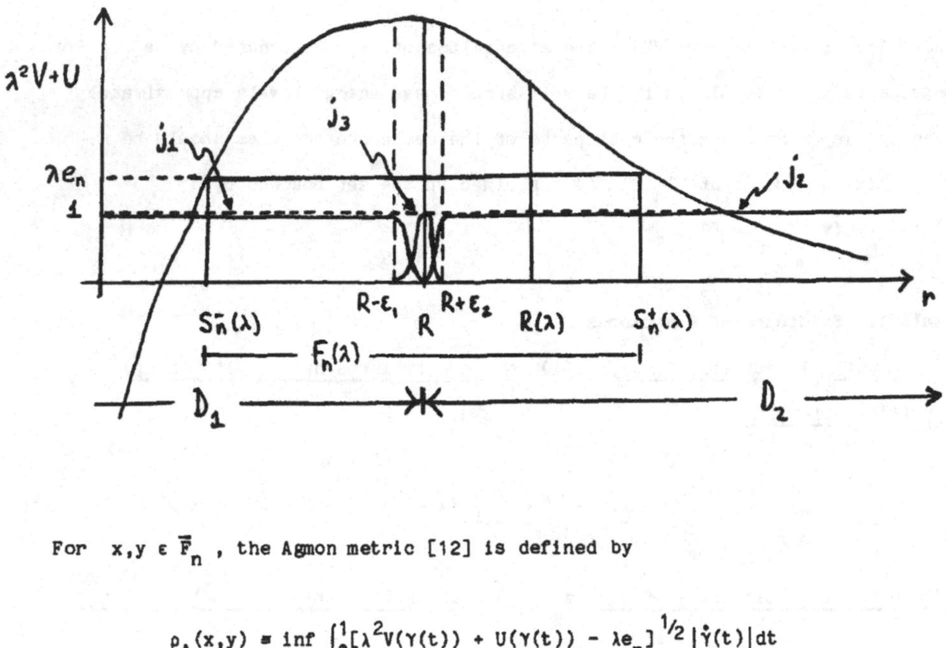

For $x, y \in \bar{F}_n$, the Agmon metric [12] is defined by

$$\rho_A(x,y) = \inf_{\gamma \in \mathcal{P}} \int_0^1 [\lambda^2 V(\gamma(t)) + U(\gamma(t)) - \lambda e_n]^{1/2} |\dot{\gamma}(t)| dt$$

where $\mathcal{P}$ is the set of all absolutely continuous paths in $\bar{F}_n$ with $\gamma(0) = x$ and $\gamma(1) = y$. Let

$$\rho_n(x) = \min_{y \in S_n^-} \rho_A(x,y) \ , \ x \in \bar{F}_n \qquad (2.1)$$

be the distance from S_n^- .

Note that $\rho_A(x,y)$ is the geodesic distance between x and y in the metric $ds^2 = (\lambda^2V+U-\lambda e_n)_+ dx^2$. We obtain the following result in most situations, see Section 5 for the details.

Result 2: Width of the Resonance

There exists a constant $c_n > 0$ such that for all λ suitably large and any $\varepsilon > 0$,

$$\Gamma_n \equiv |\mathrm{Im}\, z_n(\lambda)| \leq c_n e^{-2\beta(\lambda)(\rho_n-\varepsilon)}$$

where $\beta(\lambda) = \lambda$ for $\alpha > 2$, $\lambda \ln \lambda$ for $\alpha = 2$, $\lambda^{1/\alpha+1/2}$ for $0 < \alpha < 2$, and $\beta(\lambda)\rho_n$ is the leading asymptotic to the geodesic distance in the Agmon metric between the surfaces S_n^- and S_n^+ .

The factor ρ_n satisfies $\lim\limits_{\lambda\to\infty} \rho_n < \infty$. In the short-range case, $\alpha > 2$, we obtain the expected result: $\rho_n = \min\limits_{x\varepsilon S_n^+} \lambda^{-1}\rho_n(x)$, where $\rho_n(x)$ is given in (2.1). In the long-range case, $0 < \alpha \leq 2$, $\lambda^{-1}\rho_n(x)$ diverges as $\lambda \to \infty$. For $0 < \alpha < 2$, ρ_n is given asymptotically as $\lambda \to \infty$ by the distance from the origin to the surface $\{x|V_T(x) = e_n\}$ in the metric $ds^2 = (V_T(x)-e_n)dx^2$, where $V_T(x) = c_\alpha\delta_o|x|^{-\alpha}$. For $\alpha = 2$, the asymptotic form of ρ_n is $(1/2)(c_\alpha\delta_o)^{1/2}$ where c_α is the coefficient of $|x|^{-\alpha}$ and δ_o is given in (V3).

In the one-dimensional case and the three-dimensional case with a spherically symmetric potential, this result for $\alpha > 2$ reduces approximately to the one obtained using heuristic arguments with the JWKB approximation [13]. In these situations, the geodesics in the Agmon metric are given by straight lines.

3. Approximate and Distorted Hamiltonians and their Spectra

For simplicity, we will outline the proofs of Results (1) and (2) in the case when $U = 0$ and the potential V is short-range, i.e., $\alpha > 2$. The case $U \neq 0$ can be obtained from the proofs sketched here using standard arguments. The existence proof outlined here works for $0 < \alpha \leq 2$ cases but the method must be modified to obtain the exponential bounds of Result (2).

A. The Perturbation Problem

We define an approximate Hamiltonian $H_0(\lambda)$ by decoupling the attractive well region from the outside region using a partition of unity. Let $\{j_i\}_{i=1}^3$ be a C^∞

partition of unity with $0 \le j_i \le 1$ and supports chosen as follows. Let R be given in (V3) and let $R(\lambda) = \lambda^{1/\alpha}(\ln \lambda)^{-1}R$ with $\alpha > 2$. Let $D_1 = B_R(0)$ and $D_2 = \mathrm{Int}(\mathbb{R}^3 \setminus D_1)$. We take $\mathrm{supp}(j_1) = \bar{D}_1$ and $\mathrm{supp}(j_2) = \bar{D}_2$ so that $j_1(x) = 0 = j_2(x)$ for $|x| = R$. Let $\varepsilon_1, \varepsilon_2 > 0$ and choose j_3 such that $\mathrm{supp}(j_3) \subset B_{R+\varepsilon_2}(0) \setminus B_{R-\varepsilon_1}(0)$ and choose ε_1, ε_2 small enough such that $\mathrm{supp}(j_3) \subset F_n(\lambda)$ for all large λ. Finally, we normalize the partition of unity by requiring $\sum_{i=1}^{3} j_i^2 = 1$. The turning surface $S_n^+(\lambda)$ is $O(\lambda^{1/\alpha})$ so $B_{R(\lambda)}(0) \setminus B_{R-\varepsilon_1}(0) \subset F_n(\lambda)$ for all $\lambda > 1$ (see Figure 2.1).

By the IMS localization formula [14] - [16], we have:

$$H(\lambda) = \sum_{i=1}^{3} (j_i p^2 j_i - |\nabla j_i|^2) + \lambda^2 V$$
$$\equiv H_0(\lambda) + W$$

where

$$H_0(\lambda) \equiv H_{01}(\lambda) \oplus H_{02}(\lambda)$$

is the approximate Hamiltonian. Each term is given by

$$H_{0i}(\lambda) \equiv j_i p^2 j_i + \lambda^2 V \chi_{D_i} - J \chi_{D_i}$$

acting on $L^2(D_i)$ where χ_{D_i} is the characteristic function for D_i, $i = 1,2$, and $J \equiv \sum_{i=1}^{3} |\nabla j_i|^2$. W is the localized perturbation:

$$W \equiv j_3 p^2 j_3 .$$

Our first result characterizes the spectrum of $H(\lambda)$ and $H_0(\lambda)$:

Lemma 3.1.

(i) $\sigma_{ess}(H_{02}(\lambda)) = [0,\infty)$, $\sigma_{sc}(H_{02}(\lambda)) = \phi$, <u>and if</u> $e(\lambda) \in \sigma_{pp}(H_{02}(\lambda))$ <u>then</u> $e(\lambda) = O(\lambda^2)$;

(ii) $\sigma_{ess}(H_{01}(\lambda)) \subset [c\lambda^2,\infty)$, <u>some</u> $c > 0$; $\sigma_d(H_{01}(\lambda))$ <u>is non-empty and finite and if</u> $e_n(\lambda) \in \sigma_d(H_{01}(\lambda))$ <u>then</u> $\lim_{\lambda\to\infty} \lambda^{-1} e_n(\lambda) = e_n$;

(iii) $\sigma_{ac}(H(\lambda)) \subset [0,\infty)$, $\sigma_{sc}(H(\lambda)) = \phi$, <u>and</u> $\sigma_{pp}(H(\lambda)) \cap (0,\infty) = \phi$.

Sketch of the Proof:

(i) The estimate on $e(\lambda) \in \sigma_{pp}(H_{02}(\lambda))$ follows from a numerical range argument (see (iii) in the proof of Lemma 3.2). That $\sigma_{sc}(H_{02}(\lambda)) = \phi$ follows from the type A analyticity discussed in the next Section.

(ii) We use the Weyl criterion to prove that $\sigma_{ess}(H_{01}(\lambda)) \subset [c\lambda^2,\infty)$. Let $\sigma > 0$ be such that $B_\sigma(0) \cap (\mathrm{supp}|\nabla j_1|) = \phi$. We prove, using a local compactness argument, that if $\omega \in \sigma_{ess}(H_{01}(\lambda))$ then $\exists$ a Weyl sequence $\{\psi_n\}$ such that $\mathrm{supp}(\psi_n) \subset \bar{D}_1 \setminus B_\sigma(0)$. Then

$$\|(H_{01}(\lambda)-\omega)\psi_n\| \geq \lambda^2 V_\sigma - \omega$$

where $V_\sigma \equiv \min\{V(x)|x\in\bar{D}_1\setminus B_\sigma(0)\} > 0$, so $\omega = O(\lambda^2)$. The second part of the statement follows from the harmonic approximation (see, e.g., [15]).

(iii) The absence of positive eigenvalues follows from the Kato-Agmon-Simon theorem [17] and the fact that $\sigma_{sc}(H(\lambda)) = \phi$ is due to the type A analyticity discussed below. ∎

The approximate Hamiltonian $H_0(\lambda)$ has positive eigenvalues embedded in its continuous spectrum. We expect that these dissolve into resonances as W is added. As there is no natural perturbation parameter, we will compare the resolvents of $H_0(\lambda)$ and $H(\lambda)$ for λ large, after these operators have be distorted by U_θ and analytically continued in θ .

B. The Distorted Hamiltonians

We define a unitary group $U_\theta, \theta \in \mathbb{R}$, by

$$(U_\theta f)(x) = J(x,\theta)f(\Phi_\theta(x))$$

where $x \to \Phi_\theta(x)$ is a flow on $\mathbb{R}^n$ and $J(x,\theta)^2$ is the determinant of the Jacobian matrix. We construct the flow Φ_θ from a spherically symmetric vector field determined by a function $f:\mathbb{R}_+ \to \mathbb{R}_+$ satisfying:

(1) $f \in C^2$, $f(|x|) = 0$ for $|x| \leq R$ (where R is given in (V3));

(2) $\lim_{|x|\to\infty} (|x|-R)^{-1}f(|x|) < \infty$, so f is linear at infinity;

(3) f is the restriction to the half-line $|x| > R$ of a function $\tilde{f}$ analytic in a sector containing this half-line.

Without being more specific about growth conditions, we give an example:

$$\tilde{f}(z) = (z-R)(1-(a^2+b^2)(a^2+(z-R+b)^2)^{-1})^2$$

for $a,b > 0$. We now consider a family of flows Φ_θ^λ and groups U_θ^λ obtained by replacing R by $R(\lambda)$, $\lambda > 1$, in the above definitions.

We let $H_\mu(\lambda), \mu = 0,1$, denote $H_0(\lambda)$ or $H(\lambda)$ and define

$$H_\mu(\lambda,\theta) = U_\theta^\lambda H(\lambda)(U_\theta^\lambda)^{-1}, \quad \theta \in R .$$

Lemma 3.2.

(i) $H_\mu(\lambda,\theta)$ are analytic families of type A with domains $D(-\Delta)$ and $D(-j_1\Delta j_1 \oplus -j_2\Delta j_2)$, respectively, on the strip

$$S_{\theta_0} = \{\theta\in\mathbf{C} \mid |\mathrm{Im}\ \theta|<\theta_0\}$$

for some $\theta_0 > 0$;

(ii) $\sigma_{ess}(H(\lambda,\theta)) = e^{-2\theta}\mathbb{R}_+$, and $\sigma_{ess}(H_0(\lambda,\theta)) \subset e^{-2\theta}\mathbb{R}_+ \cup [c\lambda^2,\infty)$, some $c > 0$;

(iii) if $z \in \sigma_d(H_{02}(\lambda,\theta))$, then $z = O(\lambda(\ln \lambda)^\alpha)$ or z belongs to a sector bounded away from $\mathbb{R}_+$.

Sketch of the Proof:

(i) Let $p_\theta^2 = - U_\theta^\lambda\Delta(U_\theta^\lambda)^{-1}$ for $\theta \in \mathbb{R}$. Then we can write

$$p_\theta^2 = e^{-2\theta}p^2 + p_i a_{ij}(x,\theta)p_j + b_k(x,\theta)p_k + c(x,\theta)$$

where a_{ij}, b_k, c vanish as $|x| \to \infty$ and as $\theta \to 0$, are $O(1)$ in λ, and repeated indices are summed. The type A property follows from relative bound estimates between p_θ^2 and p^2 for $\mathrm{Im}\ \theta$ chosen suitably small.

(ii) One shows that the difference between the resolvents of p_θ^2 and $e^{-2\theta}p^2$ is compact using the fact that a_{ij}, b_k and c vanish as $|x| \to \infty$. The invariance of σ_{ess} then follows from a generalized Weyl theorem [18]. $\sigma_{ess}(H_{02}(\lambda,\theta))$ is computed using a local compactness argument to show that it is determined by Weyl sequences supported either near $\mathrm{supp}|\nabla j_2|$ where $V > \kappa > 0$ or near infinity.

(iii) $z \in \sigma_d(H_{02}(\lambda,\theta))$ is estimated using a numerical range estimate based on the facts that if $\mathrm{supp}(u) \subset B_{R(\lambda)}(0)\setminus B_R(0)$, then $(u,H_{02}(\lambda,\theta)u) \geq \lambda(\ell n\lambda)^{\alpha}\delta$, $\delta > 0$, and if $\mathrm{supp}(u) \subset \mathbb{R}^n\setminus B_{R(\lambda)}(0)$, then $(u,H_{02}(\lambda,\theta)u)$ lies in a sector in $\mathbb{C}^-$ bounded away from $\mathbb{R}_+$. ∎

As a consequence, we obtain a picture of the spectra of $H_\mu(\lambda,\theta)$ as given in Figure 3.1.

<u>Figure 3.1.</u> <u>Spectrum for</u> $\theta \in S^+_{\theta_0}$

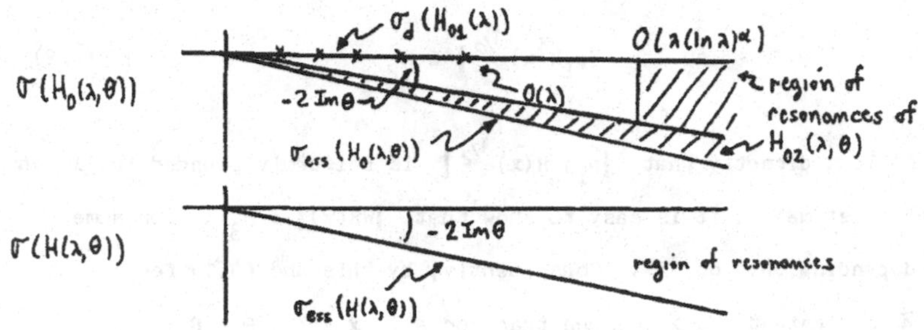

4. Outline of the Existence Proof

Let $F_\lambda(z,\theta)$ be the difference of the resolvents:

$$F_\lambda(z,\theta) = (z-H_0(\lambda,\theta))^{-1} - (z-H(\lambda,\theta))^{-1}$$

defined for $\theta \in S_{\theta_0}$ and $z \in \mathbb{C}\setminus\{z|\mathrm{Re}\, z > 0,\ |\arg z| < 2\theta_0\}$. We want to show that for $\theta \in S^+_{\theta_0} = \{\theta \in S_{\theta_0}|\theta_0>\mathrm{Im}\,\theta>0\}$ and z in a complex neighborhood of $e_n(\lambda) = \lambda e_n$, $\|F_\lambda(z,\theta)\| < \varepsilon$ for $\varepsilon > 0$, any λ sufficiently large. Let $R_0(z) = (z-H_0(\lambda,\theta))^{-1}$ and $R(z) = (z-H(\lambda,\theta))^{-1}$.

Lemma 4.1. <u>Let</u> $\theta \in \mathbb{R}$ <u>and</u> $z \in \mathbb{C}\setminus[0,\infty)$. <u>Then</u> $\lim_{\lambda\to\infty}\|F_\lambda(z,\theta)\| = 0$ <u>and the convergence is uniform on compact subsets of</u> $\mathbb{C}\setminus[0,\infty)$ <u>and in</u> $\theta \in \mathbb{R}$.

Sketch of the Proof: By the second resolvent formula, we have;

$$F_\lambda(z,\theta) = -R_0(z)WR(z) \ .$$

First, consider $x \in K \subset (-\infty,0)$, K compact. Using the Schwarz inequality and positivity, one shows that there exist constants M_1 , $M_2 > 0$ and independent of λ such that

$$|j_3R_\mu(x)^{1/2}| \leq \lambda^{-1}M_1 \ , \ \mu = 0,1 \tag{4.1}$$

and

$$|p_iR(x)^{1/2}| \leq M_2 \tag{4.2}$$

It then follows directly that $|p_ij_3R(x)^{1/2}|$ is uniformly bounded in λ on K . Given these estimates, it is easy to show that $|WR(x)| < M_3$, for some $M_3 > 0$ depending only on K . Consequently, by this and (4.1) for $\mu = 0$, $\exists$ a constant $M_K > 0$ such that for all $x \in K$, $\theta \in \mathbf{R}$:

$$|F_\lambda(x,\theta)| \leq \lambda^{-1}M_K \ . \tag{4.3}$$

Next, let $\tilde{K} \subset \mathbf{C}\backslash[0,\infty)$ be compact. Then $\exists$ a constant $M_{\tilde{K}} > 0$ such that for all $z \in \tilde{K}$:

$$|F_\lambda(z,\theta)| \leq \lambda^{-1}M_{\tilde{K}} \tag{4.4}$$

by the first resolvent identity, (4.3) and the fact that $H_\mu(\lambda)$ are self-adjoint. Hence, $\lim_{\lambda\to\infty}|F_\lambda(z,\theta)| = 0$, $z \in \tilde{K}$, and the covergence is uniform on $\tilde{K}$. ∎

To extend these estimates, we need uniform bounds on the resolvents of $H(\lambda,\theta)$ and $H_0(\lambda,\theta)$. The proof of these bounds is the most difficult part of the work.

Lemma 4.2. <u>For any compact</u> $\mathcal{C}_\eta \subset \{\theta\varepsilon C \,|\, -\eta < \mathrm{Im}\,\theta < \theta_0 \;,\; \eta < \theta_0\}$ <u>and compact</u> $K_{\eta,\mu} \subset \cap\, \rho(H_\mu(\lambda,\theta))$ <u>for all</u> λ <u>large and</u> $\theta \,\varepsilon\, \mathcal{C}_\eta$, <u>there exist constants</u> δ_μ , $\mu = 0$, 1 , <u>depending only on</u> $\mathcal{C}_\eta$ <u>and</u> $K_{\eta,\mu}$ <u>such that for all</u> $u \,\varepsilon\, C_0^\infty(\mathbf{R}^3)$:

$$\|(z-H_\mu(\lambda,\theta))u\| \geq \lambda\delta_\mu \|u\| \,.$$

Sketch of the Proof: We introduce another partition of unity $\{g_k\}_{k=1}^3$ with $\mathrm{supp}(g_1) \subset \mathrm{supp}(j_1)$, $\mathrm{supp}(g_3) \subset \mathbf{R}^n \setminus B_{R(\lambda)}(0)$, and $\sum_{k=1}^{3} g_k^2 = 1$. By a slight variation of the IMS formula, we have:

$$\|(H_\mu(\lambda,\theta)-z)u\|^2 \geq \sum_{k=1}^{3} \|(H_\mu(\lambda,\theta)-z)g_k u\|^2 - |(u,Ru)| \qquad (4.5)$$

where R is the remainder from the localization. We prove that $\exists \varepsilon_1$, $\varepsilon_2 > 0$ such that:

$$|(u,Ru)| = O(\lambda^{-\varepsilon_1}) + \lambda^{-\varepsilon_2} \|(H_\mu(\lambda,\theta)-z)u\|^2 \,.$$

Each piece of the sum on the right side of (4.5) is estimated as follows:

<u>k=1</u> one uses the harmonic approximation, i.e., one compares

$K(\lambda) = p^2 + \lambda^2 \sum_{i,j} A_{ij} x_i x_j$ with $H_\mu(\lambda,\theta)$ on $\mathrm{supp}(g_1)$ and uses the fact that $\mathrm{dist}(z,\sigma(K(\lambda))) > \delta_1 > 0$

<u>k=2</u> one uses the fact that $\lambda^2 V|\mathrm{supp}(g_2) > \lambda(\ln\lambda)^\alpha$

<u>k=3</u> one uses a numerical range estimate since

$H_\mu(\lambda,\theta)$ on $\mathrm{supp}(g_3) = e^{-2\theta} p^2 + \lambda^2 V_\theta$, and the numerical range of the latter operator lies in a sector bounded away from $\mathbf{R}_+$.

These calculations are facilitated by first scaling the Hamiltonians by the global dilation group which implements $x \to \lambda^{-1/2}x$. The result now follows from these estimates. ∎

Corollary 4.3. Let $K \subset \mathbb{C}^+$ be compact. Then $\lim_{\lambda\to\infty} |F_\lambda(z,\theta)| = 0$ for $\theta \in S^+_{\theta_0}$.

Proof: Let $\eta_K > 0$ be chosen such that the ray $R_K = \{re^{2i\eta_K} | r\in R_+\} \cap K = \phi$. Then $F_\lambda(z,\theta)$ is analytic on $K \times \{\theta | -\eta_K < \text{Im}\ \theta < \theta_0\}$ and uniformly bounded on compact subsets of the θ-domain by Lemma 4.2. Since this domain includes $\mathbb{R}$ on which $\lim_{\lambda\to\infty} |F_\lambda(z,\theta)| = 0$ by Lemma 4.1, it follows from Vitali's Theorem that $\lim_{\lambda\to\infty} |F_\lambda(z,\theta)| = 0$ for $\theta \in \{\theta | -\eta_K < \text{Im}\ \theta < \theta_0\}$. ∎

Proof of Result 1: Existence

Let $e_n(\lambda) \in \sigma_d(H_{01}(\lambda))$, $e_n(\lambda) \sim \lambda e_n$ and $S_R(0) \subset F_n(1)$. We first scale $H_\mu(\lambda,\theta)$ by the global dilation group: define $\tilde{H}_\mu(\lambda,\theta) = \lambda^{-1} U_\lambda H(\lambda,\theta) U_\lambda^{-1}$ where $(U_\lambda f)(x) = \lambda^{-3/4} f(\lambda^{-1/2}x)$. Then $\tilde{e}_n(\lambda) = \lambda^{-1} e_n(\lambda) \sim e_n$. Suppose $\exists$ an annular region $A_{n,\varepsilon} = \{z | \varepsilon < |z-\tilde{e}_n(\lambda)| < 2\varepsilon\}$ about $\tilde{e}_n(\lambda)$ and that $A_{n,\varepsilon} \subset \cap\ \rho(\tilde{H}_\mu(\lambda,\theta))$, for $\text{Im}\ \Theta = \theta_0$ and λ sufficiently large. Let γ be a circle about $\tilde{e}_n(\lambda)$ in $A_{n,\varepsilon}$ and construct

$$P_0^n = (2\pi i)^{-1} \oint_\gamma (z-\tilde{H}_0(\lambda,\theta))^{-1} dz$$

$$P^n = (2\pi i)^{-1} \oint_\gamma (z-\tilde{H}(\lambda,\theta))^{-1} dz$$

so that

$$|P^n - P_0^n| \le (2\pi)^{-1} \oint_\gamma |\tilde{F}_\lambda(z,\theta)| dz .$$

Now $\tilde{F}_\lambda(z,\theta)$ is analytic on $A_{n,\varepsilon}$, uniformly bounded there, and for $z \in A_{n,\varepsilon} \cap \mathbb{C}^+$, $\lim_{\lambda\to\infty} |\tilde{F}_\lambda(z,\theta)| = 0$. Hence, by applying Vitali's Theorem, it follows that for all λ large, $P^n \neq 0$. Hence, $H(\lambda)$ has spectral resonances near $e_n(\lambda)$. If no $\varepsilon > 0$ can be found such that $A_{n,\varepsilon} \subset \cap \rho(\tilde{H}_\mu(\lambda,\theta))$ for all λ large, then $H(\lambda)$ still has a resonance near $e_n(\lambda)$. By combining these two arguments, it is clear that for all λ suitably large, $H(\lambda)$ has resonances near $e_n(\lambda)$ and the total multiplicity of these resonances is at least as large as $\dim(\mathrm{Ran}\, P_0^n)$. In the case that P^n , constructed as above, satisfies $\|P^n - P_0^n\| < 1$, the multiplicity of the $z_{n(i)}(\lambda)$ is the same as $\dim(\mathrm{Ran}\, P_0^n)$. ∎

5. Outline of the Proof of the Width Estimate

We estimate $\Gamma_n = |\mathrm{Im}\, z_n(\lambda)|$ (writing n for $n(i)$ for simplicity) for those resonances for which the projection P^n defined in Section 4 exists, i.e., $\exists\ \lambda_n > 0$ such that $\lambda > \lambda_n$ implies that $\|P^n - P_0^n\| < 1/2$. The proof of Result 2 consists of two parts: (1) an expression for $|z_n(\lambda) - e_n(\lambda)|$; (2) an Agmon-type estimate [12] on the exponential decay in λ of the eigenfunctions of $H_{01}(\lambda)$ restricted to the classically forbidden region $\overline{F}_n \cap D_1$. Similar ideals have been used by Simon [19]. Note that due to Lemma 3.1, $\mathrm{Im}\, z_n(\lambda) < 0$.

Lemma 5.1. Let χ_3 be the characteristic function for $\mathrm{supp}(j_3) \cap \overline{D}_1$ and let ψ_n satisfy $H(\lambda,\theta)\psi_n = z_n(\lambda)\psi_n$ for some $\theta \in S^+_{\theta_0}$. Then $\exists\ c > 0$ (independent of λ) such that:

$$|z_n(\lambda) - e_n(\lambda)| \le c\lambda^{5/2} \|\chi_3 P_0^n \psi_n\|^2 . \qquad (5.1)$$

Proof: We observe that there exists a bounded operator W_1 such that $W\psi_n = W_1\psi_n$, that W_1 has the same localization as W , and that $\|W_1\| < M$, M independent of λ . We let $h_\lambda = H_0(\lambda,\theta) + W_1$ and apply the Feshbach projection method to h_λ using P_0^n [20-22]. As a result, we obtain

$$|z_n(\lambda) - e_n(\lambda)| \le \lambda \|P_0^n \psi_n\|^{-2} [\,|(P_0\psi_n, W_1 P_0 \psi_n)| + \|R(z_n)\| \|W_1 P_0 \psi_n\|^2] \qquad (5.2)$$

where $R(z_n)$ is the resolvent of $\overline{P}_0 h_\lambda \overline{P}_0$ on $\mathrm{Ran}\, \overline{P}_0$ at $z_n(\lambda)$, and $\overline{P}_0 = 1 - P_0$. Using an analysis similar to that sketched in Lemma 4.2, we show that $\|R(z_n)\|$ is bounded uniformly in λ . Finally, it follows from the assumption on the projections that $\|P_0\psi_n\| \geq 1/2$ for all large λ . The result now follows by applying the Schwarz inequality to (5.2) and using the fact that $\chi_3 W_1 = W_1$. ■

In the next step, the Agmon method is applied to eigenfunctions of $H_{01}(\lambda)$ in order to estimate $\|\chi_3 P_0\psi_n\|$. Let ϕ_n satisfy $H_{01}(\lambda)\phi_n = e_n(\lambda)\phi_n$ and $\|\phi_n\| = 1$. The distance $\rho_n(x)$, defined in (2.1), from S_n^- to $x \in \overline{F}_n$ is differentiable almost everywhere and satisfies

$$|\nabla\rho_n|^2 \leq \lambda^2 V - \lambda e_n \quad \text{a.e.} \; .$$

Choose $\delta > 0$ small and let $F_\delta(\lambda) \equiv \{x|\lambda V(x)-e_n>\delta\}$. Then $(\lambda^2 V-\lambda e_n)|F_\delta(\lambda) \geq \lambda\delta$ and $F_\delta(\lambda) \subset F_n(\lambda)$.

Lemma 5.2. Let $\eta \in C^\infty$ be such that $\eta \geq 0$, $\text{supp}(\eta) \subset F_\delta(\lambda)$ and $\eta|\text{supp}(j_3) = 1$. There exists $\lambda_n > 0$ and a constant $c_n > 0$ such that for all $\lambda > \lambda_n$:

$$\|e^{\rho_n}\eta\phi_n(\lambda)\|^2 \leq c_n\lambda e^{2\lambda\rho_n^0}$$

where $\rho_n^0 \equiv \sup\{\lambda^{-1}\rho_n(x)|x\varepsilon(\text{supp}|\nabla\eta|) \cap \overline{D}_1\}$.

Proof: Let $\phi \equiv \eta e^{\rho_n}\phi_n$. By a standard argument, one proves that for each $\kappa > 0$:

$$\begin{aligned} &\text{Re}(e^{(1-\kappa)\rho_n}\phi,[H_{01}(\lambda)-e_n(\lambda)]e^{-(1-\kappa)\rho_n}\phi) \geq \\ &\kappa(\phi,(\lambda^2 V-\lambda e_n-O(\lambda^{4/5}))\phi) \geq \lambda\kappa\delta c\|\phi\|^2 \end{aligned} \tag{5.3}$$

for $\lambda > \lambda_n$ chosen suitably large and $c > 0$. We choose $\kappa = O(\lambda^{-1})$ so $\lambda\kappa = O(1)$ and we can take $\kappa = 0$ in the left side of (5.3) since $\lim_{\lambda\to\infty} \lambda^{-1}\rho_n(x) < \infty$. Using the definition of ϕ , the left side of (5.3) is evaluated by commuting $H_{01}(\lambda)$ through η . The result is

$$c\delta\|e^{\rho_n}\eta\phi_n\|^2 \le \mathrm{Re}(e^{2\rho_n}\eta\phi_n,(\Delta\eta+2\nabla\eta\cdot\nabla)\phi_n) .$$

The term involving $\nabla\phi_n$ is evaluated using the divergence theorem (this contributes the prefactor λ) and the result now follows. ∎

We now take $\mathrm{diam}(\mathrm{supp}|\nabla\eta|) = O(\lambda^{-1})$ so $\rho_n^0 = O(\lambda^{-1})$. The penalty for this is simply an additional power of λ in the prefactor.

Proof of Result 2: Resonance Width

From Lemmas 5.1 - 5.2, we have

$$\|\chi_3 P_0^n\psi_n\|^2 \le \|e^{\rho_n}\eta P_0^n\psi_n\|^2 e^{-2\lambda\tilde{\rho}_n} \le \lambda^2 c_n e^{-2\lambda\tilde{\rho}_n} \tag{5.4}$$

using the fact that $\eta^2 \ge \chi_3^2$, and where $\tilde{\rho}_n \equiv \min\{\lambda^{-1}\rho_n(x)|x \in \mathrm{supp}(j_3)\}$. By the triangle inequality, $\tilde{\rho}_n \ge \rho_n - \varepsilon$ where ε depends on ε_1 and $R^{1-\alpha/2}$ and ρ_n is as defined in Result 2. By taking ε_1 small and R large, we can make ε arbitrarily small. As a result of this and (5.4), we get for any $\varepsilon > 0$:

$$\|\chi_3 P_0^n\psi_n\|^2 \le \lambda^2 c_n' e^{-2\lambda(\rho_n-\varepsilon)} . \tag{5.5}$$

Combining this and (5.1), we obtain $|z_n(\lambda)-e_n(\lambda)| \le c\lambda^5 e^{-2\lambda(\rho_n-\varepsilon)}$ for any $\varepsilon > 0$. This yields Result 2. ∎

Acknowledgements. We would like to thank W. Hunziker and especially B. Simon for many valuable remarks.

REFERENCES

1. Gamow, G.: Zs.f. Phys. 51, 204 (1928); Gurney, R.W., Condon, E.U.: Nature 122, 439 (1928).

2. Aguilar, J., Combes, J.M.: Commun. Math. Phys. 22, 269-279 (1971); Balslev, E., Combes, J.M.: Commun. Math. Phys. 22, 280-294 (1971).

3. Simon, B.: Ann. Math. 97, 247-274 (1973); Phys. Letts. 71A, 211-214 (1979); Int. J. Quant. Chem. 14, 529-542 (1978).

4. Hunziker, W.: ETH preprint (1986), to be published in Ann. Inst. Henri Poincaré.

5. Sigal, I.M.: Ann. Inst. Henri Poincaré 41 103-114 (1984), Addendum, ibid, 41, 333 (1984).

6. Cycon, H.L.: Helv. Phys. Acta 53, 969-981 (1985).

7. Combes, J.M.: In: Proc. Nato Inst. on Scattering Theory, J.A. LaVita, J.P. Marchand, ed., 243-272 (1974).

8. Babbitt, D., Balslev, E.: J. Math. Anal. Appl. 54, 316-347 (1976); Jensen A.: J. Math. Anal. Appl. 59, 505-513 (1977).

9. Ashbaugh, M., Harrell, E.: Commun. Math. Phys. 83, 151-170 (1982).

10. Helffer, B., Sjöstrand, J.: Univèrsité de Nantes preprint, (1985).

11. Combes, J.M., Duclos, P., Klein, M., Seiler, R.: Marseille preprint, CPT-85/P1797 (1986), and Proceedings of this conference.

12. Agmon, S.: Lectures on Exponential Decay of Solutions of Second-Order Elliptic Equations. Princeton N.J.: Princeton University Press 1982.

13. Landau, L.D., Lifshitz, E.M.: Quantum Mechanics, Oxford: Pergamon Press 1977.

14. Sigal, I.M.: Commun. Math. Phys. 85, 309-324 (1982).

15. Simon, B.: Ann. Inst. Henri Poincaré 13, 295-307 (1983).

16. Cycon,H.L., Froese, R., Kirsch, W., Simon, B.: Lectures on the Schrödinger Equation, to appear.

17. Reed, M., Simon, B., Methods of Modern Mathematical Physics. IV. New York: Academic Press 1978.

18. Sigal, I.M.: J. Op. Th. 13, 119-129 (1985).

19. Simon, B.: Ann. Math 120, 89-118 (1984).

20 Horwitz, L.P., Sigal, I.M.: Helv. Phys. Acta, 51, 685-715 (1978).

21. Combes, J.M.: In: The Schrödinger Equation, W. Thirring and P. Urban ed., Vienna: Springer-Verlag 1977.

22. Howland, J.S.: In Proc. Nato Inst. on Scattering Th., J.A. LaVita and J.M. Marchand, ed. (1974).

Random Perturbation Theory and Quantum Chaos

James S. Howland[1]
University of Virginia, Charlottesville, VA 22903

Abstract

The Floquet Hamiltonian of a time-periodic rank one perturbation of a discrete quantum system has a pure point spectrum for a.e. value of the coupling constant.

In his classic paper [1], P. W. Anderson argued that a discrete one-dimensional Schrödinger operator with a random potential would have a pure point spectrum almost surely. This phenomenon, which has come to be known as <u>localization</u>, has been the subject of a great deal of work, both rigorous and not, some of which is described in the lectures of Simon and Kotani at this conference.

Although most of the work on localization concerns discrete and continuous Schrödinger operators, there is another source of pure point spectrum in quantum mechanics, namely the Floquet operators of discrete systems in a time-periodic perturbing field. Let H be a Hamiltonian operator on $\mathcal{H}$ with a complete orthonormal set of eigenvectors e_k, $He_k = \lambda_k e_k$, and V(t) a bounded operator-valued function with period 2π in t. The <u>Floquet operator</u> is the operator

$$i \frac{d}{dt} + H(t) = i \frac{d}{dt} + H + V(t)$$

on $L^2([0,2\pi],\mathcal{H})$ with periodic boundary conditions. As explained, for example in [4,10], an eigenvalue μ of the Floquet operator corresponds to a solution of the Schrödinger equation

$$-i \frac{d\Psi}{dt} = H(t)\Psi$$

[1]Supported by NSF contract NSF-DMS-8500516.

satisfying $\Psi(2\pi) = e^{-2\pi i\mu}\Psi(0)$. If $V(t) \equiv 0$, then

$$K = i\,\frac{d}{dt} + H$$

is pure point, having eigenvectors $\varphi_{n,k} = e^{int}e_k$ and eigenvalues $n+\lambda_k$, ($n = 0,\pm1,\ldots,\ k = 1,2,\ldots$). For most sequences λ_k, this will be a dense set in $\mathbb{R}$, or, if not, it will be if the frequency is changed slightly.

In work discussed in his lecture at this conference, Bellisard [2] considered the Pulsed Rotor

$$K(\alpha,\beta) = -i\,\frac{\partial}{\partial t} - \alpha\,\frac{\partial^2}{\partial\theta^2} + \beta V(\theta,t)$$

where $0 < \theta < 2\pi$, $0 < t < 2\pi$ with periodic boundary conditions. Thus H is just $(-\alpha)$ times the Laplacian on the circle. If V is an analytic function, then Bellisard proves by methods related to the KAM theorem, that for $\varepsilon > 0$, there exists a set $\Omega_\varepsilon \subset [1,\infty)$ and a "critical" value $\beta_0(\varepsilon)$ so that for $\alpha \in [1,\infty) \sim \Omega_\varepsilon$, and $|\beta| < \beta_0(\varepsilon)$, $K(\alpha,\beta)$ is pure point. Moreover, $\text{meas}(\Omega_\varepsilon) \to 0$, and $\beta_c(\varepsilon) \to 0$ as $\varepsilon \downarrow 0$. For reasons that we cannot enter into here, Bellisard asks if there exists a critical value β_c such that for $0 < \beta < \beta_c$, $K(\beta)$ is pure point and for $\beta > \beta_c$, $K(\beta)$ is absolutely continuous. Such behavior could be described as a "transition from stability to chaos."

Recently, the author [5] developed a general theory of random compact perturbations of pure point operators, based on the work of Simon and Wolff [9] reported on in Simon's lecture. In this paper, we wish to illustrate the methods and results of [5] by extending them to the discussion of a simple model related to the Pulsed Rotor. Excepting a few technical points, our discussion is self-contained, and illustrates the essential ideas in a simple context. It will also say something about the possibility of the transition to the Quantum Chaos.

The model is the following. Let H be self-adjoint and pure point, with a complete orthonormal set e_k of eigenvectors, $He_k = \lambda_k e_k$. Let φ be a unit vector, $w(t)$ a 2π-periodic function of t, and

$$H(t,\beta) = H + \beta w(t)\,\langle\cdot,\varphi\rangle\varphi.$$

Thus, instead of a local potential $q(\theta,t)$, we consider an operator of rank one. Let $K(\beta) = i\frac{d}{dt} + H(t,\beta)$ be the Floquet operator, and define

$$P = \langle\cdot,\varphi\rangle\varphi.$$

__Theorem 1__. *If $w(t)$ is bounded, then the operator $K(\beta)$ is pure point for a.e. β, provided that*

$$\sum_k |\langle\varphi,e_k\rangle| < \infty.$$

Thus, if the Rotor Hamiltonian behaves as this model does, there can be no transition to absolutely continuous spectrum for β above some critical value β_c.

§2. H-finiteness.

The proof of Theorem 1 involves two basic ideas. The first is abstracted from the basic estimate of Simon and Wolff [9], which they derived from estimates of Fröhlich and Spencer [3].

__2.1 Definition__. *Let H be self-adjoint. A bounded operator A is H-finite iff $A(H-\lambda)^{-2}A^*$ is a bounded operator for a.e. λ.*

H-finiteness is a very strong condition. If the range of A is cyclic for H, then the spectral measure of H is supported by the exceptional set N where $A(H-\lambda)^{-2}A^*$ is not bounded. Since this is of measure zero, H has no absolutely continuous part [5, Theorem 2.4].

If H is a pure point operator as above, then a *sufficient* condition for H-finiteness is that A be *strongly* H-*finite*; that is,

$$\sum_n |Ae_n| < \infty.$$

[5, Proposition 4.2]. Strong H-finiteness implies that A is trace class.

__2.2 Lemma__. *Let H be pure point, A bounded, and $\Phi(\lambda)$ a nonvanishing continuous function on $\mathbb{R}$. If $A\Phi(H)$ is H-finite, then A is H-finite.*

<u>Proof</u>. Let J be a finite open interval, and $|\varphi(\lambda)| \geq \delta > 0$ on J. To prove $A(H-\lambda)^{-2}A^*$ finite for $\lambda \in J$, it is enough to observe that

$$\sum_{\lambda_k \in J} (\lambda-\lambda_k)^{-2}\langle\cdot, Ae_k\rangle Ae_k$$

$$= \sum_{\lambda_k \in J} |\varphi(\lambda_k)|^{-2}(\lambda-\lambda_k)^{-2}\langle\cdot, A\varphi(H)e_k\rangle A\varphi(H)e_k$$

$$\leq \delta^{-2} \sum_{\lambda_k \in J} (\lambda-\lambda_k)^{-2}\langle\cdot, A\varphi(H)e_k\rangle A\varphi(H)e_k$$

$$\leq \delta^{-2}A\varphi(H)(H-\lambda)^{-2}\overline{\varphi(H)}A^*.$$ ■

<u>2.3 Theorem</u>. <u>There exists a set</u> N <u>of Lebesgue measure zero, which does not depend on</u> β, <u>such that the singular continuous part of</u> $K(\beta)$ <u>is supported by</u> N. <u>The absolutely continuous part of</u> $K(\beta)$ <u>vanishes</u>.

<u>2.4 Lemma</u>. $P(K+i)^{-2}$ <u>is strongly</u> K-<u>finite</u>.

<u>Proof</u>. We have

$$\sum_{nk} |P(K+\lambda)^{-2}\varphi_{nk}| = \sum_{nk} |n+\lambda_k+\lambda|^{-2}|Pe_k|$$

$$= \sum_k |\langle\varphi, e_k\rangle| \sum_n ((n+\lambda_k)^2+1)^{-1} \leq (2+\pi) \sum_k |\langle\varphi, e_k\rangle| < \infty$$

since

$$\sum_n ((n+\lambda_k)^2+1)^{-1} \leq 2 + \int_{-\infty}^{+\infty} \frac{1}{(x+\lambda_k)^2+1}\, dx = 2+\pi.$$ ■

By Lemma 2.2, P is therefore H-finite.

<u>2.5 Lemma</u>. <u>The operator</u> $Q(z) = P(K-z)^{-1}P$ <u>is compact and</u> $Q(\lambda \pm i0)$ <u>exists in operator norm for a.e.</u> λ.

<u>Proof</u>. For compactness, take the Fourier transform in t, and write Q(z) on $\ell_2(\mathcal{H})$ as the diagonal operator

$$Q(z)x_n = Q_n x_n = P(H+n-z)^{-1}Px_n.$$

For each n, Q_n is rank one, and

$$\lim \|Q_n\| = \lim_{n\to\infty} \left|\sum_k \frac{|\langle\varphi,e_k\rangle|^2}{(\lambda_k+n-z)^{-1}}\right| = 0.$$

Hence Q(z) is compact.

Fix a bounded open interval J of the real axis, let E be the spectral resolution of K, R(z) its resolvent, and write

$$PR(z)P = PE[J^c]R(z)P$$
$$+ [(K+i)^{-2}P]^*[(K+i)^2]^*E[J]R(z)E[J](K+i)^2[(K+i)^{-2}P].$$

The first term is norm analytic across J. The second is trace class, with positive imaginary part, and therefore has a.e. limits in Hilbert-Schmidt norm by a theorem of De Branges [7, p. 149]. ■

<u>Proof of Theorem 2.3</u>. Write

$$K(\beta) = K+\beta PWP$$

where $Wu(t) = w(t)u(t)$ is bounded. Let N_1 be the exceptional set where

$$F(\lambda;K) = P(K-\lambda)^{-2}P$$

does not exist. Note that for $\lambda \notin N_1$,

$$F(\lambda;K) = \lim_{\varepsilon\downarrow 0} F_\varepsilon(\lambda;K) = \lim_{\varepsilon\downarrow 0} P[(H-\lambda)^2+\varepsilon^2]^{-1}P$$

Now $\varepsilon F_\varepsilon(\lambda;K) = \pi\delta_\varepsilon(K-\lambda)$, so the usual scattering theory formulas [6,7] yield

(2.1) $$F_\varepsilon(\lambda;K(\beta)) = [\Delta^*(\lambda+i\varepsilon,\beta)]^{-1}F_\varepsilon(\lambda;K)[\Delta(\lambda+i\varepsilon,\beta)]^{-1}$$

where $\Delta(z,\beta) = I+\beta WPR(z)P$.

By Lemma 2 there is a null set N_2, independent of β, such that $\Delta(\lambda+i0,\beta)$ exists in norm for $\lambda \notin N_2$. Let

$$N_3(\beta) = \{\lambda \notin N_2 : \Delta(\lambda+i0,\beta) \text{ is not invertible}\}.$$

From (2.1) it follows that $F(\lambda;K(\beta))$ is bounded except for $\lambda \in N_1 \cup N_2 \cup N_3(\beta)$.

Let $\lambda \in N_3(\beta)$, but $\lambda \notin N_1 \cup N_2$. We claim that λ is an eigenvalue of $K(\beta)$. In fact, by compactness, there exists u with

$$u+\beta WPR(\lambda+i0)Pu = 0.$$

Since $\lambda \notin N_1$, the vector $\Psi = (K-\lambda)^{-1}Pu$ is well defined because

$$|\Psi|^2 = |(K-\lambda)^{-1}Pu|^2 = \langle P(K-\lambda)^{-2}Pu,u\rangle < \infty.$$

Hence, $Pu = (K-\lambda)\Psi$ and $u+\beta WP\Psi = 0$ which implies $K(\beta)\Psi = \lambda\Psi$.

Thus, $N_3(\beta)$ is countable and the singular continuous part of $K(\beta)$ is supported by $N = N_1 \cup N_2$. Since P is $K(\beta)$-finite, the absolutely continuous part of $K(\beta)$ must vanish.

§3. Absolute continuity of multiplication operators.

The second ingredient in the proof is the recognition of the importance of the absolute continuity of the multiplication operator

$$\mathbb{K}u(\beta) = K(\beta)u(\beta)$$

on the larger space $L^2([-1,1],\mathcal{X})$.

Actually, we shall change the parametrization slightly. We want to prove that $K(\beta) = K+\beta PWP$ is pure point for a.e. β in $(-1,1)$. This is equivalent to proving that

$$K(x) = K+(\tanh x)(PWP+\gamma I)$$

is pure point for a.e. x in $\mathbb{R}$. Here we choose $\gamma > \sup|w(t)|$, so that $(PWP+\gamma I) > 0$. Define

$$\mathbb{K}u(x) = K(x)u(x)$$

on $L^2(\mathbb{R};\mathcal{X})$.

3.1 Proposition. The operator $\mathbb{K}$ is spectrally absolutely continuous.

Proof. If $p = -i\,\frac{d}{dx}$ is the momentum operator then the commutator

$$C = i[\arctan(p/2), \tanh x] > 0$$

is positive definite on $L^2(\mathbb{R})$. This may be proved by expressing arctan(p/2) explicitly as a convolution. (See [5].) Hence, on $L^2(\mathbb{R},\mathcal{X})$,

$$i[\arctan(p/2),\mathbb{K}] = C \otimes (PWP+\gamma I) > 0.$$

The result now follows from the Kato-Putnam positive commutator theorem [8, p. 157].

Proof of Theorem 1. Let $\mathbb{E}$ and E_x be the spectral resolutions of $\mathbb{K}$ and $K(x)$. We have for $u = \chi(x)y$,

$$0 = |\mathbb{E}[N]u|^2 = \int |E_x[N]y|^2 \chi(x)^2 dx.$$

Since this holds for every test function $\chi(x)$,

$$E_x[N]y = 0 \quad \text{a.e.}$$

The exceptional set depends on y, but letting y range over a countable dense set gives

$$E_x[N] = 0 \quad \text{a.e.}$$

But N supports $K_{sc}(x)$, so $K_{sc}(x)$ vanishes for a.e. x. It follows that K(x) is pure point a.e.

The author is indebted to I. W. Herbst for the proof of Lemma 2.2, and for the trick of adding γI which eliminated an earlier assumption that w(t) be positive.

REFERENCES

1. Anderson, P. W. Absence of diffusion in certain random lattices. Phys. Rev. 109 (1958), 1492-1505.

2. Bellisard, J. Stability and instability in quantum mechanics, preprint.

3. Fröhlich, J. and T. Spencer. Absence of diffusion in the Anderson tight bonding model for large disorder or low energy. Comm. Math. Phys. 88 (1983), 151-189.

4. Howland, J. S. Scattering theory for Hamiltonians periodic in time. Indiana J. Math. 28 (1979), 471-494.

5. Howland, J. S. Perturbation theory of dense point spectra. J. Functional Analysis (to appear).

6. Kato, T. Wave operators and similarity for some non-selfadjoint operators. Math. Ann. 162 (1966), 258-279.

7. Kato, T. and S. T. Kuroda. The abstract theory of scattering. Rocky Mountain J. Math. 1 (1971), 127-171.

8. Reed, N. and B. Simon. Methods of Modern Mathematical Physics, v. IV. Academic Press, New York, 1978.

9. Simon, B. and T. Wolff. Singular continuous spectrum under rank one perturbations and localization for random Hamiltonians. Comm. Pure Appl. Math. 39 (1986) 75-90.

10. Yajima, K. Scattering theory for Schrödinger equations with potential periodic in time. J. Math. Soc. Japan 29 (1977), 729-743.

PATH INTEGRAL FOR A WEYL QUANTIZED RELATIVISTIC HAMILTONIAN AND THE NONRELATIVISTIC LIMIT PROBLEM

Takashi Ichinose
Department of Mathematics, Kanazawa University
920-Kanazawa, Japan

Abstract. A path integral formula is given which represents the solution of the Cauchy problem for the pure-imaginary-time Schrödinger equation

$$\partial_t \psi^c(t,x) = -[H^c - mc^2]\psi^c(t,x), \quad t>0, \quad x \in R^d .$$

Here H^c is the relativistic quantum Hamiltonian associated, via the Weyl correspondence, with the relativistic classical Hamiltonian

$$h^c(p,x) = \sqrt{c^2(p-eA(x))^2+m^2c^4} + e\Phi(x)$$

for a spinless particle in an electromagnetic field. The nonrelativistic limit $c\to\infty$ is also discussed in both pure-imaginary and real times. The problem is connected with a time-homogeneous Lévy process.

1. Description of the Main Result

Consider the Schrödinger equation

$$(1.1) \qquad \partial_t \phi^c(t,x) = -i[H^c - mc^2]\phi^c(t,x), \quad t \in R, \quad x \in R^d,$$

and also the pure-imaginary-time Schrödinger equation

$$(1.2) \qquad \partial_t \psi^c(t,x) = -[H^c - mc^2]\psi^c(t,x), \quad t>0, \quad x \in R^d,$$

for the relativistic Hamiltonian H^c, with the rest energy mc^2 subtracted, of a spinless particle of mass $m>0$ and charge e in an electromagnetic field. c is the light velocity, and Planck's constant h is taken to equal 2π so that $\hbar=h/2\pi=1$. For H^c we adopt the Weyl quantized Hamiltonian (e.g. [1], [4])

$$(1.3) \qquad H^c = H^c_A + e\Phi(x),$$

$$(H^c_A g)(x) = (2\pi)^{-d}\iint_{R^d\times R^d} e^{i(x-y)p}\, h^c_A\left(p,\frac{x+y}{2}\right)g(y)\,dy\,dp, \quad g \in \mathcal{S}(R^d),$$

corresponding to the relativistic classical Hamiltonian (e.g. [11])

$$(1.4) \quad h^c(p,x) = h^c_A(p,x) + e\Phi(x) \equiv \sqrt{c^2(p-eA(x))^2+m^2c^4} + e\Phi(x), \quad p \in R^d, x \in R^d,$$

of the particle. Here we assume for simplicity that $A(x)$ (resp. $\Phi(x)$), the vector (resp. scalar) potential of the field, is an $\mathbf{R}^d$ (resp. $\mathbf{R}$)-valued C^∞ function in $\mathbf{R}^d$ which together with their derivatives up to a sufficiently high order is bounded. It is shown ([14],[7]) that both H_A^c and H^c are essentially selfadjoint on $C_0^\infty(\mathbf{R}^d)$, and so define selfadjoint operators in $L^2(\mathbf{R}^d)$ which have domain $H^1(\mathbf{R}^d)$ and are bounded from below. Note that H_A^c differs from the square root of the nonnegative selfadjoint operator $c^2(-i\partial-eA(x))^2+m^2c^4$.

The aim of this note is first to give a path integral formula representing the solution of the Cauchy problem for Eq.(1.2). It has a close analogy with the Feynman-Kac-Itô formula(e.g. [15]) which represents the solution for the pure-imaginary-time Schrödinger equation, i.e. the heat equation

$$(1.5) \qquad \partial_t\psi(t,x) = -H\psi(t,x), \quad t>0, \quad x\in\mathbf{R}^d,$$

for the nonrelativistic quantum Hamiltonian $H = \frac{1}{2m}(-i\partial-eA(x))^2+e\Phi(x)$ of the same particle. Then this path integral representation is used to discuss the nonrelativistic limit $c\to\infty$ of the solution for Eq.(1.2) and also, as a by-product, for Eq.(1.1). For this purpose the usual factor $1/c$ in front of $eA(x)$ in H^c and H is omitted (cf.[5]) so that it can be kept fixed in the limit $c\to\infty$. Our derivation of the path integral formula is nothing but a rigorous application of the <u>phase space path integral</u> or <u>Hamiltonian path integral</u> method with the "midpoint" prescription([12], [13]; [3]).

The main result is the following three propositions. Statement of Theorem 1 needs some notions on a time-homogeneous Lévy process, which will be explained at the end of this section.

<u>THEOREM 1.(Path integral formula for the solution $\psi^c(t,x)$ of Eq.(1.2))</u>. There exists a probability measure λ_x^c on the space $D_x([0,\infty)\to\mathbf{R}^d)$ of the right-continuous paths $X:[0,\infty)\to\mathbf{R}^d$ having left-hand limits and satisfying $X(0)=x$ such that for every initial data $\psi^c(0,x)=g(x)$ in $L^2(\mathbf{R}^d)$ and for all A and Φ,

$$(1.6) \qquad \psi^c(t,x) = \left(e^{-t[H^c-mc^2]}g\right)(x) = \int e^{-V(t,0;X)}g(X(t))d\lambda_x^c(X)$$

with

$$(1.7) \qquad V(t,0;X) = i\int_0^{t+}\int_{|y|>0} eA(X(s-)+y/2)y\,\tilde{N}_X(dsdy)$$
$$+ i\int_0^t\int_{|y|>0} e[A(X(s)+y/2)-A(X(s))]y\,\hat{N}(dsdy)$$
$$+ \int_0^t e\Phi(X(s))ds.$$

From Theorem 1 it follows, in particular, that $H_A^c - mc^2$ is nonnegative selfadjoint. As we use a general theory of pseudo-differential operators, so we have assumed sufficient regularity and boundedness of the vector and scalar potentials A and Φ. However, in applications, nonregular and unbounded A and Φ appear. It is shown that the formula (1.6) is true if $\Phi(x)$ is measurable in $\mathbf{R}^d$ and satisfies $e\Phi=e\Phi_+-e\Phi_-$ with $e\Phi_+\geq 0$, $e\Phi_-\geq 0$, where $\Phi_+ \in L^1_{loc}(\mathbf{R}^d\setminus G)$ with G a closed subset of measure zero, and $e\Phi_- \leq \delta(-c^2\Delta+m^2c^4)^{1/2}+C$, as the quadratic forms, with some constants $0<\delta<1$ and $C\geq 0$. As for A, it will suffice to assume, as $V(t,0;X)$ in (1.7) suggests, that A is Hölder-continuous of order α, $0<\alpha\leq 1$.

<u>THEOREM 2</u>.(<u>Nonrelativistic limit for Eq.(1.2)</u>). $e^{-t[H^c-mc^2]}$ is strongly convergent to e^{-tH} on $L^2(\mathbf{R}^d)$ as $c\to\infty$, uniformly on every finite t-interval in $[0,\infty)$.

<u>Remark</u>. Theorem 2 is also true on $C_\infty(\mathbf{R}^d)$, the Banach space of the bounded continuous functions in $\mathbf{R}^d$ which tend to zero as $|x|\to\infty$, instead of $L^2(\mathbf{R}^d)$.

<u>COROLLARY</u>.(<u>Nonrelativistic limit for Eq.(1.1)</u>). $e^{-it[H^c-mc^2]}$ is strongly convergent to e^{-itH} on $L^2(\mathbf{R}^d)$ as $c\to\infty$, uniformly on every finite t-interval in $(-\infty,\infty)$.

The nonrelativistic limit problems for the Dirac and/or Klein-Gordon equations are discussed by [5] and [2].

Finally, we come to the notions, used in Theorem 1, on a time-homogeneous Lévy process (e.g. [8],[9]).

The path space measure λ_x^c in (1.6), dependent on the light velocity c, has the characteristic function

$$(1.8)\quad \exp\{-t[\sqrt{c^2p^2+m^2c^4} - mc^2]\} = \int e^{ip(X(t)-X(0))}\, d\lambda_x^c(X), \quad t>0.$$

Notice that the left-hand side of (1.8) is a function of positive type in p. If $\Psi(X) = F(X(t_0),\cdots,X(t_n))$ with a finite partition $0=t_0<t_1<\cdots<t_n=t$ of the interval $[0,t]$ and with a bounded continuous function $F(x^{(0)},\cdots,x^{(n)})$ in $\mathbf{R}^{d(n+1)}$, then

$$(1.9)\quad \int \Psi(X)d\lambda_x^c(X) = \overbrace{\int_{\mathbf{R}^d}\cdots\int_{\mathbf{R}^d}}^{n} \prod_{j=1}^{n} k_0(t_j-t_{j-1},x^{(j-1)}-x^{(j)})$$

$$\times F(x^{(0)},\ldots,x^{(n)})dx^{(1)}\cdots\cdots dx^{(n)},$$

with $x^{(0)}=x$, where $k_0(t,x)$ is the fundamental solution of the Cauchy problem for the free equation

$$(1.10)\qquad \partial_t\psi^c(t,x) = -[(-c^2\Delta+m^2c^4)^{1/2}-mc^2]\psi^c(t,x),\quad t>0,\ x\in R^d,$$

to Eq. (1.2). $k_0(t,x)$ is a positive function in $t>0$ and $x\in R^d$ with $k_0(t,x) = k_0(t,-x)$ and $\int k_0(t,x)dx = 1$. Each X in $D_x([0,\infty)\to R^d)$ has at most finitely many points s at which the jump $|X(s)-X(s-)|$ exceeds a given positive constant. In particular, X(s) has at most countably many discontinuities. It is bounded on every finite interval [0,t], but not of bounded variation there. $N_X(dsdy)$ is a counting measure on $(0,\infty)\times(R^d\setminus\{0\})$:

$$N_X((t,t']\times B) = \#\{s;\ t<s\le t',\ 0\ne X(s)-X(s-)\in B\},$$

where $0<t<t'$ and B is a Borel set in $R^d\setminus\{0\}$, and

$$\tilde{N}_X(dsdy)=N_X(dsdy)-\hat{N}(dsdy),\quad \hat{N}(dsdy)\equiv\int N_X(dsdy)d\lambda_x^c(X)=dsn(dy).$$

Here n(dy) is the Lévy measure which is a σ-finite measure on $R^d\setminus\{0\}$ satisfying $\int_{|y|>0}[y^2/(1+y^2)]n(dy) < \infty$, and the Lévy-Khinchin formula holds :

$$\sqrt{c^2p^2+m^2c^4} - mc^2 = -\int_{|y|>0}[e^{ipy}-1-ipyI_{\{|y|<1\}}(y)]\ n(dy),$$

where $I_B(y)=1$ if $y\in B$, and $=0$ if $y\notin B$ (In [8], N_X, $\tilde{N}_X$ and $\hat{N}$ are denoted by N_p, $\tilde{N}_p$ and $\hat{N}_p$). The Lévy-Itô theorem gives the representation of X(t) :

$$X(t) = X(0) + \int_0^{t+}\int_{|y|>0} y\tilde{N}_X(dsdy).$$

2. Proof

We give a brief outline of the proof. Full details will be published elsewhere ([7], [6]).

Proof of Theorem 1. We mimic the proof of the Feynman-Kac-Itô formula in [15].

Define a bounded linear operator T(t), t>0, by

$$(2.1)\quad (T(t)g)(x) = \int_{R^d} k_0(t,x-y)\exp\left\{ieA(\tfrac{x+y}{2})(x-y)-e\phi(\tfrac{x+y}{2})t\right\}g(y)dy,$$

for $g\in L^2(R^d)$. Then we have

$$(2.2)\quad (T(t/n)^n g)(x) = \overbrace{\int_{R^d}\cdots\int_{R^d}}^{n} \prod_{j=1}^{n} k_0(t/n, x^{(j-1)}-x^{(j)})$$

$$\times \exp[-V_n(x^{(0)},\cdots,x^{(n)})]g(x^{(n)})dx^{(1)}\cdots\cdots dx^{(n)},$$

where

$$(2.3)\quad V_n(x^{(0)},\cdots,x^{(n)})$$

$$= i\sum_{j=1}^{n} eA\left(\frac{x^{(j-1)}+x^{(j)}}{2}\right)(x^{(j)}-x^{(j-1)}) + \sum_{j=1}^{n} e\Phi\left(\frac{x^{(j-1)}+x^{(j)}}{2}\right)(t/n)$$

with $x^{(0)}=x$. It can be shown that as $n\to\infty$, the left-hand side of (2.2) converges to $\exp[-t(H^c-mc^2)]g$ in L^2. On the other hand, the right-hand side of (2.2) is in virtue of (1.9) equal to the integral

$$(2.4)\quad \int \exp[-V^{(n)}(X)]g(X(t))d\lambda_x^c(X)$$

with $t_j=jt/n$ and $V^{(n)}(X)=V_n(X(t_0),\cdots,X(t_n))$. Rewriting it by use of Itô's formula [8, Chap.II,5, Theorem 5.1], we can show it approaches the last member of (1.6) as $n\to\infty$.

<u>Proof of Theorem 2</u>. Since $\exp[-t(H^c-mc^2)]$ is, by Theorem 1, uniformly quasi-contractive in $c\geq 1$, we may assume g is in $\mathcal{S}(R^d)$. Since $\psi^c(t,x)$ and $\psi(t,x)$ are small in L^2 where $|x|$ is sufficiently large, uniformly both on every finite t-interval and in $c\geq 1$, it suffices to show $\psi^c(t,x)$ converges to $\psi(t,x)$ in L^∞ as $c\to\infty$, uniformly on every finite t-interval. To this end we utilize the path integral representation (1.6) for $\psi^c(t,x)$ to prove its convergence to the Feynman-Kac-Itô formula representing the solution $\psi(t,x)$ of Eq.(1.5) with the same initial data $\psi(0,x)=g(x)$:

$$(2.5)\quad \psi(t,x) = \left[e^{-tH}g\right](x) = \int e^{-W(t,0;X)}g(X(t))d\mu_x(X)$$

with

$$(2.6)\quad W(t,0;X) = i\int_0^t eA(X(s))dX(s) + \frac{i}{2}\int_0^t (\mathrm{div}\, eA)(X(s))ds + \int_0^t e\Phi(X(s))ds.$$

Here μ_x is the Wiener measure on the space $C_x([0,\infty)\to R^d)$ of the continuous paths $X: [0,\infty)\to R^d$ with $X(0)=x$ whose characteristic function is $\exp[-tp^2/2m]$. The first integral on the right of (2.6) is the Itô integral. Notice that the integral (2.4) with μ_x in place of λ_x^c approaches, as $n\to\infty$, the Feynman-Kac-Itô formula, i.e. the last member of (2.5). The point of proof is to justify commutativity of the procedures of taking the limits $n\to\infty$ and $c\to\infty$. Needless to say, the measure λ_x^c is weakly convergent to the Wiener measure μ_x as $c\to\infty$, both as the probability measures on $D_x([0,\infty)\to R^d)$.

Proof of Corollary. Since both selfadjoint operators H^c and H are, by Theorem 1, bounded from below by $M = \inf_{x\in R^d} e\Phi(x)$, it follows from Theorem 2 with the Trotter-Kato theorem [10, Chap.IX,Theorem 2.16] that as $c\to\infty$, H^c-mc^2 converges to H in strong resolvent sense, and so does $i[H^c-mc^2]$ to iH. Then use again the Trotter-Kato theorem to get the desired assertion.

In a subsequent paper the zero-mass limit problem will be discussed.

REFERENCES

[1] F.A.Berezin and M.A.Šubin, Symbols of operators and quantization, Coll. Math. Soc. Janos Bolyai 5, Hilbert Space Operators, 21-52, Tihany, 1970.

[2] R.J.Cirincione and P.R.Chernoff, Dirac and Klein-Gordon equations : Convergence of solutions in the nonrelativistic limit, Commun. Math. Phys. 79, 33-46(1981).

[3] C.Garrod, Hamiltonian path-integral methods, Rev. Mod. Phys. 38, 483-493(1966).

[4] A.Grossmann, G.Loupias and E.M.Stein, An algebra of pseudo-differential operators and quantum mechanics in phase space, Ann. Inst. Fourier(Grenoble) 18, 343-368(1968).

[5] W.Hunziker, On the nonrelativistic limit of the Dirac theory, Commun. Math. Phys. 40, 215-222(1975).

[6] T.Ichinose, The nonrelativistic limit problem for a relativistic spinless particle in an electromagnetic field, to appear in J. Functional Analysis.

[7] T.Ichinose and H.Tamura, Imaginary-time path integral for a relativistic spinless particle in an electromagnetic field, Commun. Math. Phys. 105, 239-257(1986).

[8] N.Ikeda and S.Watanabe, Stochastic Differential Equations and Diffusion Processes, North-Holland/Kodansha, Amsterdam, Tokyo, 1981.

[9] K.Itô, Stochastic Processes, Lecture Notes Series, 16, Aårhus University, 1969.

[10] T.Kato, Perturbation Theory for Linear Operators, 2nd ed., Springer, Berlin-Heidelberg-New York, 1976.

[11] L.D.Landau and E.M.Lifschitz, Course of Theoretical Physics, Vol.2, The Classical Theory of Fields, 4th revised English ed., Pergamon Press, Oxford, 1975.

[12] M.M.Mizrahi, The Weyl correspondence and path integrals, J. Math. Phys. 16, 2201-2206(1975).

[13] M.Sato, Operator ordering and perturbation expansions in the path integral formalism, Prog. Theor. Phys. 58, 1262-1270 (1970).

[14] M.A.Shubin, Essential self-adjointness of uniformly hypo-elliptic operators, Vestnik Moskov. Univ. Ser.I Mat. Meh. 30, 91-94(1975); English transl. Moscow Univ. Math. Bull. 30, 147-150(1975).

[15] B.Simon, Functional Integration and Quantum Physics, Academic Press, New York, 1979.

Scattering with Penetrable Wall Potentials

Teruo Ikebe and Shin-ichi Shimada

Department of Mathematics, Faculty of Science
Kyoto University, Kyoto 606, Japan

§1. Introduction

In this paper we shall consider the Schrödinger operator with a penetrable wall potential formally of the form $H_{formal} = -\Delta + q(x)\times \times\delta(|x| - a)$, where $q(x)$ is real and continuous on $S_a = \{x \in R^3 ; |x| = a\}$, $a > 0$, and δ denotes the delta function. (This operator is said to provide a simple model for the α-decay [1].) The form counterpart of H_{formal} is $h[u,v] = (H_{formal}u,v) = (\nabla u,\nabla v) + (qu,v)_a$ with domain $Dom[h] = H^1(R^3)$, where $(\, , \,)$ means the $L_2(R^3)$ inner product, $(\, , \,)_a$ the $L_2(S_a)$ inner product, and $H^m(G)$ the Sobolev space of order m over G. h is shown to be a lower semibounded closed form, and thus determines a lower semibounded selfadjoint operator H. We should note here that while h is a "small" perturbation of h_o ($h_o[u,v] = (\nabla u,\nabla v)$, $Dom[h_o] = H^1(R^3)$) via an infinitesimally h_o-bounded form, $H - H_o$ is not H_o-bounded, where $H_o = -\Delta$, $Dom(H_o) = H^2(R^3)$, is the selfadjoint operator associated with h_o. We shall show, however, that the difference of the resolvents $R(z)$ and $R_o(z)$ of H and H_o lies in the trace class. This implies by a well-known theorem that the wave operators intertwining H and H_o exist and are complete.

§2. The Definition of H

Consider the form

$$h[u,v] = (\nabla u,\nabla v) + (q\gamma u,\gamma v)_a, \quad Dom[h] = H^1(R^3),$$

where γ is the trace operator from $H^1(R^3)$ to $L_2(S_a)$. h is a well-defined symmetric, lower semibounded, closed form, which follows from the fact that γ is infinitesimally form-bounded with respect to h_o, i.e., for any $\varepsilon > 0$ there exists a constant C_ε such that

$$\|\gamma u\|_a^2 \le \varepsilon\|\nabla u\|^2 + C_\varepsilon\|u\|^2 \quad \text{for } u \in H^1(R^3), \tag{2.1}$$

where $\|u\| = (u,u)^{\frac{1}{2}}$ and $\|u\|_a = (u,u)_a^{\frac{1}{2}}$, a fact obtainable in the standard manner. (2.1) also shows that γ is bounded from $H^1(R^3)$ to $L_2(S_a)$. Now by virtue of Theorem 2.1 of [2, Chap. IV] there is a unique lower semibounded selfadjoint operator H such that

(2.2) $\mathrm{Dom}(H) \subset \mathrm{Dom}[h]$; $(Hu,v) = h[u,v]$ for $u \in \mathrm{Dom}(H)$, $v \in \mathrm{Dom}[h]$.

Theorem 1. $u \in \mathrm{Dom}(H)$ if and only if $u \in H^1(R^3)$, $u \in H^2(|x| < a)$, $u \in H^2(|x| > a)$ and $\{q(x)u(x) + (\partial u/\partial n_+)(x) + (\partial u/\partial n_-)(x)\}|_{S_a} = 0$, where n_+ (n_-) denotes the outward (inward) normal to S_a.

Proof. Assume $u \in \mathrm{Dom}(H)$. Then by (2.2) we have for any $v \in C_0^\infty(|x| < a)$ $\int_{|x|<a} Hu\bar{v}dx = -\int_{|x|<a} u\overline{\Delta v}dx$, which implies $Hu = -\Delta u$ in $\{x ; |x| < a\}$ in the distribution sense and $u \in H^2(|x| < a)$. Similarly, $u \in H^2(|x| > a)$. Therefore, it makes sense to speak of $\partial u/\partial n_\pm|_{S_a}$. Thus we can integrate by parts to obtain for any $v \in C_0^\infty(R^3)$

$$(Hu,v) = -(\Delta u,v) + (q\gamma u + \{\partial u/\partial n_+ + \partial u/\partial n_-\}|_{S_a},v)_a,$$

and so $\{qu + \partial u/\partial n_+ + \partial u/\partial n_-\}|_{S_a} = 0$.

Conversely, let u satisfy the conditions of the theorem. Define $w \in L_2(R^3)$ by $w = -\Delta u$ (except on S_a). Then for any $v \in \mathrm{Dom}(H)$ we have, noting that $Hv = -\Delta v$, $(u,Hv) = (u,-\Delta v) = (w,v) + (q\gamma u + \{\partial u/\partial n_+ + \partial u/\partial n_-\}|_{S_a},v)_a = (w,v)$. This implies $u \in \mathrm{Dom}(H^*) = \mathrm{Dom}(H)$ (* means adjoint). q.e.d.

§3. The resolvent Equation

Consider the following integral operator T_κ depending on a complex parameter κ defined by

$$(T_\kappa f)(x) = -\frac{1}{4\pi}\int_{S_a} \frac{e^{i\kappa|x-y|}}{|x-y|} q(y)f(y)\, dS_y \quad (x \in R^3).$$

Lemma 1. Let $\mathrm{Im}\,\kappa > 0$. Then T_κ is a Hilbert-Schmidt operator from $L_2(S_a)$ to $L_2(R^3)$.

Proof. Put $b = \mathrm{Im}\,\kappa$. We compute the Hilbert-Schmidt norm $\|T_\kappa\|_2$ of T_κ:

$$\|4\pi T_\kappa\|_2^2 = \int_{S_a} dS_y|q(y)|^2 \int_{R^3} dx \frac{e^{-2b|x-y|}}{|x-y|^2} = \frac{2\pi}{b}\|q\|_a^2 < \infty,$$

from which follows the assertion. q.e.d.

Let us recall that $(4\pi|x-y|)^{-1} e^{i\sqrt{z}|x-y|}$ is the integral kernel of the resolvent $R_0(z)$, $z \notin [0,\infty)$, where $\sqrt{z}$ is such that $\mathrm{Im}\sqrt{z} \geq 0$.

It is easy to see that $T_\kappa^* = -q\gamma R_o(\overline{\kappa^2})$, which maps from $L_2(R^3)$ to $L_2(S_a)$. So, as a corollary to Lemma 1 with $q(x) \equiv 1$ we have

Lemma 2. Let $z \notin [0,\infty)$. Then $\gamma R_o(z)$ is a Hilbert-Schmidt operator from $L_2(R^3)$ to $L_2(S_a)$.

Theorem 2. Let $z \in \rho(H_o) \cap \rho(H)$, where ρ denotes the resolvent set. Then $\gamma R(z)$ is a bounded operator from $L_2(R^3)$ to $L_2(S_a)$, and the following resolvent equation holds:

$$R(z) - R_o(z) = T_{\sqrt{z}}\gamma R(z).$$

Proof. That $\gamma R(z)$ is bounded from $L_2(R^3)$ to $L_2(S_a)$ follows from the relation $\mathrm{Ran}(R(z)) = \mathrm{Dom}(H) \subset H^1(R^3)$ (see Theorem 1) (Ran = range) and the closed graph theorem. Let $u \in \mathrm{Dom}(H)$ and $v \in \mathrm{Dom}(H_o)$. In view of Theorem 1 and $\mathrm{Dom}(H_o) = H^2(R^3)$ we have

$$((H - z)u,v) = h[u,v] - (u,\bar{z}v) = (u,(H_o - \bar{z})v) + (q\gamma u,\gamma v)_a,$$

and hence, on putting $u = R(z)\phi$ and $v = R_o(\bar{z})\psi = R_o(z)^*\psi$,

$$(R_o(z)\phi,\psi) = (R(z)\phi,\psi) + (q\gamma R(z)\phi,\gamma R_o(\bar{z})\psi)_a = (R(z)\phi - T_{\sqrt{z}}\gamma R(z)\phi,\psi),$$

where we have used the relation $T^*_{\sqrt{z}} = -q\gamma R_o(\bar{z})$ remarked before Lemma 2. The required resolvent equation follows immediately. q.e.d.

§4. The Wave Operators

The wave operators $W_\pm$ which intertwine H and H_o are defined as

$$W_\pm = \operatorname*{strong\ limit}_{t \to \pm\infty} e^{itH}e^{-itH_o}$$

if they exist. In this § we shall prove the following

Theorem 3. $W_\pm$ exist and are complete.

Our proof of the above theorem will be based on Theorem 4.8 of [2, Chap. X] which states that the wave operators exist and are complete if the difference of the resolvents is a trace-class operator. Thus, in view of Lemma 1, Theorem 2 and the well-known fact that an operator is in the trace class if and only if it is a product of two Hilbert-Schmidt operators, the proof is reduced to that of the next

Lemma 3. $\gamma R(-b^2)$ is a Hilbert-Schmidt operator from $L_2(R^3)$ to

$L_2(S_a)$ for a sufficiently large $b > 0$.

Proof. On operating γ from left on the resolvent equation, we have by Theorem 2 $(1 - \gamma T_{ib})\gamma R(-b^2) = \gamma R_o(-b^2)$. If we show that $1 - \gamma T_{ib}$ has a bounded inverse for a suitable $b > 0$, then the lemma follows, for $\gamma R_o(-b^2)$ is a Hilbert-Schmidt operator by Lemma 2. Now γT_{ib} is an integral operator with kernel

$$g(x,y;b) \equiv (4\pi|x - y|)^{-1} e^{-b|x - y|}q(y) \quad (x,y \in S_a),$$

and both the integrals $\int_{S_a}|g(x,y;b)|dS_x$ and $\int_{S_a}|g(x,y;b)|dS_y$ are are finite and tend to 0 as b tends to $+\infty$. Therefore, according to a well-known result on integral operators (see, e.g., Example 2.4 of [2, Chap. III]), the operator norm of γT_{ib} is less than unity, which makes possible the Neumann series inversion of $1 - \gamma T_{ib}$. q.e.d.

The proof of Theorem 3 is now complete.

A more detailed discussion of the results obtained here together with further development will be given elsewhere.

References

[1] Petzold, J.: Wie gut gilt das Exponentialgesetz beim α-Zerfall? Zeits. f. Phys. 155 (1959), 422-432.

[2] Kato, T.: Perturbation Theory for Linear Operators. Berlin-Heidelberg New York, Springer, 1976.

Commutator Methods and Asymptotic Completeness for a New Class of Stark Effect Hamiltonians

Arne Jensen *

Matematisk Institut
Aarhus Universitet
DK - 8000 Aarhus C
Denmark

Department of Mathematics
University of Kentucky
Lexington, KY 40 506 - 0027
USA

1. Introduction.

In this note we present a generalization to dimension $n > 1$ of some recent results on asymptotic completeness for the Stark effect Hamiltonian [7,8]. The free Stark effect Hamiltonian is given by

$$H_0 = -\Delta + x_1$$

in $L^2(\mathbb{R}^n)$. We have taken the electric field in the direction of the negative x_1-axis, and normalized constants. We introduce the following assumptions:

Assumption 1.1. Let $U(x_1)$ be a realvalued function which satisfies $U(x_1) = W''(x_1)$ where $W \in C^\infty(\mathbb{R})$ is bounded with all derivatives bounded.

Assumption 1.2. Let V be a symmetric operator on $L^2(\mathbb{R}^n)$ with domain $\mathcal{D}(V) \supset \mathcal{D}(H_0)$. Assume $V(H_0 + i)^{-1}$ is compact and for some $\delta > \frac{1}{2}$ $(H_0 + i)^{-1} V (1 + x_1^2)^{\delta/2} \chi_{(-\infty,0)}(x_1)$ extends to a bounded operator on $L^2(\mathbb{R}^n)$.

We define the full Hamiltonian

$$H = H_0 + U(x_1) + V$$

in $L^2(\mathbb{R}^n)$, where U and V satisfy the above assumptions. The main results is:

---

* Partially supported by NSF - grant DMS - 8401748.

Theorem 1.3. Let H be as above. Then the wave operators

$$W_{\pm}(H,H_0) = s-\lim_{t\to\pm\infty} e^{itH} e^{-itH_0}$$

exist and are asymptotically complete. Furthermore, the singular continuous spectrum of H is empty, and the point spectrum of H is discrete in $\mathbb{R}$.

This result was obtained in [7,8] for $n=1$ and V a multiplication operator. For $U \equiv 0$ the result is well known, see e.g. [1, 2, 5, 10, 11, 12]. Results on existence and completeness of wave operators $W_{\pm}(H,H_0+U)$ have been obtained in [3,4], in case $U(x_1)$ is periodic.

The proof is based on commutator methods [6,9], and is somewhat more complicated than the one given for $n=1$ in [7,8].

2. Propagation Estimates.

The proofs are based on propagation estimates with respect to the conjugate operator

$$A = -p_1 = i\frac{\partial}{\partial x_1}$$

on $L^2(\mathbb{R}^n)$. We have on the Schwartz space $S(\mathbb{R}^n)$

$$i[H_0,A] = I \,. \tag{2.1}$$

Let P_A^+ (P_A^-) denote the spectral projections of A corresponding to the interval $(0,\infty)$ ($(-\infty,0)$).

Proposition 2.1. Let $s>0$. Then

$$\|(1+A^2)^{-s/2} e^{-itH_0} (1+A^2)^{-s/2}\| \le c(1+|t|)^{-s} \tag{2.2}$$

for all $t\in\mathbb{R}$, and

$$\|(1+A^2)^{-s/2} e^{-itH_0} P_A^{\pm}\| \le c(1+|t|)^{-s} \tag{2.3}$$

for all t, $\pm t>0$.

Proof. We note that (2.1) implies

$$e^{itH_0} A e^{-itH_0} = A+t$$

on $S(\mathbb{R}^n)$. From this relation (2.2) and (2.3) easily follow. □

Proposition 2.2. Let U satisfy Assumption 1.1, and let $H_1 = H_0 + U(x_1)$. Let $g \in C_0^\infty(\mathbb{R})$. Then for $0 < s' < s$ we have

$$\|(1+A^2)^{-s/2} e^{-itH_1} g(H_1)(1+A^2)^{-s/2}\| \le c(1+|t|)^{-s'} \tag{2.4}$$

for all $t \in \mathbb{R}$, and

$$\|(1+A^2)^{-s/2} e^{-itH_1} g(H_1) P_A^{\pm}\| \le c(1+|t|)^{-s'} \tag{2.5}$$

for all t, $\pm t > 0$.

Proof (sketch): To prove (2.4) and (2.5) we first note on $S(\mathbb{R}^n)$

$$i[H_1, A] = I + U'(x_1) .$$

Let $a \in \mathbb{R}$. We claim that there exist $\varepsilon > 0$, $\alpha > 0$, such that with $J = (a-\varepsilon, a+\varepsilon)$

$$E_{H_1}(J)\, i[H_1, A] E_{H_1}(J) \ge \alpha E_{H_1}(J) , \tag{2.6}$$

where E_{H_1} denotes the spectral measure of H_1. The estimate (2.6) is the Mourre estimate. The result in [6,9] and the assumption on U now imply that (2.4) and (2.5) hold. The proof of (2.6) requires the use of a direct integral decomposition and estimates obtained in [7], since $U'(x_1)$ is not compact relative to H_1 (in the case $n > 1$). Details will be given elsewhere. □

We need to replace the localization $(1+A^2)^{-s/2}$ by a localization in x_1; see the discussion in [6]. Let $\chi_+ \in C^\infty(\mathbb{R})$ satisfy $\chi_+(x_1) = 1$, $x_1 \ge 2$, $\chi_+(x_1) = 0$, $x_1 \le 1$, and let $\chi_-(x_1) = 1 - \chi_+(x_1)$. Define

$$\rho(x_1) = \chi_+(x_1) + \chi_-(x_1)(1+x_1^2)^{-\frac{1}{2}} .$$

Lemma 2.3. For δ, $0 \le \delta \le 1$, the operator

$$(A^2+1)^{\delta}(H_1+i)^{-1}\rho(x_1)^{\delta}$$

extends to a bounded operator on $L^2(\mathbb{R}^n)$.

Proof. The result is obtained by interpolation. Thus it suffices to prove that

$$p_1^2 (H_1+i)^{-1} \rho(x_1)$$

extends to a bounded operator on $L^2(\mathbb{R}^n)$.

This is done by commutator computations, using results on U and p_1 obtained in [7]. □

Proposition 2.4. Let U satisfy Assumption 1.1 and let $H_1 = H_0 + U(x_1)$. For $0 \leq s' < s \leq 1$, $g \in C_0^\infty(\mathbb{R})$, we have

$$\|\rho(x_1)^s e^{-itH_1} g(H_1)\rho(x_1)^s\| \leq c(1+|t|)^{-2s'} \tag{2.7}$$

for all $t \in \mathbb{R}$, and

$$\|\rho(x_1)^s e^{-itH_1} g(H_1) P_A^{\pm}\| \leq c(1+|t|)^{-2s'} \tag{2.8}$$

for all t, $\pm t > 0$.

Proof. The results follow from Proposition 2.2 and Lemma 2.3. Note that the restriction $s \leq 1$ comes from Lemma 2.3. □

3. Wave operators.

Throughout this section it is assumed that $U(x_1)$ satisfies Assumption 1.1 and that V satisfies Assumption 1.2. As above we let $H_1 = H_0 + U(x_1)$ and $H = H_0 + U(x_1) + V = H_1 + V$. The proof of Theorem 1.3 is based on the chain rule for wave operators:

$$W_\pm(H,H_0) = W_\pm(H,H_1)W_\pm(H_1,H_0) . \tag{3.1}$$

Proposition 3.1. The wave operators

$$W_\pm(H,H_1) = s - \lim_{t\to\pm\infty} e^{itH} e^{-itH_1}$$

exist and are asymptotically complete. Furthermore, the singular continuous spectrum of H is empty, and the point spectrum of H is discrete in $\mathbb{R}$.

Proof. The result follows from the Enss method in the version given by Mourre. See e.g. [9,10] for similar proofs. The propagation estimates needed in this method are given in Proposition 2.4, and the assumption on V implies that it is a short range perturbation of H_1. □

<u>Remark 3.2</u>. Results on $W_\pm(H,H_1)$ were obtained in [3,4] for $U(x_1)$ periodic. We note that our result can be extended to $U = W''$, $W \in B^5(\mathbb{R})$, if we take $\delta \geq 1$ in Assumption 1.2. This follows from [9; Theorem 4.4].

<u>Proposition 3.3</u>. The wave operators

$$W_\pm(H_1,H_0) = \underset{t\to\pm\infty}{s-\lim}\, e^{itH_1} e^{-itH_0}$$

exist and are unitary.

<u>Proof</u>. The function $U(x_1)$ depends only on x_1. Thus the problem is reduced to the one-dimensional case, and the result follows from those in [7]. □

<u>Proof of Theorem 1.3</u>. This result follows from (3.1), Proposition 3.1, and Proposition 3.3. □

4. Examples of Potentials.

In this section we briefly give some examples of potentials satisfying our assumptions. It is clear that a trigonometric polynomial (without constant term) satisfies Assumption 1.1. More generally, a large class of periodic, sums of periodic, and almost-periodic functions satisfy this assumption. A large class can be obtained by the following construction: Let U be a realvalued function which can be represented as

$$U(x_1) = \int_{-\infty}^{\infty} e^{ix_1\omega}\, d\mu(\omega)$$

where μ is a Borel measure satisfying

$$\int_{-\infty}^{\infty} \omega^{-2} d|\mu|(\omega) < \infty$$

and

$$\int_{-\infty}^{\infty} \omega^{2\ell} d|\mu|(\omega) < \infty$$

for all $\ell = 0,1,2,\dots$. It suffices to take

$$W(x_1) = \int_{\infty}^{\infty} (-\omega^{-2}) e^{i x_1 \omega} \, d\mu(\omega) .$$

For V multiplication by a realvalued function V(x) the following conditions given in [12] imply that V satisfies Assumption 1.2.

$$V(x) = \{(1+x_1^2)^{-\delta/2}\chi_-(x_1) + (1+x_1^2)^{\frac{1}{2}}\chi_+(x_1)\}\cdot(V_1(x)+V_2(x))$$

with $\delta > \frac{1}{2}$,

$$V_1 \in L^\infty(\mathbb{R}^n) \ , \quad \lim_{|x|\to\infty} V_1(x) = 0 \ ,$$

and $V_2 \in L^2_{loc}(\mathbb{R}^n)$ such that for some ν , $0 < \nu < 4$,

$$\lim_{|x|\to\infty} (1+x_1^2) \int_{|x-y|\le 1} |V_2(y)|^2 |x-y|^{-n+\nu} dy = 0 .$$

5. References

1. Avron, J. E., Herbst, I. W.: Spectral and scattering theory for Schrödinger operators related to the Stark effect. Commun. Math. Phys. 52 (1977), 239-254.

2. Ben-Artzi, M.: Remarks on Schrödinger operators with an electric field and deterministic potentials. J. Math. Anal. Appl. 109 (1985), 333-339.

3. Ben-Artzi, M., Devinatz, A.: Resolvent estimates for a sum of tensor products with applications to the spectral theory of differential operators. J. Analyse Math. 43 (1983/84), 215-250.

4. Ben-Artzi, M., Devinatz, A.: The limiting absorption principle for partial differential operators. Preprint, 1985.

5. Herbst, I. W.: Unitary equivalence of Stark effect Hamiltonians. Math. Z. 155 (1977), 55-70.

6. Jensen, A.: Propagation estimates for Schrödinger-type operators. Trans. Amer. Math. Soc. 291 (1985), 129-144.

7. Jensen, A.: Asymptotic completeness for a new class of Stark effect Hamiltonians. Commun. Math. Phys., to appear.

8. Jensen, A.: Commutator methods and asymptotic completeness for one-dimensional Stark-effect Hamiltonians. Proc. Conf. Schrödinger Operators, Aarhus Univ.,1985. (E. Balslev, ed.). Springer Lecture Notes in Mathematics, to appear.

9. Jensen, A., Mourre, E., Perry, P.: Multiple commutator estimates and resolvent smoothness in quantum scattering theory. Ann. Inst. H. Poincaré, Sect.A, 41 (1984), 207-224.

10. Perry, P.: Scattering theory by the Enss method. Math. Reports 1, part 1, 1983.

11. Simon, B.: Phase space analysis of simple scattering systems: Extensions of some work of Enss. Duke Math. J. 46 (1979), 119-168.

12. Yajima, K.: Spectral and scattering theory for Schrödinger operators with Stark-effect. J. Fac. Sci. Univ. Tokyo, Sect. IA, 26 (1979), 377-390.

ASYMPTOTICS OF THE TITCHMARSH-WEYL m-COEFFICIENT FOR INTEGRABLE POTENTIALS, II

Hans G. Kaper and Man Kam Kwong
Mathematics and Computer Science Division,
Argonne National Laboratory, Argonne, IL 60439-4844, U.S.A.

Abstract

This article is concerned with the Titchmarsh-Weyl m-coefficient for the differential equation $y'' + (\lambda - q(x))y = 0$ on $[0,\infty)$, where the potential q is integrable near 0. The m-coefficient, which is a function of the complex variable $\lambda \in C$, depends on an initialization parameter $\alpha \in (-\pi, \pi]$. A uniform as well as various non-uniform asymptotic expansions of m are derived, which are valid as $\lambda \to \infty$ in any sector of the complex plane that does not contain the real axis. The results reported here extend those of Kaper and Kwong [*Proc. Roy. Soc. Edinburgh Sect. A*, to appear].

1. The Titchmarsh-Weyl m-coefficient

Consider the singular differential equation

$$y'' + (\lambda - q(x))y = 0, \quad x \geq 0, \tag{1}$$

where q is real-valued and $q \in L^1(0, X)$ for some $X > 0$; $\lambda \in C$, with $\text{Im}\,\lambda \neq 0$. Let ϕ and θ be the solutions of (1) that satisfy the initial conditions

$$\phi(0) = \cos\alpha, \quad \phi'(0) = -\sin\alpha, \tag{2a}$$

$$\theta(0) = \sin\alpha, \quad \theta'(0) = \cos\alpha, \tag{2b}$$

where α is fixed, $\alpha \in (-\pi, \pi]$. The functions ϕ and θ, which depend parametrically on λ and α, define a family of functions $\{M_x(\cdot) : C \to C\}_{x \geq 0}$,

$$M_x(z) = \frac{\theta(x)z + \theta'(x)}{\phi(x)z + \phi'(x)}, \quad z \in C. \tag{3}$$

If $\text{Im}\,\lambda > 0$, M_x maps the lower half of the complex plane onto the interior of the *Weyl disc*, $D_x = M_x(\{z \in C : \text{Im}\, z \leq 0\})$, the diameter of which is

$$\text{diam}\, D_x = \left| \text{Im}\,(\lambda) \int_0^x |\phi(t)|^2 dt \right|^{-1}. \tag{4}$$

As $x \to \infty$, the boundary ∂D_x of D_x contracts to a limit point or limit circle. Let m denote this limit point or, in the limit-circle case, a point on the limit circle; m, the *Titchmarsh-Weyl m-coefficient*, is a function of α and λ.

The properties of m have been studied by several authors [1, 2, 3]. In particular, the asymptotic behavior of m as $\lambda \to \infty$ has been the subject of extensive investigations [4, 5, 6, 7, 8, 9, 10, 11, 12]. In [12], which is Part I of this investigation, we considered the case $\alpha = 0$ and obtained the asymptotic expansion of m as $\lambda \to \infty$. In the present note we generalize this result and derive an asymptotic expansion of m that is valid uniformly in the initialization parameter α. The expansion exhibits (singular) boundary layer behavior near $\alpha = \pm\frac{1}{2}\pi$.

2. Asymptotic expansion of $m(\alpha, \lambda)$

In this section we derive a uniform, as well as various non-uniform asymptotic expansions of the Titchmarsh-Weyl coefficient $m(\alpha, \lambda)$ as $\lambda \to \infty$ in a sector in the upper half of the complex plane.

Let ε be fixed, $0 < \varepsilon < \frac{1}{2}\pi$, and let S_ε be the following sector in the upper half of the complex plane:

$$S_\varepsilon = \{\lambda \in \mathbf{C} : \varepsilon \le \arg \lambda \le \pi - \varepsilon\}. \tag{5}$$

We assume throughout this investigation that $\lambda \in S_\varepsilon$. We use the notation

$$k = -i\lambda^{1/2} = a - ib. \tag{6}$$

Thus, a and b are strictly positive for all $\lambda \in S_\varepsilon$. More precisely,

$$a = \mathrm{Im}\,(\lambda^{1/2}) \ge |\lambda|^{1/2} \sin(\tfrac{1}{2}\varepsilon), \quad b = \mathrm{Re}\,(\lambda^{1/2}) \ge |\lambda|^{1/2} \sin(\tfrac{1}{2}\varepsilon). \tag{7}$$

The first lemma shows that the diameter of the Weyl disc D_X becomes asymptotically small as $\lambda \to \infty$.

LEMMA 1. *The diameter of the Weyl disc D_X satisfies the asymptotic estimate*

$$\mathrm{diam}\, D_X \le C|\lambda|^{1/2}\, e^{-2X \sin(\frac{1}{2}\varepsilon)\, |\lambda|^{1/2}}, \quad \lambda \to \infty. \tag{8}$$

The constant C depends on ε, but not on α or λ; $C \to \infty$ as $\varepsilon \to 0$.

Proof. Let the function u be defined by the identity

$$u(x) = \phi(x)e^{-kx}, \quad x \ge 0. \tag{9}$$

Using the variation-of-constants method, we find that u satisfies the equation

$$u(x) = \frac{1}{2k}\,(k\cos\alpha - \sin\alpha)\left\{1 + \frac{k\cos\alpha + \sin\alpha}{k\cos\alpha - \sin\alpha}e^{-2kx}\right\} + \frac{1}{2k}\int_0^x (1 - e^{-2k(x-t)})\, q(t)u(t)\, dt. \tag{10}$$

Consider the second term inside the braces. The pre-exponential factor is O(1) as $k \to \infty$, uniformly in α, so the exponential forces down the size of this term as $|k|$ increases. By taking $|k|$ sufficiently large, we can certainly achieve that the inequality

$$\left|\frac{k\cos\alpha + \sin\alpha}{k\cos\alpha - \sin\alpha}e^{-2ikx}\right| \le \frac{1}{4} \tag{11}$$

is satisfied for all $\alpha \in (-\pi, \pi]$. Eqns. (10) and (11) then yield the following integral inequality:

$$|u(x)| \le \frac{1}{|k|}|k\cos\alpha - \sin\alpha| + \frac{1}{|k|}\int_0^x |q(t)|\,|u(t)|\,dt. \tag{12}$$

Applying Gronwall's lemma [13, Section I.1.8], we thus find that

$$|u(x)| \le \frac{1}{|k|}|k\cos\alpha - \sin\alpha| \exp\left\{\frac{1}{|k|}\int_0^x |q(t)|\,dt\right\}, \qquad x \in [0, X]. \tag{13}$$

Since the integral is bounded uniformly by $\|q\|$, the L^1-norm of q on $[0, X]$, the inequality (13) yields the following upper bound for $u(x)$:

$$|u(x)| \le \frac{1}{|k|}|k\cos\alpha - \sin\alpha|\, e^{\|q\|/|k|}, \qquad x \in [0, X]. \tag{14}$$

We now derive a lower bound. From (10) we obtain

$$|u(x)| \ge \frac{1}{2|k|}|k\cos\alpha - \sin\alpha|\left\{1 - \left|\frac{k\cos\alpha + \sin\alpha}{k\cos\alpha - \sin\alpha}e^{-2ikx}\right|\right\} - \frac{1}{|k|}\int_0^x |q(t)|\,|u(t)|\,dt. \tag{15}$$

Hence, because of (14),

$$|u(x)| \ge \frac{1}{2|k|}|k\cos\alpha - \sin\alpha|\left\{1 - \left|\frac{k\cos\alpha + \sin\alpha}{k\cos\alpha - \sin\alpha}e^{-2ikx}\right| - 2\frac{e^{\|q\|/|k|}}{|k|}\int_0^x |q(t)|\,dt\right\}. \tag{16}$$

By choosing $|k|$ sufficiently large, we can certainly achieve that the last term inside the braces is less than 1/4. The middle term inside the braces is already less than 1/4, uniformly in α, according to (11). Therefore,

$$|u(x)| \ge \frac{1}{4|k|}|k\cos\alpha - \sin\alpha|, \qquad x \in [0, X]. \tag{17}$$

We can replace this lower bound by one that is independent of α. As α varies from $-\pi$ to π (k fixed), the quantity $|k\cos\alpha - \sin\alpha|$ achieves its minimum value at $\alpha = \alpha_0 (\mathrm{mod}\ \pi)$, where

$$\alpha_0 = \frac{1}{2}\tan^{-1}\frac{-2a}{|k|^2 - 1}, \qquad \alpha_0 \in (0, \tfrac{1}{2}\pi). \tag{18}$$

The minimum value can be computed and estimated:

$$|k\cos\alpha_0 - \sin\alpha_0| = \frac{|k|}{2^{1/2}}\left\{\left(1 + \frac{1}{|k|^2}\right) - \left[\left(1 + \frac{1}{|k|^2}\right)^2 - \frac{4b^2}{|k|^4}\right]^{1/2}\right\}^{1/2}$$

$$\ge \frac{|k|}{2^{1/2}}\left\{\left(1 + \frac{1}{|k|^2}\right) - \left[\left(1 + \frac{1}{|k|^2}\right)^2 - \frac{4\sin^2(\tfrac{1}{2}\varepsilon)}{|k|^2}\right]^{1/2}\right\}^{1/2}. \tag{19}$$

Here, the lower bound is $(1 + O(|k|^{-2}))\sin(\tfrac{1}{2}\varepsilon)$ as $k \to \infty$, so by choosing $|k|$ sufficiently large we can certainly achieve that

$$|k\cos\alpha - \sin\alpha| \geq \frac{1}{2}\sin(\tfrac{1}{2}\varepsilon), \qquad \alpha \in (-\pi, \pi]. \tag{20}$$

Consequently,

$$|u(x)| \geq \frac{\sin(\tfrac{1}{2}\varepsilon)}{8|k|}, \qquad x \in [0, X]. \tag{21}$$

for $|k|$ sufficiently large. We use this lower bound to estimate the integral in the expression (4) for the diameter of the Weyl disc:

$$\int_0^x |\phi(t)|^2\, dt \geq \frac{\sin^2(\tfrac{1}{2}\varepsilon)}{64|k|^2}\int_0^x e^{2at}\, dt \geq \frac{e^{2ax}\sin^2(\tfrac{1}{2}\varepsilon)}{256a|k|^2}. \tag{22}$$

The assertion of the lemma follows from (4), given the identity $\operatorname{Im}\lambda = 2ab$ and the inequalities (7) and (22). ■

Let z be a solution of (1) on $[0, X]$ and let r be defined by the ratio

$$r(x) = -\frac{z'(x)}{z(x)}, \qquad x \in [0, X]. \tag{23}$$

One readily verifies that r satisfies the Riccati equation

$$r' + (k^2 + q(x)) - r^2 = 0, \qquad x \in [0, X]. \tag{24}$$

The following lemma relates $r(x)$ to $r(0)$.

LEMMA 2. *The value of the function M_x, defined in (3), at $r(x)$ is*

$$M_x(r(x)) = \frac{r(0)\sin\alpha + \cos\alpha}{r(0)\cos\alpha - \sin\alpha}. \tag{25}$$

Proof. Taking $z = r(x)$ in (3), we see that $M_x(r(x))$ is the ratio of the Wronskians $W(z,\theta)$ and $W(z,\phi)$. ■

LEMMA 3. *If r is a solution of (24) and* $\operatorname{Im} r(X) < 0$, *then*

$$m(\alpha, \lambda) = \frac{r(0)\sin\alpha + \cos\alpha}{r(0)\cos\alpha - \sin\alpha} + R(\lambda), \tag{26}$$

where the remainder R satisfies the asymptotic estimate

$$|R(\lambda)| \leq C|\lambda|^{1/2}\, e^{-2X\sin(\frac{1}{2}\varepsilon)\,|\lambda|^{1/2}}, \qquad \lambda \to \infty. \tag{27}$$

The constant C depends on ε, but not on α or λ; $C \to \infty$ as $\varepsilon \to 0$.

Proof. The lemma is an immediate consequence of Lemmas 1 and 2. ■

Let the (complex-valued) functions q_n be defined recursively on $[0, X]$ by the expression

$$q_{n+1}(x;k) = -k\int_x^X e^{-2k(t-x)}\exp\left\{-2\int_x^t \sum_{i=1}^{n} \frac{q_i(s;k)}{k^{2^{i-1}-1}} ds\right\}(q_n(t;k))^2\, dt, \qquad n = 1, 2, \cdots \tag{28}$$

where

$$q_1(x;k) = \int_x^X e^{-2k(t-x)} q(t)\, dt. \tag{29}$$

In [12] we showed that the functions q_n are $o(1)$ as $k \to \infty$, uniformly in x, and proved the following lemma.

LEMMA 4. *Let $r(\cdot\,;k)$ be the solution of the Ricatti differential equation (24) that satisfies the boundary condition $r(X;k) = k$. Then*

$$r(x;k) = k + \sum_{n=1}^{N} \frac{q_n(x;k)}{k^{2^{n-1}-1}} + R_N^{\#}(x;k), \qquad x \in [0, X], \tag{30}$$

where the remainder $R_N^{\#}$ satisfies the asymptotic estimate

$$R_N^{\#}(x;k) = o(|k|^{-(2^N-1)}), \qquad k \to \infty, \tag{31}$$

uniformly in x on the interval $[0, X]$.

Because $\operatorname{Im} k = -b < 0$, the function $r(\cdot\,;k)$ satisfies the condition of Lemma 3. Its value at the origin can therefore be used in the expression (26) to obtain an approximation that is exponentially close to $m(\alpha, \lambda)$ as $\lambda \to \infty$.

THEOREM 5. *Let q be real-valued, $q \in L^1(0, X)$ for some $X > 0$. Let S_ε be the sector defined in (5). Then*

$$m(\alpha,\lambda) = \frac{\left(-i\lambda^{1/2} + \sum_{n=1}^{N} \frac{q_n(0;-i\lambda^{1/2})}{(-i\lambda^{1/2})^{2^{n-1}-1}}\right)\sin\alpha + \cos\alpha}{\left(-i\lambda^{1/2} + \sum_{n=1}^{N} \frac{q_n(0;-i\lambda^{1/2})}{(-i\lambda^{1/2})^{2^{n-1}-1}}\right)\cos\alpha - \sin\alpha} + R_N(\alpha,\lambda), \quad \alpha \in (-\pi, \pi], \quad \lambda \in S_\varepsilon, \tag{32}$$

where the remainder R_N satisfies the asymptotic estimate

$$R_N(\alpha,\lambda) = o(|\lambda|^{-(2^N-1)}), \qquad \lambda \to \infty, \tag{33}$$

uniformly in α.

Proof. Let $r(\cdot\,;k)$ be the solution of (24) that satisfies the boundary condition $r(X;k) = k$. A straight-forward calculation shows that

$$R_N(\alpha,\lambda) = m(\alpha,\lambda) - \frac{r(0;k)\sin\alpha + \cos\alpha}{r(0;k)\cos\alpha - \sin\alpha}$$

$$+ R_N^{\#}(0;k)\,\big(r(0;k)\cos\alpha - \sin\alpha\big)^{-1}\Big(\big(k + \sum_{n=1}^{N}\frac{q_n(0;k)}{k^{2^n-1}}\big)\cos\alpha - \sin\alpha\Big)^{-1}, \qquad (34)$$

where $R_N^{\#}$ is the remainder in (30).

The difference between the first two terms in the right member of (34) is exponentially small, according to Lemma 3. We claim that the last term is $o(|k|^{-(2^N-1)})$ as $k \to \infty$, uniformly in α.

Using arguments similar to the ones that led to the inequality (20), one shows that

$$|r(0;k)\cos\alpha - \sin\alpha| \geq \frac{1}{2}\sin(\tfrac{1}{2}\varepsilon), \qquad \alpha \in (-\pi, \pi], \qquad (35)$$

and

$$\Big|\big(k + \sum_{n=1}^{N}\frac{q_n(0;k)}{k^{2^n-1}}\big)\cos\alpha - \sin\alpha\Big| \geq \frac{1}{2}\sin(\tfrac{1}{2}\varepsilon), \qquad \alpha \in (-\pi, \pi], \qquad (36)$$

for $|k|$ sufficiently large. It follows that the modulus of the factor which multiplies $R_N^{\#}(0;k)$ is less than or equal to $4(\sin(\tfrac{1}{2}\varepsilon))^{-2}$. This upper bound is $O(1)$ as $k \to \infty$, so the asymptotic behavior of the last term of (34) is the same as that of $R_N^{\#}(0;k)$, which is $o(|k|^{-(2^N-1)})$, according to (33). This proves the claim. ■

Because the estimate (33) holds uniformly in α, we have the following result:

$$m(\alpha,\lambda) \sim \frac{\Big(-i\lambda^{1/2} + \sum_{n=1}^{\infty}\frac{q_n(0;-i\lambda^{1/2})}{(-i\lambda^{1/2})^{2^n-1}}\Big)\sin\alpha + \cos\alpha}{\Big(-i\lambda^{1/2} + \sum_{n=1}^{\infty}\frac{q_n(0;-i\lambda^{1/2})}{(-i\lambda^{1/2})^{2^n-1}}\Big)\cos\alpha - \sin\alpha}, \qquad \lambda \to \infty, \qquad (37)$$

uniformly in α, $\alpha \in (-\pi, \pi]$.

3. Non-uniform asymptotic expansions of $m(\alpha, \lambda)$

By fixing α and expanding the ratio in (37) in inverse powers of $\lambda^{1/2}$, we can derive various (non-uniform) expansions of m.

If α is bounded away from $\pm\tfrac{1}{2}\pi$, we find that the leading terms in the asymptotic expansion of $m(\alpha, \lambda)$ are

$$m(\alpha,\lambda) \sim \frac{\sin\alpha}{\cos\alpha} + \frac{1}{\cos^2\alpha}\Big(\frac{i}{\lambda^{1/2}} + \frac{1}{\lambda}\int_0^X e^{2it\lambda^{1/2}}q(t)\,dt\Big), \qquad \lambda \to \infty. \qquad (38)$$

On the other hand, if $\alpha = \pm\tfrac{1}{2}\pi$, then the leading terms in the asymptotic expansion of $m(\alpha, \lambda)$ are

$$m(\alpha, \lambda) \sim i\lambda^{1/2} - \int_0^X e^{2it\lambda^{1/2}} q(t)\, dt, \qquad \lambda \to \infty, \tag{39}$$

These leading terms were given earlier in [3].

The results (38) and (39) show that the asymptotic behavior of $m(\alpha, \lambda)$ for large λ is singular near $\alpha = \pm\frac{1}{2}\pi$: As $\alpha \to \pm\frac{1}{2}\pi$, (38) loses its meaning and the asymptotic behavior of $m(\alpha, \lambda)$ suddenly changes from $O(1)$ to $O(|\lambda|^{1/2})$. This phenomenon points to the existence of boundary layers about $\alpha = \pm\frac{1}{2}\pi$.

We obtain the asymptotic behavior of $m(\alpha, \lambda)$ inside these boundary layers by treating $\cos\alpha$ as a small parameter, such that $\lambda^{1/2}\cos\alpha = O(1)$ as $\lambda \to \infty$. Thus,

$$m(\alpha, \lambda) \sim \frac{i\lambda^{1/2}}{1 + i\lambda^{1/2}\cos\alpha} - \frac{1}{(1 + i\lambda^{1/2}\cos\alpha)^2}\int_0^X e^{2it\lambda^{1/2}} q(t)\, dt, \qquad \lambda \to \infty. \tag{40}$$

This result reduces to (38) if α is kept fixed and $\lambda \to \infty$, while it reduces to (39) if λ is kept fixed and $\alpha \to \pm\frac{1}{2}\pi$.

If $\alpha = 0$, the leading term in (38) vanishes, so the asymptotic behavior of $m(\alpha, \lambda)$ changes from $O(1)$ to $O(|\lambda|^{-1/2})$. However, the transition is regular, in the sense that the behavior at 0 follows from (38) by taking the limit $\alpha \to 0$. Details of the asymptotic expansion at $\alpha = 0$ are given in [12].

References

1 Weyl, H. Ueber gewöhnliche Differentialgleichungen mit Singularitäten und die zugehörigen Entwicklungen willkürlicher Funktionen. *Math. Ann.* 68 (1910), 220-269.

2 Titchmarsh, E. C. *Eigenfunction expansions associated with second order differential equations*, Part I, 2nd ed. (Oxford University Press, 1962).

3 Everitt, W. N. and C. Bennewitz. Some remarks on the Titchmarsh-Weyl *m*-coefficient. In: *Tribute to Ake Pleijel*, pp. 49-108 (Dept. of Mathematics, University of Uppsala, Sweden, 1980).

4 Hille, E. *Lectures on ordinary differential equations* (Addison-Wesley, London, 1969).

5 Everitt, W. N. On a property of the *m*-coefficient of a second-order linear differential equation. *J. London Math. Soc.* 4 (1972), 443-457.

6 Everitt, W. N. and S. G. Halvorsen. On the asymptotic form of the Titchmarsh-Weyl *m*-coefficient. *Applicable Anal.* 8 (1978), 153-169.

7 Atkinson, F. V. On the location of the Weyl circles. *Proc. Roy. Soc. Edinburgh Sect. A* 88 (1981), 345-356.

8 Atkinson, F. V. and C. T. Fulton. The Titchmarsh-Weyl *m*-coefficient for non-integrable potentials (preprint, 1985).

9 Harris, B. J. On the Titchmarsh-Weyl *m*-function. *Proc. Roy. Soc. Edinburgh Sect. A* 95 (1983), 223-237.

10 Harris, B. J. The asymptotic form of the Titchmarsh-Weyl *m*-function. *J. London Math. Soc. (2)* 30 (1984), 110-118.

11 Harris, B. J. The asymptotic form of the Titchmarsh-Weyl *m*-function associated with a second order differential equation with locally integrable coefficient (preprint, 1985).

12 Kaper, H. G. and Man Kam Kwong. Asymptotics of the Titchmarsh-Weyl *m*-coefficient for integrable potential functions. *Proc. Roy. Soc. Edinburgh Sect. A* (to appear).

13 Piccinini, L. C., G. Stampacchia and G. Vidossich. *Ordinary Differential Equations in R^n* (Springer-Verlag, New York, 1984).

Notes

This work was supported by the Applied Mathematical subprogram of the Office of Energy Research, U.S. Department of Energy, under contract W-31-109-Eng-38.

The permanent address of the second author (M.K.K.) is: Department of Mathematical Sciences, Northern Illinois University, DeKalb, IL 60115.

ON THE DIFFERENCE BETWEEN EIGENVALUES OF STURM-LIOUVILLE OPERATORS AND THE SEMI-CLASSICAL LIMIT [1]

Werner Kirsch
Institut für Mathematik
Ruhr-Universität Bochum
D-4630 Bochum, West Germany

1. Introduction: Consider a Sturm-Liouville operator L given by

$$L\,u := -\,(p\,u')' + V\,u \tag{1}$$

where p and V are bounded functions on an interval $[a,b]$.
We suppose that p is nonnegative and absolutely continuous and that $1/p$ is integrable on $[a,b]$ (see Neumark [6]). L is considered as an operator on $L^2(a,b)$, $-\infty < a < b < \infty$ with boundary conditions:

$$u(a) = u(b) = 0 \tag{2}$$

L has purely discrete spectrum (see [6]). We denote by $E_n = E_n(L)$ the eigenvalue of L where

$$E_o < E_1 \quad \dots \quad E_{n-1} < E_n < \quad \dots$$

All eigenvalues of L are nondegenerate. The purpose of this paper is to give estimates from below on the differences $E_n - E_{n-1}$. Quantum mechanical tunneling shows that the difference between E_n and E_{n-1} may be exponentially small (Harrel [2]) so that an exponentially small lower bound is the best we can hope for.

In [4] Simon and the present author gave estimates of this kind for the Schrödinger case, i.e. for $p \equiv 1$, including operators on the whole real line. We extend their methods to the above situation essentially by allowing their parameter λ to be a function rather than a constant. The same idea was used independently by S. Nakamura [5] to obtain estimates for one dimensional Schrödinger operators in the semiclassical limit. We will discuss this work also in the Sturm-Liouville case.

The method presented here combined with arguments from [4] can be used to treat the operator L on the whole real line. However, to avoid technicalities which may obscure

---

1 Research partially supported by the Deutsche Forschungsgemeinschaft (DFG)

the elementary main argument we will restrict ourselves to the case of a finite interval and leave the extension for a subsequent paper.

I am indebted to Shu Nakamura and to Barry Simon for very stimulating discussions.

2. General setup and Prüfer variables: From the equation

$$- (p\, u')' + V u = E u \tag{3}$$

we easily obtain the Ricatti equation:

$$\left(\frac{pu'}{u}\right)' = (V-E) - \frac{1}{p}\left(\frac{pu'}{u}\right)^2 \tag{4}$$

Introducing the Prüfer variables $r = r(x)$ and $\theta = \theta(x)$ by

$$u(x) = r(x) \cos \theta(x) \tag{5a}$$

$$p\, u'(x) = - \lambda(x)\, r(x) \sin \theta(x) \tag{5b}$$

we obtain first order differential equations for θ and r via (4) which are essentially equivalent to equation (3). In particular we get for the Prüfer angle θ:

$$\theta' = \frac{E-V}{\lambda} \cos^2 \theta + \frac{\lambda}{p} \sin^2 \theta - \frac{\lambda'}{\lambda} \sin \theta \cos \theta \tag{6}$$

Note that our definition of the Prüfer variables differs from the usual one by the factor $\lambda(x)$. The function λ remains arbitrary until we choose it in concrete situations. We may even allow λ to depend on the energy parameter E. Of course, to avoid problems with the definition (5) we will always take $\lambda(x) > 0$ if $p(x) > 0$.

For fixed E we choose a solution u of (3) satisfying

$$u(a) = 0 \tag{7}$$

Up to normalization this defines a unique function u. E is an eigenvalue of L and u is an eigenfunction iff $u(b) = 0$. Condition (7) and definition (5) determine the Prüfer angle θ only mod π. For definiteness we set

$$\theta(a) = - \frac{\pi}{2} \tag{8}$$

A solution $\theta = \theta(x,E)$ of (6) and (8) defines via (5) a solution u of (3), (7) up to normalization. E is an eigenvalue iff $\theta(b,E) = \pi/2 \mod \pi$. Moreover by the Sturm oscillation Theorem the function $\theta(x,E_n)$ satisfies:

$$\theta(b,E_n) = \frac{\pi}{2} + n\pi \tag{9}$$

in particular we get:

$$\theta(b,E_n) - \theta(b,E_{n-1}) = \pi \tag{10}$$

We set $\phi(x,E) = \frac{\partial\theta}{\partial E}(x,E)$. Obviously:

$$\theta(b,E_n) - \theta(b,E_{n-1}) = \int_{E_{n-1}}^{E_n} \phi(b,E)\, dE$$

Combining this with (10) we arrive at:

$$\pi \leq (E_n - E_{n-1}) \sup_{E \in [E_{n-1}, E_n]} \phi(b,E)$$

which gives

$$E_n - E_{n-1} \geq \pi\Big(\sup_{E \in [E_{n-1}, E_n]} \phi(b,E)\Big)^{-1} \tag{11}$$

Inequality (11) is our starting point for estimating $E_n - E_{n-1}$. It suggests to search for estimates of $\phi(b,E)$ from above.

3. Estimates on the energy derivative of the Prüfer angle:

Differentiation of equation (6) with respect to E gives a differential equation for $\phi = \dot{\theta}$. We adopt the notation that $\dot{}$ denotes derivative with respect to E while $'$ means derivative with respect to x. Supposing that λ does not depend on the energy E we get:

$$\phi' = \{(\frac{\lambda}{p} + \frac{V-E}{\lambda}) \sin 2\theta - \frac{\lambda'}{\lambda} \cos 2\theta \} \phi + \frac{\cos^2\theta}{\lambda} \tag{12}$$

To have a shorthand notation we call the expression in curly brackets on the right side of (12) f(x) and set $g(x) = \lambda^{-1} \cos^2\theta$. So (12) becomes:

$$\phi' = f\phi + g \tag{12'}$$

If λ is E-dependent we have to replace g with

$$\tilde{g} = g + \{ \frac{V-E}{\lambda^2} \cos^2\theta + \frac{1}{p} \sin^2\theta \} \dot{\lambda} + \frac{\dot{\lambda}'\lambda - \lambda'\dot{\lambda}}{\lambda^2} \cos\theta \sin\theta \tag{13}$$

Since $\theta(a,E) \equiv 0$ we may estimate $\phi(x,E)$ from (12) resp. (13) using Gronwall's Lemma. In case of (12) we get:

$$\phi(x) \leq \int_a^x |g(y)|\, dy\; e^{\int_a^x |f(y)|dy} \tag{14}$$

Again, if λ is E-dependent we replace g with $\tilde{g}$.

Inserting in (14) the expressions for f and g we arrive at the following result:

$$\phi(x,E) \leq \int_a^x \frac{1}{\lambda(y)}\, dy \;\; \exp \{ \int_a^x (\frac{|V(y)-E|}{\lambda(y)} + \frac{\lambda(y)}{p(y)} + \frac{|\lambda'(y)|}{\lambda(y)} dy\} \tag{15}$$

for any function $\lambda = \lambda(x) > 0$.

4. The general estimate for $E_n - E_{n-1}$

Putting together the estimates (11) and (15) we have proven the following result: For any function $\lambda(x) > 0$ we have:

$$E_n - E_{n-1} \geq \pi \left(\int_a^b \frac{1}{\lambda(x)} dx \right)^{-1} \exp \left\{ - \int_a^b \frac{\mu(x)}{\lambda(x)} + \frac{\lambda(x)}{p(x)} + \frac{|\lambda'(x)|}{\lambda(x)} \, dx \right\} \tag{16}$$

where we used the abbreviation $\mu(x) = \sup_{E \in [E_{n-1}, E_n]} |V(x) - E|$.

By an easy trick we improve (16) slightly. Let us set:

$$h(x) = \frac{\mu(x)}{\lambda(x)} + \frac{\lambda(x)}{p(x)} + \frac{|\lambda'(x)|}{\lambda(x)}$$

We choose c a,b in such a way that

$$\int_a^c h(x)\, dx = \int_c^b h(x)\, dx \tag{17}$$

Instead of solving the differential equation (3) "from the left to the right" we could as well solve it "backwards" which corresponds to a change of variables $x \to a+b-x$. Let us call the "backwards" solution θ. Then we have:

$$\theta(c,E_n) - \theta(c,E_{n-1}) \geq \pi/2 \text{ or } \tilde{\theta}(c,E_n) - \tilde{\theta}(c,E_{n-1}) \geq \pi/2.$$

Without loss of generality we may assume that the first alternative holds. This implies that:

$$E_n - E_{n-1} \geq \frac{\pi}{2} \left(\sup_{E \in [E_{n-1}, E_n]} \phi(c,E) \right)^{-1} \tag{18}$$

a formula in complete analogy to (11). By (17) and (15) we finally obtain:

$$E_n - E_{n-1} \geq \frac{\pi}{2} \left(\int_a^b \frac{1}{\lambda(y)} dy \right)^{-1} \exp \left\{ -\frac{1}{2} \int_a^b \frac{\mu(x)}{\lambda(x)} + \frac{\lambda(x)}{p(x)} + \frac{|\lambda'(x)|}{\lambda(x)} \, dx \right\} \tag{19}$$

for any function $\lambda > 0$.

There are various points to improve the above estimates. Especially, if we wish to consider the operator L on the whole line we have to use the Gronwall Lemma in a more clever way; e.g. by looking at the sign of $\sin 2\theta$ etc. This will be done elsewhere (see also [4]).

5. Applications of the general estimate:

Now, we show how to apply our estimate (19) by choosing the function λ. First we choose λ to be a constant. Then the exponent $1/\lambda \int \mu(x)\, dx + \lambda \int p(x)^{-1} dx$ is minimized by $\lambda = \left(\int \mu(x) dx \right)^{1/2} \left(\int p(x)^{-1}\, dx \right)^{-1/2}$.

So, if

$$\frac{1}{b-a} \int_a^b \mu(x)\, dx \leq \mu_1 \text{ and } \frac{1}{b-a} \int_a^b p(x)^{-1}\, dx \leq p_1^{-1}$$

we get by setting $\lambda = (\mu_1 p_1)^{1/2}$

$$E_n - E_{n-1} \geq \frac{\pi}{2} \frac{\mu_1 p_1}{b-a} e^{-(b-a)\left(\frac{\mu_1}{p_1}\right)^{1/2}} \tag{20}$$

For example we may choose $\mu_1 \geq \sup \mu(x)$, $p_1 \leq \inf p(x)$. It is usually not difficult to get such estimates in concrete situations.

Now we will allow λ to depend on x. The above discussion suggests to try $\lambda(x) = \mu(x)^{1/2}\, p(x)^{1/2}$. With this choice, however, λ is not differentiable, even if V is smooth. Hence the right hand side of (19) is not well defined. Supposing that V has a continuous derivative we may, however, take the distributional derivative of μ which is a bounded function. With this interpretation (19) makes perfect sence and can be verified by an approxiamtion argument.
Hence we obtain:

$$E_n - E_{n-1} \geq \frac{\pi}{2} \left(\int \frac{1}{\lambda(x)}\, dx\right)^{-1} e^{-\frac{1}{2}\alpha}\, e^{-\int \mu(x)^{1/2}\, p(x)^{-1/2}\, dx} \tag{21}$$

where we set $\alpha = \int \chi(x)\, dx$ and $\chi(x) = |p(x)\, V'(x) + p'(x)\, \mu(x)| \cdot (\mu(x)p(x))^{-3/2}$.
Similar as above we can also take $\mu_1(x) \geq \mu(x)$ and $p_1(x) \leq p(x)$ and we get:

$$E_n - E_{n-1} \geq \frac{\pi}{2} \left(\int \frac{1}{\mu_1 p_1}\right)^{-1} e^{-\int \beta(x) dx}\, e^{-\int \frac{\mu_1(x)^{1/2}}{p_1(x)^{1/2}}\, dx} \tag{22}$$

where $\beta(x) = |((\mu_1 p_1)^{1/2})'\, |(\mu_1 p_1)^{-1/2}$.
Note that formula (22) can be used for nondifferentiable V.

6. The semiclassical limit:

Nakamura [5] used estimates of the above type in the context of the semiclassical limit
To explain this idea let us consider the Schrödinger operator:

$$H(k) = -\frac{d^2}{dx^2} + k^2 V(x) \quad \text{on } L^2(\mathbb{R}).$$

Let us suppose that V is nonnegative and continuous, that $V(a) = V(b) = 0$ and $V(x) > 0$ otherwise, $\underline{\lim}_{x \to \pm\infty} V(x) > 0$.

The problem of the semiclassical limit is to determine the behavior of the eigenvalues $E_n(k)$ of $H(k)$ as k tends to infinity. This corresponds to sending h, the Planck constant, to zero, hence the name semiclassical limit.

It is known that under some additional hypotheses on V

$$\underline{\lim}_{k \to \infty} \frac{1}{k} \ln (E_n(k) - E_{n-1}(k)) \leq -\int_a^b \sqrt{V(x)}\, dx \tag{23}$$

(see Harrell [2], Combes, Duclos, Seiler [1], Simon [7], Helffer, Sjöstrand [3]).

Nakamura used the approach of [4] to prove the converse of inequality (23). We can read off such an estimate in the present context from (22). One chooses a function $\tilde{\lambda}$ with continuous derivative such that $V(x) + \varepsilon \leq \tilde{\lambda} \leq V(x) + 2\varepsilon$ and takes as a trial function $\lambda_k(x) = k\,(\tilde{\lambda}(x))^{1/2}$.

Inserting this into (22) for $p \equiv 1$ gives the converse to (23) (for operators on the interval). Moreover this inequality with the obvious changement can be obtained for Sturm-Liouville operators as well, both in the cases of an interval and of the line. Let us finally remark that estimates on $E_1 - E_o$ can be obtained in the case of multi-dimensional Schrödinger operators by quite different methods (Kirsch-Simon, in preparation). Again, as expected, those estimates are exponentially small in certain geometric quantities connected with the potential.

References:

[1] Combes J.M., Duclos P., Seiler R.: Convergence expansion for tunneling; Commun. Math. Phys. 92, 229-245 (1983)

[2] Harrell E.: Double wells; Commun. Math. Phys. 75, 239-261 (1980)

[3] Helffer B., Sjöstrand J.: Multiple wells in the semiclassical limit I; Commun. in PDE 9, 337-408 (1984)

[4] Kirsch W., Simon B.: Universal lower bounds on eigenvalue splittings for one dimensional Schrödinger operators; Commun. Math. Phys. 97, 453-460 (1985)

[5] Nakamura S.: A remark on eigenvalue splittings for one-dimensional double-well Hamiltonians, University of Tokyo, Preprint

[6] Neumark M.A.: Lineare Differentialoperatoren; Akademie-Verlag, Berlin

[7] Simon B.: Semiclassical analysis of low lying eigenvalues II. Ann. Math. 120, 89-118 (1984); IV. The flea on the elephant, J. Funct. Anal. 63, 123-136 (1985)

FINITE ELEMENT APPROXIMATION TO SINGULAR MINIMIZERS, AND APPLICATIONS TO CAVITATION IN NON-LINEAR ELASTICITY

Greg Knowles
Dept. of Electrical Engineering
Imperial College
London

Recently several classical examples in the calculus of variations [6], [7] have aroused new interest due to certain analogies to cavitation-type fracture in non-linear elasticity. Specifically consider the following example due to Mania [7], minimize

$$I(u) = \int_0 (u^3 - x)^2 (u')^6 \, dx \qquad (1)$$

in the set of admissible functions

$$\mathcal{A} = \{u \in W^{1,1}(0,1) : u(0) = 0, u(1) = 1\}. \qquad (2)$$

It is easily seen that the unique minimimum of I in $\mathcal{A}$ is $u^*(x) = x^{1/3}$, and that $I(u^*) = 0$. Remarkably, it was shown by Mania that

$$\inf\{I(u): u \in \mathcal{A},\ u \text{ is Lipschitz continuous}\} > \inf_{u \in \mathcal{A}} I(u) = 0. \qquad (3)$$

This property is known as the *Lavrentiev effect* after Lavrentiev [6], see also Cesari [3]. In view of (3) it is clear that standard numerical techniques (eg finite element or finite differences) *cannot simultaneously approximate* u^* and the *minimum value* of I, since they inherently involve the use of Lipschitz functions. This is borne out by numerical experiments. For instance, a simple finite element method is to approximate u by $u^h \in S^h$, the space of piecewise linear splines in $\mathcal{A}$ on a uniform mesh on [0,1], with mesh spacing $h = 1/N$, and to minimize $I(u^h)$ in S^h. If the internal nodal values of u^h are $\{a_1, \dots, a_{N-1}\}$, then

$$I(u^h) \triangleq I_N(a_1, \dots, a_{N-1}), \tag{4}$$

so we are left with the programming problem

$$\underset{\mathbf{a} \in \mathfrak{R}^{N-1}}{\text{minimise}}\ I_N(\mathbf{a}). \tag{5}$$

this has been done for a sequence of values $h \downarrow 0$ and a plot of the numerical minimizer u^h of (5) is shown as the lower curve in Figure 1(the upper curve is $x^{1/3}$). As predicted above the scheme does not "see" the true minimizer u^* but converges to a completely different function. (See [2] for a further discussion).

The connection with non-linear elasticity occurs in the experimentally observed phenomenon of cavitation (Gent & Lindley [4]). Consider deformations of an elastic body occupying a reference configuration in a bounded domain $\Omega \subset \mathfrak{R}^n$, (n = 2,3). Suppose the material is homogeneous with stored-energy function $W: M^{n\times n}_+ \to \mathfrak{R}$, where $M^{n\times n}_+$ denotes the set of real n x n matricies with positive determinant. We seek a deformation $x: \Omega \to \mathfrak{R}^n$ minimizing

$$I(x) = \int_\Omega W(Dx(X))dX \tag{6}$$

in the set of admissible functions

$$\mathcal{A} = \{x \in W^{1,1}(\Omega;\mathfrak{R}^n) : x|_{\partial\Omega} = AX\} \tag{7}$$

where $A \in M^{n\times n}_+$ is given. It was shown in [1] that if W is quasi-convex ([8]) the absolute minimum of I over those deformations in $\mathcal{A}$ which are Lipschitz continuous is attained by the linear deformation $x(X) = AX$. However, for certain A & W the *Lavrentiev effect*

$$\inf \{I(u) : u \in \mathcal{A}, u \text{ is Lipschitz continuous}\} = W(A).\text{vol}\,\Omega > \inf_{\mathcal{A}} I$$

holds. For example, in the case Ω is a ball, and $A = aI$ then for certain W there is a critical value of a, a_{cr}, such that for $a < a_{cr}$, the minimizer of I on the set of radial deformations $x(X) = (r(R)/R)X$, $R = |X|$, $x \in \mathcal{A}$ is attained by the linear deformation, $r(R) = aR$. However, for $a > a_{cr}$, the absolute minimum is attained by a function with $r(0) > 0$, so a cavity forms at the origin. The existence of cavitating minimizers is considered a possible explanation for fracture.

The numerical proceedure given below avoids the Lavrentiev phenomenon and can detect singular minimizers. It is important to emphasize that the location, or order of the singularity <u>need not be known</u> in advance. The basic idea is to decouple the unknown function u from its gradient in a manner reminiscent of control theory (or, in another context, in certain mixed finite element methods).

In section 1 we describe the numerical method and sketch a convergence proof., (For further details see [2].) In section 2 we apply the method to several numerical examples. In this note the effects of quadrature are not considered - for singular integrals these can be surprising, see [2] for further details.

1. THE METHOD

We consider the problem of minimizing

$$I(u) = \int_0^1 f(x, u(x), u'(x))dx \qquad (9)$$

over the set of admissible functions

$$\mathcal{A} \;=\; \{u \in W^{1,p}(0,1) : u(0) = \alpha,\ u(1) = \beta\}, \qquad (10)$$

where α, β are constants and $p > 1$. (the case $p = 1$ is slightly more technical, and is considered in [2]). We replace (9), (10) by numerically minimising the decoupled integral

$$I(u,v) = \int_0^1 f(x, u(x), v(x))dx \qquad (11)$$

among all pairs of functions $(u,v) \in \mathcal{A} \times L^p(0,1)$ satisfying the constraint

$$\int_0^1 |u'(x) - v(x)|^p dx \leq \epsilon, \qquad (12)$$

and let $\epsilon \to 0$. This is similar to the standard way a problem in the Calculus of Variations is written in control form, (eg [5]), except that the usual state constraint

$$u'(x) = v(x) \quad \text{a.e. } x \in [0,1] \qquad (13)$$

is replaced by the inequality constraint (12). When $\epsilon = 0$ the two problems are equivalent.

To simplify the proof we assume f satisfies the following,

(i) $f: [0,1] \times R \times R \to R$ is continuous

(ii) there exist constants $k_0 > 0$, k_1 such that

$$f(x,u,v) \geq k_0 |v|^p + k_1 \quad \text{for all } x \in [0,1],\ u,v \in R$$

(iii) $f(x,u,.)$ is convex for all $x \in [0,1]$, $u \in R$.

More general results are given in [2]. Under these assumptions the problems (9), (10), and (11), (12) attain their minima. The proofs are standard applications of the direct method of the calculus of variations.

To approximate (9), (10) we suppose that finite dimensional affine subspaces $\{S^h\}$ of $\mathcal{A}$ and $\{V^h\}$ of $L^\infty[0,1]$, are given such that

(i) for any $u \in \mathcal{A}$ there exist functions $u^h \in S^h$ such that $u^h \to u$ in $W^{1,p}(0,1)$, and

(ii) for any $v \in L^\infty[0,1]$, there exist $v^h \in V^h$ with $|v^h| \leq K$ for a.e. $x \in [0,1]$, all h and some constant K, and $v^h(x) \to v(x)$, a.e. x, as $h \to 0$.

Typical examples are S^h the space of piecewise linear splines in $\mathcal{A}$ and V^h the space of piecewise constants, on a uniform mesh on $[0,1]$, with mesh size h.

The natural finite dimensional approximation of (9)-(10) is to minimise

$$I(u,v) = \int_0^1 f(x, u(x), v(x))dx$$

among pairs $(u,v) \in S^h \times V^h$ satisfying

$$\int_0^1 |u'(x) - v(x)|^p \, dx \leq \epsilon$$

Since the constraint set of this problem is closed for h sufficiently small, $I(u,v) \to \infty$ as $\|(u,v)\|_{\mathcal{A}^h \times V^h} \to \infty$ (by (ii)) and f is continuous, the approximating problem has a minimum attained by, say, $(u^h, v^h) \in \mathcal{A}^h$. Denote the minimum value by I^h.

<u>Theorem 1.</u> Under the above assumptions there exists a nondecreasing function $\gamma: (0, \infty) \to (0, \infty)$ such that,

$$\lim_{\substack{h,\epsilon \to 0 \\ 0<h<\gamma(\epsilon)}} I_\epsilon^h = \min_{u \in \mathcal{A}} I(u)$$

Let $h_j \to 0$, $\epsilon_j \to 0$ be sequences with $0 < h_j < \gamma(\epsilon_j)$, and let $(u_{\epsilon_j}, v_{\epsilon_j})$ minimise $I(u,v)$ in $\mathcal{A} \times V$ subject to the constraint

$$\int_0^1 |u'(x) - v(x)|^p \, dx < \epsilon_j$$

Then there exists a subsequence (h_μ, ϵ_μ) of (h_j, ϵ_j) and a minimiser u^* of $I(u)$ in $\mathcal{A}$ such that as $\mu \to \infty$

$$u_{\epsilon_\mu}^{h_\mu} \to u^* \text{ uniformly in } [0,1]$$

$$v_{\epsilon_\mu}^{h_\mu} \to (u^*)' \text{ weakly in } L^p(0,1).$$

<u>Proof</u> (Outline)

Let $\epsilon > 0$, and u^* minimize I on A. Obviously if we didn't know about the Lavrentiev effect at this stage, we would be tempted to approximate u in $W^{1,p}(0,1)$ by functions in S^h, and pass to the limit in the integral. However u' can be unbounded so we cannot apply dominated convergence. To continue we make use of the extra leeway provided by (12).

Since u is continuous, it follows from (ii), (iii) that there exists a $M>0$ such that if $|v|>M$

$$f(x, u(x), v) \geq f(x, u(x), 0) \quad \text{for all } x \in [0,1].$$

For any $\delta > M$ define

$$v_\delta(x) = \begin{cases} 0 & \text{if } |u'(x)| > \delta \\ u'(x) & \text{otherwise.} \end{cases}$$

Then,

$$I(u,v_\delta) \le I(u), \tag{14}$$

and, by choosing δ sufficiently large,

$$\int_0^1 |u' - v_\delta|^p \, dx = \int_{E_\delta} |u'|^p \, dx < \epsilon, \tag{15}$$

since $|E_\delta| = |\{x\in[0,1]: |u'(x)| > \delta\}| \to 0$, as $\delta \to \infty$. Now the method of proof becomes clear. By truncating u' (ie replacing it by v_δ) the function $f(x, u(x), v_\delta(x))$ is bounded in x on [0,1], so that if we choose functions $u^h \to u$ in $W^{1,p}[0,1]$, and $v^h \to v_\delta$ ae, with $|v^h| \le K$ (guaranteed by (i),(ii)), then

$$\lim I(u^h, v^h) = I(u,v_\delta) \tag{16}$$

by the dominated convergence theorem. So by using the inequality constraint (12), instead of the equality constraint (13) the approximation of the integral (9) becomes benign.

Summarising, we can find a function $\gamma:(0,\infty) \to (0,\infty)$, (which can be chosen non-decreasing), so that for $h < \gamma(\epsilon)$,

$$I(u^h, v^h) < I(u, v_\delta) + \epsilon \le I(u) + \epsilon$$

and

$$\int |(u^h)' - v^h|^p \, dx \le \epsilon .$$

$$\text{Thus,} \quad I^h_\epsilon \le \min_{\mathcal{A}} I + \epsilon . \tag{17}$$

The remainder of the proof follows by the direct method of the Calculus of Variations. Let $h_j \to 0, \epsilon_j \to 0$ with $0 < h_j < \gamma(\epsilon_j)$, and set $u_j = u^{h_j}_{\epsilon_j}$, $v_j = v^{h_j}_{\epsilon_j}$, by (ii)

$$\sup_j \int^1 \|v_j\|^p \, dx < \infty,$$

so that we can extract a weakly convergent subsequence v_μ of v_j with

$$v_\mu \rightarrow v^* \text{ in } L^p(0,1),$$

for some v^*. Since

$$\int^1 |u'_\mu - v_\mu|^p \, dx \leq \epsilon_\mu,$$

it follows that $u'_\mu \rightarrow v^*$ in $L^p(0,1)$, and hence that $u_\mu \rightarrow u^*$ uniformly on $[0,1]$, where,

$$u^*(x) \triangleq \alpha + \int_0^x v^*(t)dt,$$

and thus $u^* \in \mathcal{A}$.

Since f is convex in v, we deduce from standard lower semi-continuity results that

$$\inf I \leq I(u^*) \leq \liminf_{\mu\to\infty} I(u_\mu, v_\mu) = \liminf_{\mu\to\infty} I^{h_\mu}_{\epsilon_\mu}. \qquad (18)$$

Combining (17), (18) gives that u^* minimizes I, and the result follows.

2. NUMERICAL RESULTS

The first example considered is the Mania example (1). S^h, V^h were chosen to be piecewise linears and constants respectively, and the mid-point rule was used for the calculation of $I(u^h, v^h)$. The finite dimensional optimisation was carried out using a modified penalty method and a quasi-Newton unconstrained optimisation algorithm. The starting values for the optimisation routine were taken on the straight line $u(x) = x$. Numerical results for various values of $h = 1/N$ can be seen in Fig 2. It can be seen that u_ϵ^h does approximate the minimizer $x^{1/3}$, the effect of the mid-point rule can be seen - the graph of u_ϵ^h intersects $x^{1/3}$ close to the mid-point of each subinterval.

The second example computes cavitation solutions to the problem of uniformly stretching a disc of compressible elastic material of unit radius, B out to a radius of a. That is we seek a radially symmetric deformation $x: B \to R^2$ minimising the energy

$$I(x) = \int_B \{ \|\nabla x(x)\|^p + \alpha (\det(\nabla x(x)) - 1)^2 \} \, dx \qquad (19)$$

for $\det \nabla x(x) \geq 0$ ae, and $\alpha > 0, p > 1$ given constants. Setting

$$x(X) = \frac{r(R)}{R} X, \quad R = |X|,$$

allows this to be treated as a regular minimization problem for radial solutions . Now, we seek to minimise

$$I(r) = 2\pi \int_0^1 R\,[((r')^2 + r^2/R^2)^{p/2} + \alpha (r\,r'/R - 1)^2] dR$$

on the set

$$\mathcal{A}_a = \{ r \in W^{1,1}(0,1): r(0) \geq 0, r'(R) > 0 \text{ ae}, r(1) = a, I(r) < \infty \}. \qquad (20)$$

(eg [1]). Solutions cavitating at the origin will have $r(0) > 0$, the smooth minimizer is $r(R) = aR$, the uniform extension. One way to reduce the number of constraints in (20) is to solve for $\sqrt{r'}$ on each mesh interval, and integrate backwards from $R=1$, $r(1)=a$, to find the nodal values of r.

Sample numerical results for the case $p = 3/2$ (for $p>2$ cavitation cannot occur, see[1]), and $\alpha = 0{\cdot}5$, and various values of a are given in Fig 3. As expected for $a < a_{cr} \approx 1{\cdot}6$, the absolute minimiser is $r(R) = aR$, but for a values greater than the critical point the cavitation solution shown has less energy.

Finally, we remark that work is proceeding on the application of these ideas to multi-dimensional problems in elasticity, with some preliminary success. The main disadvantage seems to be the extra number of unknowns introduced in the decoupled problem (in n space dimensions v will be an nxn matrix.) This necessitates modifying the fully decoupled system so that less unknowns are needed, and tailoring the unconstrained optimisation algorithm to take account of the sparseness inherent in such problems.

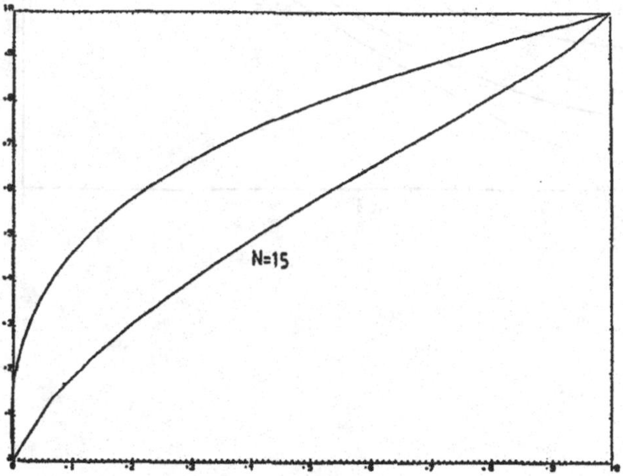

Fig.1

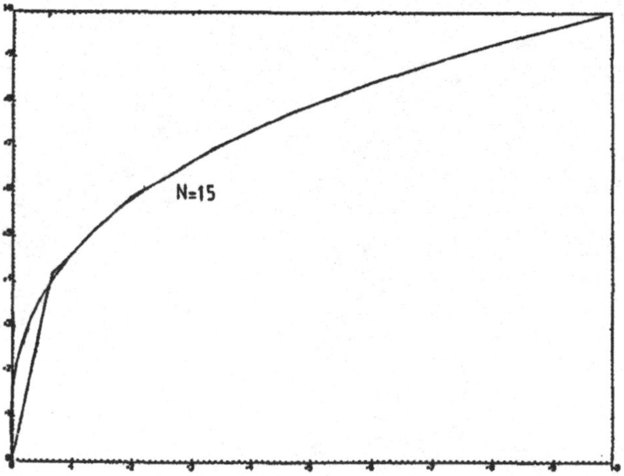

Fig.2

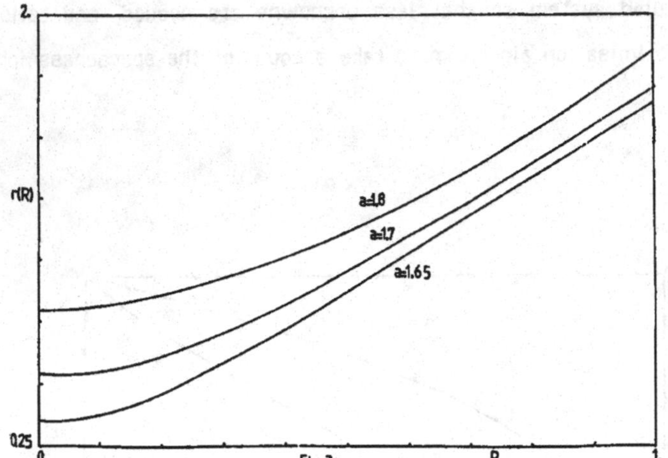

Fig. 3

REFERENCES

[1] J.M.BALL *Discontinuous solutions and cavitation in nonlinear elasticity*, Phil. Trans. Royal Soc. London A 306,577-611,1982

[2] J.M.BALL & G.KNOWLES *A numerical method for detecting singular minimizers*, Numerische Math.,1986.

[3] L.CESARI *Optimization-theory and applications*, Springer-Verlag,New York,1983.

[4] A.N.GENT & P.B.LINDLEY *Internal rupture of bonded rubber cylinders in tension*, Proc. Roy. Soc. London Ser.A,249,195-205,1958.

[5] G.KNOWLES *Introduction to Applied Optimal Control*, Academic Press,1981.

[6] M.LAVRENTIEV *Sur quelques problemes du calcul des variations*, Ann. Mat. Pure Appl. 4,7-28,1926.

[7] B. MANIA *Soppa un esempio di Lavrentieff*, Bull. Unione Mat. Ital., 13,147-153,1934.

[8] C.B. MORREY *Multiple integrals in the Calculus of Variations*, Springer-Verlag, Berlin,1966.

ON RELATING GENERALIZED EXPANSIONS TO FOURIER INTEGRALS

M. A. Kon*, Columbia University and Boston University
L. A. Raphael**, Howard University
J. E. Young***, Massachusetts Institute of Technology

Eigenfunction expansions are compared (with respect to pointwise and function space convergence) for pairs of uniformly elliptic operators with the same leading term. For two such operators, eigenfunction expansions behave in essentially the same way under analytic summation techniques. This shows, for example, that in this sense expansions in eigenfunctions of a class of elliptic operators (including Schrödinger operators) behave like the Fourier transform on $\mathbf{R}^n$.

In 1923, Marshall Stone [St1] answered some questions on the relationship of expansions in eigenfunctions of Sturm-Liouville operators (i.e., Schrodinger operators on intervals), to ordinary Fourier series. He studied so-called Birkhoff series (higher-order analogs of Sturm-Liouville series) on an interval, and proved that for a given function, these were equiconvergent with ordinary Fourier series. That is, the set of points where one series converges is identical with that where the other does, and the limits are the same when they exist.

Stone later showed [St2] that equisummability (i.e., identical convergence under a summability method) is sometimes the correct replacemant for equiconvergence. Equisummability has since been studied by Titschmarsh [Ti], Levitan and Sargsjan [LS], and Benzinger [Be1], among others. Benzinger [Be2] also used equisummability to study convergence to initial values in the heat equation. In [LS], it is also shown that equisummability is a useful concept in the study of convergence in $C^k(\mathbf{R}^+)$ of eigenfunction expansions. In higher dimensions, Il'in has investigated equisummability of expansions in eigenfunctions of the Laplacian on domains of $\mathbf{R}^n$, and the Fourier transform on all of $\mathbf{R}^n$. There has been recent interest in the equiconvergence and equisummability properties of Fourier series with irregular frequencies [Se,Be3]. See [Go] for a survey of Soviet equisummability results up to 1982.

A motivation for the study of analytic multipliers in equisummability arises form the fact that ordinary summation of multidimensional elliptic eigenfunction expansions is known to fail in L^p $(p \neq 2)$ [F,KST], and that the Riesz method improves matters only for a limited range of p values [KST].

* Research partially supported under N.S.F. grant DMS-8509458
** Research supported by Army Research Office grant number DAAG-29-84-G0004
*** Research supported in part by the Energy Research and Development Administration under contracts AT(11-1)-3069 and AT(40-1)-3992

Use of the class of analytic multipliers in summation methods avoids such convergence problems. Indeed, it will be shown that if A_1, A_2 are uniformly elliptic operators (with possibly variable leading coefficients) on $\mathbf{R}^n$ and share the same leading term, their eigenfuntion expansions are equisummable (under analytic multipliers). The most immediate example of this is equisummability of the Fourier transform and generalized expansions in eigenfunctions of operators with constant coefficient leading terms (see the remarks on applications at the end).

The basic results on uniformly elliptic operators which we use are due to D. Gurarie [G].

Let A, B be linear operators on $L^p(\mathbf{R}^n)$, $1 \le p \le \infty$. Let $f \in L^p(X)$, and ϕ be a function analytic on the spectrum of A, B, with $\phi(0) = 1$. According to the Dunford operator calculus, we define

$$\phi(\epsilon A) = \frac{1}{2\pi i} \int_\Gamma \phi(\epsilon\zeta)(\zeta - A)^{-1} d\zeta, \tag{1}$$

where $\Gamma \subset \mathbf{C}$ is a contour enclosing the spectrum $\sigma(A)$ of A.

Note that $\phi(\epsilon A)f$ represents the ϕ-summability means of the expansion of f with respect to eigenfunctions of A. That is, formally, if $\{u_\gamma\}_{\gamma \in \sigma(A)}$ is a complete set of eigenfunctions of A, then if

$$f(x) \sim \int_{\sigma(A)} u_\gamma(x) F(\gamma) d\rho(\gamma)$$

is a spectral expansion, we have

$$\phi(\epsilon A) f(x) \sim \int_{\sigma(A)} \phi(\epsilon\gamma) u_\gamma(x) F(\gamma) d\rho(\gamma).$$

The question of equisummability is: When does convergence (as $\epsilon \to 0$) of $\phi(\epsilon A)f$ imply that of $\phi(\epsilon B)f$? Specifically, we say that eigenfunction expansions with respect to A and B are ϕ-equisummable from L^p to L^q if for $f \in L^p$

$$\phi(\epsilon A)f - \phi(\epsilon B)f \underset{\epsilon \to 0}{\longrightarrow} 0$$

in L^q. If $q = \infty$, the expansions are ***pointwise equisummable.***

The above construction of eigenfunction expansions is hard in general, and the present operator claculus approach makes our results independent of the detailed properties of eigenfunction expansions. Agmon [Ag] describes two approaches to constructing eigenfunction expansions for Schrödinger operators based on expansions for the Laplacian; such techniques might be used to expilcitly prove the present results for Schrödinger eigenfunction expansions.

We begin by presenting basic definitions and notation. Let $\alpha = (\alpha_1, \ldots, \alpha_n)$ be a multiindex and $D^\alpha = i^{-|\alpha|}\partial^\alpha \equiv i^{-|\alpha|} \frac{\partial^{\alpha_1}}{\partial x_1^{\alpha_1}} \cdots \frac{\partial^{\alpha_n}}{\partial x_n^{\alpha_n}}$, $\xi^\alpha = \xi_1^{\alpha_1}, \ldots, \xi_n^{\alpha_n}$, $|\alpha| = \sum_i \alpha_i$. An operator $A = \sum_{|\alpha| \le m} a_\alpha(x) D^\alpha$ is ***uniformly elliptic*** if its leading symbol $a(x,\xi) = \sum_{|\alpha|=m} a_\alpha(x)\xi^\alpha$ satisfies

$$c_1|\xi|^m \le a(x,\xi) \le c_2|\xi|^m \qquad (\xi \in \mathbf{R}^n)$$

with constants $c_1, c_2 > 0$ independent of $x \in \mathbf{R}^n$. Let $A_0 = \sum_{|\alpha|=m} a_\alpha(x)D^\alpha$ be a uniformly elliptic homogeneous operator. Let $CB^\infty(\mathbf{R}^n)$ be the class of C^∞ functions f on $\mathbf{R}^n$ which are bounded with all their partial derivatives, i.e., $\sup_{x\in\mathbf{R}^n} |\partial^\alpha f(x)| < \infty$ for all α. We assume the coefficients a_α of A_0 are in CB^∞, and that the symbol of A_0 is real-valued.

The operators we consider are perturbations of A_0. Namely, we study $A = A_0 + B$ where $B = \sum_{|\alpha|<m} b_\alpha(x)D^\alpha$, with $b_\alpha \in L^{r_\alpha} + L^\infty$ on $\mathbf{R}^n$. We must require (see [GK2]) that

$$d \equiv \max_\alpha\left(\frac{n}{r_\alpha} + |\alpha|\right) < m.$$

More generally $\{b_\alpha\}$ can be defined on quotient spaces $\mathbf{R}^n/U_\alpha$ where U_α is a linear subspace of $\mathbf{R}^n$, with $b_\alpha \in (L^{r_\alpha} + L^\infty)(\mathbf{R}^n/U_\alpha)$ (see [GK1]). The allowable singularities in b_α are important in applications involving Schrödinger operators.

For the sequel we define the exponentially decaying radial function

$$H_{s,t,\gamma}(|z|) = \begin{cases} 1+|z|^{-s} \quad (-\ln|z| \text{ if } s=0); & |z|\le 1 \\ |z|^{-t}e^{-\gamma|z|}; & |z|>1 \end{cases} \qquad (z\in\mathbf{R}^n). \tag{2}$$

Let $A = A_0 + B$ be as above. Denote

$$R_\zeta \equiv (\zeta - A)^{-1}; \qquad R_\zeta^0 \equiv (\zeta - A_0)^{-1}, \tag{3}$$

when the resolvents exist. Define the parabolic domain (about $\mathbf{R}^+ \in \mathbf{C}$)

$$\Omega_{k,\tau} \equiv \{\zeta = \rho e^{i\theta} : \rho^{\frac{1}{m}} < k|\theta|^{-\tau}\}, \tag{4}$$

for $k, \tau > 0$, and

$$C(\zeta) = C|\theta|^{-\tau}(1 - C|\theta|^{-\tau}\rho^{-1/m})^{-1} \tag{5}$$

for some $C > 0$, $\tau > 0$. Here, $\zeta = re^{i\theta}$.

We will require the following characteristics of our class of uniformly elliptic operators. The first two parts of the following theorem are due to Gurarie [G].

Theorem 1: *Let A_0 be a uniformly elliptic homogeneous differential operator of order m as above, and D be any homogeneous differential operator of order $m' < m$, with coefficients in CB^∞. Then*

(i) there exists a parabolic domain $\Omega_{k,\tau}$ of the form (4) containing the L^p-spectrum of A_0, and of A, for $1 \le p \le \min_\alpha r_\alpha$.

(ii) in the complement of $\Omega_{k,\tau}$, the kernel of $K_\zeta = D(\zeta - A_0)^{-1}$ is bounded by L^1-dilations of a radial function $H_{s,t,\gamma}$:

$$|K_\zeta(x,y)| \le C_1(\zeta)\rho^{-1+\frac{m'+n}{m}} H_{s,t,\gamma}(\rho^{\frac{1}{m}}|x-y|) \qquad (\zeta = \rho e^{i\theta}), \tag{6}$$

where $C_1(\zeta)$ has the form (5) for some C, $s = n - m + m'$, $t > n$ and $\gamma = \gamma_0 \sin\frac{\theta}{m}$.

(iii) if $A = A_0 + B$ is as above, there exists a domain $\Omega_{k',r'}$, outside of which the kernel of $L_\zeta^{(1)} = R_\zeta - R_\zeta^0$ is estimated by

$$|L_\zeta^{(1)}(x,y)| \leq C(\zeta)\rho^{\frac{d+n}{m}-2} H_{s,t,\gamma}(\rho^{\frac{1}{m}}|x-y|) \qquad (\zeta = \rho e^{i\theta}) \tag{7}$$

where $C(\zeta)$ is as in (5), $s = n - 2m + d$, $t > n$, and $\gamma = \gamma_0 \sin\frac{\theta}{m}$.

Proof of (iii): We begin with a standard perturbation approach, representing the resolvent of A by the series

$$R_\zeta = R_\zeta^0 \sum_{k=0}^{\infty} (BR_\zeta^0)^k, \tag{8}$$

where $R_\zeta^0 = (\zeta - A_0)^{-1}$, so that

$$R_\zeta - R_\zeta^0 = R_\zeta^0 \sum_{k=1}^{\infty} (BR_\zeta^0)^k. \tag{9}$$

The operator BR_ζ^0 has a kernel which, by part (*ii*), can be bounded by convolution kernels of the form (2), multiplied and scaled by powers of $|\zeta|$. To be precise, let us look at the term $R_\zeta^0(BR_\zeta^0)^k$, $k \geq 1$. This can be expanded to an expression which contains terms of the form

$$R_\zeta^0 b_{\alpha^1} D^{\alpha^1} R_\zeta^0 b_{\alpha^2} D^{\alpha^2} \cdots R_\zeta^0 b_{\alpha^k} D^{\alpha^k} R_\zeta^0, \tag{10}$$

whose kernel is $L_{\alpha^1,\ldots,\alpha^k}(x,y)$, with $\{\alpha^i\}_{i=1}^k$ the multiindices in B. The part $D^\alpha R_\zeta^o = K_\alpha$ has a kernel which is smaller than

$$|K_\alpha(x,y)| \leq C_\alpha(\rho,\theta)\rho^{\frac{n}{m}-1} H_{s,t,\gamma}(\rho^{\frac{1}{m}}|x-y|), \tag{11}$$

where $\zeta = \rho e^{i\theta}$, and $C_\alpha, s < n$, $t > n, \gamma$ are constants (see [G]). Without loss of generality, we can assume that

$$b_\alpha(x) \in L^{r_\alpha} \qquad (|\alpha| < m). \tag{12}$$

Choose p_α such that $\frac{1}{p_\alpha} + \frac{1}{r_\alpha} = 1$. Using a multiple Hölder's inequality on the kernel of (10), we have

$$\left| L_{\alpha^1,\ldots,\alpha^k}(x,y) \right| \leq \prod_{i=1}^{k} \|b_{\alpha^i}\|_{r_{\alpha^i}} |K_0| * |K_{\alpha^1}| * \cdots * |K_{\alpha^k}|, \tag{13}$$

where the ith $*$ denotes a p_{α^i}-convolution.

Applying the bounds of (11) to the above equation, we obtain the proper bounds on each term of the sum, and then through summation on $R_\zeta - R_\zeta^0$ itself. ∎

Equisummability Under Analytic Multipliers and Some Applications

Theorem 1 together with the Dunford calculus will provide the desired results on equisummability of elliptic expansions under analytic multipliers. The advantage of the operator calculus approach to equisummability is its insensitivity to detailed spectral considerations, or even to the existence of a complete spectrum. For example, the approach addresses non-self-adjoint problems as readily as self-adjoint ones.

Let A be a uniformly elliptic operator as above. We now examine analytic functions $\phi(A)$, defined by (1). The unbounded contour Γ encloses the spectrum $\sigma(A) \subset \Omega_{k,r}$ and we require

$$\int_\Gamma |\phi(\zeta)| d\mu(\zeta) < \infty, \tag{14a}$$

where the measure is $d\mu(\zeta) = \frac{1}{\rho} C(\zeta) d\zeta$, with $C(\zeta)$ given by (5). We also assume for well-definedness that

$$\int_{\{|\zeta|=c\}\cap\Gamma^0} |\phi(\zeta)| d\mu \underset{c\to\infty}{\longrightarrow} 0 \tag{14b}$$

where Γ^0 is the interior of Γ.

For $\theta > 0$, let $W_\theta = \{\zeta \in \mathbf{C}: |\arg\zeta| < \theta\}$; note W_θ contains the positive real axis. Given operators A_0, A and $1 \le p, q \le \infty$, the eigenfunction expansions with respect to A and A_0 are *resolvent equisummable* from L^p to L^q (in $\Omega \subset \mathbf{C}$) if for $f \in L^p(\mathbf{R}^n)$

$$\|\zeta(\zeta - A)^{-1} f - \zeta(\zeta - A_0)^{-1} f\|_q \to 0$$

as $\zeta \to \infty$ (in Ω). If $q = \infty$, they are *pointwise* equisummable.

We now state an optimal equisummability theorem. Our strategy is to study first the case when $\phi(\epsilon A)$ represents the resolvent, i.e. $\phi(\epsilon A) = \zeta(\zeta - A)^{-1}$, $(\zeta = -\frac{1}{\epsilon})$. In this case we employ a sharp bound on the integral kernel of the difference $(\zeta - A)^{-1} - (\zeta - A_0)^{-1}$. The rest of the argument follows directly from the integral representation (1), where $\Gamma \subset \mathbf{C}$ is a contour enclosing the spectrum $\sigma(A)$ of A.

For $r > 0$, let D_r be the open disk of radius r, centered at 0.

Theorem 2: *Let A_0 and A be as in Theorem 1, and let $\theta > 0$. Let $\sigma(A) \cup \sigma(A_0) \subset D_r \cup W_\theta \equiv \mathcal{D}_{r,\theta}$, and ϕ be analytic in $\mathcal{D}_{r,\theta}$, with $\phi(0) = 1$ and $\phi(z) = O(z^{-\delta})$ for some $\delta > 0$. Let*

$$\frac{1}{p} - \frac{1}{q} < \frac{m-d}{n} \tag{15}$$

and $f \in L^p + L^\infty$. Then the expansions of f with respect to A and A_0 are ϕ-equisummable, that is,

$$\phi(\epsilon A) f - \phi(\epsilon A_0) f \longrightarrow 0$$

in L^q as $\epsilon \to 0$.

Proof: By the Dunford operator calculus, it suffices to prove the result for the resolvent $\phi(\epsilon A) = \frac{1}{\epsilon A}$ as $\epsilon \to 0$ (in a complex domain), or, equivalently, using $\zeta = \frac{-1}{\epsilon}$, as $\zeta \to \infty$ in the complement of a domain $W_\theta = \{z \in \mathbf{C} : |\arg z| \le \theta\}$. To this end, then, note that if $\zeta \in \sim W_\theta$ is large and R_ζ, R^0_ζ are as in (3), then for $H_{s,t,\gamma}$ as above,

$$\begin{aligned}\|\zeta(R_\zeta - R^0_\zeta) f\|_q &\le C(\zeta)\rho^{\frac{d+n}{m}-1} \|H_{s,t,\gamma}(\rho^{\frac{1}{m}} x) * f(x)\|_q \\ &\le C(\zeta)\rho^{\frac{d+n}{m}-1} \|H_{s,t,\gamma}(\rho^{\frac{1}{m}} x)\|_r \|f\|_p.\end{aligned} \tag{16}$$

Here, $\frac{1}{r}+\frac{1}{p}=1+\frac{1}{q}$. We have $H_{s,t,\gamma}\in L^r$ if $\frac{1}{r}>\frac{n-2m+d}{n}$; this occurs if $p>\frac{n}{m-d}$. With the standard scaling identity

$$\|g(ax)\|_r=a^{-\frac{n}{r}}\|g\|_r, \tag{17}$$

we obtain

$$\|\zeta(R_\zeta-R_\zeta^0)f\|_q\leq C(\zeta)\rho^{(d+n-\frac{n}{r})\frac{1}{m}-1}\|H_{s,t,\gamma}\|_r\|f\|_p. \tag{18}$$

The exponent of ρ above is easily shown to be negative, so that the right side of (18) converges to 0 as $\zeta\to\infty$ in $\sim W_\theta$.∎

We remark that the scale of L^p-spaces given in (15) is the best possible, in that there exist A_0 and A for which equisummability fails when (15) is violated. In constructing such A_0 and A, the leading term can be arbitrary, and the lower order terms (as before) are first chosen so that the coefficients $b_\alpha(x)\in L^{r_\alpha}+L^\infty$. Let $d=\max\limits_{|\alpha|<m}(\frac{n}{r_\alpha}+|\alpha|)$. We construct B so that the latter maximum is assumed for only one multiindex α^1, and that $b_{\alpha^1}=\chi_1|x|^{-\frac{n}{r_{\alpha^1}}+\delta}$, for a $\delta>0$; thus $b_{\alpha^1}\in L^{r_{\alpha^1}}$. Choose $\phi(\epsilon A)=\zeta R_\zeta$, with $\zeta=-\frac{1}{\epsilon}$, and $f(x)=\chi_1|x|^{d-m-\frac{n}{q}-\epsilon}$, with $\epsilon>0$ sufficiently small that $f\in L^p$.

It can be shown through technical arguments that for the above A, ϕ, and f, the expansions of f with respect to eigenfunctions of A and of its leading term are not ϕ-equisummable under the above hypotheses. This proves sharpness of Theorem 2.

Remarks: The above results extend to uniformly elliptic systems, including the Dirac operator

$$D=\left[i\sum_{j=1}^{3}\gamma_j(\partial^j+\mathbf{A}^j(x))-\beta\Phi(x)\right],$$

where γ_j,β are 4×4 gamma matrices, $\partial^j=\frac{\partial}{\partial x_j}$, and $\mathbf{A},\Phi$ denote vector and electromagnetic potential. This gives nice bounds, for example, on the ergodic $t\to0$ limit [HP] of

$$e^{-tD}-e^{-tD_0},$$

where $D_0=i\sum_{j=1}^{3}\gamma_j\partial^j$. The opposite ergodic limit $t\to\infty$, is useful in index theory of the Dirac operator; see [Ca].

We remark that the above applies to study of convergence to initial values for solutions of parabolic equations. This includes, for example, the initial value problem for the heat equation in a moving medium with non-uniform conductivity:

$$Au\equiv\left(-a(x)\Delta+\vec{b}(x)\cdot\vec{\nabla}+q(x)\right)u=-\frac{\partial u}{\partial t}, \tag{19a}$$

$$u(x,0)=u_0(x)\in(L^p+L^\infty)(\mathbf{R}^n). \tag{19b}$$

Here $a(x)>0$ represents a position dependent on heat conductivity, and $\vec{b}(x)$ and $q(x)$ are drift and dissipation terms. We must assume $a(x)\in CB^\infty(\mathbf{R}^n)$, $\vec{b}(x)\in L^{n+\epsilon}+L^\infty$, and

$q(x) \in L^{\frac{n}{2}+\epsilon} + L^{\infty}$, for some $\epsilon > 0$. We can conclude from Theorem 2 that If $u_0 \in L^{\frac{n}{2}+\epsilon} + L^{\infty}$, then $u(x,t) \underset{t\to 0}{\longrightarrow} u_0(x)$ at exactly the same set of points as the solution of the unperturbed problem $-a(x)\Delta u = -\frac{\partial u}{\partial t}$. This extends some results of Benzinger [Be2]; it follows from the fact that $e^{-tA}u_0$ solves (19a), and the semigroup falls into the class of multipliers ϕ in Theorem 2.

We remark also that Theorem 2 shows, for example, equisummability of Sturm-Liouville expansions on $\mathbf{R}^1$ and the ordinary Fourier transform. More generally, since an expansion in eigenfunctions of a constant coefficient elliptic operator is essentially a Fourier transform, we have showed equisummability of the Fourier transform with expansions in eigenfunctions of a large class of elliptic operators with constant coefficient leading terms.

Acknowledgement: The authors would like to thank P. Gilkey for his encouragement to study this problem.

Bibliography

[Ag] Agmon, S., Spectral theory of Schrödinger operators on Euclidean and on non-Euclidean spaces, to appear, Frontiers in the Mathematical Sciences, Courant Institute.

[Be1] Benzinger, H., Green's function for ordinary differential operators, *J. Differential Equations* **7** (1970), 478.

[Be2] Benzinger, H.E., Perturbation of the heat equation, *J. Differential Equations* **32** (1979), 398.

[Be3] Benzinger, H.E., Non-harmonic Fourier series and spectral theory, preprint.

[Ca] Callias, C., Axial anomalies and the index theorem on open spaces, *Comm. in Math. Physics*, **62** (1978), 213.

[DS] Dunford, N. and J.T. Schwartz, *Linear Operators*, Interscience, New York, 1959.

[F] Fefferman, C., The new multiplier theorem for the ball, Ann. of Math. 94 (1971), 330-336.

[Go] Golubov, B.I., Multiple series and Fourier integrals. *Mathematical Analysis*, Vol. 19, 3-54, Akad. Nauk SSSR, Moscow 1982.

[G] Gurarie, D., Kernels of Elliptic Operators: Bounds and Summability, *J. Differential Equations* **55** (1984), 1.

[GK1] Gurarie, D., and Kon, M., Radial bounds for perturbations of elliptic operators, *J. Functional Analysis* **56** (1984), 99.

[GK2] Gurarie, D., and Kon, M., Resolvents and regularity properties of elliptic operators. *Operator Theory: Advances and Applications*, 151-162, Birkhaüser, Basel, 1983.

[HP] Hille, E. and R. Phillips, Functional Analysis and Semigroups, American Mathematical Society, Providence, 1957.

[Il] Il'in, V.A., The Riesz equisummability of expansions in eigenfunctions and in the n-dimensional Fourier integral. *Trudy Mat. Inst. Steklov.* **128** (1972), 151.

[KST] Kenig, C.E., R.J. Stanton, and P.A. Tomas, Divergence of eigenfunction expansions, J. Functional Analysis 46 (1982), 28-44.

[LS] Levitan, B., and Sargsjan, I., *Introduction to Spectral Theory; Self-Adjoint Differential Operators*, A.M.S. Translations, Providence, 1975.

[Ra] Raphael, L.A., Equisummability of eigenfunction expansions under analytic multipliers, *J. Math. Anal. and Appl.* (to appear).

[Se] Sedletskiǐ, A.M., The equiconvergence and equisummability of nonharmonic Fourier expansions and ordinary trigonometric series. *Mat. Zametki* **18** (1975), 9 (Russian), *Math. Notes* **18** (1975), 586 (English).

[St1] Stone, M.H., A comparison of the series of Fourier and Birkhoff, *Trans. Amer. Math. Soc.* **28** (1926), 695.

[St2] Stone, M.H., Irregular differential systems of order two and related expansion problems. *Trans. Amer. Math. Soc.* **29** (1927), 23.

[Ti] Titchmarsh, E.C., *Eigenfunction Expansions*, Vol. I and II, Oxford, 1962.

LINK BETWEEN PERIODIC POTENTIALS AND RANDOM POTENTIALS IN ONE-DIMENSIONAL SCHRÖDINGER OPERATORS

Shinichi Kotani
Department of Mathematics
Kyoto University, Kyoto JAPAN

§1. Introduction to random Schrödinger operators.

First we make clear our framework in considering random Schrödinger operators. Let

$$\Omega = L^2_{real}(R, \frac{dx}{1+|x|^3})$$

$$\Omega_d = \{(a_n, b_n) \in R_+^Z \times R^Z;\ \sum a_n^{-1} = \infty\},$$

where $R_+ = \{a \in R,\ a > 0\}$. For $q \in \Omega$ $(q = (a_n, b_n) \in \Omega_d)$, define a symmetric operator $L(q)$ $(L_d(q))$ on $L^2(R,dx)$ $(\ell^2(Z))$ by

$$L(q)u(x) = -u''(x) + q(x)u(x)$$

$$(L_d(q)u_n = a_n u_{n+1} + a_{n-1}u_{n-1} + b_n u_n)$$

respectively. It is known that they have unique self-adjoint extensions in $L^2(R,dx)$ or $\ell^2(Z)$ (see [6],[20]). In the sequel, we mainly discuss $L(q)$. $L_d(q)$ can be treated similarly. Introduce a one-parameter group $\{T_x\}$ on Ω by

$$T_x q(\cdot) = q(\cdot + x) \qquad x \in R.$$

Let $\mathbb{B}$ be the topological Borel field of Ω. Consider a probability measure P on $(\Omega, \mathbb{B})$ satisfying the following:

(1.1) $\{T_x\}$-invariance: $P(T_x A) = P(A)$ for any $x \in R$ and any $A \in \mathbb{B}$.

(1.2) integrability: $\int_\Omega \{\int q(x)^2 dx\} P(dq) < \infty$.

(1.3) $\{T_x\}$-ergodicity: $P(T_x A \Delta A) = 0$ for any $x \in R$ implies $P(A) = 0$ or 1.

Set $\mathcal{P}$ = {the set of all probability measures on $(\Omega, \mathbb{B})$ satisfying (1.1) and (1.2)},

$\mathcal{P}_e = \{P \in \mathcal{P};\ P$ satisfies also (1.3)$\}$.

Let us now give three examples of P's belonging to $\mathcal{P}_e$.

Example 1. (periodic potential) Let q_0 be a continuous periodic function on R with period $T > 0$. Then q_0 is considered to be a continuous function on R/TZ. Define a map π from R/TZ to Ω by

$$\pi(\theta) = q_0(\cdot + \theta).$$

Then P associated with q_0 can be defined by $P = \pi^{-1}\mu$, where μ is the normalized Lebesgue measure on R/TZ. It should be noted here that a family of self-adjoint operators $\{L(\theta) = -\Delta + q_0(\cdot + \theta)\}$ are unitarily equivalent, hence their spectral properties are the same.

Example 2. (almost periodic potential) Let q_0 be a continuous almost periodic function in Bohr's sense. Then $\{q_0(\cdot + x); x \in R\}$ becomes a precompact set in $C_b(R)$ with sup-norm. Let X be the closure. Then X turns to be a compact abelian group with pointwise multiplication. Let μ be the normalized Haar measure on X. The shift operation on $\{q_0(\cdot + x); x \in R\}$ can be uniquely extended to X and we denote it by τ_x. Since q_0 can be considered as a continuous function on X, we can define

$$\pi: X \to \Omega \quad \text{by} \quad \pi(\theta) = q_0(\tau.\theta).$$

Then we can introduce $P = \pi^{-1}\mu$. It is needless to say that periodic potentials are a special case of almost periodic potentials.

Example 3. (random potential induced by Brownian motion on compact manifold) Let M be a compact Riemannian manifold and $\{X_x(\omega); x \in R\}$ be the Brownian motion on M with the Riemannian volume as its marginal distribution. Let F be a continuous function on M. Then we can define a map π from a probability space defining the Brownian motion to Ω by

$$\pi(\omega) = F(X.(\omega)).$$

P is introduced by $\pi^{-1}\mu$, where μ is the probability measure describing the Brownian motion. This type of random potentials was first considered by I. Ja. Goldseid, S. A. Molchanov, L. A. Pastur [13]. They proved that $L(\omega) = -\Delta + F(X.(\omega))$ has only point spectrum and their eigenfunctions decay exponentially fast under a non-degeneracy condition on F.

Since the beginning of the eighties, lots of papers have been devoted to the study of almost periodic potentials and it is known that they exhibit all possible features of spectrum. We give references for this only J. Avron, B. Simon [1],[2], V. A. Chulaevsky [7], W. Craig [9], J. Pöschel [31] and J. Moser [26]. Random potentials such as in

Example 3 have received a special interest by physicist. Although the original proof given by G-M-P was rather difficult to understand, R. Carmona [4],[5] has succeeded to simplify the proof. Recently by the works S. Kotani [19], F. Delyon, Y. Lévy, B. Souillard [12], B. Simon [33] and S. Kotani, B. Simon [21], it can be said that the mechanism of the exponential localization in one-dimensional random systems is almost completely analyzed.

On the other hand, periodic potentials have been investigated from various aspects. Especially, periodic potentials with finite band spectrum have been studied intensively in relation to the KdV-equation (see S. P. Novikov [27] and H. P. McKean, P. van Moerbeck [24]). Recently their works have been extended to several cases of almost periodic potentials by V. A. Marchenko, I. V. Ostrovsky [23], B. M. Levitan [22] and L. A. Pastur, B. A. Tkachenko [29]. Therefore it is preferable to treat their ideas from a more general point of view. Fortunately, as were known in [17],[18],[20], several general theorems for spectral properties of a random system $(L(q),\Omega,P)$ already exist. This note is an attempt of further extension of the above three papers by the author. We present two problems. One is a relationship between spectrum and the logarithmic potential theory first noticed by R. Johnson, J. Moser [15]. The second is an infinite dimensional Hamiltonian system first introduced by J. Moser [25] in finite band case.

Notations.

Any integral on the whole line R will be written simply as $\int$. We use notations: $R_+ = (0,\infty)$, $R_- = (-\infty,0)$, $C_+ = \{\lambda \in C;\ \mathrm{Im}\,\lambda > 0\}$, $C_- = \{\lambda \in C;\ \mathrm{Im}\,\lambda < 0$. The set of all Herglotz functions on C_+ will be denoted by $\mathcal{H}$: $\mathcal{H} = \{h;\ h$ is holomorphic on C_+ and maps C_+ into $C_+\}$. A subclass $\mathcal{W}$ of $\mathcal{H}$ is introduced: $\mathcal{W} = \{w;\ w, w', -w \in \mathcal{H}\}$.

§2. Floquet exponent and spectrum.

In this section we summarize known facts needed for later use. The details can be found in [20]. Let $q \in \Omega$. For $\lambda \in C$, let ϕ_λ, ψ_λ be unique solutions of

$$L(q)\phi_\lambda = \lambda\phi_\lambda,\quad \phi_\lambda(0) = 1,\quad \phi_\lambda'(0) = 0$$

$$L(q)\psi_\lambda = \lambda\psi_\lambda,\quad \psi_\lambda(0) = 0,\quad \psi_\lambda'(0) = 1.$$

Then it is known that there exist unique $h_{\pm}(\lambda,q) \in C_+$ such that

$$f_{\lambda}^{\pm}(x,q) = \phi_{\lambda}(x,q) \pm h_{\pm}(\lambda,q)\psi_{\lambda}(x,q) \in L^2(R\pm,dx). \tag{2.1}$$

Then the Green function $g_{\lambda}(x,y,q)$ of $L(q) - \lambda$ can be represented by $f_{\lambda}^{\pm}$:

$$g_{\lambda}(x,y,q) = g_{\lambda}(y,x,q) = -\big(h_+(\lambda,q) + h_-(\lambda,q)\big)^{-1} f_{\lambda}^{+}(x,q) f_{\lambda}^{-}(y,q) \tag{2.2}$$

for $x \geq y$ and $\lambda \in C_+$. Another expression of g_{λ} is possible. Let

$$H(\lambda) = \begin{pmatrix} \dfrac{-1}{h_+(\lambda) + h_-(\lambda)} , & \dfrac{-h_+(\lambda)}{h_+(\lambda) + h_-(\lambda)} + \dfrac{1}{2} \\ \dfrac{-h_+(\lambda)}{h_+(\lambda) + h_-(\lambda)} + \dfrac{1}{2} , & \dfrac{h_+(\lambda)h_-(\lambda)}{h_+(\lambda) + h_-(\lambda)} \end{pmatrix} . \tag{2.3}$$

Here we omitted q. It is easy to see that

$$\big(H(\cdot)z,z\big) \in \mathcal{H} \quad \text{for any} \quad z \in C^2.$$

Hence there exist a positive definite 2×2 matrix valued measure $\Sigma(d\xi)$ and a self-adjoint 2×2 matrix A such that

$$H(\lambda) = A + \int \Big(\frac{1}{\xi - \lambda} - \frac{\xi}{1 + \xi^2} \Big) \Sigma(d\xi).$$

By using this $\Sigma(d\xi)$, $g_{\lambda}(x,y)$ can be expressed as

$$\operatorname{Im} g_{\lambda}(x,y) = \int \frac{\operatorname{Im}\lambda}{|\xi - \lambda|^2} \big(\Sigma(d\xi)\Phi_{\xi}(x),\Phi_{\xi}(y)\big), \tag{2.4}$$

where $\Phi_{\lambda}(x) = {}^t\big(\phi_{\lambda}(x),\psi_{\lambda}(x)\big)$. Since $\Sigma(d\xi)$ is positive definite, setting $\sigma(d\xi) = \operatorname{tr}\Sigma(d\xi)$, we see that every component of $\Sigma(d\xi)$ is absolutely continuous with respect to σ. Therefore, making a diagonalization of the density matrix, we have

$$\operatorname{Im} g_{\lambda}(x,y) = \operatorname{Im}\lambda \int \frac{\{f_1(x,\xi)f_1(y,\xi) + f_2(x,\xi)f_2(y,\xi)\}}{|\xi - \lambda|^2} \sigma(d\xi), \tag{2.5}$$

where $f_{1,2}$ are solutions of $L(q)f = \xi f$ (which may be dependent).

$g_{\lambda}(x,y,q)$ has an asymptotic behaviour:

$$g_{\lambda}(x,y,q) = \exp\{w(\lambda)|x - y| + o(|x - y|)\} \tag{2.6}$$

as $|x - y| \to \infty$ for $\lambda \in C_+$. This is valid only for almost every $q \in \Omega$ with respect to P. $w(\lambda)$ is a function holomorphic on C_+ independent of $q \in \Omega$ and possesses the following relations ([15]):

$$w(\lambda) = -\frac{1}{2} \int_{\Omega} g_{\lambda}(0,0,q)^{-1} P(dq), \tag{2.7}$$

$$w'(\lambda) = \int_{\Omega} g_{\lambda}(0,0,q) P(dq). \tag{2.8}$$

Hence we can define a w-function by (2.7) also for $P \in \mathcal{P}$ not satisfying (1.3). If necessary, we denote it by w_P. Since $g_\lambda(0,x,q) \in L^2(R_+,dx)$, Re $w(\lambda)$ should be negative on C_+. Hence this combined with (2.7) and (2.8) implies $w \in \mathcal{W}$. On the other hand, from the asymptotic expansion of $g_\lambda(0,0,q)$ near $\lambda = \infty$, it follows that

$$-i\sqrt{\lambda}\, w(\lambda) = \lambda + m + O(|\lambda|^{-1}), \tag{2.9}$$

where $m = -\int_\Omega \{\int_0^1 q(x)dx\}P(dq) \in R$. Let us introduce a subclass of $\mathcal{W}$ by

$$\mathcal{W}_0 = \{w \in \mathcal{W};\ w \text{ satisfies } (2.9)\}.$$

w is known to have a finite limit at $\xi \in R$, which we denote by

$$w(\xi + i0) = -\gamma(\xi) + in(\xi). \tag{2.10}$$

γ and n are called the Lyapounov exponent and the integrated density of states for $(L(q),\Omega,P)$. From the properties of w it is easy to see that

$$\gamma(\xi), n(\xi) \geq 0 \text{ and } n(\xi) \text{ is non-decreasing.}$$

Moreover (2.10) implies

$$\gamma(\xi) = -\text{Re } w(i) + \frac{1}{\pi}\int \log\left|\frac{\eta - \xi}{\eta - i}\right| n(d\eta) \qquad \text{(Thouless formula).}$$

From this we see that $n(\xi)$ has at least log-Hölder continuity (W. Craig, B. Simon [11]).

Denote by $\sum(q)$, $\sum_{a.c.}(q)$ the spectrum and the absolutely continuous spectrum of $L(q)$ respectively. Then we have

Theorem 2.1. Let $P \in \mathcal{P}_e$. Then for a.e. $q \in \Omega$ with respect to P

$\sum(q) = \text{supp } dn$ (L. A. Pastur [28]).

$\sum_{a.c.}(q) = \overline{\{\xi \in R;\ \gamma(\xi) = 0\}}^{\mu}$ (K. Ishii [14], L. A. Pastur [28], S. Kotani [17])

where for a Borel set $A \in \mathcal{B}(R)$,

$$\bar{A}^{\mu} = \{x \in R;\ \mu(U \cap A) > 0 \text{ for any neighborhood } U \text{ of } x\}.$$

μ is Lebesgue measure on R.

Theorem 2.2. (S. Kotani [17]) Let $P \in \mathcal{P}$. Suppose for an $A \in \mathcal{B}(R)$ $\gamma(\xi) = 0$ a.e. on A. Then for any $q \in \text{supp } P$ (with respect to the sequentially weak convergence in Ω)

$$h_+(\xi+i0,q) = -\overline{h_-(\xi+i0,q)} \quad \text{a.e. on } A. \tag{2.11}$$

Hence for any $x \in R$ and any $q \in \text{supp } P$

$$\text{Re } g_{\xi+i0}(x,x,q) = 0 \quad \text{a.e. on } A \tag{2.12}$$

Especially if $A = I$ an open interval, then for any $x \in R$ and $q \in \text{supp } P$, $g_\lambda(x,x,q)$ can be analytically continuable down to C_- through I.

In this way, the Floquet exponent $w(\lambda)$ plays an essential role in considering a random system $(L(q),\Omega,P)$.

§3. The logarithmic potential theory and spectrum.

Before stating the relationship between the potential theory and spectrum, we mention two results on almost periodic potentials obtained by the Russian school.

Let

B_2^+ = {the set of all L^2-almost periodic functions q in Besicovitch's sense whose $L(q)$ is non-negative definite}.

In B_2^+, we define

$$\|q\|^2 = \lim_{a\to\infty} \frac{1}{2a} \int_{-a}^{a} |q(x)|^2\, dx.$$

Let $A = \{a_n\}_1^\infty$ be a positive sequence satisfying $a_{n+1}/a_n \in Z$. Set

$$Q_\infty^+(A) = \{q \in B_2^+;\ \text{there exist } a_n\text{-periodic } q_n \text{ such that}$$
$$\|q - q_n\| \exp(ca_{n+1}) \to 0 \quad \text{as } n \to \infty \quad \text{for any } c > 0\}.$$

Moreover set

$$R_n = \{\pi k/a_n;\ k \in Z\}, \qquad R(A) = \bigcup_n R_n.$$

Then the spectrum of q as an almost periodic function is contained in $R(A)$. For $\theta = (\theta_r) \in C^{R(A)}$, define

$$\|\theta\|^2 = |\theta_0|^2 + \sum r^2 |\theta_r|^2.$$

Introduce

$$\Theta_\infty^+(A) = \{(\theta_r) \in C^{R(A)};\ \theta_r = \bar{\theta}_{-r},\ \text{Im } \theta_0 = 0 \text{ and}$$
$$\Big(\sum_{r\in R(A)\setminus R_n} r^2|\theta_r|^2\Big)\exp(ca_{n+1}) \to 0 \quad \text{for any } c > 0\}.$$

For a given $\theta = (\theta_r) \in \Theta_\infty^+(A)$, we can construct $w \in W_0$ in the following way. Let

$$\Pi^\theta = \{\lambda \in C;\ \text{Im } \lambda > 0,\ \text{Re } \lambda < 0\} \setminus \bigcup_{r\in R(A)} [-|\theta_r| + i|r|,\ i|r|],$$

where $[\alpha,\beta]$ denotes the line connecting α and β in C. Since

Π^{θ} is simply connected, by the Riemann mapping theorem, there exists uniquely a conformal mapping w from C_+ onto Π^{θ} satisfying the normalization condition (2.9) and $w(0) = 0$. Then it is not difficult to see $w \in \mathcal{W}_0$. L. A. Pastur, B. A. Tkachenko [29] have obtained the following theorem as a generalization of V. A. Marchenko, I. V. Ostrovsky [23]:

Theorem 3.1. There exists a unique correspondence between $Q_{\infty}^{+}(A)$ and $\Theta_{\infty}^{+}(A)$ as an extension of the correspondence in periodic case. Moreover the above $w(\lambda)$ gives the Floquet exponent.

It is known that in the above case dn turns to be absolutely continuous and the spectrum of $L(q)$ consists only of absolutely continuous component. Now let us state the second result by B. M. Levitan [22]. Let F be a perfect set in R_+ and set

$$F^c = \bigcup_{n=1}^{\infty} (\alpha_n, \beta_n).$$

We arrange (α_n, β_n) in the decreasing order of their length. Let (λ_j, μ_j) be a subinterval of (α_j, β_j) satisfying $\lambda_j - \alpha_j = \beta_j - \mu_j$ and set

$$0 < q_j = \frac{\mu_j - \lambda_j}{\beta_j - \alpha_j} < 1.$$

Impose the following condition on them:

$$\sup_j \frac{\mu_j}{\lambda_j} < \infty, \quad \sum q_j < \infty, \quad \sum \sqrt{\lambda_j} q_j < \infty.$$

Then we have

Theorem 3.2. There exists a sequence of quasi periodic potentials $\{q_N\}$ whose spectrum is $[0,\lambda_1] \cup [\mu_1,\lambda_2] \cup \cdots \cup [\mu_{N-1},\lambda_N] \cup [\mu_N,\infty)$ such that $\{q_N\}$ congerges to q uniformly on R and $L(q)$ has F as its spectrum.

This q is obviously almost periodic, however it is not known if $L(q)$ has absolutely continuous spectrum or not. In the above context, it seems interesting to establish a general theorem to guarantee the existence of $P \in \mathcal{P}$ or $\mathcal{P}_e$ for a given $w \in \mathcal{W}_0$. In this respect we have

Theorem 3.3. (S. Kotani [18],[20]) For any $w \in \mathcal{W}_0$, we have $P \in \mathcal{P}$ such that $w = w_P$. If $w \in \mathcal{W}_0$ satisfies

$$\int \gamma(\xi) dn(\xi) = 0, \tag{3.1}$$

then we can choose $P \in \mathcal{P}_e$.

The condition (3.1) implies that $w \in \mathcal{W}_0$ is an extreme point in $\mathcal{W}_0$. It is not known if the converse statement is also true or not.

Here we mention briefly the case of a discrete system $(L_d(q), \Omega_d, P)$. Since we can define a shift operator on Ω_d, the properties (1.1) and (1.3) can be discussed also for P. Instead of (1.2) we impose

(1.2)' integrability:

$$\int_{\Omega_d} \{|\log a_0| + a_0^2 + b_0^2\} P(dq) < \infty.$$

We introduce $\mathcal{P}^d$, $\mathcal{P}_e^d$, similarly as $\mathcal{P}$, $\mathcal{P}_e$. Then we have Theorems 2.1 and 2.2 also in this case (see B. Simon [32], R. Carmona, S. Kotani [6]), if we define w, γ, n in the following way. Let

$$g_\lambda(0,0,q) = \left((L_d(q)-\lambda)^{-1}\delta_0, \delta_0\right). \tag{3.2}$$

Then

$$w'(\lambda) = \int_{\Omega_d} g_\lambda(0,0,q) P(dq) = \int \frac{dn(\xi)}{\xi - \lambda}, \quad w(\lambda) = \log \frac{1}{\xi - \lambda} dn(\xi). \tag{3.3}$$

The Lyapounov exponent $\gamma(\xi)$ satisfies

$$\gamma(\xi) = -\operatorname{Re} w(\lambda) - \int_{\Omega_d} (\log a_0) P(dq) \geq 0. \tag{3.4}$$

The integrated density of states fulfills

$$\int dn(\xi) = 1, \quad \int \xi dn(\xi) = \int_{\Omega_d} b_0 P(dq),$$

$$\int \xi^2 dn(\xi) \leq \int_{\Omega_d} (2a_0^2 + b_0^2) P(dq) < \infty.$$

We introduce

$$\mathcal{W}_0^d = \{w;\ w(\lambda) = \int \log \frac{1}{\xi - \lambda} dn(\xi) \text{ with a measure } dn \text{ on } R$$

$$\text{satisfying } \int dn(\xi) = 1, \quad \int \xi^2 dn(\xi) < \infty \text{ and} \tag{3.5}$$

$$\operatorname{Re} w(\lambda) \text{ is bounded from above on } C_+\}. \tag{3.6}$$

Then we have analogous result of Theorem 3.3.

<u>Theorem 3.3'</u>. (R. Carmona, S. Kotani [6]) For any $w \in \mathcal{W}_0^d$ we have $P \in \mathcal{P}^d$ such that $w = w_P$. If w satisfies

$$\int \gamma(\xi) dn(\xi) = 0 \tag{3.7}$$

also with $\gamma(\xi) = c - \operatorname{Re} w(\lambda)$, $c = \sup_{\lambda \in C_+} \operatorname{Re} w(\lambda)$, then we can choose

$P \in P_e^d$ and $c = -\int_{\Omega_d} (\log a_0) P(dq)$.

Now we are in a stage to mention the relationship of the spectrum to the logarithmic potential theory. Let us recall several notions of the potential theory. Let K be a compact set of C. Set

$$M(K) = \{\text{the set of all probability measures on } K\}.$$

For $\nu \in M(K)$ define

$$I(\nu) = -\iint_{K\times K} \log|z_1 - z_2| \ \nu(dz_1)\nu(dz_2) \quad \text{and}$$

$$\mathrm{Cap}(K) = \sup_{\nu\in M(K)} \exp\{-I(\nu)\}.$$

Then $0 \leq \mathrm{Cap}(K) < \infty$. For any Borel set A of C, $\mathrm{Cap}(A)$ can be defined naturally. It should be noted that $\mathrm{Cap}(A) = 0$ implies $\mu(A) = 0$. μ is Lebesgue measure on R.

Potential theoretic arguments were first employed by R. Johnson, J. Moser [15]. Later W. Craig, B. Simon [10] also used the subharmonicity of the Lyapounov exponent to show the coincidence of the Lyapounov exponent and $-\mathrm{Re}\, w(\lambda)$ on R. In their paper [15], R. Johnson, J. Moser essentially proved

<u>Theorem 3.4</u>. Let $P \in P_e$. Then Theorem 2.1 says $\sum(q) = \mathrm{supp}\, dn$ a.e. $q \in \Omega$. However $\sum = \mathrm{supp}\, dn$ satisfies $\bar{\sum}^{\mathrm{Cap}} = \sum$. The definition of $\bar{A}^{\mathrm{Cap}}$ is analogous to that of $\bar{A}^{\mu}$.

This implies automatically $\mathrm{Cap}(\sum) > 0$ and $\sum$ has no isolated points. This theorem comes immediately from the Thouless formula if we notice that dn has a locally finite energy and any such measure does not charge capacity zero sets.

In Theorems 3.3 and 3.3', we have introduced the conditions (3.1) and (3.7). These conditions are interesting from the potential theoretic point of view. For $w \in W_0$ or W_0^d, set

$$N(w) = \{\xi \in R;\ \gamma(\xi) = 0\}.$$

If w satisfies (3.1) or (3.7), then $\overline{N(w)} = \mathrm{supp}\, dn$. We denote

$$A_1 = A_2 \quad (\mathrm{mod.\ cap})$$

if $\mathrm{Cap}(A_1 \Delta A_2) = 0$. Then we have

<u>Theorem 3.5</u>. Assume $w_1, w_2 \in W_0$ satisfy (3.1). If $N(w_1) = N(w_2)$

(mod. cap), then $w_1 = w_2$. The analogous theorem is valid also for $\mathcal{W}_0^d$.

Proof. First note that for $w_1, w_2 \in \mathcal{W}_0$

$$\int_{C_+} |w_1'(\lambda) - w_2'(\lambda)|^2 \, dxdy = -\pi \int \{\gamma_1(\xi) - \gamma_2(\xi)\} d\{n_1(\xi) - n_2(\xi)\} \quad (3.8)$$

See for the proof S. Kotani [20]. From (3.1) we have

$$0 = \int \gamma_1(\xi) dn_1(\xi) = \int_{N(w_1)^c} \gamma_1(\xi) dn_1(\xi),$$

hence $dn_1(N(w_1)^c) = 0$. However, since dn_1 does not charge capacity zero sets, we see $dn_1(N(w_2)^c) = 0$. Therefore

$$\int \gamma_2(\xi) dn_1(\xi) = \int_{N(w_2)^c} \gamma_2(\xi) dn_1(\xi) = 0.$$

Similarly we have $\int \gamma_1(\xi) dn_2(\xi) = 0$. This combined with (3.8) concludes the proof.

As a next step, it will be interesting to give a condition for a Borel set A of R to be an $N(w)$ for a $w \in \mathcal{W}_0$ satisfying (3.1). We can solve this problem for the discrete systems. To see this, we introduce the notion of regular points for $A \in \mathcal{B}(R)$. $z \in C$ is called a regular point for A, if a two-dimensional Brownian motion starting at z reaches instantaneously A. See the details S. C. Port, C. J. Stone [30]. Set

$$A^r = \{z \in C; \ z \text{ is a regular point for } A\}.$$

Then $\mathring{A} \subset A^r \subset \bar{A}$ and $\mathrm{Cap}(A \setminus A^r) = 0$.

Theorem 3.6. Let A be a bounded Borel set of R. Then $A = N(w)$ (mod. cap) for a $w \in \mathcal{W}_0^d$ satisfying (3.7) if and only if $A^r = A$ (mod. cap) and $\mathrm{Cap}(A) > 0$. In this case dn turns out to be the equilibrium measure of A, which is characterized by

$$\int \log \frac{1}{|\eta - \xi|} \, dn(\eta) = \text{const. (mod. cap) on } A, \quad dn(A^c) = 0. \quad (3.9)$$

The above const. = $-\log\{\mathrm{Cap}(A)\}$ $\left(= \int_{\Omega_d} \log a_0 P(dq) \right)$.

Proof. We have only to prove that for $w \in \mathcal{W}_0^d$ satisfying (3.7), $N(w)^r \subset N(w)$. The potential

$$u(\xi) = \int \log \frac{1}{|\eta - \xi|} \, dn(\eta)$$

satisfies $u(\xi) = \text{const.}$ on $A \equiv N(w)$. Moreover (3.7) implies $dn(A) = 1$. These two properties show that dn coincides with ν_A the equilibrium measure of A. In this case, it is known that

$$u(\xi) = \text{const}.(= -\log\{\text{Cap}(A)\}) \quad \text{on} \quad A^r.$$

Therefore $A^r \subset A$. $\text{Cap}(A) > 0$ follows from that dn does not charge capacity zero sets.

This theorem shows that the condition (3.7) does not necessarily imply the presence of absolutely continuous spectrum. Especially if we take a compact set K of R with $\text{Cap}(K) > 0$, $K = K^r$ and $\mu(K) = 0$, then the corresponding $P \in P_e^d$ induces a random system $(L_d(q), \Omega_d, P)$ whose spectrum $\sum(q) = K$ a.e. $q \in \Omega_d$, and the Lyapounov exponent vanishes exactly on K. Hence this system has no absolutely continuous spectrum and the spectrum is very thin. This kind of random system was considered by J. Bellisard, B. Simon [3] and V. A. Chulaevsky, S. A. Molchanov [8].

For the continuous systems, a similar result to Theorem 3.6 will be discussed elsewhere. In this case, since $N(w)$ becomes an unbounded Borel set of R, we have to modify the logarithmic potential theory.

§4. Absolutely continuous spectrum and Hamiltonian system.

In this final section, we remark a way to study random systems with non-degenerate absolutely continuous spectrum by extending the argument done by J. Moser [25].

Let $P \in P$ and $w = w_P$. For simplicity we assume

$$[c_1, \infty) \subset N(w) \subset [c_2, \infty) \tag{4.1}$$

for $c_1 > c_2$. Then set

$$u(\lambda) = \text{const.} \exp\{J(\lambda)\} \text{ with } J(\lambda) = \frac{1}{2}\int_{N(w)} \left(\frac{1}{\xi - \lambda} - \frac{\xi}{1 + \xi^2} \right) d\xi. \tag{4.2}$$

In view of (4.1), it is easy to see

$$u(\lambda) = \text{const.} \frac{i}{\sqrt{\lambda}} + o(\sqrt{|\lambda|}^{\,-1}) \tag{4.3}$$

near $\lambda = \infty$. The const. in (4.2) should be chosen so that the const. in (4.3) equals 2. Now notice

$$0 < \text{Im } j(\lambda) < \pi \quad \text{on} \quad C_+ \tag{4.4}$$

$$\text{Im } J(\xi + i0) = \begin{cases} 0 & \text{a.e. on } N(w)^c \\ \pi/2 & \text{a.e. on } N(w). \end{cases} \tag{4.5}$$

On the other hand, from Theorem 2.2 it follows that for any $q \in \operatorname{supp} P$,

$$\operatorname{Im} \log\bigl(H(\xi+i0,q)z,z\bigr) = \begin{cases} \pi/2 & \text{a.e. on } N(w) \\ 0 \text{ or } \pi & \text{a.e. on } N(w)^c. \end{cases} \tag{4.6}$$

Therefore we can conclude that

$$\bigl(u(\cdot)H(\cdot,q)z,z\bigr) \in H, \quad \text{for any} \quad z \in C^2. \tag{4.7}$$

Since $H(\lambda,q)$ behaves like $\frac{i}{2\sqrt{\lambda}}I$ at $\lambda = \infty$, by (4.3) and (4.7), we see that with a positive definite 2×2 matrix valued measure Ξ

$$u(\lambda)H(\lambda,q) = \int \frac{1}{\xi - \lambda}\, \Xi(d\xi,q), \quad \int \Xi(d\xi,q) = I. \tag{4.8}$$

Now note the following fact: Let $f(\lambda)$ be a real entire function and $h \in H$ with representing measure σ. Then

$$\lim_{\varepsilon \downarrow 0} \int_I \operatorname{Im}\bigl(f(\xi + i\varepsilon)h(\xi + i\varepsilon)\bigr)d\xi = \int_I f(\xi)\sigma(d\xi). \tag{4.9}$$

Applying (4.9) to $f(\lambda) = \Phi_\lambda(x)$ and $h(\lambda) = u(\lambda)H(\lambda,q)$, (2.1) and (2.2) imply

$$u(\lambda)g_\lambda(x,x,q) = \int \frac{1}{\xi - \lambda}\bigl(\Xi(d\xi,q)\Phi_\xi(x),\Phi_\xi(x)\bigr). \tag{4.10}$$

Setting $\nu(d\xi,q) = \operatorname{tr}\Xi(d\xi,q)$, we see by a similar argument to that in §2,

$$u(\lambda)g_\lambda(x,x,q) = \int \frac{f_1(x,\xi)^2 + f_2(x,\xi)^2}{\xi - \lambda}\,\nu(d\xi,q) \tag{4.11}$$

where $f_{1,2}$ are solutions (which may be dependent) of $L(q)f = \xi f$. We summarize the above argument in

<u>Theorem 4.1</u>. Assume the condition (4.1). Then for any $q \in \operatorname{supp} P$ there exist a finite singular measure ν on $[c_2,c_1]$ and two solutions $f_{1,2}$ of $L(q)f = \xi f$ on R for a.e. $\xi \in R$ with respect to ν such that (4.11) is valid for all $x \in R$. Moreover they satisfy

$$\int \{f_1(x,\xi)^2 + f_2(x,\xi)^2\}\nu(d\xi) = 1. \tag{4.12}$$

<u>Proof</u>. From the condition (4.1) we easily see $\operatorname{supp} \nu \subset [c_2,c_1]$. The singularity of the measure ν comes from (4.5) and (4.6). The property (4.12) is an easy consequence of (4.3) and the fact that $g_\lambda(x,x,q)$ behaves like $\frac{i}{2\sqrt{\lambda}}$ at $\lambda = \infty$.

In [25], J. Moser discussed a case that

$$N(w) = [\alpha_1,\beta_1] \cup [\alpha_2,\beta_2] \cup \cdots \cup [\alpha_{n-1},\beta_{n-1}] \cup [\alpha_n,\infty).$$

He chose as $u(\lambda)$

$$u(\lambda) = 2\left(-\frac{(\lambda-\beta_1)\cdots(\lambda-\beta_{n-1})}{(\lambda-\alpha_1)\cdots(\lambda-\alpha_n)}\right)^{1/2},$$

which coincides with our $u(\lambda)$. In this case

$$\nu(d\xi) = \sum_{j=1}^{n} \delta_{\alpha_j}(d\xi).$$

Moreover he proved that either f_1 or f_2 vanishes. In this sense ν is simple. However in our general case we do not know if this is valid or not. Theorem 4.1 can be rewritten in terms of a certain infinite dimensional Hamiltonian system. Let

$$\begin{cases} H = L^2(R,\nu) \times L^2(R,\nu) \\ Af(\xi) = \left(\xi f_1(\xi), \xi f_2(\xi)\right) \quad \text{on} \quad H. \end{cases}$$

Then

$$\begin{cases} -f''(x) + q(x)f(x) = Af(x) \\ \|f(x)\| = 1 \qquad \text{on} \quad R, \end{cases} \tag{4.14}$$

where $f(x) = \left(f_1(x,\xi), f_2(x,\xi)\right) \in H$. Here A is a bounded self-adjoint operator on H. If ν is of form (4.13), then it is known that the Hamiltonian system (4.14) is completely integrable. Therefore it seems interesting to study an infinite dimensional system (4.14). Here it should be remarked that R. Johnson [16] pointed out that if a random system $\left(L(q),\Omega,P\right)$ satisfies (4.1) then every $q \in \text{supp } P$ is of Segal-Wilson type and hence q is a holomorphic function on C.

REFERENCES

1. Avron, J. and Simon, B. "Almost periodic Schrödinger operators I, Limit periodic potentials," Comm. Math. Phys. 82(1982), 101-120.
2. ______, "Singular continuous spectrum for a class of almost periodic Jacobi matrices," Bull.(New Series) of the AMS 6(1982), 81-85.
3. Bellisard, J. and Simon, B. "Cantor spectrum for the almost Mathieu equation," J. Funct. Anal. 48(1982), 408-419.
4. Carmona, R. "Exponential localization in one-dimensional disordered systems," Duke Math. J. 49(1982), 191-213.
5. ______, "One-dimensional Schrödinger operators with random or deterministic potentials," New Spectral Types, J. Funct. Anal. 51(1983), 229-258.
6. Carmona, R. and Kotani, S. "On inverse problem for random Jacobi matrices," in preparation.
7. Chulaevsky, V. A. "On perturbations of a Schrödinger operator with periodic potential," Russian Math. Survey 36(1981), 143-144.
8. Chulaevsky, V. A. and Molchanov, S. A. "Structure of the spectrum of a lacunary-limit periodic Schrödinger operator," Funct. Anal. and its Appl. 18(1984), 90-91.

9. Craig, W. "Pure point spectrum for discrete almost periodic Schrödinger operators," Comm. Math. Phys. 88(1983), 113-131.
10. Craig, W. and Simon, B. "Subharmonicity of the Ljapunov index," Duke Math J. 50(1983), 551-559.
11. _____, "Log-Hölder continuity of the integrated density of states for stochastic Jacobi matrices," Comm. Math. Phys. 90(1983), 207-218.
12. Delyon, F., Lévy, Y. and Souillard, B. "Anderson localization for one- and quasi one-dimensional systems," J. of Stat. Phys. 41(1985), 375-388.
13. Goldseid, I. Ja., Pastur, L. A. and Molchanov, S. A. "A pure point spectrum of the stochastic and one-dimensional equation," Funct. Anal. and its Appl. 11(1977), 1-10.
14. Ishii, K. "Localization of eigenstates and transport phenomena in one-dimensional disordered systems," Supp. Prog. Theor. Phys. 53 (1973), 77-138.
15. Johnson, R. and Moser, J. "The rotation number of almost periodic potentials," Comm. Math. Phys. 84(1982), 403-438.
16. _____, Some remarks on a paper of Segal-Wilson, Universität Heidelberg Preprint, 1986.
17. Kotani, S. "Ljapounov indices determine absolutely continuous spectra of stationary random one-dimensional Schrödinger operators," Proc. Taniguchi Symp. S. A. Katata (1982), 225-247.
18. _____, "On an inverse problem for random Schrödinger operators," AMS series of Contemporary Math. 41(1985), 267-280.
19. _____, "Lyapounov exponents and spectra for one-dimensional random Schrödinger operators," AMS series of Contemporary Math. 50(1986), 277-286.
20. _____, "One-dimensional random Schrödinger operators and Herglotz functions," Proc. Taniguchi Symp. on Probabilistic Methods in Mathematical Physics, Katata, to appear.
21. Kotani, S. and Simon, B. Localization in general one-dimensional systems II. Continuous Schrödinger operators, in preparation.
22. Levitan, B. M. "On the closure of the set of finite-zone potentials," Math. USSR Sb. 51(1985), 67-89.
23. Marchenko, V. A. and Ostrovsky, I. V. "A characterization of the spectrum of Hill's operator," Math. USSR Sb. 26(1974), 493-554.
24. McKean, H. P. and van Moerbeck, P. "The spectrum of Hill's operator," Invention Mat. 30(1975), 217-274.
25. Moser, J. Integrable Hamiltonian Systems and Spectral Theory, Lezioni Fermiane, Pisa-1981.
26. _____, "An example of Schrödinger operator with almost periodic potential and nowhere dense spectrum," Comment. Math. Helv. 56 (1981), 198-224.
27. Novikov, S. P. "The periodic problem for the KdV equation I," Funct. Anal. and its Appl. 8(1974), 236-246.
28. Pastur, L. A. "Spectral properties of disordered systems in one-body approximation," Comm. Math. Phys. 75(1980), 107-196.
29. Pastur, L. A. and Tkachenko, B. A. "On the spectral theory of the one-dimensional Schrödinger operator with a limit periodic potential," Dokl. Acad. Nauk SSSR 279(1984), 1050-1054.
30. Port, S. C. and Stone, C. J. "Brownian motion and classical potential theory," Academic Press.
31. Pöschel, J. "Examples of discrete Schrödinger operators with pure point spectrum," Comm. Math. Phys. 88(1983), 447-463.
32. Simon, B. "Kotani theory for one-dimensional stochastic Jacobi matrices," Comm. Math. Phys. 89(1983), 227-234.
33. _____, "Localization in general one-dimensional systems I. Jacobi matrices," Comm. Math. Phys. 102(1985), 327-336.

Undressing of Odd Pseudodifferential Operators

B. A. Kupershmidt

The University of Tennessee Space Institute

Tullahoma, Tennessee 37388, USA

Abstract. For $\mathbf{Z}_2$-homogeneous pseudodifferential operators in an odd space, a differential analog is found of the classical condition of conservation of $(N-1)$th coefficient of Lax equaitons. The vanishing of this differential expression is the necessary and sufficient condition for the conjugation of the corresponding odd pseudodifferential operator into its principal term.

§0. Introduction

Lax equations constitute the most important and the best understood class of integrable systems. These are equations of the form [11]

(0.1) $L_t = [P_+, L] = [-P_-, L]$,

where

(0.2) $L = \sum_{0 \leq i \leq N} u_i \xi^i, \quad \xi = \dfrac{d}{dx}$,

is an $\ell \times \ell$ matrix differential operator satisfying the conditions

(0.3*i*) u_N is a constant invertible diagonalizable matrix,

(0.3*ii*) $u_{N-1} \in Im\, ad\,(u_N)$;

P runs over elements of the centralizer $Z(L)$ of L in the ring of matrix pseudodifferential operators over the differential ring $Mat_\ell(C)$,

(0.4) $C = K\left[u^{(j)}_{i,\alpha\beta}\right], \quad 1 \leq \alpha, \beta \leq \ell, \quad j \in \mathbf{Z}_-, \quad 0 \leq i < N,$

is a differential ring over a differential ring K, with a derivation $\partial : K \to K$.

The semisimplicity conditon (0.3) is crucial: it enables one to "undress" L, i.e., to find a Volterra operator

(0.5) $K = 1 + \sum_{i>0} \chi_i \xi^{-i}$

which conjugates L into its highest term:

(0.6) $K^{-1} L K = u_N \xi^N$.

Possessing the dressing operator K allows one to establish a great many nontrivial results, e.g.,: to classify Lax equations [11], to derive their Hamiltonian forms [7; 6], to analyze various constructions of conservation laws and formal eigenfunctions [12], to construct τ-function formulae [1; 4, Ch. IX], etc. The situation is similar in the super case [3; 5; 6], i.e., when one has $\mathbf{Z}_2$-gradings imposed upon the matrices u_i's in (0.2). Notice that in the scalar case $\ell = 1$, (0.3) becomes

(0.7i) $u_N = \text{const} \neq 0,$

(0.7ii) $u_{N-1} = 0.$

The condition (0.7ii) is routinely accepted in various constructions of scalar Lax equations on the grounds that for $\ell = 1$, (0.1) implies $u_{N-1,t} = 1$, so that u_{N-1} is time-independent and can be made to vanish by an invertible change of the x-variable. However, in most of constructions, the condition $\{u_{N-1}$ is arbitrary$\}$ serves just as well [7]; it is precisely the undressing property (0.6) that forces (0.7ii) and (0.3ii).

Recently Manin and Radul [10], motivated by constructions in supergeometry, introduced a radically new object into the area of integrable systems: an odd space with an odd derivation. They also began to study corresponding scalar Lax equations associated with odd Lax operators. Their method follows the classical scalar approach [9] of fractional powers of pseudodifferential operators, bypassing the problem of an existence of the dressing operator K. As explained above, the latter problem is, however, of much importance and I take it up in this paper. The main result, Theorem 2.2, states that K exists iff the condition (2.9) (a differential generalization of (0.7ii)) is satisfied. As in the classical case, the condition (2.9) is compatible with respect to power raising and root taking (Corollary 1.7).

§1. Pseudodifferential Operators

Let $K = K_0 + K_1$ be a commutative superalgebra with a $\mathbf{Z}_2$-grading p, and let $\theta : K \to K$ be an odd derivation of K (see, e.g., [2; 8] for basic superalgebra). The associative ring $\mathcal{O}_K$ of pseudodifferential operators over K is the space $\left\{ \sum\limits_{i<\infty} f_i \varsigma^i | f_i \in K \right\}$, with the natural $\mathbf{Z}_2$-grading p defined as

(1.1) $p(f\varsigma^i) = p(f) + ip(\theta),$

and with the multiplication

(1.2) $g\varsigma^i f\varsigma^j = \sum_{k\geq 0}\left\{\begin{matrix} i \\ k \end{matrix}\right\}(-1)^{(i-k)p(f)p(\theta)} g f^{(k)}\varsigma^{i-k+j}, \quad \forall g,f\in K, \quad \forall i,j\in \mathbf{Z},$

where $f^{(k)} := \theta^k(f)$ and $\left\{\begin{matrix} i \\ k \end{matrix}\right\}$ are the superbinomial coefficients [10]:

(1.3) $\left\{\begin{matrix} i \\ 0 \end{matrix}\right\} = 1; \left\{\begin{matrix} i \\ k \end{matrix}\right\} = \prod_{j=1}^{k}(1-q^{i+1-j}) / \prod_{j=1}^{k}(1-q^j), \quad k>0, q := (-1)^{p(\theta)}.$

If $Y = \sum^N f_i\varsigma^i \in O_K$ and $f_N \neq 0$ then $N = ord(Y)$ is called the order of Y. Denote by $O_K(I)$ the multiplicative subsemigroup of O_K consisting of $\mathbf{Z}_2$-homogenious pseudodifferential operators Y with the highest term $\mu\varsigma^{ord(Y)}, \mu \in K_c := Ker\theta|_K$. Define, for $\mathbf{Z}_2$-homogeneous $Y = f_N\varsigma^N + f_{N-1}\varsigma^{N-1} + f_{N-2}\varsigma^{N-1} + \cdots \in O_K, ord(Y) \leq N$,

(1.4) $Con\,(Y) := (-1)^{p(f_{N-2})} f_{N-2} + \frac{1}{2} f^{(1)}_{N-1}.$

1.5. Theorem. If $Y \in O_K(I)$ and $p(Y) = ord(Y)\,mod\,2$, then, for any $r\in \mathbf{Z}_+$,

(1.6) $Con\,(Y^r) = r\,Con\,(Y).$

1.7. Corollary. Let $O_K(I_0) = \{Y \in O_K(I) | \,Con\,(Y) = 0, p(Y) = ord(Y)\,mod 2.\}$. Then $O_K(I_0)$ is closed with respect to power raising.

1.8. Remark. $O_K(I_0)$ is *not closed* under multiplication, while in the even case $\left\{Y = \sum^N f_i\varsigma^{2i} | f_i \in K_0, f_N = const \neq 0\right\}$ ($\partial = \theta^2$ is an even derivation of K_0), the condition $Con\,(Y) = 0$ collapses into $f_{N-1} = 0$ and the latter conditon (exactly (0.7*ii*)) is, obviously, *closed* under multiplication.

Proof of Theorem. For $r = 0, 1$, (1.6) is evident. For $r \geq 2$, we can take $Y = \varsigma^N + f_{N-1}\varsigma^{N-1} + f_{N-2}\varsigma^{N-2} + \cdots$. We have, remembering that f_{N-1} is odd and f_{N-2} is even: $Y^r = \varsigma^{Nr} + \sum_{i=0}^{r-1}\varsigma^{Ni}\left(f_{N-1}\varsigma^{N-1} + f_{N-2}\varsigma^{N-2}\right)\varsigma^{N(r-i-1)} + O(\varsigma^{Nr-3})$. Now $\sum \varsigma^{Ni} f_{N-2}\varsigma^{N-2}\varsigma^{N(r-i-1)} = r f_{N-2}\varsigma^{Nr-2} + \cdots$, and since $\varsigma^{Ni} f_{N-1} = (-1)^{Ni} f_{N-1}\varsigma^{Ni} \cdot (-1)^{Ni-1}\left\{\begin{matrix} Ni \\ 1 \end{matrix}\right\} f^{(1)}_{N-1}\varsigma^{Ni-1} + \cdots$, we obtain $Y^r = \varsigma^{Nr} + F_{rN-1}\varsigma^{Nr-1} + F_{rN-2}\varsigma^{Nr-2} + \cdots$, where $F_{rN-1} = f_{N-1}\sum_{i=0}^{r-1}(-1)^{Ni}$,

$F_{rN-2} = r f_{N-2} + f^{(1)}_{N-1}\sum_{i=0}^{r-1}(-1)^{Ni-1}\left\{\begin{matrix} Ni \\ 1 \end{matrix}\right\}$. By (1.3), $\left\{\begin{matrix} Ni \\ 1 \end{matrix}\right\} = \frac{1-(-1)^{Ni}}{2}$. Thus, for N even, $F_{rN-1} = r f_{N-1}, F_{rN-2} = r f_{N-2}$, and (1.6) follows. For N odd, $F_{rN-1} = f_{N-1}\frac{1+(-1)^{r+1}}{2}, F_{rN-2} = r f_{N-2} + \frac{1}{2} f^{(1)}_{N-1}\left[r - \frac{1+(-1)^{r+1}}{2}\right]$, and (1.6) follows again. ∎

1.9. Proposition. If $R \in O_K$ with $ord\,(R) < 0$, and $Y \in O_K(I)$ with $ord\,(Y) = 1\,mod\,2$, then

(1.10) $Con\,([Y, R]) = 0,$

provided we consider $[Y, R]$ starting with $G\varsigma^{ord(Y)} + \cdots$.

1.11. <u>Remark</u>. For the case $ord(Y) = p(Y) = 1$, an observation equivalent to (1.10) was made in [10, p. 67].

1.12. <u>Remark</u>. If $ord(Y) = 0 \, mod\, 2$, (1.10) fails.

<u>Proof of Proposition</u>. Let $Y = \mu\varsigma^N + f_{N-1}\varsigma^{N-1} + f_{N-2}\varsigma^{N-2} + \cdots$, $\mu \in K_c$. If $ord(R) < -1$, then $ord([Y, R]) < N - 2$. Indeed, $[\mu\varsigma^N, g\varsigma^{-2}] = \mu\varsigma^N g\varsigma^{-2} - (-1)^{[p(\mu)+N]p(g)} g\varsigma^{-2}\mu\varsigma^N = \left(\mu(-1)^{Np(g)}g - (-1)^{[p(\mu)+N]p(g)}g\mu\right)\varsigma^{N-2} + \cdots = 0\varsigma^{N-2} + \cdots$, since $\mu g = (-1)^{p(\mu)p(g)} g\mu$. Thus, it is enough to consider the case $R = g\varsigma^{-1}$. Remembering that $N \in 1 + 2\mathbf{Z}$, we have $[Y, R] = [\mu\varsigma^N + f_{N-1}\varsigma^{N-1} + \cdots, g\varsigma^{-1}] = \mu\varsigma^N g\varsigma^{-1} - (-1)^{[p(\mu)+1][p(g)+1]} g\varsigma^{-1}\mu\varsigma^N + f_{N-1}g\varsigma^{N-2} - (-1)^{p(f_{N-1})[p(g)+1]} g\varsigma^{-1} f_{N-1}\varsigma^{N-1} + \cdots = 2\mu g(-1)^{p(g)}\varsigma^{N-1} + \mu g^{(1)}\left\{ {N \atop 1} \right\}\varsigma^{N-2} + \cdots = 2\mu g(-1)^{p(g)}\varsigma^{N-1} + \mu g^{(1)}\varsigma^{N-2} + \cdots$. Hence, by (1.4), $Con([Y, R]) = (-1)^{p(\mu)+p(g)+1}\mu g^{(1)} + \frac{1}{2}2(-1)^{p(g)}(-1)^{p(\mu)}\mu g^{(1)} = 0$. ∎

§2. <u>Conjugation</u>

Intersecting properties of operators Y figuring in Theorem 1.5 and Proposition 1.9, we consider $\mathcal{O}_K(I') = \left\{ Y = \varsigma^{ord(Y)} + \cdots \in \mathcal{O}_K(I) \mid ord(Y) = 1\, mod\, 2 \right\}$. Let us fix some $L \in \mathcal{O}_C(I')$:

$$(2.1) \quad L = \varsigma^N + \sum_{i<N} u_i\varsigma^i, \quad u_i \in C, \quad p(u_i) = i + 1, \quad N = 1\, mod\, 2,$$

where $C = K\left[u_i^{(j)}\right]$, $j \in \mathbf{Z}_+$, is a commutative differential superalgebra generated over K by independent generators $u_i^{(j)}$, with $p\left(u_i^{(j)}\right) = i + j + 1$. Denote by $\overline{C}$ a supercommutative differential extension of C on which θ acts and $\partial := \theta^2$ acts surjectively, and on which the L-grading $rk\left(u_i^{(j)}\right) = j + N - i$ is extended from C retaining the property $rk(\theta) = 1$ [11].

2.2. <u>Theorem</u>. (*i*) There exists a $\mathbf{Z}_2$-homogeneous operator $\mathcal{K}$ in $\mathcal{O}_{\overline{C}}$,

$$(2.3) \quad \mathcal{K} = 1 + \sum_{s>0} \chi_s\varsigma^{-s}, \quad \chi_s \in \overline{C},$$

conjugating L into its highest term:

$$(2.4) \quad \mathcal{K}^{-1}L\mathcal{K} = \varsigma^N,$$

iff $Con(L) = 0$. (*ii*) Such $\mathcal{K}$ is unique provided $rk(\mathcal{K}) = 0$.

Proof. Uniqueness is obvious. Rewriting (2.4) in the form $LK = K\varsigma^N$:

(2.5) $\sum u_i\varsigma^i + \sum \varsigma^N \chi_s \varsigma^{-s} + \sum u_i \varsigma^i \chi_s \varsigma^{-s} = \sum \chi_s \varsigma^{N-s}$,

we get

(2.6) $u_{N-1} = \chi_1 \left[1 - (-1)^N\right] = 2\chi_1$,

(2.7) $u_{N-2} = -\chi_1^{(1)}$,

(2.8) $(-1)^s \chi_s + \chi_{s-1}^{(1)} + (-1)^s \frac{N-1}{2} \chi_{s-2}^{(2)} + u_{N-1}\chi_{s-1} + u_{N-2}(-1)^s \chi_{s-2} + R_s = \chi_s$,

where R_s stands for a differential polynomial which does not contain χ_j's with $j \geq s-2$. From (2.6), (2.7) we see that (2.4) implies

(2.9) $Con\,(L) = u_{N-2} + \frac{1}{2} u_{N-1}^{(1)} = 0.$

Suppose now that $Con\,(L) = 0$. Then we find $\chi_1 = \frac{1}{2} u_{N-1} \in C$ from (2.6), and then proceed to define χ_s's by induction on s. Suppose we have already determined $\chi_1, \ldots, \chi_{2r-1}$ from the equations (2.8) for $s = 1, \ldots, 2r$. For $s = 2r+1$ and $s = 2r+2$, (2.8) yields, respectively

(2.10) $\chi_{2r+1} = \frac{1}{2}(u_{N-1} + \theta)(\chi_{2r}) + R_{2r+2}$,

(2.11) $(u_{N-1} + \theta)(\chi_{2r+1}) + (u_{N-2} + \frac{N-1}{2}\theta^2)(\chi_{2r}) = R_{2r+2}$.

Since R_{2r+2} in (2.10) contains χ_j's only for $j < 2r$, (2.10) simply tells us how to compute χ_{2r+1} in terms of $\chi_1, \ldots, \chi_{2r}$. Substituting (2.10) into (2.11) we obtain

(2.12) $S(\chi_{2r}) = R_{2r+2}$,

where $S = \frac{1}{2}(u_{N-1} + \theta)(u_{N-1} + \theta) + u_{N-2} + \frac{N-1}{2}\theta^2 = \frac{1}{2}\left(u_{N-1}^{(1)} + \theta^2\right) + u_{N-2} + \frac{N-1}{2}\theta^2 = Con\,(L) + \frac{N}{2}\theta^2 = \frac{N}{2}\theta^2$. Since θ^2 is invertible in $\overline{C}$, we find χ_{2r} from (2.12) and then χ_{2r+1} from (2.10). This completes the induction step. ∎

Acknowledgement

It is a pleasure to thank Ian Knowles and Roger Lewis for their hospitality during the conference. This work was supported in part by the National Science Foundation.

References

[1] Flaschka, H., "Construction of Conservation Laws for Lax Equations: Comments on Paper by G. Wilson", Quart. J. Math. Oxford (2), 34 (1983), 61–65.

[2] Kac, V. G., "Lie Superalgebras", Adv. Math. 26 (1976), 8–96.

[3] Kupershmidt, B. A., "Superintegrable Systems", Proc. Nat-l Acad. Sci. USA 81, 6562–6563.

[4] Kupershmidt, B. A., "Discrete Lax Equations and Differential-Difference Calculus", Asterisque, Paris, 1985.

[5] Kupershmidt, B. A., "A Review of Superintegrable Systems", Lect. Appl. Math. 23 (I), (1986), 83–121.

[6] Kupershmidt, B. A., "Elements of Superintegrable Systems" (to appear).

[7] Kupershmidt, B. A., and Wilson, G., "Modifying Lax Equations and the Second Hamiltonian Structure", Inventiones Math. 62 (1981), 403–436.

[8] Lietes, D. A., "Introduction to Supermanifolds", Russ. Math. Surveys 35 (1980), 1–64.

[9] Manin, Yu. I., "Algebric Aspects of Nonlinear Differential Equations", J. Sov. Math. 11 (1979), 1–122.

[10] Manin, Yu, I., and Radul, A. O., "A Supersymmetric Extention of the Kadomtsev Petviashvili Hierarchy", Comm. Math. Phys. 98 (1985), 65–77.

[11] Wilson, G., "Commuting Flows and Conservation Laws for Lax Equations", Math. Proc. Camb. Phil. Soc. 86 (1979), 131–143.

[12] Wilson, G., "On Two Constructions of Conservation Laws for Lax Equations", Quart. J. Math. Oxford (2), 32 (1981), 491–512.

SOME MATHEMATICAL ASPECTS OF THE WAVEMAKER THEORY

K. G. Lamb(*) and G. Tenti
Department of Applied Mathematics
University of Waterloo
Waterloo, Ontario, Canada
N2L 3G1

ABSTRACT

The wavemaker theory, like most topics in Fluid Mechanics, offers some challenging mathematical problems. Even in the linearized version, where the regulating equation is the Laplace equation, the full initial-boundary value problem is non-standard and offers serious difficulties to a completely rigorous treatment. In this paper we show how a reformulation of the theory allows for a solution of both the linear and the nonlinear problems by the same technique, which is essentially like the one introduced by Kennard to treat the traditional linear theory.

I. INTRODUCTION

A wavemaker is any device whose prescribed motion produces surface waves in a fluid with a free surface. The classical example, of course, is the piston-type wavemaker, in which one of the end walls of a long rectangular water tank performs oscillations of a prescribed frequency, and one is interested in understanding the behaviour of the water waves so generated as they travel along the tank. To date most laboratory testing of floating or bottom-mounted structures and studies of beach profiles and related phonomena have utilized wave tanks; and even waves caused by earthquake excitations of the seafloor or human-made structures can be studied by wavemaker theory.

The classical mathematical formulation of the theory of water waves[1-3] assumes that water is inviscid, incompressible, and irrotational, so that the motion may be described in terms of a velocity potential satisfying the Laplace equation, supplemented by the kinematic and dynamic boundary conditions at the free surface, and by the condition of no flow through the bottom of the container. In the wavemaker theory, this already complicated problem is rendered even more difficult by the fact that the velocity potential must also satisfy a further boundary condition at the wavemaker itself and the initial condition appropriate to the experimental situation being analyzed. As a result, even the case of infinitesimal waves -- where the nonlinearity of the surface boundary conditions may be disregarded -- poses some serious mathematical questions which have not yet been fully answered.

(*) Present Address: Department of Mathematics, Program in Applied Mathematics, Princeton University, Princeton, NJ 08544

Although we cannot give here a fair review of the literature on the subject, it should be mentioned that accounts of the linear theory -- which goes back to Havelock(4) are given in the classical references above, and have been critically re-examined by Ursell et al.(5). Extensions to wavemakers of various shapes have been given by Biésel and collaborators(6), and nonlinear effects have been considered more recently by Fontanet(7), Madsen(8), Flick and Guza(9), and many other investigators. These works share the common feature of being concerned with the permanent regime resulting from the long-time periodic motion of the wavemaker, while neglecting completely its transient motion; mathematically, the solutions thus obtained are asymptotic in nature.

From a physical point of view, these solutions are not completely satisfactory, as they can only predict the behavior of the waves far enough from the wavemaker since they incorporate the so-called radiation condition(5); this in turn makes it difficult to account for the influence of reflected waves from the far end of the tank. To overcome the problem, one may use the "burst method". in which the wavemaker generates waves for a short period of time and is then stopped; in this case, however, transient effects must be considered(8), with the result that the full initial value problem has to be solved. For the linear case this has been done first by Kennard(10) and his solution has been used by Madsen(8); however, it is not easy to see how Kennard's method could be used to estimate the (finite amplitude) nonlinear corrections, and this may explain why his work has not received over the years the attention that it deserves. It is the purpose of this paper to outline a new approach to the wavemaker theory which is capable of overcoming such difficulties.

II. FORMULATION OF THE PROBLEM

The starting point of our approach is the so-called pressure formulation of the water wave theory(11), whose main feature is the treatment of the pressure as an independent variable of the problem. Using dimensionless quantities, and taking the x-axis along the undisturbed free surface, with the y-axis upwardly oriented, the regulating equations for the stream function $\psi(x,p,t)$ and the height $y(x,p,t)$ are complicated nonlinear partial differential equations; however, since the free surface is one of one of constant pressure, the boundary conditions become exactly linear. Consequently, once a solution is sought by perturbation theory, the full problem is reduced to a sequence of linear problems.

The details of the derivation of the first and second order problems are, unfortunately, rather lengthy and will be published elsewhere. Using superscripts to denote the perturbation order and subscripts to denote partial derivatives, we shall simply state below the results of such a calculation. The governing equations of the linear theory are:

$$\psi^{(1)}_{xx} + \psi^{(1)}_{pp} = 0 , \tag{1}$$

$$\psi^{(1)}_{pt} - \delta^{(0)} y^{(1)}_x = 0 , \tag{2}$$

$$\psi^{(1)}_{xt} + \delta^{(0)} y^{(1)}_p = 0 , \tag{3}$$

on the domain $x,t \geq 0$, $p \in (0,h)$. Here h is the dimensionless water depth and $\delta^{(0)}$ is the leading order in the expansion of the quantity g/c^2k , where g is the acceleration of gravity, c is the phase speed and k the wave number of the wave. The boundary conditions, on the other hand, are simply

$$\psi^{(1)}_p = -F_t(-p,t) \quad \text{at} \quad x = 0 , \tag{4}$$

$$\psi^{(1)}_x = 0 \quad \text{at} \quad p = h , \tag{5}$$

$$y^{(1)}_t + \psi^{(1)}_x = 0 \quad \text{at} \quad p = 0 . \tag{6}$$

Here we have used the information that the position of the wavemaker is given by the prescribed motion $x = \epsilon F(y,t)$, with ϵ denoting the dimensionless amplitude, which is assumed small enough to be useful as the perturbation parameter. Eq. (5) is just the statement of no flow through the bottom of the tank, while Eq. (6) represents the boundary condition at the free surface. Finally, to complete the formulation of the linear problem, we have the initial conditions

$$y^{(1)}(x,0,0) = \psi^{(1)}(x,0,0) = 0 . \tag{7}$$

The second order problem embodying the nonlinear effects, has a similar mathematical structure, but is far more complicated. The salient feature is represented by the fact that, except for the free-surface boundary condition which remains homogeneous to all orders, all equations become inhomogeneous. Specifically, the regulating equations turn out to be the following:

$$\psi^{(2)}_{xx} + \psi^{(2)}_{pp} = 2(y^{(1)}_p \psi^{(1)}_{xx} - y^{(1)}_x \psi^{(1)}_{px}) , \tag{8}$$

$$\psi^{(2)}_{pt} - \delta^{(0)} y^{(2)}_x = \delta^{(1)} y^{(1)}_x - \psi^{(1)}_{pp}(\psi^{(1)}_x + y^{(1)}_t) + \psi^{(1)}_p(\psi^{(1)}_{px} - y^{(1)}_{pt}) , \tag{9}$$

$$\begin{aligned}\psi^{(2)}_{xt} + \delta^{(0)} y^{(2)}_p = -\delta^{(1)} y^{(1)}_p - \delta^{(0)}(y^{(1)}_p)^2 - \psi^{(1)}_{px}(\psi^{(1)}_x + y^{(1)}_t) + \\ + \psi^{(1)}_p(\psi^{(1)}_{xx} - y^{(1)}_{xt}) - y^{(1)}_x \psi^{(1)}_{pt} .\end{aligned} \tag{10}$$

These equations must be solved subject to following boundary conditions:

$$\psi^{(2)}_p = y^{(1)}_p F_t(-p,t) - y^{(1)} F_{ty}(-p,t) + \psi^{(1)}_x F_y(-p,t) - \psi^{(1)}_{px} F(-p,t) \quad \text{at} \quad x = 0, \tag{11}$$

$$\psi^{(2)}_x = y^{(1)}_p \psi^{(1)}_{xp} - y^{(1)}_x \psi^{(1)}_p \quad \text{at} \quad p = h , \tag{12}$$

$$y_t^{(2)} + \psi_x^{(2)} = 0 \quad \text{at} \quad p = 0 \ , \tag{13}$$

while the initial conditions are simply

$$y^{(2)}(x,0,0) = \psi^{(2)}(x,0,0) = 0 \ . \tag{14}$$

It is worth noting that although the inhomogeneous terms are all known functions at this stage, the explicit form of $\psi^{(1)}$ and $y^{(1)}$ is rather complicated; consequently, the integration of Eqs. (8)-(10) subject to (11)-(14) appears to represent an intimidatingly complex task. It is a remarkable fact, therefore, that by an appropriate change of variables the second order regulating equations may be reduced to the same mathematical form as in the first order problem. More precisely, it can be easily verified that, after setting

$$\xi = \psi^{(2)} + \frac{1}{\delta^{(0)}} \psi_t^{(1)} \psi_x^{(1)} \ , \tag{15}$$

$$w = y^{(2)} + \frac{1}{\delta^{(0)}} [\delta^{(1)} y^{(1)} + \frac{1}{2} (\psi_x^{(1)})^2 + \frac{1}{2} (\psi_p^{(1)})^2 + y_x^{(1)} \psi_t^{(1)} + + y_t^{(1)} \psi_x^{(1)} + \frac{1}{\delta^{(0)}} \psi_{tt}^{(1)} \psi_p^{(1)}] \ , \tag{16}$$

Eqs. (8)-(10) become

$$\xi_{xx} + \xi_{pp} = 0 \ , \tag{17}$$

$$\xi_{pt} - \delta^{(0)} w_x = 0 \ , \tag{18}$$

$$\xi_{xt} + \delta^{(0)} w_p = 0 \ . \tag{19}$$

Of course, the boundary and initial conditions must now be expressed in terms of ξ and w ; but this offers no difficulty, and in the end they take on the form

$$\xi_p = C(p,t) \quad \text{at} \quad x = 0 \ , \tag{20}$$

$$\xi_x = G(x,t) \quad \text{at} \quad p = h \ , \tag{21}$$

$$w_t + \xi_x = H(x,t) \quad \text{at} \quad p = 0 \ , \tag{22}$$

$$w(x,0,0) = f(x) \ , \tag{23}$$

where the functions on the right-hand side are all known in terms of the first order solution, although their explicit form is not important for our purposes in this paper. The important point here is that the first order problem is a special case of the auxiliary problem above; in fact, identifying ξ with $\psi^{(1)}$ and w with $y^{(1)}$, and setting $C(p,t) = -F_t(-p,t)$, $G = H = f = 0$, reduces Eqs. (17)-(23) to Eqs. (1)-(7). Thus we may cencentrate on finding the solution of the auxiliary problem only, which will be done in the next section.

III. SOLUTION OF THE PROBLEM

As a first step in the solution procedure, we take advantage of the linearity of the problem to break it up in four sub-problems for the variables $\xi^{(i)}$ and $w^{(i)}$, $i = 1,2,3,4$, satisfying the following set of equations:

$$\xi^{(i)}_{xx} + \xi^{(i)}_{pp} = 0 \,, \quad (i = 1,2,3,4) \tag{24}$$

$$\xi^{(i)}_{pt} - \delta^{(0)} w^{(i)}_{x} = 0 \,, \tag{25}$$

$$\xi^{(i)}_{xt} + \delta^{(0)} w^{(i)}_{p} = 0 \,, \tag{26}$$

$$\xi^{(i)}_{p} = C(p,t)\delta_{i1} \quad \text{at} \quad x = 0 \,, \tag{27}$$

$$\xi^{(i)}_{x} = G(x,t)\delta_{i2} \quad \text{at} \quad p = h \,, \tag{28}$$

$$w^{(i)}_{t} + \xi^{(i)}_{x} = H(x,t)\delta_{i3} \quad \text{at} \quad p = 0 \,, \tag{29}$$

$$w^{(i)}(x,0,0) = f(x)\delta_{i4} \,, \tag{30}$$

where δ_{ij} is the Kronecker delta. Each sub-problem can then be solved by the same mathematical technique, and the solution of the original problem follows by superposition. We shall limit ourselves to outlining the solution procedure for the first sub-problem.

We let $i = 1$ in Eqs. (24)-(30), and since $\xi^{(1)}$ must be harmonic change variables according to

$$\phi_x = -\xi^{(1)}_{p} ; \quad \phi_p = \xi^{(1)}_{x} \,. \tag{31}$$

Then Eqs. (25)-(26) may be integrated once, and the problem becomes

$$\phi_{xx} + \phi_{pp} = 0 \,, \tag{32}$$

$$\phi_x = -C(p,t) \quad \text{at} \quad x = 0 \,, \tag{33}$$

$$\phi_p = 0 \quad \text{at} \quad p = h \,, \tag{34}$$

$$\phi_p = \frac{1}{\delta^{(0)}} \phi_{tt} \quad \text{at} \quad p = 0 \,, \tag{35}$$

$$\phi_t(x,0,0) = 0 \,, \tag{36}$$

with the wave profile determined by

$$\eta^{(1)}(x,t) = w^{(1)}\Big|_{p=0} = -\frac{1}{\delta^{(0)}} \phi_t\Big|_{p=0} \tag{37}$$

From a mathematical point of view it is interesting to observe that this problem is not a standard one in the theory of elliptic equations on the domain $x \geq 0$ and $p \in [0,h]$; it fails to be the Neumann problem because the normal derivative of ϕ at $p = 0$ is not given, but rather is related to ϕ_{tt} via Eq. (35). However, aside from the different role played by the variable p, it is the very same problem solved by Kennard[10], and we may thus use the same procedure to solve our problems. The reader is referred to Kennard's paper from the details, and we conclude that the solution for ϕ is given by his Eq. (28) with appropriate nondimensionalization and re-interpretation of the symbols; from this it is easy to obtain $\xi^{(1)}$ and $w^{(1)}$, and hence the contribution of the first sub-problem to the wave shape via our Eq. (37). Finally, it is not difficult, albeit lengthy, to show that the other three sub-problems can be solved by perfectly analogous procedures.

IV. SUMMARY AND CONCLUDING REMARKS

Most topics in Fluid Mechanics have traditionally played the dual role of being of great engineering interest and, at the same time, of being a rich source of interesting mathematical problems. The theory of the wavemaker is no exception, and in this paper we have reported on some recent progress which has a bearing on both aspects. The main feature of our theory is that both the linear (infinitesimal amplitude) waves and the nonlinear (finite amplitude) corrections may be found with the same basic technique, which is essentially the procedure used by Kennard[10] to solve the linear case. Recognizing the non-standard nature of the problem, the procedure consists of constructing the solution in two stages. First one disregards the effect of gravity, and by an appropriate extension of the domain this part of the problem is shown to reduce to the Neumann problem for the Laplace equation, for which the Green function is straightforwardly found. Then one takes into account gravity by using the same mathematics which explains the evolution of an initial heap of water. The resulting solution can be easily checked to satisfy the Laplace equation and all the prescribed boundary and initial conditions; nevertheless, the procedure is not completely rigorous, nor is there an existence and uniqueness theorem which would be highly desirable.

REFERENCES

1. H. Lamb, "Hydrodynamics", 6th Ed., Cambridge Univ. Press (1932).
2. J. J. Stoker, "Water Waves", Interscience, N.Y. (1957).
3. J. V. Wehausen and E. V. Laitone, in Handbuck der Physik, Bd. IX. (1960).
4. T. H. Havelock, Phil. Mag., Series 7,8, 569-76 (1929).
5. F. Ursell, G. R. Dean, and Y. S. Yu, J. Fluid Mech. 7, 33-52 (1960).
6. F. Biésel and F. Suquet, Houille Blanche 6, 475-96, 723-37 (1951).
7. P. Fontanet, Houille Blanche 16, 3-31, 174-197 (1961).
8. O. S. Madsen, J. Geo. Res. 76, 8672-83 (1971).
9. R. E. Flick and R. T. Guza, J. Waterway Port, Coastal, and Ocean Engin. 106, 80-96 (1980).
10. E. H. Kennard, Q. Appl. Math. 7, 303-312 (1949).
11. W. H. Hui and G. Tenti, in "The Ocean Surface", Eds. Y. Toba and H. Mitsuyasu, pp. 17-24, Reidel (1985).

INTEGRO-DIFFERENTIAL EQUATIONS ASSOCIATED WITH PIECEWISE DETERMINISTIC PROCESSES

Suzanne M. Lenhart
Department of Mathematics
University of Tennessee
Knoxville, TN 37996

Yu-Chung Liao*
Prime Computer, Inc.
500 Old Connecticut Path
Framingham, MA 01701

1. Introduction

We consider integro-differential equations associated with piecewise deterministic (PD) processes. A PD process follows deterministic dynamics between random jumps. We consider equations of the form:

$$- \sum_{i=1}^{n} g_i(x)u_{x_i}(x) + \alpha(x)u(x) - \lambda(x) \int_{\mathbb{R}^n} (u(y) - u(x))Q(dy, x) = f(x) \qquad (1.1)$$

where for each fixed x, Q(dy, x) is a probability measure. Our main results are $W^{1,\infty}$ existence results and probabilistic representations for the solution of (1.1). See [1] for the definition of viscosity solution for first order partial differential equations used in section 3.

Davis [2] has recently developed the probabilistic side of PD process. See [2] also for references on applications in queueing, storage, and renewal processes. See [3, 5, 7, 8] for optimal control of PD processes.

2. Brief Stochastic Background

To construct a sample path of the PD process, let T_n be the nth jump time of the PD process x_t. For $t < T_1$, x_t follows deterministic dynamics,

$$\dot{x}_t = [g_1(x_t), \ldots, g_n(x_t)] , \; x_0 = x .$$

Let $\lambda(x)$ be the jump rate and

$$Q: \mathbb{R}^n \to P(\mathbb{R}^n) , \text{ probability measures on } \mathbb{R}^n .$$

---

*This work was begun while Liao was at the University of Kentucky.

The probability distribution of T_1 is

$$P_x(T_1 > t) = \exp(- \int_0^t \lambda(x_s)ds) .$$

The conditional distribution of x_{T_1}, given T_1, is specified by the transition measure Q:

$$P_x[x_{T_1} \in A | T_1] = Q(A; x_{T_1^-}) .$$

The process restarts at x_{T_1} and we assume $T_m \to \infty$ as $m \to \infty$.

Davis [2] shows that x_t is a strong Markov process and characterizes the extended generator as

$$Au(x) = \sum_{i=1}^n g_i u_{x_i} + \lambda \int_{\mathbb{R}^n} (u(y) - u(x))Q(dy, x) .$$

3. Some PDE Estimates

We need some preliminary estimates for the iterative scheme in the next section. Consider the following equation:

$$- \sum_{i=1}^n g_i u_{x_i} + cu + h = 0 \quad \text{in } \mathbb{R}^n \tag{3.1}$$

with the assumptions,

$$c(x) \geq \alpha_0 > 0 \text{ and } c, h, g_i \in W^{1,\infty}(\mathbb{R}^n) , i = 1, \ldots, n .$$

From standard results [4, 6], the unique viscosity solution of (3.1) satisfies

$$\|u\|_{L^\infty(\mathbb{R}^n)} \leq \frac{1}{\alpha_0} \|h\|_{L^\infty(\mathbb{R}^n)} \tag{3.2}$$

and for α_0 sufficiently large,

$$\|\nabla u\|_{L^\infty(\mathbb{R}^n)} \leq C(\|h\|_{W^{1,\infty}(\mathbb{R}^n)} + \|u\|_{L^\infty(\mathbb{R}^n)}) . \tag{3.3}$$

(C depends on the coefficients)

4. Problems with State Space $\mathbb{R}^n$

In this section we consider equation (1.1) for the PD process with state space $\mathbb{R}^n$, with the following assumptions.

$$\alpha(x) \geq \alpha_o > 0 \text{ for all } x \in \mathbb{R}^n \tag{4.2}$$

where α_o is sufficiently large from section 3,

$$\lambda(x) \geq 0 \text{ for all } x \in \mathbb{R}^n \tag{4.3}$$

$$\alpha, f, \lambda, g_i \in W^{1,\infty}(\mathbb{R}^n) , i = 1, \ldots, n \tag{4.4}$$

and

$$|\int_{\mathbb{R}^n} \phi(y)Q(dy, a) - \int_{\mathbb{R}^n}\phi(y)Q(dy, b)| \leq C_Q|b-a| \tag{4.5}$$

with C_Q constant, for all ϕ with $\|\phi\|_{L^\infty(\mathbb{R}^n)} \leq 1$ and a, b $\in \mathbb{R}^n$.

We need an extended definition of viscosity solution for integro-differential equations like (1.1).

<u>Definition</u>: Let $u \in B \cup C(\mathbb{R}^n)$ (bounded, uniformly continuous)

u is called a <u>viscosity</u> <u>supersolution</u> (<u>subsolution</u>) of (1.1) on $\mathbb{R}^n$, if

$$-g_i(x_o)\phi_{x_i}(x_o) + \alpha(x_o)u(x_o) - \lambda(x_o)\int(u(y) - u(x_o))Q(dy, x_o)$$

$$-f(x_o) \underset{(\leq)}{\geq} o ,$$

whenever $\phi \in C\,'(E)$ and $u - \phi$ has a global minimum (maximum) at x_o, relative to $\mathbb{R}^n$.

Let x_t be the PD process defined in section 2 on $\mathbb{R}^n$. We now define a sequence of approximating processes. Let $x_t^o(x)$ be the solution of this ordinary differential equation

$$\dot{x}_t = [g_1(x_t), \ldots, g_n(x_t)] \tag{4.6}$$

$$x_o = x .$$

Let τ be a random variable with distribution

$$P(\tau>t) = \exp(-\int_0^t \lambda(x_s^{\upsilon})ds) .$$

We define the process x_t^1 with initial state x by

$$x_t^1(x) = x_t^0(x) \text{ for } t<\tau$$

$$x_\tau^1 \text{ has distribution } Q(dy, x_\tau^o)$$

$$x_{\tau+s}^1(x) = x_s^o(x_\tau^1)$$

(x_t^1 has one jump at time τ.) For m<l, we define x_t^m by

$$x_t^m(x) = x_t^1(x),\ t\leq\tau$$

$$x_{\tau+s}^m(x) = x_s^{m-1}(x_\tau^1(x)),\ s\geq 0 .$$

The process x_t^m satisfies

$$x_t^m = x_t \text{ for } t\leq T_m \tag{4.7}$$

where T_m is the mth jump time of x_t. After T_m, x_t^m follows the deterministic dynamics (4.6). The process x_t^m is deterministic before T_1 and then "begins afresh" with initial state x_{T_1}. Hence x_t^m is Markovian in the following sense:

$$E[h(x^m_{T_1+s}) \mid x^m_t,\ t \leq T_1] = E[h(x_s^{m-1}(x_s^{m-1}(x^1_{T_1})))] \tag{4.8}$$

for all bounded measurable functions h on $\mathbb{R}^n$.

We now define a sequence of functions u_m to approximate the solution of (1.1):

$$u_0(x) = \int_0^\infty e^{-\alpha t} f(x_t^0(x))\,dt$$

$$u_m(x) = E\int_0^\infty e^{-\alpha t} f(x_t^m(x))\,dt$$

(α is taken to be constant for simplicity).

It is well-known that u_0 satisfies

$$g_i(u_o)_{x_i}(x) - \alpha u_0 + f = 0 \text{ in } \mathbb{R}^n\ ,\ u_0 \in W^{1,\infty}(\mathbb{R}^n) \tag{4.9}$$

We now derive a relationship between u_m and u_{m-1}. Let $F_\tau = \sigma\{x_s^m : s \leq \tau\}$. Then we have

$$u_m(x) = E_x\int_0^\infty e^{-\alpha t} f(x_t^m)\,dt$$

$$= E_x[\int_0^\tau e^{-\alpha t} f(x_t^m)\,dt + E_{F_\tau}\int_\tau^\infty e^{-\alpha t} f(x_t^m)\,dt]$$

$$= E_x[\int_0^\tau e^{-\alpha t} f(x_t^m)\,dt + E_{F_\tau}\int_0^\infty e^{-\alpha(s+\tau)} f(x^m_{\tau+s})\,ds]$$

$$= E_x[\int_0^\tau e^{-\alpha t} f(x_t^1)\,dt + e^{-\alpha\tau} E_{F_\tau}\int_0^\infty e^{-\alpha s} f(x_s^{m-1}(x_\tau^1))\,ds]\ .$$

By (4.8),

$$u_m(x) = E_x[\int_0^\tau e^{-\alpha t} f(x_t^1)\,dt + e^{-\alpha\tau} u_{m-1}(x_\tau^1)]\ .$$

Since the probability density of τ is

$$\lambda(x_t^0)\exp(-\int_0^t \lambda(x_s^0)\,ds)$$

$x_t^m = x_t^0$ for $t < \tau$, and x_τ^m has distribution $Q(dy, x_\tau^0)$,

$$u_m(x) = \int_0^\infty \lambda(x_t^0)\exp(-\int_0^t \lambda(x_s^0)ds)[\int_0^t e^{-\alpha s} f(x_s^0)ds + e^{-\alpha t}\int u_{m-1}(y)Q(dy, x_t^0)]dt + \exp(-\int_0^t \lambda(x_s^0)ds)\int_0^\infty e^{-\alpha t} f(x_s^0)ds.$$

Using integration by parts,

$$u_m(x) = \int_0^\infty \exp - \int_0^t (\alpha+\lambda)(x_s^0)ds[f(x_t^0) + \lambda(x_t^0)\int u_{m-1}(y)Q(dy, x_t^0)]dt \quad (4.10)$$

Proposition 4.1: Under assumption (4.2) - (4.5), the sequence u_m, $m \geq 1$ satisfies:

u_m is the unique viscosity solution of

$$-g_i(u_m)_{x_i} + (\alpha+\lambda)u_m - f - \lambda(x)\int_{\mathbb{R}^n} u_{m-1}(y)Q(dy, x) = 0 \text{ in } \mathbb{R}^n \quad (4.11)$$

(implicit summation)

$$\|u_m\|_{L^\infty(\mathbb{R}^n)} \leq \frac{1}{\alpha_0} \|f\|_{L^\infty(\mathbb{R}^n)} \quad (4.12)$$

and

$$\|\nabla_x u_m\|_{L^\infty(\mathbb{R}^n)} \leq C \quad (4.13)$$

where C is independent of m.

Proof: The fact that the cost function u_m in (4.10) is the unique viscosity solution of (4.11) is well documented [6]. The bound (4.12) follows from the definition of u^m. The bound (4.13) follows from (3.2) and (3.3) with

$$h = -\lambda\int_{\mathbb{R}^n} u_{m-1}(y)Q(dy, x) - f \quad \blacksquare$$

Remark 4.1: Notice that x_t^m is not a Markov process, but u_m still has the analytical property (4.10).

Now we consider the convergence of u_m as $m \to \infty$.

Theorem 4.1: Under assumptions (4.2) - (4.5), the sequence u_m converges uniformly on compact subsets of $\mathbb{R}^n$ to u, where u is the viscosity solution of (4.1), satisfying $u \in W^{1,\infty}(\mathbb{R}^n)$, and

$$u(x) = E_x\int_0^\infty e^{-\alpha t} f(x_t)dt \quad . \quad (4.14)$$

Proof: The bounds on u_m from Proposition 4.1 imply the existence of u such that

$$u_m \to u \text{ uniformly on compact subsets of } \mathbb{R}^n$$

and

$\nabla u_m \to \nabla u$ weak * L^∞_{loc}.

Because, for $m \leq m^1$,

$$|u_m(x) - u_{m'}(x)| \leq 2\|f\|_{L^\infty(\mathbb{R}^n)} E\int_{T^m}^\infty e^{-\alpha t}dt \to 0 \text{ as } m \to \infty ,$$

the whole sequence converges, not just a subsequence. To show u solves (4.1), we must show

$$\int_{\mathbb{R}^n} u_m(y)Q(dy, x) \to \int_{\mathbb{R}^n} u(y)Q(dy, x) \tag{4.15}$$

uniformly on compact subsets of $\mathbb{R}^n$.

Let $\varepsilon > 0$ and K be a compact set. For any $x \in K$, there is a compact set K(x) such that

$$Q(K(x),x) > 1 - \varepsilon/2.$$

By (4.5), Q(A,x) is continuous in x for any measurable set A. Hence, there is a neighborhood, B(x), of x such that

$$Q(K(x),y) > 1 - \varepsilon \text{ for all } y \in B(x) .$$

Since K is compact, there is a finite set $\{x_i\}_{i=1}^k$ in K such that

$$K \subset \bigcup_{i=1}^k B(x_i) .$$

Let

$$\hat{K} = \bigcup_{i=1}^k K(x_i) .$$

Then we have

$$Q(\hat{K},x) > 1 - \varepsilon \text{ for all } x \in K.$$

For m large, we have

$$|u_m(y) - u(y)| < \varepsilon \text{ for all } y \in \hat{K}, \text{ compact.}$$

Hence for m large,

$$|\int_{\mathbb{R}^n} u_m(y)Q(dy, x) - \int_{\mathbb{R}^n} u(y)Q(dy, x)|$$

$$\leq \int_{\hat{K}} |u_m-u|(y)Q(dy, x) + \int_{\mathbb{R}^n\setminus\hat{K}}|u_m-u|(y)Q(dy, x)$$

$$\leq \varepsilon + \frac{2}{\alpha_0}\varepsilon\|f\|_{L^\infty(\mathbb{R}^n)}$$

for all $x \in K$, m large. This completes the proof of (4.15).

Thus u solves (1.1) by letting $m \to \infty$ in (4.11). It is clear that u is in $D(A)$. Hence by the representation formula in [2, Thm. 6.3]

$$u(x) = E_x[-\int_0^t e^{-\alpha s}(Au(x_s) + \alpha u(x_s))ds$$
$$+ e^{-\alpha t}u(x_t)] .$$

Letting $t \to \infty$, by the boundedness of u,

$$u(x) = E_x \int_0^\infty e^{-\alpha t}f(x_t)dt .$$

To show u is a viscosity solution, suppose $u - \phi$ attains its maximum at x_o. There exists x^m such that $u_m - \phi$ attains its maximum at x^m and

$x^m \to x_o$ as $m \to \infty$. Since u_m is a viscosity solution,

$$-g_i\phi_{x_i} + \alpha u_m - \lambda\int_{\mathbb{R}^n}(u_m(y) - u_m(x^m))Q(dy,x^m) - f \leq 0 \text{ at } x^m.$$

As $m \to \infty$, we have u is a viscosity subsolution. Similarly for supersolution. ■

<u>Remark</u>: We can show a similar existence and representation formula on a bounded domain. The PD process jumps immediately upon hitting the boundary, which results in the following boundary condition: $u(x) = \int_\Omega u(y)Q(dy, x)$ for $x \in \partial\Omega$.

REFERENCES

1. Crandall, M. G., L. C. Evans, and P. L. Lions, Some properties of viscosity solutions of Hamilton-Jacobi Equations, Trans. Amer. Math. Soc., 282(1984), 487-502.
2. Davis, M. H. A., Piecewise-Deterministic Markov processes: a general class of non-diffusion stochastic models, J. Royal Stat. Soc. 46 (1984), 353-388.
3. Davis, M. H. A., Control of Piecewise-deterministic processes via Discrete Time Dynamics Programming, Proceedings of 3rd Bad Hannel Symposium on <u>Stochastic Differential Systems</u>, 1985.

4. Lenhart, S. M., Semilinear Approximation Technique for Max-Min Type Hamilton-Jacobi Equations over finite Max-Min set, Journal of Nonlinear Analysis, 8(1984), 407-415.
5. Lenhart, S. M. and Y. C. Liao, Integro-Differential Equations associated with optimal stopping time of a piecewise-deterministic process, Stochastics, 15(1985), 183-297.
6. Lions, P. L., Generalized solutions of Hamilton-Jacobi Equations, Pitman, Boston, 1982.
7. Soner, M., Optimal Control with State-Space Constraint II, SIAM J. of Optimal Control, to appear.
8. Vermes, D., Optimal Control of piecewise deterministic processes, Stochastics, 14(1984), 165-208.

AMBROSETTI-PRODI TYPE RESULTS IN NONLINEAR BOUNDARY VALUE PROBLEMS

Jean Mawhin
Université de Louvain
Institut Mathématique
B-1348 Louvain-la-Neuve, Belgium

1. INTRODUCTION

An Ambrosetti-Prodi type result for an equation of the form

$$G(u,s) = 0 \tag{1}$$

depending upon a real parameter s consists in finding some s_o such that (1) has no solution, at least one solution or at least two solutions according to $s < s_o$, $s = s_o$ or $s > s_o$. The terminology will be explained later when references to the pioneering literature will be given. Such a situation occurs for example for the simple equation

$$f(u) = s \tag{2_s}$$

when $f : \mathbb{R} \to \mathbb{R}$ is continuous and such that

$$f(u) \to +\infty \text{ if } |u| \to \infty. \tag{3}$$

To prove an Ambrosetti-Prodi type result for (2_s) one can proceed as follows :

a) by (3) and Weierstrass theorem, f achieves its minimum on $\mathbb{R}$, i.e. if

$$s_1 = \inf_{\mathbb{R}} f,$$

there exists $u_o \in \mathbb{R}$ such that

$$f(u_o) = s_1.$$

Clearly, (2_s) *has no solution if* $s < s_1$ *and at least one solution* u_o *if* $s = s_1$.

b) if $s > s_1$, we can use (3) to obtain $R_- < u_o < R_+$ such that

$$f(R_-) > s,\ f(R_+) > s.$$

Then Bolzano's theorem will imply the existence of at least two distinct solutions $u_- \in]R_-,u_o[$ and $u_+ \in]u_o,R_+[$, so that (2_s) *has at least two solutions if* $s > s_1$.

Notice that u_o and u_-, u_+ will be the only solutions for $s = s_1$ or $s > s_1$ if f is decreasing on $]-\infty,u_o[$ and increasing on $]u_o,+\infty[$.

This elementary result can be phrased in terms of differential equations by considering the periodic problem

$$(4_s) \qquad \begin{aligned} &u'(x) + f(u(x)) = s \\ &u(0) = u(2\pi) \end{aligned}$$

with f satisfying condition (3). If u is a solution of (4_s), then, integrating (4_s) after multiplication by u' and using the periodicity immediately implies that $\|u'\|_{L^2} = 0$ and hence that u is a constant which satisfies (2). It is therefore natural to consider the less trivial non-autonomous situation.

2. THE PERIODIC PROBLEM FOR A SCALAR FIRST ORDER ORDINARY DIFFERENTIAL EQUATION

Let us consider the problem

$$(5_s) \qquad \begin{aligned} &F(u)(x) \equiv u'(x) + f(x,u(x)) = s \\ &u(0) = u(2\pi) \end{aligned}$$

where $f : [0,2\pi] \times \mathbb{R} \to \mathbb{R}$ is continuous, $s \in R$ and

(6) $$f(x,u) \to +\infty \text{ as } |u| \to \infty$$

uniformly on $I = [0,2\pi]$.

Recall that $\alpha \in C^1(I)$ (resp. $\beta \in C^1(I)$) is called a *lower solution* (resp. *upper solution*) for (5_s) if $\alpha(0) = \alpha(2\pi)$ (resp. $\beta(0) = \beta(2\pi)$) and α (resp. β) satisfies the inequality

(7) $$F(\alpha)(x) \geqslant s \text{ (resp. } F(\beta)(x) \leqslant s)$$

for all $x \in I$. Such a lower or upper solution will be called *strict* if the strict inequality holds everywhere in (7). This concept allows to state and prove a *Bolzano-type theorem* for (5_s), namely if (5_s) has a lower (resp. upper) solution α and an upper (resp. lower) solution β such that

$$\alpha(x) \leqslant \beta(x),\ x \in I,$$

then (5_s) has a solution u such that

$$\alpha(x) \leqslant u(x) \leqslant \beta(x),\ x \in I.$$

Moreover, if α and β are strict, then $\alpha(x) < u(x) < \beta(x)$, $x \in I$. This theorem was proved in [17] for f Lipschitzian and in [18] for f continuous.

We can now prove the following

Theorem 1. *If f satisfies (6), then there exists $s_1 \in \mathbb{R}$ such that (5_s) has zero, at least one or at least two solutions according to $s < s_1$, $s = s_1$ or $s > s_1$.*

Proof. Let

$$S_j = \{s \in \mathbb{R} : (5_s) \text{ has at least } j \text{ solutions}\}\ (j \geqslant 1).$$

a) $\underline{S_1 \neq \phi}$. Take $s^* > \max_{x\in I} f(x,0)$ and use then (6) to obtain $\underline{R}^* < 0$ such that $f(x,\underline{R}^*) > s^*$ for all $x \in I$, so that $\underline{R}^*$ is a strict lower solution and 0 a strict upper solution to (5_{s^*}) and hence $s^* \in S_1$.

b) *if* $\tilde{s} \in S_1$ *and* $s > \tilde{s}$, *then* $s \in S_2$.
Let $\tilde{u}$ be a solution of $(5_{\tilde{s}})$ and let $s > \tilde{s}$; then $\tilde{u}$ is a strict upper solution for (5_s). Take now $R_- < \min_I \tilde{u}$ and $R_+ > \max_I \tilde{u}$ such that

$$f(x,R_-) > s \text{ and } f(x,R_+) > s$$

for all $x \in I$, so that R_- and R_+ are strict lower solutions for (5_s). Then (5_s) has two solutions u_-, u_+ such that

$$R_- < u_-(x) < \tilde{u}(x) < u_+(x) < R_+,\ x \in I.$$

c) $s_1 = \inf S_1$ *is finite and* $S_2 \supset]s_1,+\infty[$.
By (6), we have $f(x,u) \geq C$ for some $C \in \mathbb{R}$ and all $(x,u) \in I \times \mathbb{R}$, and hence, we see by integrating (5_s) over I that it has no solution if $s < C$. Thus $s_1 > -\infty$ and b) provides the second conclusion.

d) *for each* $s_2 > s_1$, *the set of all possible solutions of* (5_s) *with* $s \leq s_2$ *is a priori bounded.* Let $R_2 > 0$ be such that $f(x,u) > s_2$ whenever $|u| \geq R_2$ and let $y \in I$ be such that $u(y) = \max_I u$ for a possible solution u of (5_s) with some $s \leq s_2$; then, if $y \in]0,2\pi[$,

$$0 = u'(y) = s - f(y,u(y)) \leq s_2 - f(y,u(y))$$

so that $\max_I u = u(y) < R_2$. If $y = 0$ or 2π, then $\max_I u = u(0) = u(2\pi)$ and hence $u'(0) \leq 0$ and $u'(2\pi) \geq 0$, so that

$$0 \leq u'(2\pi) = s - f(2\pi,u(2\pi)) \leq s_2 - f(2\pi,u(2\pi))$$

and $\max_I u = u(2\pi) < R_2$. Similarly, $\min_I u > -R_2$.

e) $s_1 \in S_1$. Taking a decreasing sequence (σ_k) in $]s_1,\infty[$ converging to s_1, a corresponding sequence (u_k) of solutions of (5_{σ_k}) and using the a priori bound of (d) and Ascoli-Arzela theorem, we obtain a subsequence (u_{j_k}) which converges uniformly on I to a solution u of (5_{s_1}). The proof is complete.

<u>Remark 1</u>. When

$$f(x,u) \to -\infty \text{ as } |u| \to \infty$$

uniformly for $x \in I$, there exists $s_1 \in \mathbb{R}$ such that (5_s) has zero, at least one or at least two solutions according to $s > s_1$, $s = s_1$ or $s < s_1$.

<u>Remark 2</u>. The lower bound two when $s \geqslant s_1$ is sharp as shown by the example $f(x,u) = u^2$ and the end of Section 1. The number of solutions can be infinite as shown by the example

$$f(x,u) = \begin{cases} -1-u & \text{if } u < -1 \\ 0 & \text{if } |u| \leqslant 1 \\ u-1 & \text{if } u > 1 \end{cases}$$

for which problem (5_0) admits the solutions $u(x) = c$, $|c| \leqslant 1$. However, if $f(x,.)$ is analytic for each $x \in I$, it follows from a result of [5] that, for each $s \in \mathbb{R}$, (5_s) has at most finitely many solutions.

<u>Remark 3</u>. When $f(x,.)$ is strictly convex on $\mathbb{R}$ for each $x \in I$, then (6) implies that (5_s) has exactly zero, one or two solutions according to $s < s_1$, $s = s_1$ or $s > s_1$. See [21] for the corresponding proofs.

Problem (5_s) or special cases have been studied by Vidossich [24] through Ambrosetti-Prodi's original differential topological approach, by Scovel [23] through singularity theory and by Nkashama [22] through degree theory.

3. THE PERIODIC AND NEUMANN PROBLEMS FOR SECOND ORDER DIFFERENTIAL EQUATIONS

Let us now consider the second order periodic problem

$$(8_s)\qquad \begin{aligned} F(u)(x) &\equiv u''(x) + f(x,u(x)) = s \\ u(0) - u(2\pi) &= u'(0) - u'(2\pi) = 0 \end{aligned}$$

where $f : [0,2\pi] \times \mathbb{R} \to \mathbb{R}$ is continuous, $s \in \mathbb{R}$ and

$$(9)\qquad f(x,u) \to +\infty \text{ as } |u| \to \infty$$

uniformly on $I = [0,2\pi]$.

Recall that $\alpha \in C^2(I)$ (resp. $\beta \in C^2(I)$) is called a *lower solution* (resp. *upper solution*) for (8_s) if

$$\alpha(0) - \alpha(2\pi) = \alpha'(0) - \alpha'(2\pi) = 0$$
$$(\text{resp. } \beta(0) - \beta(2\pi) = \beta'(0) - \beta'(2\pi) = 0)$$

and satisfies the inequality

$$(10)\qquad F(\alpha)(x) \geq s \quad (\text{resp. } F(\beta)(x) \leq s)$$

for all $x \in I$. Such a lower or upper solution will be called *strict* if the strict inequality holds everywhere in (10). We also have a *Bolzano-type theorem* for (8_s), namely if (8_s) has a lower solution α and an upper solution β such that

$$\alpha(x) \leq \beta(x),\ x \in I,$$

then (8_s) has a solution u satisfying

$$\alpha(x) \leq u(x) \leq \beta(x),\ x \in I.$$

Moreover, if α and β are strict, then

$$\alpha(x) < u(x) < \beta(x), \ x \in I$$

and

$$|D_L(F(.)-s, \Omega_{\alpha,\beta})| = 1$$

where D_L denotes the degree relative to L (see e.g. [20]) of $F(.)-s$ in $\Omega_{\alpha,\beta}$, where $D(L) = \{u \in C^2(I) : u(0) - u(2\pi) = u'(0) - u'(2\pi) = 0\}$, $L : D(L) \subset C(I) \to C(I)$, $u \mapsto u''$, $\Omega_{\alpha\beta} = \{u \in C(I) : \alpha(x) < u(x) < \beta(x), x \in I\}$. We refer to [20] for those results and we notice that, in contrast to the first order case, no Bolzano's like theorem holds for the case where $\beta(x) \leqslant \alpha(x)$, $x \in I$.

A result similar to Theorem 1 holds for (8_s).

<u>Theorem 2</u>. *If* f *satisfies* (9), *then there exists* $s_1 \in \mathbb{R}$ *such that* (8_s) *has zero, at least one or at least two solutions according to* $s < s_1$, $s = s_1$ *or* $s > s_1$.

<u>Proof</u>. Let

$$S_j = \{s \in \mathbb{R} : (8_s) \text{ has at least j solutions}\} \quad (j \geqslant 1).$$

a) $S_1 \neq \phi$. Proceed exactly like in part (a) of the proof of Theorem 1.

b) *if* $\tilde{s} \in S_1$ *and* $s > \tilde{s}$, *then* $s \in S_1$.
Proceed like in part (b) of the proof of Theorem 1 except that one solution only is obtained through the pair R_-, $\tilde{u}$ of lower and upper solutions.

c) $s_1 = \inf S_1$ *is finite and* $S_1 \supset]s_1,+\infty[$.
Proceed like in part (c) of the proof of Theorem 1.

d) *for each* $s_2 > s_1$, *the set of all possible solutions of* (8_s) *with* $s \leqslant s_2$ *is a priori bounded.*

Let $u = \bar{u} + \tilde{u}$, with

$$\bar{u} = (1/2\pi) \int_0^{2\pi} u(x)dx, \int_0^{2\pi} \tilde{u}(x)dx = 0$$

be a possible solution of (8_s) for some $s \leq s_1$. Then

$$(11) \qquad (1/2\pi) \int_0^{2\pi} f(x,u(x))dx = s \leq s_2$$

and, as $f(x,u) \geq C$ for some $C \in \mathbb{R}$ and all $(x,u) \in I \times \mathbb{R}$ we have

$$\int_I u'^2(x)dx = \int_I f(x,u(x))\tilde{u}(x)dx = \int_I (f(x,u(x))-C)\tilde{u}(x)dx \leq$$

$$\leq \|\tilde{u}\|_{C^0} \int_I (f(x,u(x))-C)dx = 2\pi\|\tilde{u}\|_{C^0}(s-C) \leq 2\pi\|\tilde{u}\|_{C^0}(s_2-C).$$

Hence by a well known inequality

$$\|u'\|_{L^2} \leq 3^{-1/2}\pi(s_2-C) = R_1.$$

Moreover, if $R_2 > 0$ is such that

$$f(x,u) > s_2$$

whenever $|u| > R_2$, it follows from (11) that there exists $y \in I$ such that

$$|u(y)| \leq R_2.$$

Hence, for $x \in I$,

$$|u(x)| = |u(y) + \int_y^x u'(t)dt| \leq R_2 + \sqrt{2\pi}\, \|u'\|_{L^2} \leq$$

$$\leq R_2 + \sqrt{2\pi}\, R_1 = R(s_2),\ x \in I.$$

e) *for each* $s \leq s_2$ *and each* $R > R(s_2)$, $D_L(F(.)-s, B(R)) = 0$, *where* $B(R) = \{u \in C^o(I) : \|u\|_{C^o} < R\}$.

By (d) and the invariance of degree, $D_L(F(.)-s, B(R))$ is independent of s and is therefore equal to zero, as $F(.)-s$ has no zero whenever $s < s_1$.

f) $S_2 \supset]s_1,+\infty[$.
Let $s \in]s_1,+\infty[$, $s_2 \in]s_1,s[$ and u_2 a solution of (8_{s_2}). Then u_2 is a strict upper solution for (8_s) and, by (9) there will exist $R_2 < \min_I u_2$ such that $f(x,R_2) > s$ for all $x \in I$, so that R_2 is a strict lower solution for (8_s). Consequently, (8_s) has a solution in $\Omega_{R_2u_2}$ and

$$|D_L(F(.)-s, \Omega_{R_2u_2})| = 1.$$

Taking $R > \max(|R_2|, \|u_2\|_{C^o}, R(s))$, we deduce from (e) and from the additivity of degree that

$$|D_L(F(.)-s, B(R) \setminus \bar{\Omega}_{R_2u_2})| = |D_L(F(.)-s, B(R)) -$$

$$- D_L(F(.)-s, \Omega_{R_2u_2})| = |D_L(F(.)-s, \Omega_{R_2u_2})| = 1$$

so that (8_s) has a second solution in $B(R) \setminus \bar{\Omega}_{R_2u_2}$.

g) $s_1 \in S_1$. Proceed exactly like in part (e) of the proof of Theorem 1. The proof is complete.

Remark 4. When

$$f(x,u) \to -\infty \text{ as } |u| \to \infty$$

uniformly on I, there exists $s_1 \in \mathbb{R}$ such that (8_s) has zero, at least one or at least two solutions according to $s > s_1$, $s = s_1$ or $s < s_1$. We refer to [13] for more general results concerning the periodic problem for a larger class of second order equations.

Remark 5. The example of nonlinearity given in Remark 2 shows that (8_s) may admit infinitely many solutions for some values of s. It is not known if the analyticity of $f(x,.)$ for each $x \in I$ implies that (8_s) has at most finitely many solutions (see [5]). Another open problem is to find sharp conditions on f (in the non-differentiable case) which imply that there are exactly zero, one or two solutions for (8_s) according to $s < s_1$, $s = s_1$ or $s > s_1$.

Remark 6. The same result as in Theorem 2, with an entirely similar proof, holds for the *Neumann problem*

$$F(u)(x) \equiv u''(x) + f(x,u(x)) = s$$
$$u'(0) = u'(\pi) = 0$$

where $f : [0,\pi] \times \mathbb{R} \to \mathbb{R}$ is continuous, $s \in \mathbb{R}$ and

$$f(x,u) \to +\infty \text{ as } |u| \to \infty$$

uniformly on $I = [0,\pi]$. In this case, when $f(x,.)$ is analytic for each $x \in I$, the corresponding Neumann problem has at most finitely many solutions (see [5]).

It is then natural to consider the extension of Theorem 2 to the Neumann problem for an elliptic partial differential equation, say

$$(12_s)\qquad \begin{aligned} F(u)(x) &\equiv \Delta u(x) + f(x,u(x)) = s \quad \text{in } \Omega \\ \frac{\partial u}{\partial \nu}(x) &= 0 \quad \text{in } \partial\Omega \end{aligned}$$

where $\Omega \subset \mathbb{R}^N$ is a bounded smooth domain, $f : \bar{\Omega} \times \mathbb{R} \to \mathbb{R}$ is Hölder continuous, Δ is the Laplacian and $\partial/\partial\nu$ the normal derivative. We shall assume again that

$$(13)\qquad f(x,u) \to +\infty \text{ as } |u| \to \infty$$

uniformly in $\bar{\Omega}$ and moreover that

(14) $$\frac{f(x,u)}{u^{\sigma}} \to 0 \text{ as } u \to +\infty$$

uniformly in $\bar{\Omega}$, where $\sigma = \frac{N}{N-2}$ if $N \geq 3$ and σ is finite in $N = 2$. Recall that $\alpha \in C^2(\Omega) \cap C^1(\bar{\Omega})$. (resp. β) will be called a lower (resp. upper) solution for (12_s) if $\frac{\partial\alpha}{\partial\nu} = 0$ (resp. $\frac{\partial\beta}{\partial\nu} = 0$) on $\partial\Omega$ and

(15) $$F(\alpha)(x) \geq s \text{ (resp. } F(\alpha)(x) \leq s)$$

on Ω. Such a lower or upper solution will be called strict if the strict inequality sign holds in (15). A Bolzano-type theorem similar to the one of the periodic case holds in this situation (see e.g. [20] for statements, proofs and references).

<u>Theorem 3</u>. *If* f *satisfies* (13) *and* (14), *then there exists* $s_1 \in \mathbb{R}$ *such that* (12_s) *has zero, at least one or at least two solutions according to* $s < s_1$, $s = s_1$ *or* $s > s_1$.

<u>Proof</u>. We shall only outline the differences with respect to the proof of Theorem 2. Let

$$S_j = \{s \in \mathbb{R} : (12_s) \text{ has at least } j \text{ solutions}\} \quad (j \geq 1).$$

One proceeds exactly like in Theorem 2 to prove that $S_1 \neq \phi$, that $\tilde{s} \in S_1$ and $s > s_1$ imply $s \in S_1$, that $s_1 = \inf S_1$ is finite and hence that $S_1 \supset]s_1,+\infty[$. The a priori bound for the possible solutions is a more delicate question.

i) *For each* $s_2 \in \mathbb{R}$ *there exists* $D = D(s_2)$ *such that each possible solution* u *of* (12_s) *with* $s \in [s_1,s_2]$ *satisfies the inequality*

(16) $$u(x) \geq D, \; x \in \bar{\Omega}.$$

Indeed, let D be such that

$$f(x,v) > s_2$$

whenever $x \in \bar{\Omega}$ and $v < D$ (assumption (13)). If there exists $s \in [s_1,s_2]$, u solution of (12_s) and $x_o \in \bar{\Omega}$ such that $u(x_o) < D$, then $D-u$ achieves a positive maximum on $\bar{\Omega}$ at some $x_1 \in \bar{\Omega}$ for which one has therefore $f(x_1,u(x_1)) > s_2$. If $x_1 \in \Omega$, then

$$0 \geq -\Delta u(x_1) = f(x_1,u(x_1)) - s > s_2 - s \geq 0,$$

a contradiction. If $x_1 \in \partial\Omega$, then $D - u(x_1) > D - u(x)$, $x \in \Omega$; moreover there exists $r > 0$ such that $f(x,u(x)) > s_2$ if $x \in \bar{\Omega} \cap B(x_1,r)$ and hence

$$\Delta(D-u(x)) = -\Delta u(x) = f(x,u(x)) - s > s_2 - s \geq 0$$

for $x \in \Omega \cap B(x_1,r)$. The boundary point lemma (see [14]) implies then that $\frac{\partial}{\partial\nu}(D-u)(x_o) = -\frac{\partial u}{\partial\nu}(x_o) > 0$, a contradiction

ii) *For each* $s_2 \in \mathbb{R}$ *there exists* $R = R(s_2)$ *such that each possible solution* u *of* (12_s) *with* $s \in [s_1,s_2]$ *satisfies the inequality*

$$\|u\|_{C^0} \leq R.$$

We shall use an argument introduced by Ward [25] in another context.

By (i), we already know that $\|u^-\|_{L^\infty} \leq |D|$. Now, if C is such that $f(x,u) \geq C$ on $\bar{\Omega} \times \mathbb{R}$, and if u is a possible solution of (12_s) with $s \in [s_1,s_2]$, then

$$\int_\Omega f(x,u(x))dx = s|\Omega|$$

and

$$\|\Delta u\|_{L^1} \leq \|f(.,u(.))\|_{L^1} + |\Omega||s_1| \leq \int_\Omega (f(x,u(x)) - C)dx +$$

$$+ |\Omega|(|s_1| + |C|) = |\Omega|(s - C + |s_1| + |C|) \leq \tag{17}$$

$$\leq 2|\Omega|(s_1^+ + C^-) \leq 2|\Omega|(|s_1| + |C|) = D_1.$$

If we write $u = \overline{u} + \tilde{u}$ with

$$\overline{u} = |\Omega|^{-1} \int_\Omega u(x)dx, \int_\Omega \tilde{u}(x)dx = 0,$$

it follows from (17) and known estimates [14] that, for each $1 \leq q < N/N-1$ there exists $C_1(q)$ such that

$$\|\tilde{u}\|_{W^{1,q}} \leq C_1(q)D_1 \tag{18}$$

and $W^{1,q}(\Omega)$ is compactly imbedded in $L^p(\Omega)$ for

$$1 \leq p < \frac{Nq}{N-q}.$$

Now if the possible solutions u of (12_s) with $s \leq s_2$ are not a priori bounded in $W^{1,q}(\Omega)$ it follows from (i), (18) and the compact imbedding that we can find a sequence (σ_k) in $[s_1,s_2]$ converging to $s_3 \in [s_1,s_2]$, a sequence (u_k) of solutions of (12_{σ_k}) and a function $\tilde{u} \in W^{1,q}(\Omega)$ such that

$$\tilde{u}_k \xrightarrow{L^p} \tilde{u}$$

$$\tilde{u}_k(x) \to \tilde{u}(x) \text{ a.e. on } \Omega$$

$$\overline{u}_k \to +\infty$$

when $k \to \infty$. Therefore, $u_k(x) \to +\infty$ a.e. on Ω and from

$$\frac{1}{|\Omega|} \int_\Omega f(x,u_k(x))dx = \sigma_k$$

(13) and Fatou's lemma, we deduce

$$s_2 \geq s_3 = \lim_{k\to\infty} \sigma_k = \lim_{k\to\infty} \frac{1}{|\Omega|} \int_\Omega f(x,u_k(x))dx \geq$$

$$\geq \frac{1}{|\Omega|} \int_\Omega \lim_{k\to\infty} f(x,u_k(x))dx = \infty,$$

a contradiction. Thus $\|u\|_{W^{1,q}} \leq D_2$ and hence

$$(19) \qquad \|u\|_{L^p} \leq D_3(p)$$

for all $1 \leq p \leq \frac{Nq}{N-q}$. As the function $h : r \mapsto \frac{Nr}{N-r}$ is strictly increasing on $[1, \frac{N}{N-1}]$ and $h(1) = \frac{N}{N-1}$, $h(\frac{N}{N-1}) = \frac{N}{N-2}$, we see that we can take any $1 \leq p < \frac{N}{N-2}$ in (19). Now conditions (14), (16), (19) and a usual bootstrap argument imply that

$$\|u\|_{L^\infty} \leq R$$

for some $R > 0$.

Now the degree argument leading to the existence of a second solution in $]s_1,\infty[$ and the limit process showing that $s_1 \in S_1$ are very similar to the ordinary case (except that one deals with generalized continuous solutions in the degree argument and those solutions are shown afterwise to be classical ones by a usual regularization technique). The proof is therefore complete.

Remark 7. Theorem 3 generalizes in various directions earlier results of Kazdan-Warner, Amann-Hess and Berestycki-Lions. Kazdan and Warner [16] were the first to inject the technique of upper and lower solutions in Ambrosetti-Prodi type problems when f is smooth and

$$(20) \qquad \limsup_{u\to-\infty} \frac{f(x,u)}{u} < 0 < \liminf_{u\to+\infty} \frac{f(x,u)}{u}$$

uniformly in $\bar{\Omega}$ and Amann-Hess [1] and Berestycki-Lions [3] have proved the multiplicity result when f is smooth, satisfies (20) and either $\limsup_{u \to +\infty} u^{-1} f(x,u) < +\infty$ in [1] or the more general condition (14) in [3]. Thus, Theorem 3 weakens the smoothness assumptions and replaces (20) by the more general condition (13).

Remark 8. The Laplacian in (12_s) could be replaced by a more general class of elliptic operators. Also, in problems (5_s), (8_s) and (12_s) the right hand member could be replaced by $s\psi(x)$ where ψ is positive on I or $\bar{\Omega}$. Finally a result analogous to that of Remark 4 holds for (12_s) when we replace $+\infty$ by $-\infty$ in (13).

4. THE DIRICHLET PROBLEM FOR SECOND ORDER DIFFERENTIAL EQUATIONS

The results of the previous section suggest that Ambrosetti-Prodi type results should be true also for the Dirichlet problems

$$(21_s) \quad \begin{aligned} F(u)(x) &\equiv u''(x) + u(x) + f(x,u(x)) = s\sqrt{2/\pi}\sin x \\ u(0) &= u(\pi) = 0 \end{aligned}$$

or

$$(22_s) \quad \begin{aligned} F(u)(x) &\equiv \Delta u(x) + \lambda_1 u(x) + f(x,u(x)) = s\varphi(x) \text{ on } \Omega \\ u(x) &= 0 \text{ on } \partial\Omega \end{aligned}$$

where λ_1 denotes the first eigenvalue of $-\Delta$ on Ω with Dirichlet boundary conditions, Ω is a smooth bounded domain in $\mathbb{R}^N$ and f is continuous in (21_s), Hölder continuous in (22_s).

Indeed, the original Ambrosetti-Prodi result [2], in his more precise formulation due to Berger-Podolak [4], dealed with the following special case of (22_s)

$$(23_s)\quad \begin{aligned} &\Delta u(x) + \lambda_1 u(x) + g(u(x)) - h(x) = s\varphi(x) \text{ on } \Omega \\ &u(x) = 0 \text{ on } \partial\Omega \end{aligned}$$

where g is of class C^2, strictly convex and such that

$$(24)\qquad -\lambda_1 < \lim_{u\to-\infty} g'(u) < 0 < \lim_{u\to+\infty} g'(u) < \lambda_2 - \lambda_1,$$

where λ_2 is the second eigenvalue of $-\Delta$ with Dirichlet boundary conditions on Ω. They proved the existence of s_1 such that (23_s) has exactly zero, one or two solutions according to $s < s_1$, $s = s_1$ or $s > s_1$. The more general situation (22_s) with (24) replaced by (20) was considered by Kazdan-Warner [16] who obtained zero or at least one solution according to $s < s_1$ or $s > s_1$ when f is sufficiently smooth. The multiplicity result (at least one solution for $s = s_1$ and at least two solutions for $s > s_1$) was then proved by Amann-Hess [1] when f is smooth, verifies (20) and

$$(25)\qquad \limsup_{u\to+\infty} \frac{f(x,u)}{u} < \infty$$

and independently by Dancer [8] when (25) is replaced by the more general condition

$$(26)\qquad \limsup_{u\to\infty} \frac{f(x,u)}{u^\sigma} = 0$$

for some $\sigma < \frac{N+1}{N-1}$ if $N \geq 3$ and $\sigma < 3$ if $N = 2$.

Finally, under the same smoothness and growth assumption (26), Kannan-Ortega [15] have replaced (20) by the conditions

$$(27)\qquad \lim_{u\to-\infty} (\lambda_1 u + f(x,u)) = +\infty,\quad \lim_{u\to+\infty} f(x,u) = +\infty$$

uniformly on $\overline{\Omega}$, which is less general than (20) at $-\infty$ and more general at $+\infty$, and Chang [6], de Figueiredo [10] and de Figueiredo-Solimini [11] have used variational methods to weaken the growth condition (26) when f is smooth and satisfies (20). See [9] for a survey of the various results and methods up to 1980.

It still seems to be an open problem to know if an Ambrosetti-Prodi type result holds for (21_s) with f continuous or (22_s) with f Hölder continuous under the mere assumption that

$$f(x,u) \to +\infty \text{ as } |u| \to \infty \tag{28}$$

uniformly in I or $\overline{\Omega}$. The case N = 1 has been recently discussed by Chiappinelli-Mawhin-Nugari [7] and provides a partial positive answer that we shall briefly describe now. Their treatment requires the introduction of the associated non-homogeneous Dirichlet problem

$$(29_s^\gamma) \quad \begin{aligned} F(u)(x) &\equiv u''(x) + u(x) + f(x,u(x)) = s\sqrt{2/\pi}\,\sin x \\ u(0) &= u(\pi) = \gamma \end{aligned}$$

for $\gamma \in [0,1]$. Recall that a *lower* (resp. *upper*) *solution* for (29_s^γ) is a C^2-function α (resp. β) on I such that

$$F(\alpha)(x) \geq 0,\ \alpha(0) \leq \gamma,\ \alpha(\pi) \leq \gamma$$
$$(\text{resp. } F(\beta)(x) \leq 0,\ \beta(0) \geq \gamma,\ \beta(\pi) \geq \gamma)$$

for all $x \in I$. It will be called *strict* if the strict inequalities hold everywhere. Again, a Bolzano-type theorem similar to the one stated in the beginning of Section 3 holds. Let

$$S_j^\gamma = \{s \in \mathbb{R} : (29_s^\gamma) \text{ has at least j solutions}\}\ (j \geq 1).$$

The result proved in [7] is the following.

Theorem 4. *If f satisfies (28), then there exist* $s_1^o \leq s_1^+$ *in* $\mathbb{R}$ *such that* (21_s) *has no solution if* $s < s_1^o$, *at least one solution if* $s \in [s_1^o, s_1^+]$ *and at least two solutions if* $s > s_1^+$.

Proof. It follows the same lines as that of Theorem 2, but some technical difficulties occur due to the nature of the Dirichlet problem.

a) $S_1^\gamma \neq \phi$ *for* $\gamma \in [0,1]$. The lower and upper solutions used to prove this result cannot be chosen constant anymore and their existence comes from some extensions of Lemmas 2.3 and 2.7 of [7] which can be stated as follows :

Lemma 1. There exists $s^* \in \mathbb{R}$ such that, for each $s \geq s^*$ and each $\gamma \in [0,1]$, (29_s^γ) has an upper solution.

Lemma 2. If (28) holds, then, for each s, (29_s^γ) has a strict lower solution which can be taken smaller than any given C^1 function with zero boundary conditions.

b) *if* $\tilde{s} \in S_1^\gamma$ *and* $s > \tilde{s}$, *then* $s \in S_1^\gamma$ $(\gamma \in [0,1])$.
This is done exactly like in part (b) of the proof of Theorem 2.

c) $s_1^\gamma = \inf S_1^\gamma$ *is finite*, $S_1^\gamma \supset]s_1^\gamma, \infty[$ *and* $\gamma \longmapsto s_1^\gamma$ *is non-decreasing* $(\gamma \in [0,1])$.

The two first assertions are proved like in part (c) of Theorem 2 and the last one is an easy lower and upper solution argument. Consequently

$$s_1^+ = \lim_{\gamma \to 0_+} s_1^\gamma \geq s_1^o.$$

d) *for each* $s_2 > s_1^o$, *the set of all possible solutions of* (21_s) *with* $s \leq s_2$ *is a priori bounded.*

Writing $u(x) = \bar{u}\varphi(x) + \tilde{u}(x)$ with $\int_I \tilde{u}(x)\sin x\, dx = 0$, we obtain like in part (ii) of the proof of Theorem 3 an estimate

$$\int_I |u''(x) + u(x)|\sin x\, dx \le D_1 = D_1(s_2).$$

By an argument of [19], this implies that

$$\|\tilde{u}\|_{C^0} \le D_2$$

and hence an estimate upon $|\bar{u}|$ is obtained from condition (28) in a way similar to that of part (ii) in Theorem 3.

e) *for each* $s \le s_2$ *and each sufficiently large* $R > 0$, $D_L(F(.)-s\varphi, B(R)) = 0$, where $\varphi(x) = \sqrt{2/\pi}\sin x$ and $B(R) = \{u \in C_0^0(I) : \|u\|_{C^0} < R\}$, $C_0^0(I) = \{u \in C^0(I) : u(0) = u(\pi) = 0\}$.

The proof is similar to that of part (e) in Theorem 2.

f) $S_2 \supset]s_1^+,+\infty[$.
Let $\gamma \in]0,1]$, $s \in]s_1^\gamma,+\infty[$, $s_2 \in]s_1^\gamma,s[$ and u_2^γ a solution of $(29_{s_2}^\gamma)$. Then u_2^γ is a strict upper solution for (21_s) and, by Lemma 2, there exists a strict lower solution α_s for (21_s) such that $\alpha_s(x) < u_2^\gamma(x)$, $x \in I$. We can then use the degree argument of part (e) of the proof of Theorem 2 with $\Omega_{R_2u_2}$ replaced by the open bounded subset of $C_0^0(I)$

$$\Omega_{\alpha_s u_2^\gamma} = \{u \in C^0(I) : \alpha_s(x) < u(x) < u_2^\gamma(x),\ x \in I\}.$$

Thus $S_2 \supset]s_1^\gamma,+\infty[$ for all $\gamma \in]0,1]$, which gives the result.

g) $[s_1^0,s_1^+] \subset S_1^0$.
This is proved by the usual limit argument from the fact that $S_1^0 \supset]s_1^0,\infty[$, and the proof is complete.

Remark 9. We do not know if $s_1^0 = s_1^+$ in general. However, it is proved in [7] that it is the case if there exists $M > 1$ and $r > 0$ such that the function $u \mapsto Mu + f(x,u)$ is non-decreasing on $[-r,r]$ for each $x \in I$ (in particular if f is Lipschitz-continuous uniformly in $x \in I$). It is also the case if f is Hölder continuous in u uniformly in $x \in I$ for some exponent $\alpha > 1/2$ (C. Fabry, personal communication). It is likely that Theorem 4 holds for (22_s). We are working presently upon the proof.

5. THE PERIODIC PROBLEM FOR HIGHER ORDER DIFFERENTIAL EQUATIONS

The results of sections 2 to 4 are based upon lower and upper solutions techniques which only exist for first and second order differential equations. An interesting question is therefore the possibility of proving Ambrosetti-Prodi type results for higher order differential equations. We describe here some results of Ding-Mawhin [12] in the simplest situation, namely the periodic problem for forced autonomous ordinary differential equations.

Let us consider the problem

$$(30_s)\qquad \begin{aligned} F(u)(x) &\equiv u^{(m)}(x) + g(u(x)) - h(x) = s \\ u^{(k)}(0) &- u^{(k)}(2\pi) = 0 \quad (0 \leq k \leq m-1) \end{aligned}$$

where $m \geq 3$, $g : \mathbb{R} \to \mathbb{R}$ is continuous, $h \in L^1(I)$, $I = [0,2\pi]$, $s \in \mathbb{R}$ and

$$(31)\qquad g(u) \to +\infty \text{ if } |u| \to \infty.$$

Without loss of generality, we can assume that $\int_0^{2\pi} h(x)dx = 0$.

<u>Theorem 5</u>. *If* m *is odd and* g *satisfies* (31), *there exists* $s_o < s_1$ *such that* (30_s) *has zero, at least one or at least two solutions according to* $s < s_o$, $s = s_1$ *or* $s > s_1$.

<u>Proof</u>. Let

$$S_j = \{s \in \mathbb{R} : (30_s) \text{ has at least } j \text{ solutions}\} \ (j \geq 1).$$

$$S_o = \{s \in \mathbb{R} : (30_s) \text{ has no solution}\}.$$

a) *If* $\gamma = \min_{\mathbb{R}} g$, *then* $s_o = \sup S_o = \inf S_1 \geq \gamma$.

Indeed, if u is a solution of (30_s), then

$$s = (1/2\pi) \int_0^{2\pi} g(u(x))dx \geq \gamma.$$

b) *There exist* $\rho, \tilde{R} > 0$ *such that each possible solution* $u = \bar{u} + \tilde{u}$ *of*

$$(30_s^\lambda) \quad \begin{aligned} &F_\lambda(u)(x) \equiv u^{(m)}(x) + \lambda g(u(x)) - \lambda h(x) = \lambda s \quad (s \in \mathbb{R},\ \lambda \in]0,1]) \\ &u^{(k)}(0) - u^{(k)}(2\pi) = 0 \quad (0 \leq k \leq m-1) \end{aligned}$$

satisfies the inequalities

$$\| u^{\frac{(m+1)}{2}} \|_{L^2} < \rho,\ \|\tilde{u}\|_{C^0} < \tilde{R}.$$

If $u = \bar{u} + \tilde{u}$, with $\bar{u} = (1/2\pi) \int_0^{2\pi} u(x)dx$, is a possible solution of (30_s^λ) then integrating over $[0,2\pi]$ after multiplying by u' and using the periodicity and integrations by parts, get

$$(-1)^{\frac{m-1}{2}} \| u^{\frac{(m+1)}{2}} \|_{L^2}^2 = \lambda \int_I h(x)u'(x)dx \leq \|h\|_{L^1} \|u'\|_{C^0},$$

and hence, using well-known inequalities,

$$\|\tilde{u}\|_{C^0}^2 \leq C^2 \| u^{\frac{(m+1)}{2}} \|_{L^2}^2 \leq C \|h\|_{L^1} \|u'\|_{C^0} \leq C^3 \|h\|_{L^1} \| u^{\frac{(m+1)}{2}} \|_{L^2}$$

for some universal constant C, which gives the result.

c) *For each* $s_2 \in \mathbb{R}$ *there exist* $r(s_2) > 0$ *such that each possible solution* $u = \bar{u} + \tilde{u}$ *of* (30_s^λ) *with* $s \leq s_2$ *satisfies the inequality*

$$|\bar{u}| < r(s_2) + \tilde{R}.$$

By (31) we can find $r(s_2)$ such that $g(u) > s_2$ whenever $|u| > r(s_2)$. Now each possible solution u of (30_s^λ) with $s \leqslant s_2$ satisfies the relations

$$(1/2\pi) \int_0^{2\pi} g(\bar{u} + \tilde{u}(x))dx = s \leqslant s_2$$

so that necessarily, for some $y \in I$, one has

$$|\bar{u} + \tilde{u}(y)| < r(s_2)$$

and hence

$$|\bar{u}| \leqslant |\bar{u} + \tilde{u}(y)| + |\tilde{u}(y)| < r(s_2) + \tilde{R}.$$

Let $D(L) = \{u \in C^m(I) : u^{(k)}(0) - u^{(k)}(2\pi) = 0, 0 \leqslant k \leqslant m-1\}$,
$L : D(L) \subset C^0(I) \to L^1(I), u \mapsto u^{(m)}$.

d) *For each open bounded subset* $\Omega \subset C^0(I)$ *such that* $\Omega \supset \{u \in C^0(I) : |\bar{u}| < r(s_2), \|\tilde{u}\|_{C^0} < \tilde{R}\}$, *one has*

$$D_L(F(.)-s, \Omega) = 0. \qquad (s \in \mathbb{R})$$

This follows from a homotopy between s_2 and some $\tilde{s} < s_0$.

e) *There exist* $s_1 \geqslant s_0$ *such that for each* $s > s_1$ *one can find an open bounded subset* $\Omega_1(s) \subset C^0(I)$ *such that*

$$|D_L(F(.)-s, \Omega_1(s))| = 1.$$

If $u_0 \in \mathbb{R}$ is such that $g(u_0) = \gamma$ and $s_1 = \max_{[u_0-\tilde{R}, u_0+\tilde{R}]} g$, the set $\Omega_0(s) = \{u \in C^0(I) : u_0 < \bar{u} < r(s) + \tilde{R}, \|\tilde{u}\|_{C^0} < \tilde{R}\}$ has the required property. Indeed, $D_L(F_\lambda(.)-s, \Omega_1(s))$ is well defined for all $\lambda \in]0,1]$ and equal in absolute value to that of the Brouwer degree $d_B(g(.)-s,]u_0, r(s) + \tilde{R}[)$.

f) $S_2 \supset]s_1,+\infty[$ *and* $s_1 \in S_1$.
Use (d) and (e) and the additivity of degree and then the usual limit argument to get $s_1 \in S_1$.

<u>Remark 10</u>. It follows from the proof of Theorem 5 that S_1 is closed. In contrast to the first and second order cases, we do not know if S_o, S_1 and S_2 are intervals.

<u>Remark 11</u>. If m is even in (30_s), one can prove the analog of Theorem 5 under the supplementary assumption

$$(-1)^{\frac{m}{2}}(g(v) - g(w))(v-w) \geqslant -\beta(v-w)^2$$

for all $v,w \in \mathbb{R}$ and some $0 \leqslant \beta < 1$.
See [12] for the details.

REFERENCES

1. H. AMANN and P. HESS, A multiplicity result for a class of elliptic boundary value problems, Proc. R. Soc. Edinburgh 84A (1979) 145-151.
2. A. AMBROSETTI and G. PRODI, On the inversion of some differentiable mappings with singularities between Banach spaces, Ann. Mat. Pura Appl. (4) 93 (1972) 231-247.
3. H. BERESTYCKI and P.L. LIONS, Sharp existence results for a class of semilinear elliptic problems, Bol. Soc. Brasil. Mat. 12 (1981) 9-20.
4. M.S. BERGER and E. PODOLAK, On the solutions of a nonlinear Dirichlet problem, Indiana Univ. Math. J. 24 (1975) 837-846.
5. L. BRÜLL and J. MAWHIN, Finiteness of the set of solutions of some boundary-value problems for ordinary differential equations, Sémin. Math. U.C.L. (NS) n° 79, 1986.
6. K.C. CHANG, Variational methods and sub- and super-solutions, Scientia Sinica A-26 (1983) 1256-1265.
7. R. CHIAPPINELLI, J. MAWHIN and R. NUGARI, Generalized Ambrosetti-Prodi conditions for nonlinear two-point boundary value problems, preprint, 1986.
8. E.N. DANCER, On the ranges of certain weakly nonlinear elliptic partial differential equations, J. Math. Pures Appl. 57 (1978) 351-366.
9. D.G. DE FIGUEIREDO, Lectures on boundary value problems of the Ambrosetti-Prodi type, in "Atas 12e Semin. Brasileiro Analise, Sao Paulo, 1980", 230-291.
10. D.G. DE FIGUEIREDO, On the superlinear Ambrosetti-Prodi

problem, J. Nonlinear Analysis 8 (1984) 655-666.
11. D.G. DE FIGUEIREDO and S. SOLIMINI, A variational approach to superlinear elliptic problems, Comm. Partial Differential Equations 9 (1984) 699-717.
12. S.H. DING and J. MAWHIN, A multiplicity result for periodic solutions of higher-order ordinary differential equations, preprint, 1986.
13. C. FABRY, J. MAWHIN and M. NKASHAMA, A multiplicity result for periodic solutions of forced nonlinear second order ordinary differential equations, Bull. London Math. Soc. 18 (1986) 173-180.
14. D. GILBARG and N.S. TRUDINGER, "Elliptic Partial Differential Equations of Second Order", Second Edition, Springer, Berlin, 1983.
15. R. KANNAN and R. ORTEGA, Superlinear elliptic boundary value problems, Czech. Math. J., to appear.
16. J.L. KAZDAN and F.W. WARNER, Remarks on some quasilinear elliptic equations, Comm. Pure Appl. Math. 28 (1975) 567-597.
17. H.W. KNOBLOCH, An existence theorem for periodic solutions of nonlinear ordinary differential equations, Michigan Math. J. 9 (1962) 303-309.
18. J. MAWHIN, Recent results on periodic solutions of differential equations in "Intern. Conf. on Differential Equations", Academic Press, New York, 1975, 537-556.
19. J. MAWHIN, Boundary value problems with nonlinearities having infinite jumps, Comment. Math. Univ. Carolinae 25 (1984) 401-414.
20. J. MAWHIN, "Points fixes, points critiques et problèmes aux limites", Sém. Math. Sup. n° 92, Presses Univ. Montréal, 1985.
21. J. MAWHIN, First order ordinary differential equations with several periodic solutions, to appear.
22. M. NKASHAMA, A generalized upper and lower solutions method and multiplicity results for periodic solutions of nonlinear first order ordinary differential equations, preprint, 1986.
23. J.C. SCOVEL, Geometry of some nonlinear differential operators, Ph. D. Thesis, Courant Inst. New York University, 1983.
24. G. VIDOSSICH, Toward a theory for periodic solutions to first order differential equations, SISSA Trieste Report n° 59/83/M, 1983.
25. J.R. WARD, Perturbations with some superlinear growth for a class of second order elliptic boundary value problems, *J. Nonlinear Anal.* 6 (1982) 367-374.

Transmutation of Analytic and Harmonic Functions

Peter A. McCoy
United States Naval Academy
Annapolis, Maryland 21402

Abstract: Reciprocal transforms are constructed that link analytic functions in complex 3-space with harmonic functions whose angles are expressed as the spherical Euler variables in Euclidean 3-space. The representations are well suited to problems with multiple symmetry patterns along axes skewed relative to the standard spherical system. Special cases include the Bergman-Whittaker and Gilbert reciprocal integral transforms.

Introduction: There is a longstanding theory for many of the partial differential equations of mathematical physics based in classical function theory. The correspondence is made through integral transforms joining analytic functions with solutions of the object equation. An integral operator generally links properties of the functions in ways that are determined qualitatively: whereas, precise quantative information requires an inverse operator too. When one is known, a reciprocal transform pair exists and function theoretic analysis is sufficient for many studies.

One of the earliest operators to appear in this context is attributed to E.T. Whittaker [1903; 8]. It is a local map from analytic functions in C^2 onto harmonic functions in E^3 that S. Bergman [1926; 1] extended and inverted to develop the Bergman-Whittaker operator B_3 and its inverse. R.P. Gilbert [1960;3] based a detailed function theory on his operators G_3 and G_3^{-1}. Like B_3, G_3 generates harmonic functions in E^3 from analytic functions in C^2. And, it also arises in the method of ascent [4] where harmonic functions are projected onto solutions of other linear partial differential equations. These examples illustrate the central role of harmonic functions in this setting.

Existing function theoretic approaches express harmonic functions in spherical coordinates and produce representations that adapt to problems with simple symmetries, see [3,4,6] and the references therein. In many applications, complicated geometries arise with multiple symmetries or periods about axes oriented relative to the reference system. Such problems do not fall naturally into the domain of the B_3-G_3 operators or their relatives. To develop sufficient formulae, we express harmonic functions with Euler angle variables as integral transforms of analytic functions in C^3 and then invert the transforms. The result is a transform pair that extends the B_3-G_3 class of operators and applies to harmonic functions with general symmetries.

Background. The working system is a rotation of the spherical coordinates through the Euler angles (ϕ,θ,ψ). When r is the distance from a point to the origin in the Eulerian system [5,7], LaPlace's equation [7] has the form

$$(1)\quad (\partial_r(r^2\,\partial_r)) + \partial_{\theta\theta} + \mathrm{ctg}\theta\partial_\theta + (\sin\theta)^{-2}\,(\partial_{\phi\phi} - 2\cos\theta\;\partial\phi\psi + \partial\psi\psi))\;H = 0,$$

$0<\theta<\pi$, $0\leq\phi<2\pi$, $0\leq\psi<2\pi$. Harmonic functions $H(r,\theta,\phi,\psi)$ are local C^2 solutions.

Development of the transforms requires that H be expressed as a series of solid harmonics. Moving in this direction, separation of variables in eqn. (1) yields the harmonic polynomials

$$H^k_{mn}\;(r,z,e^{i\phi},\,e^{i\psi}) = r^k\,P^k_{mn}(z)\,e^{-i(m\phi+n\psi)},\quad z = \cos\theta\;,$$

$-k \leq m,\ n \leq k$, as particular solutions. The generalized Legendre polynomials, or angular functions, are the solutions of the differential equation [7]

$$[(1-z^2)D_{zz} - 2zD_z - ((m^2+n^2-2mnz)/(1-z^2) + k(k+1)]\;P^k_{mn}\;(z) = 0,\ -1 < z < +1\;,$$

that are bounded at $z = \pm 1$. The angular functions P^k_{mn} coincide with the Jacobi polynomials $P^{(\alpha,\beta)}_j$ of degree j and integer parameters (α,β) when $k = j + (\alpha+\beta)/2$, $m = (\alpha+\beta)/2$ and $n = (\alpha-\beta)/2$. And, if $n=0$ they become the associated Legendre polynomials

$$P^m_k\;(z) = i^m\;\;(k+m)!\;\;P^k_{m0}\;(z)\;.$$

Symmetric addition of the polynomials H^k_{mn} defines the k-degree solid harmonics

$$H_k(r,z,e^{i\phi},e^{i\psi}) = \sum_{m=-k}^{k}\;\sum_{n=-k}^{k}\;\;a^k_{mn}\;\;H^k_{mn}\;\;(r,z,e^{i\phi},e^{i\psi}),\ n=0,1,2,\ldots\;.$$

These are inturn superimposed as formal series

$$(1a)\quad H(r,\theta,\phi,\psi) := H(r,\cos\theta,e^{i\phi},e^{i\psi}) = \sum_{k=0}^{\infty}\;\;H_k\;(r,\cos\theta,e^{i\phi},e^{i\psi})\;.$$

To state a sufficient condition for existence [7], suppose that for some $r_0 > 0$,

$$\int_0^{2\pi}\int_0^{2\pi}\int_0^{\pi}\;|H(r_0,\cos\theta,e^{i\phi},e^{i\psi})|^2\;\;\sin\theta\;d\theta\;d\phi\;d\psi < \infty\;\;.$$

This integral is converted to the convergent series

$$\sum_{k=0}^{\infty} \sum_{m=-k}^{k} \sum_{n=-k}^{k} |r_o^k \; a_{mn}^k \; /(2k+1)|^2 < \infty$$

by Parseval's equation [7]. The implication is that the expansion (1a) is majorized by a geometrically convergent series in a sphere of (sufficiently small) radius $r_* < r_o$ where it exists as a harmonic function.

<u>The Integral Operators</u>. The process begins by identifying the solid harmonic expansions with associated analytic functions. We will show that the required associates admit convergent Laurent series expansions of the form

(1b) $$f(z_1,z_2,z_3) := \sum_{k=0}^{\infty} h_k(z_1,z_2,z_3) ,$$

$$h_k(z_1,z_2,z_3) := \sum_{m=-k}^{k} z_3^m \; g_m^k (z_1,z_2), \; k=0,1,2, \ldots ,$$

$$g_m^k(z_1,z_2) := \sum_{n=-k}^{k} a_{mn}^k \; \omega_{mn}^k \; z_1^{k-n} \; z_2^{k+n} \quad , \; -k \leq m,n \leq k ,$$

$$(\omega_{mn}^k)^2 := (k-m)!(k+m)!/(k-n)!(k+n)!,$$

in a pseudoannulus

$$A_\varepsilon : = \{(z_1,z_2,z_3) \varepsilon \, C^3 : |z_1|^2 < r_*/2 \, , \, |z_2|^2 < r_*/2 \, , \, 1 - \varepsilon < |z_3| < 1+ \varepsilon\}, \varepsilon > 0.$$

The first step in the construction is to convert the rational functions h_k to solid harmonics. We notice that the angular functions [7] are written as contour integrals

$$P_{mn}^k (z) = \omega_{mn}^k /2\pi i \int_\Gamma \tau_1^{k-n} \; \tau_2^{k+n} \; s^{m-k} \; ds/s$$

of the generating variables

$$\tau_1 = s((1+z)/2)^{1/2} + i \; ((1-z)/2)^{1/2}, \; \tau_2 = i \; s \; ((1-z)/2)^{1/2} + ((1+z)/2)^{1/2} ,$$

over the circle $\Gamma : |s| = 1$. This formula permits us to express the solid harmonics as LaPlace type integrals

$$H_k(r,z,e^{i\phi},e^{i\psi}) = 1/2\pi i \int_\Gamma h_k(e^{-i\phi}(r/s)^{1/2} \; \tau_1, e^{i\phi}(r/s)^{1/2} \; \tau_2, se^{-i\psi}) ds/s,$$

$k=0,1,2,\ldots$ when the principal branches are selected.

The ascending operator is developed by termwise integration of the series (1b). The result,

(2a) $H(r,\cos\theta,e^{i\phi},e^{i\psi}) = E(f):=$

$$1/2\pi i \int_{\Gamma} f(e^{-i\phi}(r/s)^{1/2}\,\tau_1,\ e^{i\phi}(r/s)^{1/2}\,\tau_2,\ s\,e^{-i\psi})\,ds/s\ ,$$

is valid in a sufficiently small sphere.

The inverse, or descending operator, is based on a Cauchy type integral. The kernel of the operator is defined as the meromorphic function

$$K(\sigma_1,\sigma_2,\sigma_3,\sigma_4):=\sum_{k=0}^{\infty}(2k+1)/2\ \sigma_1^k\ J_k(\sigma_2,\sigma_3,\sigma_4),$$

$$J_k(\sigma_2,\sigma_3,\sigma_4):=\sum_{n=-k}^{k}\sum_{m=-k}^{k}\omega_{mn}^k\ P_{mn}^k(\sigma_2)\ \sigma_3^{-m}\ \sigma_4^{-n}\ ,$$

in a pseudoannulus. The generating function method gives the closed form for the principal branch of the kernel as

$$K(\sigma_1,\sigma_2,\sigma_3,\sigma_4) = 1/2\ \partial/\partial\omega\ \ R(\omega,\rho,\xi_1\ \xi_2),$$

$$R(\omega,\rho,\xi_1\ \xi_2) = \omega(1+\xi_1\ \xi_2\ \omega^2)/((1-\omega^2\ \xi_1\ \xi_2\ \rho)(1-\omega^2\ \xi_1\ \xi_2\ \rho^{-1})),$$

$$\omega^2 = \sigma_1,\ \rho = \xi_2/\xi_1\ \sigma_3\ \ ,$$

with generating variables

$$\xi_1 = ((1+\sigma_2)/2)^{1/2}\ \sigma_4^{1/2} + i\ ((1-\sigma_2)/2)^{1/2}\ \sigma_4^{-1/2}\ ,$$

$$\xi_2 = i((1-\sigma_2)/2)^{1/2}\ \sigma_4^{1/2} + ((1+\sigma_2)/2)^{1/2}\ \sigma_4^{-1/2}\ ,$$

$$\xi_1\ \xi_2 = \sigma_2 + i(1-\sigma_2^2)^{1/2}\ (\sigma_4^{1/2}+\sigma_4^{-1/2})/2.$$

By using the orthogonality of the angular functions [7],

$$\int_{-1}^{+1} P_{mn}^k(\xi)\ P_{pq}^{*j}(\xi)\ d\xi = (2k+1)/2\ \delta_{mp}\ \delta_{nq}\ \delta_{kj}\ ,$$

it is easy to check that the solid harmonic polynomials and rational analytic functions are connected by the transform

$$h_k(z_1,z_2,z_3) = -1/4\pi^2 \int_{\Gamma_\zeta} \int_{\Gamma_\xi} \int_{\Gamma_\varepsilon} K(z_1 z_2/r,\ \xi,\ \eta z_3,\ \zeta/z_1 z_2)\ H_k(r,\ \xi,\ \eta,\zeta) d\xi \frac{d\eta}{\eta} \frac{d\zeta}{\zeta}$$

in the pseudoannulus. The contours are Γ_ζ, Γ_ξ : $|s| = 1$ and Γ_ε = seg[-1,+1].

The descending operator is constructed by termwise integration of the harmonic function expansion in (1a). This transformation,

(2b) $f(z_1,z_2,z_3) = E^{-1}(f) :=$

$$-1/4\pi^2 \int_{\Gamma_\zeta} \int_{\Gamma_\xi} \int_{\Gamma_\varepsilon} K(z_1 z_2/r,\ \xi,\ \eta z_3,\ \zeta/z_1 z_2)\ H(r,\ \xi,\ \eta,\ \zeta)\ d\xi\ \frac{d\eta}{\eta}\ \frac{d\zeta}{\zeta}\ ,$$

is a map from harmonic functions (1a) defined in a sphere onto meromorphic functions of the type (1b) expanded in a pseudoannulus.

Each of the transforms is an identity on the initial domain of definition of the function element on which it operates. The transforms are analytically continued by contour deformation from their initial domains of definition to their domains of association by the envelope method [3,4]. The continued representations are viewed as transmutations [2]. The analysis is summarized below.

<u>THEOREM</u>. <u>The reciprocal integral operators E and E^{-1} form a transmutation pair between the analytic and harmonic function elements (2a-b) on their domains of association.</u>

Reversing the rotations that produced the Eulerian system suppresses the ψ-angle dependence and reduces the solid harmonic expansions and rational function expansion by one independent variable. The transformation from the angular functions to the associated Legendre functions effectively produces the Bergman-Whittaker operator and its inverse.

<u>References</u>

[1] S. Bergman, Integral Operators in the Theory of Linear Partial Differential Equations, Ergebnisse der Mathematik and ihrer Grenzgebiete, Band 23, Springer-Verlag, Inc., New York (1969).

[2] R. W. Carroll, Some remarks on the Bergman-Whittaker integral operator, J. Math. pures et appl., Vol. 63 (1984) 1-14.

[3] R. P. Gilbert, Function-Theoretic Methods in Partial Differential Equations, Math. in Science and Engineering, Vol. 54, Academic Press, New York (1969).

[4] R. P. Gilbert, Constructive Methods for Elliptic Equations, Lecture Notes in Mathematics, Vol. 365, Springer-Verlag, New York (1974).

[5] H. Goldstein, Classical Mechanics, Addison-Wesley, Reading (1957).

[6] P. A. McCoy, Characterization of solutions o the generalized Cauchy-Riemann system, J. Math analysis and Appl., Vol. 101 (1984) 465-474.

[7] N. J. Vilenkin, Special Functions and the Theory of Group Representations, Trans. Math. Monographs, Vol. 2, American Mathematical Society, Providence (1968).

[8] E. T. Whittaker and G. N. Watson, A Course of Modern Analysis, Cambridge University Press, 4th ed. reprinted, London (1969).

SOME SOLVED AND UNSOLVED CANONICAL PROBLEMS
OF DIFFRACTION THEORY

Erhard Meister, TH Darmstadt, Germany

ABSTRACT. A large class of canonical problems involves diffraction by several wedges of different materials. In the case of four right-angled wedges a four-part Wiener-Hopf functional equation for the double Fourier-transformed wave-functions is established. This transmission problem and some more general multimedia boundary-transmission problems with boundary conditions on two faces of a Sommerfeld half-plane or a wedge are alternately reduced to a system of integral equations for the Fourier-transformed normal derivatives.

1. BASIC PROBLEMS OF DIFFRACTION THEORY

Besides electromagnetic wave propagation phenomena (which contain, in particular all those concerning microwaves in communication engineering) problems related to acoustic and elastic waves as well lead to the wave equations $(n = 2,3)$

$$\Delta_n u - \frac{1}{c^2} u_{tt} = f(\underline{x},t) \qquad \text{for} \quad (\underline{x},t) \in \Omega \times (0,t) \tag{1}$$

for a scalar or vectorial wave-function u (where u may be interpreted as the perturbation pressure, the displacement, the electrical or magnetic intensity vectors etc.) depending on the space and time variables $(\underline{x},t) = (x,y,z,t) \in \Omega \times (0,T)$ with a region $\Omega \subset \mathbb{R}^n$, $n = 2,3$.

If we confine ourselves to the simplest scalar situation we are led to a so-called initial boundary value problem for an unbounded domain $\Omega \subset \mathbb{R}^n$:

Find $u(\underline{x},t) \in C^2(\Omega \times (0,\infty)) \cap C^{\ell}(\overline{\Omega} \times [0,\infty))$ fulfilling the differential equation (1) in $\Omega \times (0,\infty)$ for $\ell = 0$ or 1 . Besides this the initial conditions should hold

$$u(\underline{x}, +0) = u^0(\underline{x}) , \qquad \underline{x} \in \overline{\Omega} \tag{2a}$$

$$\frac{\partial u}{\partial t}(\underline{x}, +0) = u^1(\underline{x}) , \qquad \underline{x} \in \overline{\Omega} \tag{2b}$$

and one of the boundary conditions

$$u(\underline{x},t)\Big|_{\underline{x}\in\partial\Omega} = g(\underline{x},t)\ , \qquad (\underline{x},t)\in\partial\Omega\times(0,\infty) \tag{3a}$$

or with a constant $p \in C_{++}$

$$\frac{\partial u}{\partial n}(\underline{x},t) + ip\, u(\underline{x},t)\Big|_{\underline{x}\in\partial\Omega} = h(\underline{x},t)\ , \qquad (\underline{x},t)\in\partial\Omega\times[0,\infty)\ . \tag{3b}$$

For non-smooth $\partial\Omega$, when the unit normal vector n to $\partial\Omega$ does not vary continuously, the edge condition should hold

$$u = O(1)\ ,\ \nabla u \in L^2_{loc}(\Omega)\ . \tag{4}$$

For unbounded $\Omega = \Omega_a$ but bounded $\partial\Omega_a$, which are called exterior domains, there should hold additionally

$$\lim_{|\underline{x}|\to\infty} u(\underline{x},t) = 0 \quad \text{and} \quad \lim_{|\underline{x}|\to\infty} |\nabla u(\underline{x},t)| = 0 \tag{5}$$

for all $t \in (0,\infty)$ (due to the finite propagation speed c !) .

For $\Omega_a \subset \mathbb{R}^n$, $n = 2,3$, with $\partial\Omega_a \in C^{1,\alpha}$ (Ljapounov surfaces) the initial boundary value problem has been solved completely (see e.g. R. LEIS' book [66]). In optics and microwave theory also unbounded domains exterior to screens S (these being complements to open curves Γ in $\mathbb{R}^2$ and complements to smoothly bounded surfaces S in $\mathbb{R}^3$) are of great interest. In the two-dimensional case the theory has been developed by P. DURAND (1983) [3] and in the three-dimensional case by E. STEPHAN (1984) [22] who transformed the boundary value problems into pseudo-differential equations on the screens Γ or S, respectively. The asymptotic behavior of $\nabla\Phi_s$ and of the solutions to the boundary integral equations, near the end points of Γ or S play a decisive role.

In order to control these asymptotics a series of canonical boundary value problems for the Helmholtz equation have been studied. The oldest problem of this type, SOMMERFELD's half-plane problem (1896) [21] was the first mathematically rigourously treated problem of diffraction theory. Besides the near-field, i.e. the asymptotics as $r\to 0$, the far-field, i.e. the asymptotic behavior as $r\to\infty$, has been of main interest.

Now we are going to formulate Sommerfeld's half-plane problems more precisely:

Find a scattered field $\Phi_s(\underline{x})$ exterior to the semi-infinite screen $S := \{(x,y)\in\mathbb{R}^2 : x \geq 0,\ y = 0\}$ due to the primary plane wave-field $\Phi_{pr}(x,y) := e^{ik(x\cos\theta + y\sin\theta)}$ with wave-number $k = k_1 + ik_2$, $k_1 > 0$, $k_2 \geq 0$, falling upon the screen in the direction θ $(0 \leq \theta < \pi)$ and its optically reflected field $\Phi_{ref}(x,y) := Re^{ik(x\cos\theta - y\sin\theta)}$ within the sector $S_\theta := \{(x,y) \in \mathbb{R}^2 : x \geq -\cot\theta \cdot y,\ y \geq 0\}$ and a purely diffracted wave within the optical shadow region $OS_\theta := \{(x,y) \in \mathbb{R}^2 : x \geq \cot\theta \cdot y\ ;\ y \geq 0\}$, respectively. The total field $\Phi_{tot}(x,y)$ is assumed to have vanishing boundary values or those of its normal derivative on the screen S, according to the polarization of its electric, E, or magnetic, H, field vector as $(0,0,\Phi(x,y))^T$ parallel to the diffracting edge K of the screen, i.e. the z-axis.

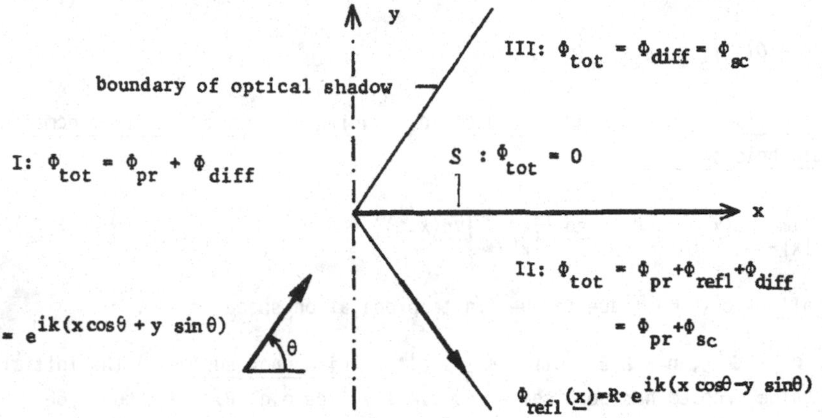

Figure 1. Decomposition of the space into different areas according to ray optics

After introducing the unknown Fourier transforms

$$\hat{E}_-(\lambda) := \int_{-\infty}^{0} e^{i\lambda x}\, \Phi_{sc}(x,0)\,dx \tag{6a}$$

and

$$\hat{V}_-(\lambda) := \int_{-\infty}^{0} e^{i\lambda x}\, \frac{\partial\Phi_{sc}}{\partial y}(x,0)\,dx \tag{6b}$$

which are holomorphic for $\operatorname{Im}\lambda = \lambda_2 < k_2$ the F-transforms of the wave-functions in both cases - the Dirichlet and Neumann one - may be represented as follows

$$\begin{aligned}\hat{\Phi}_{sc,D}(\lambda,y) &= A_D(\lambda) \cdot \exp\left[-|y|\sqrt{\lambda^2-k^2}\,\right] \\ &= (\hat{E}_-(\lambda) + [i(\lambda + k\cos\theta)]^{-1}) \exp[-|y|\sqrt{\lambda^2-k^2}]\end{aligned} \tag{7a}$$

and

$$\hat{\Phi}_{sc,N}(\lambda,y) = \text{sgny} \cdot A_N(\lambda) \cdot \exp\,[-|y|\sqrt{\lambda^2-k^2}]$$

$$= \frac{\text{sgny}}{\sqrt{\lambda^2-k^2}} \cdot (\hat{V}_-(\lambda) + \frac{k\sin\theta}{\lambda + k\cos\theta}) \exp\,[-|y|\sqrt{\lambda^2-k^2}]\ , \tag{7b}$$

respectively.

The unknown unilateral F-transforms $\hat{E}_-(\lambda)$ and $\hat{V}_-(\lambda)$ are coupled via the following <u>function-theoretic Wiener-Hopf equations</u> within the strip $-k_2\cos\theta < \text{Im}\lambda = \lambda_2 < k_2$ of the complex λ-plane:

$$\hat{E}_-(\lambda) + \frac{1}{2}\hat{J}_+(\lambda) \,/\, \sqrt{\lambda^2-k^2} = \frac{i}{\lambda + k\cos\theta} \tag{8a}$$

and

$$\hat{V}_-(\lambda) + \frac{1}{2}\hat{Q}_+(\lambda) \cdot \sqrt{\lambda^2-k^2} = \frac{-k\sin\theta}{\lambda + k\cos\theta}\ , \tag{8b}$$

respectively.

Specializing λ to the real axis (for $0 < \theta < \pi/2$) these two equations denote two (non-normal) <u>Riemann's type function-theoretic transmission problems</u> with respect to the two half-planes $H^{\pm} \subset \mathbb{C}_\lambda$ containing the boundary values $(\hat{E}_-, \hat{J}_+)$ and $(\hat{V}_-, \hat{Q}_+)$, respectively, which have to vanish as $|\lambda| \to \infty$ in their corresponding half-planes of holomorphy due to Riemann-Lebesgue's lemma. Another canonical problem in this context is the <u>mixed Sommerfeld half-plane problem</u> where different boundary conditions are prescribed on the two faces $S_\pm$ of the half-plane $S := \{(x,y) \in \mathbb{R}^2 : x \geq 0,\ y = 0\}$, e.g.

$$\lim_{y\to +0} \Phi_{sc}(x,y) = -\Phi_{pr}(x,+0) \qquad \text{for} \quad x \geq 0 \tag{9a}$$

and

$$\lim_{y\to -0} \frac{\partial\Phi_{sc}(x,y)}{\partial y} = -\frac{\partial\Phi_{pr}}{\partial y}(x,-0) \qquad \text{for} \quad x > 0\ . \tag{9b}$$

Choosing the direct approach via the F-transforms of the scattered wave-functions and their traces on $y = \pm 0$, $x \in \mathbb{R}$, one arrives at the following system instead of the scalar Wiener-Hopf functional equations (31 a,b) in $-k_2\cos\theta < \lambda_2 < k_2$:

$$\sqrt{\lambda^2-k^2} \cdot \hat{E}_-(\lambda) + \hat{V}_-(\lambda) + \hat{\Phi}'_+(\lambda,+0) = \frac{\sqrt{\lambda^2-k^2}}{\lambda + k\cos\theta} \tag{10a}$$

$$\sqrt{\lambda^2-k^2} \cdot \hat{E}_-(\lambda) - \hat{V}_-(\lambda) + \sqrt{\lambda^2-k^2} \cdot \hat{\Phi}_+(\lambda,+0) = \frac{k\sin\theta}{\lambda + k\cos\theta}\ . \tag{10b}$$

After introducing the pairs of functions holomorphic in $H^{\pm} \subset \mathbb{C}$, viz.

$$\hat{\vec{\Phi}}_-(\lambda) := \begin{bmatrix} \sqrt{\lambda-k} \cdot \hat{E}_-(\lambda) \\ \hat{V}_-(\lambda) \,/\, \sqrt{\lambda-k} \end{bmatrix} \qquad \text{for} \quad \text{Im}\,\lambda < k_2 \tag{11a}$$

and

$$\hat{\vec{\Phi}}_+(\lambda) := \frac{1}{2}\begin{bmatrix} \sqrt{\lambda+k}\cdot\hat{\Phi}_+(\lambda,-0) \\ \hat{\Phi}_+'(\lambda,+0)/\sqrt{\lambda+k} \end{bmatrix} \quad \text{for} \quad \operatorname{Im}\lambda > -k_2\cos\theta \tag{11b}$$

the WH-system (10a,b) becomes a system of Riemann transmission problems for $\lambda \in \mathbb{R}$

$$\hat{\vec{\Phi}}_-(\lambda) = \underline{K}(\lambda)\cdot\hat{\vec{\Phi}}_+(\lambda) + \hat{\vec{r}}(\lambda) \tag{12a}$$

with the function matrix

$$\underline{K}(\lambda) := \begin{bmatrix} \sqrt{\frac{\lambda-k}{\lambda+k}} , & 1 \\ -1 , & \sqrt{\frac{\lambda+k}{\lambda-k}} \end{bmatrix} \tag{12b}$$

holomorphic and non-singular for $|\operatorname{Im}\lambda| < k_2$.

The factorization of this and some related matrices, has been an open problem for a long time and has been accomplished almost at the same time, by different methods, by A.D. RAWLINS (1981) [20], A.E. HEINS (1982/83) [4] and E. MEISTER (1981/86) [14]. After splitting $\underline{K}(\lambda) = \underline{K}^-(\lambda)\cdot[\underline{K}^+(\lambda)]^{-1}$ with factors being holomorphic, non-singular and bounded in their respective half-planes $\operatorname{Im}\lambda \lessgtr \pm k_2$ the Wiener-Hopf system (12) may be solved explicitly like in the scalar case. It is of interest that the asymptotic edge behavior of $\Phi_{tot}(x,y)$ and $\nabla\Phi_{tot}(x,y)$ as $r \to 0$ now is different compared to that one of the classical Sommerfeld problems, viz.

$$\Phi_{tot}(x,-0) = O(x^{1/4}) \tag{13a}$$

and as $x \to +0$

$$\frac{\partial\Phi_{tot}}{\partial y}(x,+0) = O(x^{-3/4}) \tag{13b}$$

2. THE QUARTER-PLANE PROBLEM

Diffraction theory is also interested in the scattering behaviour of thin screens with corners in their edges. We should like to mention e.g. cracks in an elastic solid or a thin metallic plate of rectangular cross section.

The relevant canonical diffraction problem is the so-called quarter-plane problem:

> Find a scattered wave-field $\Phi_{sc}(\underline{x}) \in C^2(\mathbb{R}^3\setminus\Sigma)$ with continuous boundary values on both faces $\Sigma_\pm$ of the screen $\Sigma := \{(x,y,z) \in \mathbb{R}^3 : x \geq 0 , y \geq 0 , z = 0\}$ and continuous normal derivatives $\partial\Phi_{sc}/\partial z(x,y,0)$ off the edges $K = \partial\Sigma$ due to an incoming plane wave

$\Phi_{pr}(\underline{x}) = e^{i<\underline{k},\underline{x}>} := \exp[ik(x \sin\theta \cdot \cos\phi + y \sin\theta \sin\phi + z \cos\theta)]$ satisfying

$$(\Delta_3 + k^2)\Phi_{sc}(\underline{x}) = 0 \quad \text{in} \quad \mathbb{R}^3 \setminus \Sigma \tag{14}$$

and the boundary conditions

$$\Phi_{sc}(\underline{x})|_\Sigma = - \Phi_{pr}(\underline{x})|_\Sigma \tag{15a}$$

or

$$\frac{\partial}{\partial z}\Phi_{sc}(\underline{x})\Big|_{\overset{\circ}{\Sigma}} = -\frac{\partial}{\partial z}\Phi_{pr}(\underline{x})\Big|_{\overset{\circ}{\Sigma}} \tag{15b}$$

where $\overset{\circ}{\Sigma} := \Sigma \setminus K$, or

$$\left(\frac{\partial}{\partial z} \pm ip\right)\Phi_{sc}(\underline{x})\Big|_{\overset{\circ}{\Sigma}_\pm} = -\left(\frac{\partial}{\partial z} \pm ip\right)\Phi_{pr}(\underline{x})\Big|_{\overset{\circ}{\Sigma}_\pm} \tag{15c}$$

or with different boundary conditions on the two different faces $\Sigma_\pm$. Besides these the wave function has to fulfil the asymptotic conditions (edge conditions)

$$\Phi_{sc}(\underline{x}) = O(1) , \quad \nabla\Phi_{sc}(\underline{x}) = \begin{cases} O(\rho^{-\alpha}) & \text{as } \rho \to 0 , \ r \geq r_0 > 0 \\ O(r^{-\beta-1}) & \text{as } r \to 0 \end{cases} \tag{16a}$$

where $\rho := \text{dist}(\underline{x}, K \setminus \{0\})$, $r := |\underline{x}|$, with $0 \leq \alpha, \beta < 1$ or, in a more general form

$$\Phi_{sc}(\underline{x}) \in H^1_{loc}(\mathbb{R}^3) \tag{16b}$$

and the radiation conditions

$$\Phi_{sc}(\underline{x}) = O(e^{-k_2 r}) , \quad \left(\frac{\partial}{\partial r} - ik\right)\Phi_{sc}(\underline{x}) = o\left(\frac{e^{-k_2 r}}{r}\right) \tag{17}$$

as $r \to +\infty$ for all directions $\underline{x} / |\underline{x}|$ leaving out those belonging to the quarter-plane Σ .

J. RADLOW tried (1961) [19] to solve the canonical Dirichlet problem for the quarter-plane by means of two-dimensional Laplace transformation with respect to the x,y-variables. Sorry to say that he failed and the problem has been solved only recently by the author and F.-O. Speck [17].

3. SOME CANONICAL TRANSMISSION PROBLEMS

In this section we are going to show how an important class of diffraction problems involving different materials leads to multiple-part Wiener-Hopf operators. We shall see that there also exists a number of unsolved canonical transmission problems.

By a transmission or interface problem for Helmholtz's equation we understand the following one:

Let there be given a piecewise smoothly bounded domain $\Omega_i \subset \mathbb{R}^n$, $n = 2,3$, and its complement $\Omega_a := \mathbb{R}^n \setminus \overline{\Omega}_i$ with respective wave propagation velocities c_i, c_a and transmission coefficients $\rho_i, \rho_a \in \mathbb{C}^{++} \setminus \{0\}$. Find two wave fields $u_{i/a}(\underline{x},t) \in C^2(\Omega_{i/a} \times (0,\infty)) \cap C^{\ell}(\overline{\Omega}_{i/a} \times [0,\infty))$; $\ell = 0$ or 1 , such that $(\Delta_n - \frac{1}{c^2}\frac{\partial^2}{\partial t^2})\, u_{i/a}(\underline{x},t) = 0$ in $\Omega_{i/a} \times (0,\infty)$ fulfilling the initial value conditions

$$u_{i/a}(\underline{x},+0) = u^0_{i/a} \qquad \text{for } \underline{x} \in \overline{\Omega}_{i/a} \tag{18a}$$

$$\frac{\partial u_{i/a}}{\partial t}(\underline{x},+0) = u^1_{i/a}(x) \qquad \text{for } \underline{x} \in \Omega_{i/a} \tag{18b}$$

and the transmission conditions

$$u_i(\underline{x},t) - u_a(\underline{x},t) = g(\underline{x},t) \quad \text{for } (\underline{x},t) \in \partial\Omega \times [0,\infty) \tag{19a}$$

$$\rho_i \frac{\partial u_i}{\partial n_i}(\underline{x},t) + \rho_a \frac{\partial u_a}{\partial n_a}(\underline{x},t) = h(\underline{x},t) \quad \text{for } (\underline{x},t) \in (\partial\Omega_i \setminus K) \times (0,\infty) \tag{19b}$$

with unit normal vector $n_{i/a}$ pointing into the interior of $\Omega_{i/a}$ on $\partial\Omega_i \setminus K$.

In the time-harmonic case, i.e. $u_{i/a}(\underline{x},t) = \mathrm{Re}\{\Phi_{i/a}(\underline{x})e^{-i\omega t}\}$, one is led to two Helmholtz equations in $\Omega_{i/a}$ and has to claim additionally the edge condition $\nabla\Phi_{i/a} \in L^2_{loc}(\Omega_{i/a})$ and the radiation conditions

$$\Phi_{i/a} = o(1)\ , \quad (\frac{\partial\Phi_{i/a}}{\partial r} - ik_{i/a}\cdot\Phi_{i/a}) = o(r^{-\frac{1-n}{2}}) \tag{20}$$

as $r := |\underline{x}| \to \infty$ in $\Omega_{i/a} \cap S^2_r(0)$ where $k^2_{i/a} := \omega^2/c^2_{i/a}$.

In the two-dimensional case this problem has been studied for finite and smooth $\Gamma = \partial\Omega_i = \partial\Omega_a$ by several authors including e.g. P. WERNER (1963 , 1969) [23][24], R. KRESS & G.F. ROACH (1978) [6] and for curves Γ with vertices by M. COSTABEL & E. STEPHAN (1985) [2]. A formulation and a method of solution in the language of Sobolev spaces has been given for very general boundaries by R. MOHR (1976) [18] in his dissertation where he constructed the spectral family corresponding to the self-adjoint operator

$$A := -c^2(\underline{x})\rho(\underline{x})\nabla \cdot \frac{1}{\rho(\underline{x})} \nabla \quad \text{in} \quad H^1(\mathbb{R}^n;\Delta)$$

with piecewise continuous $c(\underline{x})$ and $\rho(\underline{x}) > 0$.

Here we shall formulate the canonical transmission and boundary transmission problems and will point out the basic ideas reporting the present state of the art. We start with the simplest optical problem of refraction and reflection of a plane wave by a plane interface F :

In connection with the problems considered in the first and second sections of this paper we are led to the corresponding boundary transmission problems for geometrically canonical boundaries, e.g. the different Sommerfeld half-plane problems when there are different materials in the two different half-planes $H^{\pm} \subset \mathbb{R}^2$:

$$(\Delta_2 + k_2^{\,2})\Phi_{sc,1}(x,y) = 0 \quad \text{in} \quad H^- \qquad (y < 0) \tag{21a}$$

$$(\Delta_2 + k_1^{\,2})\Phi_{sc,2}(x,y) = 0 \quad \text{in} \quad H^+ \qquad (y > 0) \tag{21b}$$

$$\Phi_{sc,1}(x,-0) - \Phi_{sc,2}(x,+0) = f(x) \tag{22a}$$

for $x < 0$

$$\rho_1 \frac{\partial\Phi_{sc,1}}{\partial y}(x,-0) - \rho_2 \frac{\partial\Phi_{sc,2}}{\partial y}(x,+0) = g(x) \tag{22b}$$

$$B_1[\Phi_{sc,1}](x) = F_1(x) \tag{23a}$$

for $x > 0$

$$B_2[\Phi_{sc,2}](x) = F_2(x) \tag{23b}$$

with boundary operators $B_{1,2}$ of the first, second, or third kind on both faces of the screen, $S_{\pm} := \{(x,y) \in \mathbb{R}^2 : x \geq 0 \ , \ y = \pm 0\}$. The data on the right-hand sides of eqs. (22a,b) and (23a,b) result from the corresponding homogeneous transmission and boundary conditions for the total wave-fields in $H^{\pm}$. Applying the F-transformation with respect to x we get

$$\hat{\Phi}_{sc,1/2}(\lambda,y) = A_{1/2}(\lambda) \cdot \exp[\pm y \sqrt{\lambda^2 - k_{1/2}^2}] \qquad \text{for } y \gtrless 0 \tag{24}$$

$$= \begin{cases} (\hat{F}_{1/2+}(\lambda) + \hat{E}_{1/2-}(\lambda)) \exp[\pm y \sqrt{\lambda^2 - k_{1/2}^2}] & \text{for the Dirichlet problem} \\ -(\hat{G}_{1/2+}(\lambda) + \hat{V}_{1/2-}(\lambda)) \cdot \mathrm{sgn}\, y \cdot \dfrac{\exp[\pm y \sqrt{\lambda^2 - k_{1/2}^2}]}{\sqrt{\lambda^2 - k_{1/2}^2}} & \text{for the Neumann problem} \end{cases}$$

including the unilateral left-hand F-transforms of the unknown Cauchy data on $y = \pm 0$, $x < 0$, i.e. the half-lines in front of the screen S , viz. $\hat{E}_{1/2-}(\lambda)$ and $\hat{V}_{1/2-}(\lambda)$

which are coupled via the F-transformed transmission conditions (22a,b) :

$$\hat{E}_{1-}(\lambda) - \hat{E}_{2-}(\lambda) = \hat{f}_-(\lambda) \tag{25a}$$

$$\rho_1 \hat{V}_{1-}(\lambda) - \rho_2 \hat{V}_{2-}(\lambda) = \hat{g}_-(\lambda) \quad . \tag{25b}$$

Now, in the case of the first boundary value problem on the screen, the data $F_{1/2}(x)$, $x > 0$, are known while $G_{1/2}(x)$ are to be found. Letting $y \to +0$ we obtain the following relations due to Huygens' principle among the Cauchy data: In the case of the D/D problem

$$\begin{bmatrix} \hat{E}_{1-}(\lambda) \\ \\ \hat{V}_{1-}(\lambda) \end{bmatrix} = \begin{bmatrix} \dfrac{\rho_1}{\rho_1\gamma_1 + \rho_2\gamma_2} , & \dfrac{\rho_2}{\rho_1\gamma_1 + \rho_2\gamma_2} \\ \\ \dfrac{-\rho_2\gamma_2}{\rho_1\gamma_1 + \rho_2\gamma_2} , & \dfrac{-\rho_2\gamma_1}{\rho_1\gamma_1 + \rho_2\gamma_2} \end{bmatrix} \begin{bmatrix} \hat{G}_{1+}(\lambda) \\ \\ \hat{G}_{2+}(\lambda) \end{bmatrix} + \begin{bmatrix} \hat{r}_1(\lambda) \\ \\ \hat{r}_2(\lambda) \end{bmatrix} \tag{26}$$

where the last column is known.

The explicit factorization of the 2 by 2-matrices like that one in eq. (26) or those belonging to the related vectorial Riemann problem has not yet been accomplished, so far the author is aware.

The next step in growing complexity of transmission and boundary transmission problems is reached in $\mathbb{R}^3$-space when one considers a quarter-plane or an arbitrary plane screen Σ at the interface $(z = 0)$ of two half-spaces filled with different materials such that boundary conditions are posed on the two faces $\Sigma_\pm$ of the screen and two transmission conditions off the screen : $\Sigma' := \mathbb{R}^2_{xy} \setminus \Sigma$. Since most of the one-media quarter-plane problems are still unsolved the more is true for the scalar and vectorial electro-magnetic multi-media quarter-plane problems.

4. MULTIMEDIA TRANSMISSION PROBLEMS

Remaining still in $\mathbb{R}^3$-space for the original transmission problems, wedges or quadrants seem to represent the canonical geometries. The famous dielectric wedge problem involves just two different wave-numbers $k, k' \in \mathbb{C}_{++}$ and is still unsolved with respect to all details concerning the asymptotics of the near - and far - fields and the spectral representation needed for the initial value problem.

The author investigated this problem and its natural generalization, the four-quadrant transmission problem together with N. LATZ starting in (1964) [8]. He wants to report in brief about the methods applied to it. Several authors like N.H. KUO & M.A. PLONUS (1967) [7], N. LATZ (1984), [9], E.A. KRAUT & G.W. LEHMAN (1969) [5],

A.D. RAWLINS (1977) [20], E. MEISTER & N. LATZ (1984) [11] and S. BERNTSEN (1983) [1] developed approximate solution methods, mainly based on BANACH's fixed point theorem and the two-dimensional F-transforms of the scattered wave-fields in the quadrants.

Now we shall transform firstly the transmission problem into a four-part-Wiener-Hopf problem. This method may be generalized to treat also vectorial electro-magnetic and elastodynamic transmission problems for a finite number of reasonably bounded semi-infinite domains G_j ; $j = 1,\dots,N$; $G_j \cap G_\ell = \emptyset$ for $j \neq \ell$ and $\bigcup_{j=1}^{N} \overline{G}_j = \mathbb{R}^n$, $n = 2,3$ (cf. for the electro-magnetic case N. LATZ' habilitation's thesis (1974) [10] and E. MEISTER & F.O. SPECK (1979) [13] and E. MEISTER (1981) [14] for the elastodynamic case).

Without loss of generality let us assume the electro-magnetic field to be excited by a line-source situated in the first wedge with cross-section Q_1 , the first quadrant of the x,y-plane: $(\underline{E}_{pr}(x,y), \underline{H}_{pr}(x,y))$. There exist two distinct polarizations according to the $\underline{E}$-field or the $\underline{H}$-field having a single component parallel to the z-axis, the common edge K of the four wedges. For the sake of simplicity we shall prefer here the case of electrical polarization where all the ρ_j may be assumed equal to one. This leads to the following transmission problem in $\mathbb{R}^2$:

Find $\Phi_j \in C^2(\mathring{Q}_j) \cap C^1(Q_j \setminus \{0\})$, $j = 1,\dots,4$, satisfying
$(\Delta + k_j^2)\Phi_j(x,y) = 0$ in $\mathring{Q}_j \subset \mathbb{R}^2$ with the transmission conditions

$$\Phi_j - \Phi_{j-1}\Big|_{\partial Q_j \cap \partial Q_{j-1}} = \begin{cases} -\Phi_{pr}(x,0) & \text{for } x > 0 \text{ , } j = 1 \\ \Phi_{pr}(0,y) & \text{for } y > 0 \text{ , } j = 2 \\ 0 & \text{else} \end{cases} \tag{27a}$$

$$\frac{\partial \Phi_j}{\partial n_j} + \frac{\partial \Phi_{j-1}}{\partial n_{j-1}}\Big|_{\partial Q_j \cap \partial Q_{j-1}} = \begin{cases} -i\omega\mu \cdot H_{pr1}(x,0) & \text{for } x > 0 \text{ , } j = 1 \\ i\omega\mu \cdot H_{pr2}(0,y) & \text{for } y > 0 \text{ , } j = 2 \\ 0 & \text{else} \end{cases} \tag{27b}$$

the edge conditions

$$\Phi_j = O(1) \text{ , } \quad \nabla\Phi_j = O(r^{-\alpha_j}) \qquad \underline{\text{as}} \quad r \to 0 \tag{28}$$

with some $0 \leq \alpha_j < 1$ and the radiation conditions

$$\Phi_j, \ \nabla\Phi_j = o(r^{-1/2}\, e^{-qr}) \qquad \underline{\text{as}} \quad r \to +\infty \tag{29}$$

with an appropriate $q := \min_j(\omega\sigma_j\mu) > 0$.

We are going to solve this problem by means of the two-dimensional F-transformation

$$\hat{\Phi}_j(\lambda_1,\lambda_2) := F_2\Phi_j := \int_{Q_j} e^{i\underline{\lambda}\cdot\underline{x}}\,\Phi_j(x,y)\,d(x,y) \tag{30}$$

with $\underline{\lambda} := (\lambda_1,\lambda_2)$, $\underline{x} := (x,y) \in \mathbb{R}^2$. The Helmholtz equation for Φ_j then turns into the representation

$$\hat{\Phi}_j(\lambda_1,\lambda_2) = Z_j(\lambda_1,\lambda_2)\cdot(\lambda_1^2+\lambda_2^2-k_j^2)^{-1} \tag{31}$$

holding for $\mathrm{Re}\,\lambda_1$, $\mathrm{Re}\,\lambda_2 \in \mathbb{R}$, $(\mathrm{Im}\,\lambda_1)^2 + (\mathrm{Im}\,\lambda_2)^2 < q^2$ with numerators Z_j given e.g. by

$$Z_1(\lambda_1,\lambda_2) := i\lambda_2\cdot\hat{f}_1^{(1)}(\lambda_1) + \hat{g}_1^{(1)}(\lambda_1) + i\lambda_1\cdot\hat{f}_2^{(1)}(\lambda_2) + \hat{g}_2^{(1)}(\lambda_2) \tag{32}$$

with one-dimensional, unilateral F-transforms $\hat{f}$ of the boundary values of Φ_1 and $\hat{g}$ of those of $\partial\Phi/\partial n$ on the positive x- and y-semi-axis, respectively.

Adding all four numerators Z_j and taking into account the F-transformed transmission conditions (26a,b) we arrive at

$$\sum_{j=1}^{4} Z_j(\lambda_1,\lambda_2) = Z(\lambda_1,\lambda_2) \tag{33}$$

which is a known function containing F-transforms of the Cauchy data of the primary wave-function in Q_1 .

Expressing the functions Z_j by eq. (31) and inserting into eq. (33) we obtain after subsequent division by $\lambda_1^2 + \lambda_2^2 - k^2$ with an appropriate $k \in \mathbb{C}_+$ the four-part Wiener-Hopf equation

$$\sum_{j=1}^{4} \left(1 + \frac{k^2-k_j^2}{\lambda_1^2+\lambda_2^2-k^2} \right) \hat{P}_j\,\hat{\Phi}(\lambda_1,\lambda_2) = \frac{Z(\lambda_1,\lambda_2)}{\lambda_1^2+\lambda_2^2-k^2} \tag{34}$$

holding at least for $(\lambda_1,\lambda_2) \in \mathbb{R}^2$. The four factors (...) are elliptic multiplication operators A_j on the function space $FL^p(\mathbb{R}^2)$, $1 \leq p \leq 2$, i.e. they are linear, continuous, and bijective. Furthermore we introduced the four projectors

$$\hat{P}_j := F_2\chi_j\cdot F_2^{-1}\ , \qquad j = 1,\dots,4$$

the F-transforms of the four space-projectors P_j , corresponding to the characteristic functions χ_j of the four quadrants $Q_j \subset \mathbb{R}^2$.

N. LATZ showed in (1968) [9], even for more general decompositions of $\mathbb{R}^2$-space into N (≥ 2) semi-infinite regions Q_j, that $k \in \mathbb{C}_+$ with $\mathrm{Re}\, k < 0$, $\mathrm{Im}\, k > 0$, may be chosen in such a way, depending on the $k_j \in \mathbb{C}_{++}$, that eq. (34) is uniquely and continuously solvable in $x = FL^p(\mathbb{R}^2)$ $(1 \leq p \leq \infty)$. He could also formulate the corresponding equation in the original space as a two-dimensional convolutional equation.

The question of unique solvability in appropriate weighted spaces $L^2(\mathbb{R}^2;\rho)$ is still open for positive wave-numbers k_j.

N. LATZ and the present author developed in (1984) [11] a new method to attack the canonical problem of the dielectric wedge by reducing the transmission problem (27) to (29) by means of the F_2-transformation and the one-dimensional WH-technique to a single integral equation or a system of such for the F-transformed boundary values of the normal derivatives of the wave functions along the semi-axis. Since this method seems to prove some progress in treating boundary value transmission problems where some boundary conditions are given instead of transmission conditions on some of the semi-axis in $\mathbb{R}^2$, we shall describe this one for the simplest transmission case.

Starting again from the two-dimensional F-transforms $\hat{\Phi}_j(\lambda_1,\lambda_2)$ involving the unilateral one-dimensional F-transformed Cauchy data like eq. (32) we add only the first two numerators and symmetrize with respect to λ_2 which corresponds to a Fourier cosine transform:

$$(\lambda_1^2+\lambda_2^2-k_1^2)\,\hat{\Phi}_{1,c}(\lambda_1,\lambda_2) + (\lambda_2^2+\lambda_2^2-k_2^2)\hat{\Phi}_{2c}(\lambda_1,\lambda_2) = \hat{g}_1^{(1)}(\lambda_1) + \hat{g}_3^{(2)}(\lambda_1) + \hat{r}_{1,c}(\lambda_1,\lambda_2) \tag{36}$$

where $\hat{r}_{1,c}$ is a known function. The Dirichlet data drop out while the Neumann data on $x = 0$, $y > 0$ for $\rho_1 = \rho_2$ have to be determined. Factoring the characteristic polynomials into

$$(\lambda_1^2 + \lambda_2^2 - k_j^2) = (\lambda_1 - i\sqrt{\lambda_2^2-k_j^2})(\lambda_1 + i\sqrt{\lambda_2^2-k_j^2}) \tag{37}$$

where $\mathrm{Re}\sqrt{\lambda_2^2-k_j^2} > 0$ for $\lambda_2 \in \mathbb{R}$ and dividing eq. (36) by $(\lambda_1 - i\sqrt{\lambda_2^2-k_1^2})(\lambda_1 + i\sqrt{\lambda_2^2-k_2^2})$ and finally applying the projectors $\hat{P}_1$ and $\hat{P}_2$ with respect to the F-variable λ_1 we obtain

$$\hat{\Phi}_{1/2,c}(\lambda_1,\lambda_2) = \frac{\lambda_1 \pm i\sqrt{\lambda_2{}^2-k_2{}^2}}{\lambda_1 \pm i\sqrt{\lambda_2{}^2-k_1{}^2}}\, \hat{P}_\pm([(\sigma - i\sqrt{\lambda_2{}^2-k_1{}^2})(\sigma + i\sqrt{\lambda_2{}^2-k_2{}^2}\,] \cdot \\ \cdot \{\hat{g}_1^{(1)}(\sigma) + \hat{g}_3^{(2)}(\sigma) + \hat{r}_{1,c}(\sigma,\lambda_2)\})(\lambda_1) \,. \tag{38}$$

Now we apply a second symmetrization with respect to λ_1 and get, making use of explicit forms of the projectors $\hat{P}_\pm = \frac{1}{2}(I \pm S_{\mathbb{R}})$,

$$\hat{\Phi}_{1,cc}(\lambda_1,\lambda_2) = \frac{\hat{g}^{(1)}_{1,c}(\lambda_1)}{\lambda_1{}^2+\lambda_2{}^2-k_1{}^2} - \frac{2}{\pi}\,\frac{\sqrt{\lambda_2^2-k_1^2}\cdot\sqrt{\lambda_2^2-k_2^2}}{\sqrt{\lambda_2^2-k_1^2}+\sqrt{\lambda_2^2-k_2^2}}\,(\lambda_1{}^2+\lambda_2{}^2-k_1{}^2)^{-1} \cdot \\ \cdot \left(\int_{\sigma=0}^{\infty} \frac{\hat{g}^{(1)}_{1,c}(\sigma)\,d\sigma}{\sigma^2+\lambda_2{}^2-k_1{}^2} + \int_{\sigma=0}^{\infty} \frac{\hat{g}^{(3)}_{3,c}(\sigma)\,d\sigma}{\sigma^2+\lambda_2{}^2-k_2{}^2}\right) + \frac{W_{1,cc}(\lambda_1,\lambda_2)}{\lambda_1{}^2+\lambda_2{}^2-k_1{}^2} \tag{39}$$

and a similar formula for $\hat{\Phi}_{2,cc}(\lambda_1,\lambda_2)$.

If we double symmetrized the expressions for $\hat{\Phi}_j$ in eq. (31), that are the double Fourier cosine transforms of the scattered wave-functions $\Phi_j(x,y)$, we were led to

$$\hat{\Phi}_{1,cc}(\lambda_1,\lambda_2) = \frac{\hat{g}^{(1)}_{1,c}(\lambda_1) + \hat{g}^{(1)}_{2,c}(\lambda_2)}{\lambda_1{}^2+\lambda_2{}^2-k_1{}^2} \tag{40}$$

and analogous representations for the $\hat{\Phi}_{j,cc}(\lambda_1,\lambda_2)$; $j = 2,3,4$. Since the first terms in the numerators for $\hat{\Phi}_{1,cc}(\lambda_1,\lambda_2)$ coincide in eqs. (39) and (40) there must hold the following relation among the remaining terms, viz. for $\lambda_2 > 0$:

$$\hat{g}^{(1)}_{2,c}(\lambda_2) = -\frac{2}{\pi}\,\frac{\sqrt{\lambda_2^2-k_1^2}\cdot\sqrt{\lambda_2^2-k_2^2}}{\sqrt{\lambda_2^2-k_1^2}+\sqrt{\lambda_2^2-k_2^2}} \cdot \\ \cdot \left\{\int_{\sigma=0}^{\infty} \frac{\hat{g}^{(1)}_{1,c}(\sigma)\,d\sigma}{\sigma^2+\lambda_2{}^2-k_1{}^2} + \int_{\sigma=0}^{\infty} \frac{\hat{g}^{(3)}_{3,c}(\sigma)\,d\sigma}{\sigma^2+\lambda_2{}^2-k_2{}^2}\right\} + \hat{v}_{2,c}(\lambda_2) \,. \tag{41}$$

There are three more relations of this type where the Fourier cosine transform of the normal derivative along one semi-axis is expressed, like in eq. (41), by the Fourier cosine transforms of the normal derivatives along the two neighboring half-lines. This ends up with a 4 by 4-system if integral equations of the following type

$$\hat{\vec{g}}_c(\xi) + \int_0^{\infty} \underline{\underline{K}}(\xi,\eta)\hat{\vec{g}}_c(\eta)\,d\eta = \hat{\vec{v}}_c(\xi) \,, \qquad \xi \in \mathbb{R}_+ \tag{42}$$

with $\xi = \lambda_1$ or $= \lambda_2$ and $\vec{\hat{g}}_c := (\hat{g}^{(1)}_{1,c}(\lambda_1), \hat{g}^{(1)}_{2,c}(\lambda_2), \hat{g}^{(3)}_{3,c}(\lambda_1), \hat{g}^{(3)}_{4,c}(\lambda_2))^T$,

the function vector to be found in $(F_c L^p(\mathbb{R}_+))^4$ when $\vec{\hat{v}}_c$ is prescribed there. An analogous system of integral equations turns up also in the case of H-polarization where the factors are $\rho_j \not\equiv 1$. In this case one has to replace the square rootes in eqs. (39) and (41) by weighted ones. The special structure of the alternating system of integral equations (41) allows to insert two of the expressions into the others resulting then in a 2 by 2-system with all elements $\not\equiv 0$.

Presently a detailed discussion of the system (42) is still missing when one admits an arbitrary combination of wave-numbers $k_j \in \overline{\mathbb{C}}_{++} \setminus \{0\}$ and transmission factors $\rho_j \in \overline{\mathbb{C}}_{++} \setminus \{0\}$. The goal of all investigations would be an explicit representation of the resolvent operator to the system (42).

A similar transmission problem may be formulated in $\mathbb{R}^3$-space for the eight octants with the twelve coordinate quarter-planes as their common boundary parts. It may be reduced similarly to a 12 by 12-system of two-dimensional integral equations for the two-dimensional double Fourier cosine transforms of the boundary values of the normal derivatives on the faces of the octants. For details see the author's paper [15].

In order to close this report we want to list those two-dimensional boundary value transmission problems which may be called canonical, too:

i) The generalized Sommerfeld half-plane problems and

ii) The generalized (rectangular) wedge-problems

for the 4-media case involving also their respective generalizations to the $\mathbb{R}^3$-space geometries with quarter-planes and octants and including vectorial wave-fields as well.

It seems to be the simplest way to illustrate the main situations by the following figures

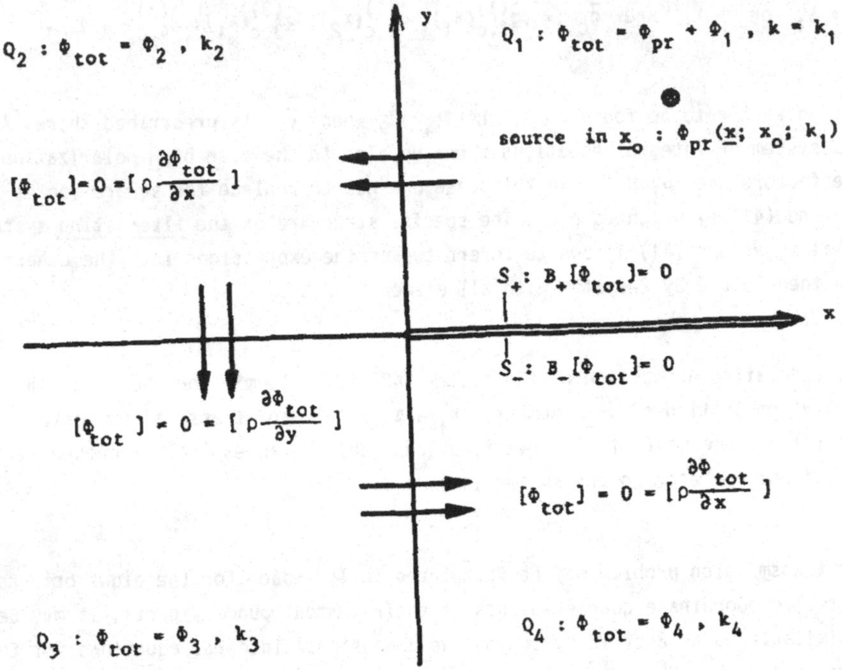

Figure 2. Multi-media Sommerfeld problems

1.a) $B_\pm[\Phi] := \Phi|_\pm$

for the electrically polarized field $\underline{E} = (0,0,\Phi)^T$

b) $B_\pm[\Phi] := \frac{\partial \Phi}{\partial y} \Big|_{S_\pm}$

for the acoustic or magnetically polarized field $\underline{H} = (0,0,\Phi)^T$

c) $B_\pm[\Phi] := \frac{\partial \Phi}{\partial y} \pm ip \cdot \Phi|_{S_\pm}$

impedance boundary conditions

d) $B_+[\Phi] := \Phi|_{S_+}$ for the mixed generalized Sommerfeld problem

$B_-[\Phi] := \frac{\partial \Phi}{\partial y} \Big|_{S-}$

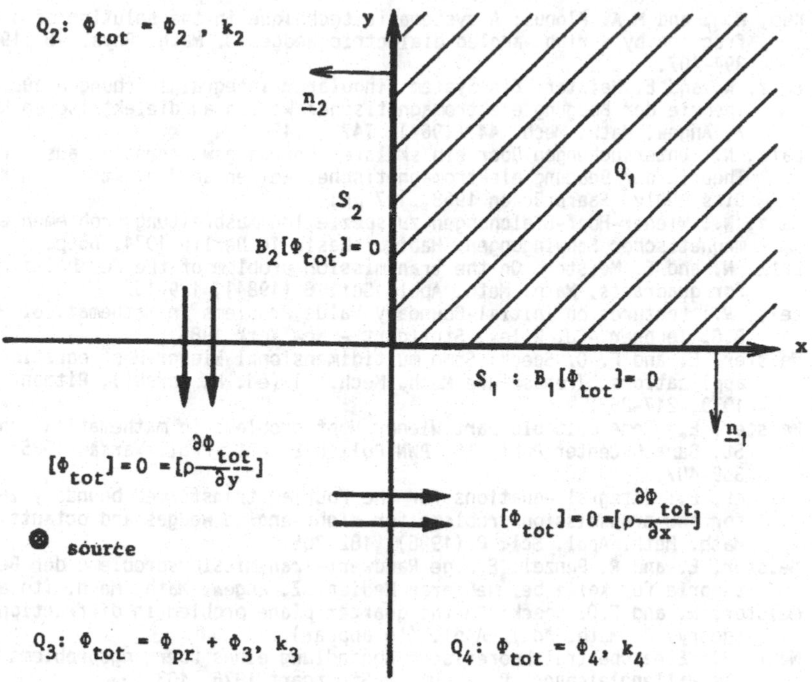

Figure 3. Multi-media wedge problem

In the meantime the author and R. Penzel derived the systems of integral equations for the unknown one-dimensional cosine-transforms of the normal derivatives on the semi-infinite boundaries of the four quadrants [16]. They have a similar form like the equations (41) for the pure transmission problem [15].

REFERENCES

[1] Berntsen, S.: Diffraction of an electric polarized wave by a dielectric wedge. SIAM J. Appl. Math. **43** (1983), 186-211.

[2] Costabel, M. and E. Stephan; A direct boundary integral equation method for transmission problems. J. Math. Anal. Appl. **106** (1985), 367-413.

[3] Durand, P.: Layer potentials and boundary value problems for the Helmholtz equation in the complement of a thin obstacle, Math. Meth. Apppl. Sci. **5** (1983), 389-421.

[4] Heins, A.E.: The Sommerfeld half-plane problem revisited II: The factorization of a matrix of analytic functions. Math. Meth. Appl. Sci. **5** (1983), 14-21.

[5] Kraut, E.A. and G.W. Lehmann: Diffraction of electromagnetic waves by a right- angle dielectric wedge. J. Math. Phys. **10** (1969), 1340-1348.

[6] Kress, R. and G.F. Roach: Transmission problems for the Helmholtz equation. J. Math. Phys. **19** (1978), 1422-1437.

[7] Kuo, N.H. and M.A. Plonus: A systematic technique in the solution of diffraction by a right-angled dielectric wedge. J. Math. Phys. **46** (1967), 394-407.

[8] Latz, N. and E. Meister: Ein System singulärer Integralgleichungen aus der Theorie der Beugung elektromagnetischer Wellen an dielektrischen Keilen. Z. Angew. Math. Mech. **44** (1964), T47 - T49.

[9] Latz, N.: Untersuchungen über ein skalares Übergangswertproblem aus der Theorie der Beugung elektromagnetischer Wellen an dielektrischen Keilen. Diss. Univ. Saarbrücken 1968, 117 pp.

[10] Latz, N.: Wiener-Hopf-Gleichungen zu speziellen Ausbreitungsproblemen elektromagnetischer Schwingungen. Habil. thesis TU Berlin 1974, 65pp.

[11] Latz, N. and E. Meister: On the transmission problem of the Helmholtz equation for quadrants. Math. Meth. Appl. Sci. **6** (1984), 129-157.

[12] Leis, R.: Lectures on Initial-Boundary Value Problems in Mathematical Physics. B.G. Teubner - J. Wiley, Stuttgart - New York 1986.

[13] Meister, E. and F.-O. Speck: Some multidimensional Wiener-Hopf equations with applications. Trends Pure Math. Mech. II (ed. H. Zorski), Pitman, London 1979, 217-262.

[14] Meister, E.: Some multiple-part Wiener-Hopf problems in mathematical physics. St. Banach Center Publ. **15**; PWN-Polish Scient. Publ. Warsaw 1985, 359-407.

[15] Meister, E.: Integral equations for the Fourier transformed boundary values for the transmission problems for right-angled wedges and octants. Math. Meth. Appl. Sci. **8** (1986), 182-205.

[16] Meister, E. and R. Penzel: Einige Randwert-Transmissionsprobleme der Beugungstheorie für Keile bei mehreren Medien. Z. angew. Math. Mech. (to appear).

[17] Meister, E. and F.O. Speck: On the quarter-plane problem in diffraction theory. J. Math. Anal. Appl. (to appear).

[18] Mohr, R.: Eine spektraltheoretische Behandlung eines Übergangsproblems für die Wellengleichung. Diss. Univ. Stuttgart 1976, 103 pp.

[19] Radlow, J.: Note on the diffraction at a corner. Arch. Rat. Mech. Anal. **19** (1965), 62-70.

[20] Rawlins, A.D.: Diffraction by a dielectric wedge. J. Inst. Math. Appl. **19** (1977), 231-279.

[21] Sommerfeld, A.: Mathematische Theorie der Diffraction. Math. Ann. **47** (1896), 317-374.

[22] Stephan, E.: Boundary integral equations for mixed boundary value problems, screen and transmission problem in $\mathbb{R}^3$. Habil. thesis = Prep. 848, FB Mathematik, TH Darmstadt, Sept. 1984, 136 pp.

[23] Werner, P.: Beugungsprobleme der mathematischen Akustik. Arch. Rat. Mech. Anal. **12** (1963), 155-184.

[24] Werner, P.: Ein Grenzübergang in der Theorie akustischer Wellenfelder. Arch. Rat. Mech. Anal. **33** (1969), 192-218.

RUNGE-KUTTA SCHEMES AND NUMERICAL INSTABILITIES: THE LOGISTIC EQUATION

Ronald E. Mickens
Department of Physics
Atlanta University
Atlanta, Georgia 30314

Abstract

We consider the Logistic differential equation and several finite difference schemes that can be used to numerically integrate the equation. We show that linear stability analysis about the fixed points of the finite difference schemes allow an understanding of how and when instabilities can occur. Some rules for modeling differential equations by finite difference schemes are presented.

Introduction

This paper summarizes certain recent results on instabilities which can arise when differential equations are modeled by difference equations for the purposes of numerical integration.

The genesis of these ideas start with the papers of Yamaguti and Ushiki [1] and Ushiki [2]. Using the Logistic differential equation,

$$\frac{dy}{dt} = y(1-y), \quad y_0 = y(0) = \text{given}, \tag{1}$$

they proved that the central difference scheme

$$\frac{y_{k+1} - y_{k-1}}{2h} = y_k(1-y_k), \tag{2}$$

gives rise to trajectories that globally depended very sensitively on the initial conditions, the time-mesh length h, and the precision of the computation used. They conclude that chaotic orbits exist for all positive time steps h.

We show that linearly stability analysis about the fixed points of a finite difference scheme allows an understanding of both how and when instabilities can occur. Such unstable schemes give rise to numerical solutions that are, in general, both qualitatively and quantitatively different from the actual solution of the differential equation. We will see in the following analysis that the time-mesh length h is restricted by numerical stability rather than by accuracy of the finite difference scheme.

The results of this paper have practial significance in the sense that a better understanding of the possible mechanisms which lead to instabilities should lead to more effective methods of constructing finite difference schemes which either suppress or eliminate them.

For reference, the exact solution to eq. (1) is

$$y(t) = y_0 e^t/[(1-y_0) + y_0 e^t]. \tag{3}$$

Note that eq. (1) has two fixed-points at $y^* = 0$ and $y^* = 1$; the first is unstable, while the second is stable.

Euler Scheme

The simplest Runge-Kutta scheme for eq. (1) is the Euler Scheme

$$\frac{y_{k+1} - y_k}{h} = y_k(1-y_k), \tag{4}$$

which has fixed-points at $y_k = 0$ and 1. Note that both the number and location of the fixed points are the same for eqs. (1) and (4). A linear stability analysis of eq. (4) about $y_k = 0$ gives, for $y_k = 0 + \varepsilon_k^{(0)}$, with neglect of higher-order terms in $\varepsilon_k^{(0)}$, the result

$$\frac{\varepsilon_{k+1}^{(0)} - \varepsilon_k^{(0)}}{h} = \varepsilon_k^{(0)}, \quad \varepsilon_k^{(0)} = \varepsilon_0^{(0)}(1+h)^k. \tag{5}$$

Thus, the fixed point at $y_k = 0$ is unstable. Likewise, considering the fixed-point $y_k = 1$, we obtain

$$\frac{\varepsilon_{k+1}^{(1)} - \varepsilon_k^{(1)}}{h} = -\varepsilon_k^{(1)} \quad \text{and} \quad \varepsilon_k^{(1)} = \varepsilon_0^{(1)}(1-h)^k \tag{6}$$

where the substitution $y_k = 1 + \varepsilon_k^{(1)}$ was made with neglect of higher-order terms. Analysis of the solution to eq. (6) gives the following results for various values of h:

(i) $0 < h < 1$: stable
(ii) $h = 1$: neutral stability,
(iii) $1 < h < 2$: oscillating, stable
(iv) $h \geq 2$: unstable.

We see that the type of stability of the fixed-point at $y_k = 1$ depends on the time-mesh length h. Therefore, depending on the magnitude of h, this difference scheme may have solution behaviors that do not correspond to the actual solution of eq. (1).

Now consider the difference scheme

$$\frac{y_{k+1} - y_k}{h} = y_k(1-y_{k+1}). \tag{7}$$

Again linear stability analysis gives

$$\varepsilon_k^{(0)} = \varepsilon_0^{(0)}(1+h)^k \quad \text{and} \quad \varepsilon_k^{(1)} = \varepsilon_0^{(1)}(1+h)^{-k}. \tag{8}$$

Consequently, for all $h > 0$, the fixed-points at $y_k = 0$ and 1 have exactly the same stability properties as the corresponding fixed-points of the differential equation. This means that for all $h > 0$, the qualitative behavior of y_k and $y(t)$ is the same; for h "small," i.e., $0 < h << 1$, we expect the quantitative behaviors to be very similar.

Central Difference Scheme

The central difference scheme for eq. (1) is [1,2]

$$\frac{y_{k+1} - y_{k-1}}{2h} = y_k(1-y_k). \tag{9}$$

This nonlinear difference equation has fixed-points at $y_k = 0$ and 1. Applying a linear stability analysis gives

$$\varepsilon_k^{(0)} = A(r_+)^k + B(r_-)^k, \quad r_\pm = h \pm \sqrt{1 + h^2}, \tag{10}$$

$$\varepsilon_k^{(1)} = C(s_+)^k + D(s_-)^k, \quad s_\pm = -h \pm \sqrt{1 + h^2}, \tag{11}$$

where (A,B,C,D) are arbitrary constants. Because $r_+ > 1$, we conclude that the fixed-point at $y_k = 0$ is unstable. Also, because $|s_-| > 1$, we conclude that the fixed-point at $y_k = 1$ is oscillating unstable. Thus, both fixed-points are unstable in constrast to the actual solution of eq. (1).

With this information on the nature of the instabilities of the fixed-points of eq. (9), we can understand the behavior of eqs. (2) or (9) as obtained by Yamaguti and Ushiki [1]. If we start with a y_0, such that $0 < y_0 < 1$, then the "trajectory," as determined by eq. (9), will be repelled from the unstable fixed-point at $y_k = 0$. As the "trajectory" approaches close to the fixed-point at $y_k = 1$, unstable oscillations will begin. The amplitude of these oscillations will build-up until y_k obtains values in the neighborhood of $y_k = 0$. At this time, the "trajectory" will be repelled and the cycle begins anew.

Runge-Kutta Method

A higher-order Runge-Kutta method for eq. (1), than the simple relation of eq. (4), is

$$y_{k+1} = \left[1 + \frac{(2+h)h}{2}\right]y_k - \left[\frac{(2+3h+h^2)h}{2}\right]y_k^2 + (1+h)h^2y_k^3 - (\frac{h^3}{2})y_k^4 . \tag{12}$$

This equation has four fixed points! They are located at

$$y_0^{(1)} = 0, \quad y_k^{(2)} = 1, \quad y_k^{(3,4)} = (\frac{1}{2})\left[\frac{2 + h \pm \sqrt{h^2-4}}{h}\right] \tag{13}$$

For $h \le 2$, the fixed points $y_k^{(3)}$ and $y_k^{(4)}$ are complex conjugates of each other; while for $h \ge 2$, all the fixed points are real. Thus, this difference equation model of eq. (1) has more fixed points than the original differential equation. However, both numerical evaluation of eq. (12) and linear stability analysis shows that instabilities will not arise until $h \ge 2$. Consequently, for small h, $0 < h << 1$, eq. (12) provides an excellent approximation to the actual solution of eq. (1). It should be clear that higher-order Runge-Kutta methods will lead to difference equations with larger numbers of fixed-points.

Discussion

Based upon the above results, as obtained from consideration of the Logistic equation, and from similar analyses of other nonlinear differential equations and their associated difference equation models, we can state the following general conclusions concerning numerical instabilities:

(a) It is clear that when the order of the difference scheme is greater than the order of the differential equation, then ghost solutions [1] or numerical instabilities will occur for any positive time-mesh size h.

(b) The time-mesh size h is restricted by numerical stability considerations rather than by the accuracy of the finite difference scheme. In general, there is a threshold value of $h = h_0$ such that for $h \ge h_0$ the difference scheme becomes unstable.

(c) In most cases, a linear stability analysis about the fixed-points of difference equations models of differential equations allow the prediction of the occurrence of numerical instabilities and the general global behavior of the solutions.

(d) Related to the ideas of this paper is the notion on "exact" finite difference schemes. These are schemes such that on the lattice points, $t_k = hk$, the solution to the differential and difference equations are <u>exactly</u> equal to each other regardless of the time-mesh size h. For systems of ordinary differential equations, one can show that such schemes always exist [3]. For example, the exact difference scheme for eq. (1) is

$$\frac{y_{k+1} - y_k}{e^h - 1} = y_k(1-y_{k+1}). \tag{14}$$

Thus, while we only have an existence theorem on exact difference schemes, in practice, a given finite difference model of a differential equation can be checked for its linear stability properties about its fixed points and those schemes that have the same stability properties as the corresponding differential equation should have solution behaviors that both qualitatively and quantitatively are close to that of the differential equation.

Finally, it should be noted that issues similar to those raised in this paper occur in the modeling of partial differential equations by difference equations [4,5].

Acknowledgement

This research was supported in part by grants from AOR, DOE, and NASA.

References

1. M. Yamaguti and S. Ushiki, "Chaos in Numerical Analysis of Ordinary Differential Equations," Physica 3D (1981), 618-626.

2. S. Ushiki, "Central Difference Scheme and Chaos," Physica 4D (1982), 407-424.

3. R. E. Mickens, "Difference Equations Models of Differential Equations Having Zero Local Truncation Errors," pps. 445-449, in Differential Equations, edited by I. W. Knowles and R. T. Lewis (North-Holland, Amsterdam, 1984).

4. A. R. Mitchell and J. C. Bruch, Jr., "A Numerical Study of Chaos in a Reaction-Diffusion Equation," Numerical Methods for Partial Differential Equations 1 (1985), 13-23.

5. J. M. Sanz-Serna, "Studies in Numerical Instability I. Why Do Leapfrog Schemes Go Unstable?," SIAM Journal of Scientific and Statistical Computing 6 (1985), 923-938.

STRUCTURE OF POSITIVE SOLUTIONS TO

$(-\Delta + V)u = 0$ IN R^n

Minoru Murata
Department of Mathematics,
Tokyo Metropolitan University
Fukazawa, Setagaya, Tokyo,
158 Japan.

1. Introduction.

Since Martin [4] gave a method for uniquely representing any positive harmonic function in a domain of R^3 by an integral on the minimal Martin boundary, several authors have extended his results to second order elliptic equations in a domain of R^n (see [9] and references in [7] and [9]). However, little attention has been paid to explicit construction of representation formula (see [2] and references in [7]). Meanwhile, Agmon [1] and Carmona-Simon [3] investigated exponential decay rate at infinity of positive L_2-eigenfunctions of Schrödinger equations. Murata [7] investigated the problem of explicitly constructing all positive solutions of a Schrödinger equation and determining their asymptotics at infinity. This problem has close connections with the existence and uniqueness of a solution for an elliptic equation on R^n and the asymptotic behavior as $t \to \infty$ of a Schrödinger semigroup (see [5],[6], and [8]). The purpose of this paper is to solve the problem for a class of potentials by applying method and results in [7].

2. Classification and minimal boundary.

Let V be a real-valued measurable function on R^n, $n \geqq 3$, of the form

$$V = V_0 + \lambda V_1 ,\ V_0(x) = a^2 \text{ for } x_n \geqq 0 \text{ and } V_0(x) = b^2 \text{ for } x_n < 0 , \tag{2.1}$$

where $0 \leqq a \leqq b$, $-\infty < \lambda < \infty$, and $V_1(x)(1+|x|)^{\rho - n/p}$ is not zero on a set of positive measure and belongs to $L_p(R^n)$ for some $p \geqq n$ and $\rho > 2$. We consider positive solutions of the equation

$$(-\Delta + V)u = 0 \text{ on } R^n . \tag{2.2}$$

We say that V is <u>subcritical</u> when there exists a positive Green's function of $-\Delta + V$ on R^n, <u>supercritical</u> when $(\nabla\phi,\nabla\phi)+(V\phi,\phi) < 0$ for some ϕ in $C_0^\infty(R^n)$, and <u>critical</u> otherwise. Recall that the equation (2.2) has a positive solution iff V is subcritical or critical.

<u>Theorem 2.1</u>. There exist positive continuous functions $\Lambda_\pm$ on $\{(a,b);\ 0 \leq a \leq b\}$ such that (i) if $V_{1,\pm}(x) = \max(\pm V_1(x),0)$ are both nonzero on a set of positive measure, then V is subcritical iff $-\Lambda_- < \lambda < \Lambda_+$, critical iff $\lambda = \pm\Lambda_\pm$, and supercritical iff $\lambda < -\Lambda_-$ or $\lambda > \Lambda_+$; and (ii) if $\pm V_1 \geq 0$, then V is subcritical iff $\pm[\lambda \pm \Lambda_\mp] > 0$, critical iff $\lambda = \mp\Lambda_\mp$, and supercritical iff $\pm[\lambda \pm \Lambda_\mp] < 0$.

<u>Proof</u>. Let $f_\pm(a,b,t)$, $t \geq 0$, be the greatest eigenvalue of a compact nonnegative operator $(V_{1,\pm})^{\frac{1}{2}}(-\Delta + V_0 \pm tV_{1,\pm})^{-1}(V_{1,\pm})^{\frac{1}{2}}$ on $L_2(R^n)$. Then the same argument as in the proof of Theorem 2.8 in [7] shows that $\Lambda_\pm$ is the unique solution of $\lambda f_\pm(a,b,\lambda) = 1$, and has the desired properties. Q.E.D.

We say that u is <u>normalized</u> when $u(0) = 1$. The set of all normalized minimal solutions equipped with the topology induced by L_∞-norms on compact sets of R^n is called <u>minimal boundary</u> for V.

<u>Theorem 2.2</u>. (i) If V is critical, then the minimal boundary for V consists of one point. (ii) If V is subcritical, then the minimal boundary is homeomorphic to the following space

$$\begin{aligned}
\sigma &= S^{n-1} \equiv \{\omega \in R^n;\ |\omega| = 1\} && \text{if } 0 < a = b, \\
&= \{\omega \in S^{n-1};\ \omega_n \geq 0 \text{ or } \omega_n \leq -(1-a^2/b^2)^{\frac{1}{2}}\} && \text{if } 0 < a < b, \\
&= \{\omega \in S^{n-1};\ |\omega_n| = 1\} && \text{if } 0 = a < b, \\
&= \{\infty\} && \text{if } 0 = a = b.
\end{aligned} \tag{2.3}$$

Furthermore, for any normalized positive solution u, there exists a unique nonnegative Borel measure μ on σ such that $\mu(\sigma) = 1$ and

$$u(x) = \int_\sigma P(x,\omega)\mu(d\omega), \tag{2.4}$$

where $P(\cdot,\omega)$ is the normalized minimal solution corresponding to ω in σ; conversely, for any such measure μ the right hand side of (2.4) represents a normalized positive solution.

<u>Proof</u>. The assertion (i) and the second half of (ii) are special cases of Theorems 2.2 and 2.3 in [7], respectively. The first half of

(ii) for the case $a=b$ is a corollary of Theorems 5.6 and 5.8 in [7]. By making use of it, we can show the first half of (ii) for the case $a<b$ along the line given in the proof of Theorem 7.1 in [7].

Q.E.D.

3. Asymptotic behaviors at infinity of minimal solutions.

In order to obtain asymptotic formulas as $x\to\infty$ for minimal solutions, we assume, for simplicity, the additional condition

$$V_1 \text{ in (2.1) has a compact support,} \tag{3.1}$$

although the following theorems hold without (3.1) at least for the case $a=b$.

For positive functions f and g, $f\sim g$ will mean in what follows that the limit

$$\lim_{r\to\infty} f(r\omega)/g(r\omega)$$

exists and is positive for any ω in a prescribed subset of S^{n-1}.

Theorem 3.1. Suppose that V is critical. Then the unique normalized positive solution u has the following asymptotics as $x\to\infty$ with $t=x_n/|x|$.

(i) When $0<a=b$, $u(x)\sim e^{-a|x|}|x|^{-(n-1)/2}$ for $-1\leqq t\leqq 1$.

(ii) When $0<a<b$,

$$\begin{aligned} u(x) &\sim e^{-a|x|}|x|^{-(n-1)/2} && \text{for } 0<t\leqq 1, \\ u(x) &\sim e^{-\psi(t)|x|}|x|^{-(n+1)/2} && \text{for } -\tau<t\leqq 0, \\ u(x) &\sim e^{-b|x|}|x|^{-n/2-1/4} && \text{for } t=-\tau, \\ u(x) &\sim e^{-b|x|}|x|^{-(n-1)/2} && \text{for } -1\leqq t\leqq -\tau, \end{aligned} \tag{3.2}$$

where $\tau=(1-a^2/b^2)^{\frac{1}{2}}$ and $\psi(t)=a(1-t^2)^{\frac{1}{2}}-(b^2-a^2)^{\frac{1}{2}}t$.

(iii) When $0=a<b$,

$$\begin{aligned} u(x) &\sim |x|^{1-n} && \text{for } t>0, \\ u(x) &\sim e^{bx_n}|x|^{-n} && \text{for } -1<t\leqq 0, \\ u(x) &\sim e^{bx_n}|x|^{-(n-1)/2} && \text{for } t=-1. \end{aligned} \tag{3.3}$$

(iv) When $0=a=b$, $u(x)\sim|x|^{2-n}$ for $-1\leqq t\leqq 1$.

<u>Proof</u>. Let $G(x,y)$ be the minimal Green's function for V_0. Then we see that

$$u(x) = - \int G(x,y)\, \lambda V_1\, u(y)\,dy\,. \tag{3.4}$$

Thus it suffices to show that for any y in a compact set $G(x,y)$ has the same asymptotics as that stated in the theorem. Elementary calculations show that

$$G(x,y) = \int_{R^{n-1}} e^{i(x'-y')\xi}\, H(|\xi|^2, x_n, y_n)\,d\xi/(2\pi)^{n-1}\,, \tag{3.5}$$

where H is defined as follows: With $A = (a^2 + |\xi|^2)^{\frac{1}{2}}$, $B = (b^2 + |\xi|^2)^{\frac{1}{2}}$, $s = x_n$, and $t = y_n$,

$$\begin{aligned} H &= (A-B)[2A(A+B)]^{-1}\exp[-A(s+t)] + (2A)^{-1}\exp[-A|s-t|] && \text{for } s,t \geqq 0,\\ H &= (A+B)^{-1}\exp[Bs-At] && \text{for } s \leqq 0 \leqq t\,,\\ H &= (A+B)^{-1}\exp[-As+Bt] && \text{for } t \leqq 0 \leqq s\,,\\ H &= (B-A)[2B(A+B)]^{-1}\exp[B(s+t)] + (2B)^{-1}\exp[-B|s-t|] && \text{for } s,t \leqq 0\,. \end{aligned} \tag{3.6}$$

Let us show that when $0 < a < b$ and $y = 0$, $G(x,y)$ has the same asymptotics as the right hand side of (3.2). We see that if $x_n \neq 0$, the stationary point of $F_c(\zeta) = -|x_n|(c^2 + \zeta^2)^{\frac{1}{2}} + ix'\zeta$, $c = a$ or b, with respect to ζ is equal to $icx'/|x|$ and nondegenerate, and that $A + B = [(b^2 + \xi\xi)^{\frac{1}{2}} - (A^2 + \xi\xi)^{\frac{1}{2}}]/(b^2 - a^2)$. When $t > 0$ or $t < -\tau$, we can deform the contour of the integral in (3.5) until the contour goes through the stationary point. Thus the stationary method yields the asymptotics. When $-\tau < t < 0$, we hit the singular point $\zeta = iax'/|x'|$ of the function $A + B$ before reaching the stationary point of $F_b(\zeta)$. We have that

$$\begin{aligned} F_b(\eta + iax'/|x'|) &= -(b^2-a^2)^{\frac{1}{2}}|x_n| - a|x'|\\ &+ ix'\eta[1-a|x_n|/(b^2-a^2)^{\frac{1}{2}}|x'|] - |\eta|^2[|x_n|/2(b^2-a^2)^{\frac{1}{2}}] + O(|\eta|^3) \end{aligned} \tag{3.7}$$

as $\eta \to 0$, and that $[a^2+(\eta+iax'/|x'|)^2]^{\frac{1}{2}} = [2iax'\eta/|x'|+\eta^2]^{\frac{1}{2}}$. Thus application of the stationary phase method to the integral (3.5) with respect to (n-2) - variables orthogonal to the variable $z = x'\eta/|x'|$ reduces the problem to the asymptotics of the integral on R^1 such as $\int e^{iz\nu}|z|^{\frac{1}{2}}dz$, $\nu \to \infty$. Hence we get the desired asymptotics for $-\tau<t<0$.

When $t=-\tau$, we hit both the stationary point and singular point. But we can obtain the desired asymptotics by the stationary method. Finally, when $t=0$, use the formula

$$\int (a^2+\xi^2)^{\frac{1}{2}} \exp(ix'\xi)\,d\xi = c(\partial/\partial x_n)^2 K_a(|x|)|_{x_n=0}\ , \tag{3.8}$$

where c is a constant and $K_a(|x-y|)$ is the Green's function for $-\Delta+a$. The formulas of G for the other cases can be shown similarly.

Theorem 3.2. If V is subcritical, then $P(x,\omega)$ in (2.4) has the following asymptotics as $x\to\infty$.

(i) When $0<a=b$, $P(x,\omega)\sim e^{ax\omega}$ on R^n.

(ii_+) When $0<a<b$ and $\omega_n \geqq 0$,

$$\begin{aligned} P(x,\omega) &\sim C_+e^{ax\omega}+C_-e^{ax\tilde{\omega}} && \text{for } x_n \geqq 0\ ,\\ P(x,\omega) &\sim 2a\omega_n \exp[ax'\omega' + x_n(b^2-a^2|\omega'|^2)^{\frac{1}{2}}] && \text{for } x_n \leqq 0\ , \end{aligned} \tag{3.9}$$

where $\tilde{\omega}=(\omega',-\omega_n)$ and $C_\pm = a\omega_n \pm (b^2-a^2|\omega'|^2)^{\frac{1}{2}}$.

(ii_-) When $0<a<b$ and $\omega_n \leqq -(1-a^2/b^2)^{\frac{1}{2}}$,

$$\begin{aligned} P(x,\omega) &\sim -2b\omega_n \exp[bx'\omega' - x_n(a^2-b^2|\omega'|^2)^{\frac{1}{2}}] && \text{for } x_n \geqq 0\ ,\\ P(x,\ ) &\sim D_+e^{bx\omega}+D_-e^{bx\tilde{\omega}} && \text{for } x_n \leqq 0\ , \end{aligned} \tag{3.10}$$

where $D_\pm = -b\omega_n \pm (a^2-b^2|\omega'|^2)^{\frac{1}{2}}$.

(iii) When $0=a<b$, with $\omega^\pm=(0,\dots,0,\pm1)$,

$$\begin{aligned} P(x,\omega^+) &\sim bx_n+1 && \text{for } x_n \geqq 0\ , \text{ and}\\ P(x,\omega^+) &\sim e^{bx_n} && \text{for } x_n \leqq 0\ , \end{aligned} \tag{3.11}$$

$$\begin{aligned} P(x,\omega^-) &\sim 2 && \text{for } x_n \geqq 0\ , \text{ and}\\ P(x,\omega^-) &\sim e^{-bx_n}+e^{bx_n} && \text{for } x_n \leqq 0\ . \end{aligned} \tag{3.12}$$

(iv) When $0=a=b$, $P(x,\infty)\sim 1$ on R^n.

Proof. (i) and (iv) follow from Theorem 5.6 and 5.8 in [7]. Let $a<b$. Then we obtain that the minimal solution $P_0(x,\omega)$ for V_0

equals a constant multiple of the right hand side of (3.9) ~ (3.12). We have

$$P(x,\omega) = Q(x,\omega)/Q(0,\omega)\ ,\ Q(x,\omega) = P_0(x,\omega) - \int G(x,z)\,\lambda V_1(z)\,P_0(z,\omega)dz\ . \tag{3.13}$$

It is easily seen that $G(x,z) = o(P_0(x,\omega))$ as $x \to \infty$ for any fixed z and ω, which yields the theorem.

Q.E.D.

References

[1] S. Agmon, Lectures on exponential decay of solutions of second-order elliptic equations, Mathematical Notes 29, Princeton University Press, 1982.

[2] S. Agmon, On positive solutions of elliptic equations with periodic coefficients in R^n, Proc. Internat. Conf. on Differential Equations, I. W. Knowles and R. T. Lewis ed., North Holland, 1984, 7-17.

[3] R. Carmona and B. Simon, Pointwise bounds on eigenfunctions and wave packets in N-body quantum systems, Commun. Math. Phys, 80 (1981), 59-98.

[4] R. S. Martin, Minimal positive harmonic functions, Trans. Amer. Math. Soc. 49 (1941), 137-172.

[5] M. Murata, Positive solutions and large time behaviors of Schrödinger semigroups, J. Funct. Anal. 56 (1984), 300-310.

[6] M. Murata, Isomorphism theorems for elliptic operators in R^n, Commun. in Partial Differential Equations, 9 (1984), 1085-1105.

[7] M. Murata, Structure of positive solutions to $(-\Delta + V)u = 0$ in R^n, to appear in Duke Math. J. 53 (1986).

[8] B. Simon, Schrödinger semi-groups, Bull. Amer. Math. Soc. 7 (1982), 447-526.

[9] M. G. Sur, The Martin boundary for a linear elliptic second-order operator, Amer. Math. Translations Ser. 2, 56, 19-35.

An Extension of Lavine's Formula for Time-Delay

Shu Nakamura
Department of Pure and Applied Sciences
University of Tokyo
3-8-1, Komaba, Meguro-ku, Tokyo 153, Japan

__Abstract__. Lavine's formula for time-delay ([1]) is extended to Schrödinger equations in n-dimensional space.

We consider Schrödinger operators:

$$H = H_0 + V(x)\cdot \quad ; \quad H_0 = -\Delta$$

on $\underline{H} = L^2(R^n)$ and we suppose that the potential V satisfies

__Assumption (V)__. There exists a constant $\varepsilon > 0$ such that the multiplication operator by $V(x)$ is a compact operator from $H^2(R^n)$ to $L^{2,2+\varepsilon}(R^n)$.

Then it is well-known that wave operators defined by

$$W_{\pm} = \underset{t \to \pm\infty}{\text{strong-limit}} \ \exp(itH)\exp(-itH_0)$$

exist and the scattering operator S defined by

$$S = W_+^* W_-$$

is unitary.

For R, X_R is a multiplication operator defined as follows:

$$X_R = X_R(x)\cdot; \quad X_R(x) = X(|x|/R);$$
$$0 \le X(x) \le 1 \quad (x \in R); \quad X \in C_0^\infty(R);$$
$$X_1(x) = 0 \quad \text{if} \quad |x| \ge 2, \quad = 1 \quad \text{if} \quad |x| \le 1.$$

For φ, $\psi \in \underline{H}$, we set $\varphi(t)$, $\psi(t)$, $\varphi_0(t)$ and $\psi_0(t)$ as

$$\varphi(t) = \exp(-itH)\, W_- \varphi \ ; \quad \varphi_0(t) = \exp(-itH_0)\, \varphi$$
$$\psi(t) = \exp(-itH)\, W_- \psi \ ; \quad \psi_0(t) = \exp(-itH_0)\, \psi.$$

Note that $\varphi(t)$ is the solution of the Schrödinger equation such that

$$\|\varphi(t) - \varphi_0(t)\| \longrightarrow 0 \quad (t \to -\infty).$$

We set $\underline{D}_0$ as

$$\underline{D}_0 = \{\varphi \in \underline{H}: E_{H_0}(\Omega)\varphi = \varphi \quad \text{for some} \quad \Omega \subset (0,\infty) \; ; \quad \Omega : \text{compact} \; ; \quad \Omega \cap \sigma_{pp}(H) : \text{empty} \; \}.$$

Then the time-delay operator is defined by the following equation:

$$(\varphi, T_R\psi) = \int_{-\infty}^{\infty} \left[\varphi(t), X_R\psi(t)\right] dt - \int_{-\infty}^{\infty} \left[\varphi_0(t), X_R\psi(t)\right] dt$$

for $\varphi \in \underline{H}$ and $\psi \in \underline{D}_0$. Since X_R is H_0-smooth and is locally H-smooth, $(\varphi, T_R\psi)$ exists for such φ and ψ. Hence, by the Riesz representation theorem, T_R is a well-defined operator on $\underline{D}_0$. T_R represents approximately the difference of the sojourn time of an interacting particle in the ball of radius R, and that of a free particle.

We define $\underline{D} \subset \underline{H}$ as follows: for arbitrarily small fixed constant δ,

$$\underline{D} = \{\varphi \in \underline{D}_0 : \varphi \in L^{2,1+\delta}(R^n), \quad \varphi \in D(\langle A\rangle^{1+\delta})\}$$

where

$$A = \frac{1}{2i}\left(x \cdot \frac{\partial}{\partial x} + \frac{\partial}{\partial x} \cdot x\right) : \quad \text{dilation generator},$$

and $\langle x\rangle = (1 + |x|^2)^{1/2}$.

Our result may be stated as follows.

<u>Theorem</u>. Suppose Assumption (V), $\varphi, \psi \in \underline{D}$ and $S\varphi, S\psi \in \underline{D}$, then

$$\lim_{R\to\infty} (\varphi, T_R H_0 \psi) = \int_{-\infty}^{\infty} \left[\varphi(t), \{V + \tfrac{i}{2}[A,V]\}\, \psi(t)\right] dt. \tag{1}$$

Of course, Theorem implies that the limit of the L.H.S. and the integral of the R.H.S. exist, and are equal.

<u>Remark 1</u>. The integrand in the R.H.S. of (1) is well-defined since $\varphi(T), \psi(t) \in H^2(R^n)$ and

$$V + \frac{i}{2}[A,V] : H^2(R^n) \longrightarrow H^{-2}(R^n) = \left[H^2(R^n)\right]^*.$$

<u>Remark 2</u>. In many cases, the set of ψ's such that $\varphi \in \underline{D}$ and $S\varphi \in \underline{D}$ is dense in $\underline{H}$. For example, (i) if V satisfies

$$V: H^2(R^n) \to L^{2,4+\epsilon}(R^n)$$

then $\varphi \in \underline{D}$ implies $S\varphi \in \underline{D}$ (Jensen [2]); (ii) if V satisfies

$$|(\frac{\partial}{\partial x})^\alpha V(x)| \leq C_\alpha \langle x\rangle^{-1-|\alpha|-\epsilon}$$

for any α, then $\hat{\varphi} \in C_0^\infty(R^n)$ and $\varphi \in \underline{D}$ imply $S\varphi \in \underline{D}$ and $S\varphi \in C_0^\infty(R^n)$ (cf. Isozaki-Kitada [4]).

Remark 3.. For one-dimensional Hamiltonian, Lavine ([1]) proved that if V satisfies

$$|V(x)| + |x\, V'(x)| \leq C\langle x\rangle^{-1-\epsilon}$$

and if $\varphi, \psi \in D(T_R H_0)$, then the formula (1) holds. On the other hand, Jensen ([2]) proved an analogous formula for n-dimensional Hamiltonian where X_R was replaced by

$$X_R = F(|A| < R).$$

Remark 4. In terms of S-matrix $\{S(\lambda)\}$, the Eisenbud-Wigner time-delay operator is defined by

$$T = \{-i\, S(\lambda)^* \frac{d}{d\lambda} S(\lambda)\}.$$

Jensen ([3]) showed

$$(\varphi, T\, H_0\, \psi) = \int_{-\infty}^{\infty} \Big[\varphi(t),\ \{V + \frac{i}{2}[A,V]\}\, \psi(t)\Big]\, dt$$

under certain conditions. Hence, as quadratic forms on $\underline{D}\, S^{-1}\underline{D}$,

$$T = \lim_{R\to\infty} T_R.$$

Sketch of the proof. We construct a psuedo-differential operator A_R such that

$$X_R\, H_0 \sim \frac{i}{2}[H_0, A_R]$$

and such that the symbol of A_R converges to $x\cdot\xi$ + constant as $R \to \infty$. Then we see

$$X_R\, H \sim \frac{i}{2}[H, A_R] + X_R\, V + \frac{i}{2}[A_R, V]$$

and

$$\begin{aligned} \Big[\varphi(t),\ X_R\, H\, \psi(t)\Big] &- \Big[\varphi_0(t),\ H_0\, X_R\, \psi(t)\Big] \\ &\sim \frac{1}{2}\frac{d}{dt}\{\Big[\varphi(t),\ A_R\, \psi(t)\Big] - \Big[\varphi_0(t),\ A_R\psi_0(t)\Big]\} \\ &\quad + \Big[\varphi(t),\ \{X_R\, V + \frac{i}{2}\{A_R, V]\}\, \psi(t)\Big]. \end{aligned} \tag{2}$$

Integrating (2), we have

$$\int_{-\infty}^{\infty} \{ \left[\varphi(t), X_R H \psi(t)\right] - \left[\varphi_0(t), X_R H_0 \psi_0(t)\right]\} dt$$

$$\sim \frac{1}{2} \lim_{\substack{t\to\infty \\ s\to-\infty}} \{\left[\varphi(t), X_R \psi(t)\right] - \left[\varphi_0(t), X_R \psi_0(t)\right]\}\Big|_s^t$$

$$+ \int_{-\infty}^{\infty} \left[\varphi(t), \{X_R V + \frac{i}{2} [A,V]\} \psi(t)\right] dt.$$

We then show that the former term vanishes, that the latter converges to

$$\int_{-\infty}^{\infty} \left[\varphi(t), \{V + \frac{i}{2} [A,V]\} \psi(t)\right] dt$$

as $R \to \infty$, completing the proof. Detailed proof will appear in a forthcoming paper.

REFERENCES

[1] R. Lavine, Commutators and local decay, in "Scattering Theory in Mathematical Physics", J. A. LaVita and J. P. Marchand (eds.), 141-156, D. Reidal, 1974.

[2] A. Jensen, Time-delay in potential scattering theory, some geometric results, Commun. Math. Phys. 82, 435-456 (1981).

[3] A. Jensen, A stationary proof of Lavine's formula for time-delay Lett. Math. Phys. 7, 137-143 (1983).

[4] H. Isozaki and H. Kitada, Scattering matrices for two-body Schrödinger operators, preprint.

SOME OPEN QUESTIONS IN MULTI-DIMENSIONAL INVERSE PROBLEMS

Roger G. Newton
Physics Department
Indiana University
Bloomington, IN 47405

For the Schrödinger equation in $\mathbb{R}^3$,

$$(\Delta + k^2)f = Vf$$

with a potential $V(x)$ that in some appropriate sense decreases to zero as $|x| \to \infty$, there are two kinds of inverse problems that have been attacked with some success: The inverse scattering problem and the inverse spectral problem. (I am restricting myself here to three dimensions for the sake of definiteness. The needed proofs have been given for two dimensions by Margaret Cheney [1-3]. In more than three dimensions the methods work formally, but no proofs exist.) In the first one wants to find $V(x)$ from a knowledge of the large-$|x|$ asymptotics of certain physically interesting "scattering solutions"; in the second, $V(x)$ is to be found from a spectral measure. The second is, in a certain sense, an analogue of Marc Kac's question "Can you hear the shape of a drum?" but here the spectrum consists not just of eigenvalues. The most direct methods of solving the inverse scattering problem are generalizations of the method first introduced by Marchenko [4] in one dimension, and those of solving the inverse spectral problem are generalizations of the method first given by Gel'fand and Levitan [5]. There is, however, also a bridge between the asymptotic, i.e., scattering, information and the spectral measure function, just as there is in one dimension, so that the inverse scattering problem can also be solved by Gel'fand-Levitan procedures. This bridge depends on the solution of a generalized Riemann-Hilbert problem, as does the Marchenko method [6].

At the present time there are four classes of solution-methods known for these problems: The first is a nonlinear equation [7-10] that utilizes backward scattering information only; the second is based on a Green's function for the Schrödinger equation introduced by Faddeev [11,12] and developed first by him [13] and also by myself [14,15]; the third is also based on Faddeev's Green's function, though arrived at and developed differently recently by Nachman and Ablowitz [16] by means of the $\bar{\partial}$-method; the fourth was found and developed over the last seven years in a series of my papers [17-20,6].

All of these solution methods contain certain open problems. I will tell you about three of them and, in conformance with the main emphasis of this conference, I will formulate two of them eventually in the language of differential equations, though they originally arise in the context of integral equations.

The first problem, which I mention only for the sake of completeness, is not strictly open, but its solutions are not as satisfactory as one would wish. It is clear a priori, from a counting of variables, that in contrast to one dimension, there have to be severe restrictions on the class of functions that are admissible as scattering amplitudes or spectral functions. (For the first-mentioned method this problem does not arise. That method, on the other hand, is known to work only for "weak" potentials). In my formulation these restrictions manifest themselves when the integral equation that leads to the potential is solved and the solution must have a surprising property which I usually refer to as the miracle. In the other formulations there are analogous conditions. They all have in common that they are extremely inconvenient to use; they do not allow us to put our hands directly on a class of admissible functions. This is particularly bothersome for any possible use of inverse methods for generating multi-dimensional analogs of the nonlinear evolution equations with soliton solutions in one dimension. Such a use was certainly very much in my mind when I developed my own procedure. Unfortunately, the miracle has so far completely blocked such a development.

I will now turn to the other two open problems and treat them in more detail. Let me start with Faddeev's Green's function [11,12,13, 14]. It is defined by

$$G_q(\lambda;x,y) = \frac{1}{(2\pi)^3}\int d^3p\, \frac{e^{i(p+iq)\cdot(x-y)}}{\lambda-(p+iq)^2}$$

where $p,q,x,y \in \mathbb{R}^3$ and $\lambda \in \mathbb{C}$. It is the kernel of the resolvent of $-\Delta$ on the Hilbert space $L^2(\mathbb{R}^3)$ with the weight function $e^{2q\cdot x}$, where $-\Delta$ is not self-adjoint. Of course, it satisfies the equation

$$(\Delta + \lambda)G_q(\lambda;x,y) = \delta(x-y).$$

We are ultimately primarily interested in the limit as $q \to 0$, but that limit will depend on the direction of q. So we will define

$q = \gamma\beta$, where γ is a unit vector, $\gamma \in \mathbb{R}^3$, $|\gamma| = 1$, $\beta \in \mathbb{R}$, $\lambda = |k + iq|^2$, $k \in \mathbb{R}^3$, $G_q(\lambda,x,y) = \tilde{G}_\gamma(|k+iq|^2; x-y)$, and decompose vectors in $\mathbb{R}^3$ along γ and orthogonal to it: $x = x_\perp + \gamma x_0$, $x_\perp \cdot \gamma = 0$, $k = k_\perp + \gamma k_0$, $k_\perp \cdot \gamma = 0$, etc., and we write $\mu^2 = k_\perp^2$, $s = k_0 + i\beta$. We find by a shift of variables of integration

$$\tilde{G}_\gamma(|k+iq|^2,x) = \frac{1}{(2\pi)^3} \int d^3p \, \frac{e^{ip\cdot x + isx_0}}{\mu^2 - p^2 - 2sp_0} = G_\gamma(\mu^2,s;x).$$

Considered as a function of the variable s, for fixed $\mu \in \mathbb{R}$ and $x \in \mathbb{R}^3$ and γ, $G_\gamma(\mu,s,x)$ is an analytic function, holomorphic in the open upper and lower half planes, $\mathbb{C}^+$ and $\mathbb{C}^-$, but not on the real axis. The limits of G_γ as s approaches $\mathbb{R}$ from $\mathbb{C}^+$ and $\mathbb{C}^-$ are in general not equal. Indeed one readily sees that for complex s

$$G_\gamma(\mu^2,s,x) = G_{-\gamma}(\mu^2,-s,x).$$

If we designate the limits as $\mathrm{Im}s \downarrow 0$ and $\mathrm{Im}s \uparrow 0$ by G_γ^+ and G_γ^- then we find

$$G_\gamma^+(\mu^2,s,x) - G_\gamma^-(\mu^2,s,x)$$
$$= i(2\pi)^{-2} \int d^3p \, \delta(\mu^2 + s^2 - |p|^2) e^{ip\cdot x} [\theta(\gamma\cdot p - s) - \theta(s - \gamma\cdot p)]$$

where

$$\theta(s) = \begin{cases} 1, & s > 0. \\ 0, & s < 0. \end{cases}$$

The function $\tilde{G}_\gamma^+(k,x) = G_\gamma^+(\mu^2,s,x)$, $k \in \mathbb{R}^3$, $k = k_\perp + \gamma s$, $|k_\perp|^2 = \mu^2$, $k_\perp \cdot \gamma = 0$, is the limit of an analytic function of the vector k as the imaginary part of k approaches zero along the direction $-\gamma$.

We now define a solution $\tilde{\psi}_\gamma^+(k,x)$ of the Schrödinger equation by the integral equation

$$\tilde{\psi}_\gamma^+(k,x) = e^{ik\cdot x} + \int d^3y \, \tilde{G}_\gamma^+(k,x-y) \, V(y) \, \tilde{\psi}_\gamma^+(k,y). \qquad (1)$$

which resembles the Lippmann-Schwinger equation, except that the usual Green's function is replaced by Faddeev's. Under similar general conditions on $V(x)$ as for the Lippmann-Schwinger equation one can convert this integral equation into a Fredholm equation by multiplying it by $|V(x)|^{1/2} \exp(-isx_0)$. Its kernel then is Hilbert-Schmidt for $\mathrm{Im}s \geq 0$ and its inhomogeneous term is in L^2 [13,14].

One may now develop the properties of $\tilde{\psi}_\gamma$, obtain its asymptotics as $|x| \to \infty$, find a relation between $\tilde{\psi}_\gamma^+$ and $\tilde{\psi}_\gamma^-$ [defined as in (1), except that $\tilde{G}_\gamma^+$ is replaced by $\tilde{G}_\gamma^-$], etc. analogously to the usual scattering solutions of the Schrödinger equation, and one can pose and solve the two kinds of inverse problems mentioned earlier [13,14]. The vector γ is an artificial construct that is convenient for the procedure, but it must ultimately disappear from the solution of the inverse problem.

Now the entire method depends on the solvability of the integral equation (1), both for real and complex s, and all directions of γ. This solvability is guaranteed unless the modified Fredholm determinant of (1) (i.e., modified for L^2-kernels) vanishes, which can happen in $\mathbb{C}^+$ only at isolated points. Such complex exceptional points would be somewhat bothersome for the solution of the inverse problem, but they could be handled, if their number is finite. However, if (1) has <u>real</u> exceptional points, no solution method is known to work. Furthermore, if such real exceptional points exist then the number of complex exceptional points may be infinite because they may accumulate on the real axis. (It has been proved that they cannot accumulate at infinity. Analytic Fredholm theory rules out accumulation points in the open $\mathbb{C}^+$.) It is therefore important for all inverse methods based on the Faddeev Green's function to be sure that (1) has no real exceptional points. Nobody has yet been able to prove this.

Suppose that (for a given value of μ) $s = s_0$ is an exceptional point of (1), with $\operatorname{Im} s_0 = 0$. Then there exists a nontrivial solution u_γ of the homogeneous form of (1). The asymptotic form of u_γ for large $|x|$ is then entirely determined by that of G^+_γ or G^-_γ. I shall concentrate on G^+_γ, since $G^-_\gamma(s) = G^+_{-\gamma}(-s)$.

We define the following two cones [13,14]:

$$C^+_\gamma(\mu,s) = \{x \mid x_0 > 0,\ \ |x_\perp|/x_0 < \mu/|s|\}$$
$$C^-_\gamma(\mu,s) = \{x \mid x_0 < 0,\ \ |x_\perp|/|x_0| < \mu/|s|\}$$

The first points in the direction of γ, the second in the direction of $-\gamma$. They have an opening angle α such that $\tan\alpha = \mu/|s|$. Then as $|x| \to \infty$ the asymptotic form of G^+_γ is known to be

$$G^+_\gamma(\mu^2,s,x) = \begin{cases} o(|x|^{-1}), & x \in C^+_\gamma(\mu,s) \\ -(2\pi|x|)^{-1}\cos k|x| + o(|x|^{-1}), & x \in C^-_\gamma(\mu,s) \\ -(4\pi|x|)^{-1}e^{ik|x|} + o(|x|^{-1}), & x \notin C^+_\gamma \cup C^-_\gamma,\ s > 0 \\ -(4\pi|x|)^{-1}e^{-ik|x|} + o(|x|^{-1}), & x \notin C^+_\gamma \cup C^-_\gamma,\ s < 0 \end{cases}$$

Thus we find for the asymptotics of a solution of the homogeneous form of eq. (1) that in the backward cone C^-_γ there are spherical waves both moving inward and outward; outside both cones there are either only incoming or only outgoing spherical waves, depending on whether s is positive or negative; and in the forward cone C_γ^+ the solution decays faster than $|x|^{-1}$. Thus, for positive s, there are incoming spherical waves in the backward cone C_γ^- only; in the forward cone C_γ^+ there are no waves at all (to order $|x|^{-1}$), and outside the forward cone there are outgoing spherical waves. (If $s < 0$ incoming and outgoing are interchanged.)

The question is: Are there reasons why such solutions of the Schrödinger equation cannot exist? Are there classes of potentials for which they can be ruled out? Of course, from a physical point of view, if they can exist, it would also be good to know what their physical significance would be.

I now turn to the inverse spectral problem for the Schrödinger equation. In the case of one dimension one defines a solution $\phi(k,x)$ by boundary conditions at a finite point and one proves the so-called completeness relation (here $k \in \mathbb{R}$)

$$\int \big(\phi(k,x),\, d\rho(k)\; \phi(k,y)\big) = \delta(x - y)$$

i.e., for $f \in L^2$

$$f(x) = \int \big(\phi(k,x),\, dg(k)\big)$$

$$dg(k) = d\rho(k) \int dy\phi(k,y)\; f(y)$$

The inner product here is on a two-dimensional Hilbert space. (Two-dimensional, because the multiplicity of the essential spectrum is two.) The crucial property of $\phi(k,x)$ that leads to the Gel'fand-Levitan integral equation for the reconstruction of $V(x)$ from the spectral measure ρ is that, for each fixed x, $\phi(k,x)$ may be regarded as an entire analytic function of k that is of exponential type $|x|$. This property implies by the Paley-Wiener theorem that the support of its Fourier transform with respect to k is confined to $(-|x|,|x|)$, and this fact leads to the Gel'fand-Levitan equation. The entire analytic nature of $\phi(k,x)$, and its exponential boundedness, are direct consequences of its being defined by boundary conditions at a finite point.

This method has been generalized to three dimensions. The multiplicity of the essential spectrum here is infinite, and it can be parametrized by a point θ on the unit sphere (or two polar angles). However, the Schrödinger equation is now a partial differential equation, and an analog of the solution ϕ that in one dimension is defined by boundary conditions at a point and therefore is an entire analytic function of k, is not readily at hand. While one may, of course, define a solution by boundary conditions on a surface, a proof that the so defined solution, as a function of k, is entire analytic of exponential type $|x|$, is lacking. However, one may proceed as follows.

Let $\psi(k,\theta,x)$, $k \in \mathbb{R}$, $\theta \in \mathbb{R}^3$, $|\theta| = 1$, $x \in \mathbb{R}^3$, be the "scattering solution" of the Schrödinger equation, i.e., the solution defined by a radiation condition, which asymptotically for large $|x|$ goes as

$$\psi(k,\theta,x) = e^{ik\theta\cdot x} + |x|^{-1}e^{ik|x|}A(k,\hat{x},\theta) + o(|x|^{-1})$$

where $\hat{x} = x/|x|$. The S matrix is defined by

$$S(k,\theta,\theta') = \delta(\theta,\theta') - \frac{k}{2\pi i} A(k,\theta,\theta').$$

We now pose the following generalized Riemann-Hilbert problem [6,17-20]. (For simplicity of formulation I will assume that the Schrödinger equation has no eigenvalues. My statements can be generalized to the case with eigenvalues.)

Find an operator family $J(k)$ from some appropriate Banach space $L^p(S^2)$ to $L^p(S^2)$, with kernel $J(k,\theta,\theta')$, such that $J(k)$ is an analytic function of k holomorphic in $\mathbb{C}^+$, such that

$$\lim_{|k|\to\infty} J(k) = \mathbb{1}$$

(where $\mathbb{1}$ is the unit operator) and such that its limit on the real axis satisfies the equation

$$J(-k,\theta,\theta') = \int_{S^2} d\theta'' \, J(k,-\theta,\theta'')S(k,-\theta'',\theta').$$

If this generalized Riemann-Hilbert problem has a solution then that solution can be found by solving a linear integral equation with compact kernel. While certain necessary conditions (such as a Levinson theorem for the Fredholm determinant of S) for the solvability of this generalized Riemann-Hilbert problem are known, there is no a priori assurance that it always has a solution.

In terms of the solution $\psi(k,\theta,x)$ of the Schrödinger equation and the solution $J(k)$ of the Riemann-Hilbert problem, we may now define a new solution $\phi(k,\theta,x)$ by [6,17-20]

$$\phi(k,\theta,x) = \int d\theta' \, J(k,\theta,\theta') \, (k,\theta',x).$$

Because of the properties of J and ψ it turns out that $\phi(k,\theta,x)$ is an entire analytic function of k, of exponential type $|x|$. As a result, the support of the Fourier transform of $[\phi - \exp(ik\theta\cdot x)]$ with respect to k is confined to the interval $(-|x|,|x|)$, by the Paley-Wiener theorem:

$$\phi(k,\theta,x) = e^{ik\theta\cdot x} - \int_{-|x|}^{|x|} dt e^{ikt} w(\theta,t,x).$$

At this point one may proceed in perfect analogy with the one dimensional case and derive a generalized Gel'fand-Levitan equation for w in terms of the spectral weight function for ϕ and thereby solve the inverse spectral problem.

While all of this is, formally, quite parallel to the situation in one dimension, there is one crucial difference. Whereas in one dimension the solution ϕ of the Schrödinger equation is well defined by a boundary condition, in three dimensions it is defined rather indirectly via the solution of a Riemann-Hilbert problem, and it may consequently not even exist. It would therefore be desirable to get at ϕ in some other manner.

Let us use the Fourier representation of $\phi(k,\theta,x)-\exp(ik\theta\cdot x)$ in terms of $w(\theta,t,x)$ in the Schrödinger equation. We then find that for $\theta\cdot x < t < |x|$ and for $-|x| < t < \theta\cdot x$ the function w must satisfy the plasma wave equation

$$(\Delta - \frac{\partial^2}{\partial t^2})w = Vw$$

and for $t = \pm|x|$ it must vanish:

$$w(\theta,\pm|x|,) = 0.$$

Furthermore, on the hyperplane $t = \theta\cdot x$ the normal derivative of w must have a discontinuity [21]:

$$\theta\cdot\nabla[w(\theta,\theta\cdot x+,x) - w(\theta,\theta\cdot x,x)] = \frac{1}{2}V(x).$$

Of course, the derivatives of w need not necessarily exist. For large classes of potentials V the function w, as a function of t, is known to be in L^2 but it is not known to be differentiable with respect to t or x. Hence the plasma wave equation may hold only in a weak sense.

The solution ϕ has the symmetry property $\phi(-k,\theta,x) = \phi(k,-\theta,x)$. As a result we must have $w(\theta,-t,x) = w(-\theta,t,x)$. Therefore w can be formed from a function $u(\theta,t,x)$,

$$w(\theta,t,x) = \begin{cases} u(\theta,t,x) & \text{for } -|x| < t < \theta\cdot x \\ u(-\theta,-t,x) & \text{for } \theta\cdot x < t < |x| \end{cases}$$

and u is subject to the Goursat problem

$$(\Delta - \frac{\partial^2}{\partial t^2})u = Vu, \quad -|x| < t < \theta\cdot x,$$

$$u(\theta,-|x|,x) = 0, \quad \theta\cdot\nabla u(\theta,\theta\cdot x,x) = -\frac{1}{4}V(x).$$

The question is: For what class of functions V(x) *does this problem have a solution?* Its solution would lead to a solution ϕ of the Schrödinger equation with the appropriate analyticity and asymptotics for the Gel'fand-Levitan procedure in the inverse spectral problem.

REFERENCES

1. Cheney, M. "Inverse scattering in dimension two," J. Math. Phys. 25(1984), 94-107.
2. _____, "Two-dimensional scattering: The number of bound states from scattering data," J. Math. Phys. 25(1984), 1449-1455.
3. _____, "Two-dimensional inverse scattering: Compactness of the generalized Marchenko operator," J. Math. Phys. 26(1985), 743-752.
4. Marchenko, V. A. "The construction of the potential energy from the phases of the scattering waves," Dokl. Skad. Nauk SSSR 104 (1955), 695-698 [Math. Rev. 17, p. 740].
5. Gel'fand, I. M. and Levitan, B. M. "On the determination of a differential equation from its spectral function," Isvest. Akad. Nauk. SSSR 15(1951), 309 [Am. Math. Soc. Transl. 1, 253-304].
6. Newton, R. G. "The Marchenko and Gel'fand-Levitan methods in the inverse scattering problem in one and three dimensions," Conference on Inverse Scattering: Theory and Application, J. B. Bednar et al., editors, SIAM, Philadelphia 1983, pp. 1-74.
7. Moses, H. E. "Calculation of the scattering potential from reflection coefficients," Phys. Rev. 102(1956), 559-567.
8. Prosser, R. T. "Formal solutions of inverse scattering problems," J. Math. Phys. 10(1969), 1819-1822.
9. _____, "Formal solutions of inverse scattering problems, II," J. Math. Phys. 17(1976), 1775-1779.
10. _____, "Formal solutions of inverse scattering problems, III, J. Math. Phys. 21(1980), 2648-2653.
11. Faddeev, L. D. "Increasing solutions of the Schroedinger equation," Dokl. Akad. Nauk. SSSR 165(1965), 514-517 [Soviet Phys.-Doklady, 10, 1033-1035 (1966)].
12. _____, "Factorization of the S-matrix for the multi-dimensional Schrödinger operator," Dokl. Akad. Mauk SSSR 167(1966), 69-72 [Soviet Phys.-Doklady 11, 209-211 (1966)].
13. _____, "Inverse problem of quantum scattering theory," Itogi Naukii Tekh. Sov. Prob. Mat. 3(1974), 93-180 [J. of Soviet Math. 5(1976), 334-396].
14. Newton, R. G. "The Gel'fand-Levitan Method in the inverse scattering problem," in Scattering Theory in Mathematical Physics, Lavita, J. A. and Marchand, J.-P., editors, Reidel Publ. Co., Dordrecht (1974), pp. 193-235.
15. _____, "A Feedeev-Marchenko method for inverse scattering in three dimensions," Inverse Prob. 1(1985), 127-132.
16. Nachman, A. I. and Ablowitz, M. J. "A multidimensional inverse scattering method," Studies in Appl. Math. 71(1984), 243-250.
17. Newton, R. G. "Inverse scattering II. Three dimensions," J. Math. Phys. 21(1980), 1698-1715; 22(1981), 631; 23(1982), 693.
18. _____, "Inverse scattering III. Three dimensions, continued," J. Math. Phys. 22(1981), 2191-2200; 23(1982), 693.

19. Newton, R. G. "Inverse scattering IV. Three dimensions: Generalized Marchenko construction with bound states," J. Math. Phys. 23(1982), 594-604.
20. ______, "On a generalized Hilbert problem," J. Math. Phys. 23(1982), 2257-2265.
21. The integrated version given on p. 49 of ref. 6 is incorrect.

Radially Symmetric Solutions of a Monge-Ampère Equation Arising in a Reflector Mapping Problem*

by

V. OLIKER**

and

P. WALTMAN

0. Introduction

In several papers Brickel, Marder, Norris, and Westcott [11], [9], [1], [8], [10] considered the following problem in geometric optics. In 3-dimensional Euclidean space R^3 fix a point O and suppose that a nonisotropic light source is positioned there. Let S be a unit sphere centered at O and Ω some fixed domain on S. Denote by F a smooth surface which projects radially in a one-to-one fashion onto Ω. The surface F is assumed to have a perfect reflection property, that is, when a light beam is issued from O and reaches F, it is reflected from it and no loss of energy occurs. Denote by y the unit vector in the direction of the reflected ray and identify it with the point on S, by translating y parallel to itself to the point O. Thus we get a mapping γ of $\Omega \subset S$ into some domain $\omega \subset S$.

The inverse problem posed in [11], roughly speaking, consists in recovering the surface F from the following data: the sets Ω, ω, and the density f(y), $y \in \omega$, of the light intensity in the reflected direction. Below we refer to this as the *reflector mapping problem.*

---

*Research supported by AFOSR Grants 83-0319 and 84-0285.

**Part of this work was done while the first author was visiting Institute for Mathematics and its Applications, University of Minnesota, 514 Vincent Hall, 206 Church Street, SE, Minneapolis, Minnesota 55455

In analytic formulation the problem reduces to solving an equation of Monge-Ampère type under certain nonlinear boundary conditions. The equations of Monge-Ampère type constitute a wide class of strongly nonlinear equations and only some special cases of such equations have been studied [4], [2]. The particular equation corresponding to the above problem has not been studied before and the known methods are not readily applicable.

In [9] the authors presented results of their numerical studies of the elliptic case in the reflector mapping problem. Because of the complexity of the problem there are no rigorous results on convergence of numerical approximations and, in fact, there are no results on existence of solutions.

In [1] the problem was reformulated in the language of complex analysis, and later Marder [8] proved a uniqueness theorem in this setting for the case where elliptic solutions of the problem are considered. An ODE approach without boundary conditions is given in Keller [7].

In this paper the equation for the reflector mapping problem is rederived (section one) using only basic concepts from differential geometry and the postulated law of reflection. Naturally the end result is the same as in [11] but the method of derivation here is within the framework of differential geometry and accessible to mathematical physicists.

The hypothesis of radial symmetry is introduced in section two and it leads to a boundary value problem for an ordinary differential equation. It is shown that this problem is overdetermined, that nontrivial solutions can exist only for special values of the parameters, (corresponding to the size of the aperture, a natural design parameter), the solutions fall naturally into two classes, and uniqueness is possible within each class up to a constant multiple. Thus, in principle, radially symmetric solutions can be constructed explicitly whenever all of the parameters are appropriate. The form of the solution actually shows how the aperture, for example, must be chosen in order to have a nontrivial radially symmetric solution. One anticipates being able to use these solutions for perturbation arguments.

This contribution should be viewed, in the tradition of mathematical physics, of presenting "known" results in rigorous and familar terms, and showing their place in a wider context of mathematical problems.

1. Derivation.

1.1. In the Euclidean space R^3 we fix a Cartesian coordinate system with origin at a point O and a sphere S of unit radius with center O. Let F be a surface of class C^∞ which projects radially from O in a one-to-one fashion onto some closed domain $\bar{\Omega} \subset S$. Denote the position vector of F by r. Then $r = \rho m\colon \bar{\Omega} \to F \subset R^3$, where

$m \in \bar{\Omega}$, $\rho = |r|$, and it is assumed that $\rho > 0$ in $\bar{\Omega}$.

On F fix a unit normal vector field n such that $0 < \langle n,m\rangle$, where $\langle , \rangle$ denotes the inner product in R^3. Such choice of n is possible because of the following observations.

Let u^1, u^2 be some local coordinates on S. Put $\partial_i = \partial/\partial u^i$, $i = 1,2$. The first fundamental form of S, that is, the metric on S induced from R^3 will be denoted by e, $e = e_{ij}du^i du^j$, where $e_{ij} = \langle \partial_i m, \partial_j m\rangle$. Here and throughout the paper the summation convention over repeated lower and upper indices is in effect. The first fundamental form of F is denoted by g, $g_{ij} = \langle \partial_i r, \partial_j r\rangle$, and by direct calculation we find:

$$\partial_i r = (\partial_i \rho)m + \rho\partial_i m,$$

$$g_{ij} = \partial_i\rho\partial_j\rho + \rho^2 e_{ij},$$

$$\det(g_{ij}) = \rho^2[|\nabla\rho|^2 + \rho^2]\det(e_{ij}),$$

where

$$|\nabla\rho|^2 = e^{ij}\partial_i\rho\partial_j\rho, \quad (e^{ij}) = (e_{ij})^{-1}.$$

Clearly, $\det(g_{ij}) > 0$ in $\bar{\Omega}$, since it was assumed that $\rho > 0$ in $\bar{\Omega}$. Hence, F is an immersed surface. In fact, it is an embedded surface, because it also projects in a one-to-one fashion onto $\bar{\Omega}$.

The normal vector field (of unit length) on F is given by

$$n = \frac{-\nabla\rho + \rho m}{\sqrt{|\nabla\rho|^2 + \rho^2}}$$

where $\nabla\rho = e^{ij}(\partial_i\rho)\partial_j m$. Since ρ is a smooth positive function in $\bar{\Omega}$, $n \neq 0$. Furthermore, $\langle n,m\rangle = \rho/\sqrt{|\nabla\rho|^2 + \rho^2} > 0$.

1.2. Define a map $\gamma: \bar{\Omega} \to S$ via F by the formula

$$y = m - 2\langle m,n\rangle n. \tag{1.1}$$

This is indeed a map into S, since $y^2 = m^2 = 1$. Obviously, $\langle y,n\rangle = -\langle m,n\rangle$, and for that reason F is called a "reflector".

Consider now the solid body defined by the map

$r(m,s) = sr(m)$: $[0,1] \times \bar{\Omega} \to R^3$, $m \in \bar{\Omega}$. The volume element of this body is

$dV = (1/3)\sqrt{\det(g_{ij})}\, h\, du_1 du_2$, where $h = \langle r,n\rangle = \rho^2/\sqrt{|\nabla\rho|^2 + \rho^2}$.

The function h has a simple geometric meaning; it represents the oriented distance from the origin 0 to the tangent plane of F at the point with normal n. It is called the *support function* of F.

The *normalized volume element* is defined as $dV_{nor} = dV/\rho^3$. In optics this quantity is called the elementary solid angle [10].

Suppose the surface F is such that the map γ is a diffeomorphism of $\bar{\Omega}$ onto some domain $\bar{\omega} \subset S$. In the following we denote by Ω and ω the corresponding domains without boundaries. Denote by $d\sigma$ the area element of S, that is,

$d\sigma = \sqrt{\det(e_{ij})}\, du_1 du_2$.

Let $m \in \Omega$ and $y = \gamma(m) \in \omega$. The *relative light intensity* in the reflected direction y is defined as the quotient

$$f(\gamma(m)) = \frac{dV_{nor}(m)}{(1/3)d\sigma(\gamma(m))}. \tag{1.2}$$

1.3. In the study of the problem of constructing the reflector from the "target" domain $\bar{\omega}$ and relative light intensity f(y), $y \in \omega$, it is convenient to look for the surface F parametrized by points in $\bar{\omega}$, that is, the position vector r of F is sought as $r = \rho m = \rho\gamma^{-1}(y)$, provided one can establish that γ is indeed a diffeomorphism. We will now derive such an expression for r. Previously, such an expression was derived in a different way in [11].

Since in (1.2) the numerator and denominator do not depend on the particular parametrization of F and S, we can write it in the form

$$\underline{f}(y) = \frac{dV_{nor(\gamma^{-1}(y))}}{(1/3)d\sigma(y)} = \frac{\sqrt{\det(e_{ij}(\gamma^{-1}(y)))}}{\sqrt{\det(e_{ij}(y))}}, \; y \in \bar{\omega}. \tag{1.3}$$

It will be always assumed that $\bar{\omega} \neq S$ and, therefore, without loss of generality we may suppose that (u^1,u^2) are coordinates on S such that $\bar{\omega}$ lies in one coordinate patch. We consider the points of ω as a function $y(u^1,u^2)$, and assume that in these parameters y is an analytic function.

Thus, let the position vector r of F be given as $\rho(y)\gamma^{-1}(y)$. We seek to find an explicit expression for it. Put $N = -2hn$. Multiplying (1.1) by ρ we get

$$\rho y = r + N. \tag{1.4}$$

By differentiating (1.4) we get

$$(\partial_i\rho)y + \rho\partial_i y = \partial_i r + \partial_i N, \; i = 1,2. \tag{1.5}$$

Multiplying (1.5) by y and taking into account $\langle\partial_i y, y\rangle = 0$, we get

$$\partial_i\rho = \langle\partial_i r, y\rangle + \langle\partial_i N, y\rangle, \; i = 1,2. \tag{1.6}$$

Because of (1.4) $\langle \partial_i r, y\rangle = \langle \partial_i r, r\rangle/\rho = \partial_i \rho$, and (1.6) implies that

$$\langle \partial_i N, y\rangle = 0,\ i = 1,2. \tag{1.7}$$

Put $p = \langle N,y\rangle$. Then, $\partial_i p = \langle N, \partial_i y\rangle$.

The vectors y, $\partial_1 y$, $\partial_2 y$ form a basis in R^3 and therefore the system of equations

$$\begin{aligned} \langle N,y\rangle &= p, \\ \langle N,\partial_i y\rangle &= \partial_i p,\ i = 1,2, \end{aligned} \tag{1.8}$$

can be solved for N at any point $y \in \bar{\omega}$. The solution is given by the formula

$$N = \nabla p + py, \tag{1.9}$$

where $\nabla p = e^{ij}\partial_i p \partial_j y$, which is the gradient of p in the metric of S.

Now from (1.4) and (1.9) we find

$$-r = \nabla p + (p-\rho)y, \tag{1.10}$$

and

$$\rho = \frac{|\nabla p|^2 + p^2}{2p}. \tag{1.11}$$

1.3.1. **Remark**. From the definition of p we obtain with the use of (1.1)

$$p = \langle N,y\rangle = -\ 2h\langle n,y\rangle = 2h\langle m,n\rangle.$$

Since $\langle m,n\rangle > 0$ and $h = \langle r,n\rangle = \rho\langle m,n\rangle > 0$ everywhere on F, p must be positive in $\bar{\omega}$. Therefore, (1.11) makes sense.

1.3.2. **Remark**. The vector function N defined above is a normal vector field on the surface F. If we translate this vector field in R^3 so that the initial point of all these vectors coincides with origin O then $N: \bar{\omega} \to R$ defines a surface T. Its tangent vectors are given by $\partial_1 N$ and $\partial_2 N$ and, in view of (1.7), the unit normal vector field on T is given by the vector field y. The function p is then the support function of T and the equations (1.8) describe T as an envelope of the family of its tangent planes given by the equations $\langle X,y\rangle = p$, where $X \in R^3$.

Note also, that for any C^1 function p in $\bar{\omega}$ the expressions (1.9) and (1.10) define surfaces in R^3, though these surfaces do not have to be immersions. A necessary and sufficient condition for N to be an immersion can be found, for example, in [6]. The question of when (1.10) is an immersion can be answered in a similar but not entirely satisfactory fashion.

1.3.3. **Remark**. The map γ^{-1} can be explicitly represented in terms of the function p. Namely, using (1.1), (1.4) and (1.9), we find

$$m = \frac{r}{\rho} = y - \frac{N}{\rho} = y - \frac{2p(\nabla p + py)}{|\nabla p|^2 + p^2}. \tag{1.12}$$

1.4. Introduce the operations of covariant differentiation in the metric e, namely, put

$$\nabla_{ij} = \partial_{ij} - \Gamma^k_{ij}\partial_k,$$

where $\partial_{ij} = \partial^2/\partial u_i \partial u_j$, $i,j = 1,2$, and Γ^k_{ij}, $i,j,k = 1,2$, are the Christoffel symbols of the second kind of the metric e. In the following we use the standard rules of covariant differentiation (see, for example, [5]).

Differentiating (1.10) covariantly, we obtain

$$-\partial_i r = [\nabla_{ij}p + (p - \rho)e_{ij}]e^{jk}\partial_k y - (\partial_i \rho)y, \tag{1.13}$$

Using (1.11), we find that

$$(\partial_i \rho)p = [\nabla_{ij}p + (p - \rho)e_{ij}]e^{jk}\partial_k p. \tag{1.14}$$

Substituting this in (1.13), we obtain

$$-p\partial_i r = [\nabla_{ij}p + (p - \rho)e_{ij}]e^{jk}[p\partial_k y - (\partial_k p)y], \quad i = 1,2.$$

The coefficients of the first fundamental form of F are $g_{ij} = \langle \partial_i r, \partial_j r\rangle$, $i,j = 1,2$. Hence,

$$p^2 g_{ij} = q_{ik}q_{j\ell}e^{ks}e^{\ell t}[p^2 e_{st} + (\partial_s p)(\partial_t p)], \quad i,j = 1,2, \tag{1.15}$$

where $q_{ik} = \nabla_{ik}p + (p - \rho)e_{ik}$.

Since $e_{ij}(\gamma^{-1}(y)) = \langle m_i, m_j\rangle = (g_{ij} - \partial_i\rho\partial_j\rho)/\rho^2$ we find, using (1.14) and (1.15),

$$e_{ij}(\gamma^{-1}(y)) = \frac{q_{ik}q_{j\ell}e^{k\ell}}{\rho^2},$$

and finally from (1.3),

$$M(p) = \frac{4p^2 \det[\nabla_{ij}p + (p - \frac{|\nabla p|^2 + p^2}{2p})e_{ij}]}{(|\nabla p|^2 + p^2)^2 \det(e_{ij})} = f. \tag{1.16}$$

1.5. Now we are in a position to formulate analytically the ***inverse reflector problem***. In that we follow essentially [10], p. 34.

Let ω, Ω be two circular domains on S, that is, both ω and Ω are intersections with S of some cones of revolution with vertices at the center of S. It is assumed that $\bar{\omega} \cap \bar{\Omega} = \phi$. Further, let f be a continuous positive function in $\bar{\omega}$. One needs to find a function p in $\bar{\omega}$ such that $p \in C^2(\omega) \cap C^1(\bar{\omega})$, $p > 0$ in $\bar{\omega}$, and satisfies the equation

$$M(p) = f \text{ in } \omega. \tag{1.17}$$

In addition, the vector function r, constructed from p, as in (1.10), must satisfy the relation

$$\langle m, \xi\rangle = \langle \frac{r}{\rho}, \xi\rangle = \phi, \text{ on } \partial\omega, \tag{1.18}$$

where ξ is a constant unit vector whose end point is the center of the domain Ω, and ϕ a given constant between zero and one.

1.6. **Remark**. The operator M, as well as the quotient r/ρ, are invariant relative to the transformation $p \to cp$, where $c = \text{const} \neq 0$.

1.7. In the following we restrict our attention to the class of functions p on which the operator M is *uniformly elliptic* (cf. [4], p. 346). The case when M is considered on functions on which it is hyperbolic is treated by entirely different methods. The elliptic and hyperbolic solutions of (1.15) are physically interpreted in [10].

1.8. Continuing the discussion in 1.4.1 we observe that if we denote by $d\mu$ the area element of S in domain Ω then it follows from (1.3) and (1.16) that

$$\text{area of } \Omega = \int_\Omega d\mu = \int_\omega M(p)\,\sqrt{\det(e_{ij})}\,du^1du^2 = \int_\omega f d\sigma.$$

Thus, a *necessary* condition for solvability of (1.17) is that

$$\text{area of } \Omega = \int_\omega f d\sigma. \tag{1.19}$$

From the point of view of geometric optics this condition represents the conservation law of energy, and it is called *energy compatibility* equation; see [10], p. 35.

2. Radially Symmetric Case in the Problem (1.17), (1.18).

2.1. In section 1.5 it was assumed that ω is a circular domain. If the function $f(y)$ is allowed to depend only on $\text{dist}(y,y_0)$, where y_0 is the center of ω, then it is natural to expect that solutions of (1.17), (1.18) will be also radially symmetric (r.s.), that is, $p = p(\text{dist}(y,y_0))$.

2.2. On the sphere S we introduce spherical coordinates α,β where $-\frac{\pi}{2} \le \alpha \le \frac{\pi}{2}$, $0 \le \beta \le 2\pi$. The metric of S in this coordinates has the form $e_{ij}du^idu^j = (d\alpha)^2 + \cos^2\alpha(d\beta)^2$. Assume that p is radially symmetric, that is, $p \equiv p(\alpha)$. Then from (1.16) it follows that

$$M(p) = \frac{4p^2\left[\ddot{p} + p - \frac{\dot{p}^2 + p^2}{2p}\right]\left[-\frac{1}{2}\dot{p}\sin 2\alpha + \left(p - \frac{\dot{p}^2 + p^2}{2p}\right)\cos^2\alpha\right]}{(\dot{p}^2 + p^2)^2\cos^2\alpha}, \tag{2.1}$$

where $\dot{p} = dp/d\alpha$, $\ddot{p} = d^2p/d\alpha^2$.

Assume further that the North Pole $(\pi/2,\beta)$ coincides with the center y_0 of the domain ω, and the angle α varies in the interval $[\bar{\alpha},\pi/2]$, where $\bar{\alpha} \in (0,\pi/2)$, that is, $\omega = \{(\bar{\alpha},\beta) | \alpha \leq \alpha \leq \pi/2,\ 0 \leq \beta \leq 2\pi\}$.

The requirement that p be a r.s. solution of class $C^1(\omega)$ then implies

$$\dot{p}\Big|_{\alpha = \pi/2} = 0. \tag{2.2}$$

2.3. Proposition. Let, as before, ω be a circular domain on S and p a positive of class $C^2(\omega) \cap C^1(\bar{\omega})$ r.s. solution of (2.1) satisfying (2.2). Then the domain $\bar{\Omega} = m(\bar{\omega})$, where m is defined by (1.12), is also circular.

2.3.1. Proof. First of all observe that the center y_o of ω is mapped into the South Pole $y_1 = (-\pi/2,\beta)$. This follows from (2.2) and (1.12).

Further, let $y \in \bar{\omega}$. Then $\langle y,y_1\rangle = -\langle y,y_o\rangle = -\cos\kappa$, $\langle \dot{y},y_1\rangle = -\sin\kappa$, where κ is the angle between y_o and y, and $\dot{y} = \partial y/\partial\alpha$.

From this and (1.12) we have

$$\langle m,y_1\rangle = \langle y - \frac{2p\,(\dot{p}\dot{y} + py)}{\dot{p}^2 + p^2},\ y_1\rangle =$$

$$= -\left(1 - \frac{2p^2}{\dot{p}^2 + p^2}\right)\cos\kappa + \frac{2p\dot{p}}{\dot{p}^2 + p^2}\sin\kappa,$$

which proves the proposition.

2.3.2. Remark. Similarly one shows that the surface $r(\bar{\omega})$ defined by (1.10) is radially symmetric with respect to the axis of direction y_1.

2.4. Now we can set up the boundary condition (1.18). From the proof of the Proposition 2.3 it follows that y_1 is the center of Ω, that is, $y_1 = \xi$. Further $\langle y,\xi\rangle\Big|_{\alpha = \bar{\alpha}} = -\sin\bar{\alpha}$ and $\langle \dot{y},\xi\rangle\Big|_{\alpha = \bar{\alpha}} = -\cos\alpha$. Consequently (1.18) assumes the form

$$(\phi + \sin\bar{\alpha})\dot{p}^2 - 2p\dot{p}\cos\bar{\alpha} + (\phi - \sin\bar{\alpha})p^2\Big|_{\alpha = \bar{\alpha}} = 0. \tag{2.3}$$

3. Radially Symmetric Solutions of the Reflector Mapping Problem.

3.1. Equation (2.1) may be written as

$$\frac{[2p\ddot{p} + p^2 - \dot{p}^2][-2p\dot{p}\tan\alpha + p^2 - \dot{p}^2]}{(\dot{p}^2 + p^2)^2} = f(\alpha),\ \alpha \in (\bar{\alpha}, \tfrac{\pi}{2}). \tag{3.1}$$

At this point we assume that $f \in C([\bar{\alpha},\pi/2))$ *and positive.* We are interested only in solutions that are of class C^1 and positive in $\bar{\omega}$ (cf. 1.3.1 and (2.2), (2.3)). Hence, we may set $\tan\theta(\alpha) = \dot{p}(\alpha)/p(\alpha)$. Then $\theta(\alpha)$ satisfies (using 3.1))

$$(2\dot{\theta} + 1)(\cos 2\theta - \sin 2\theta\tan\alpha) = f.$$

The further substitution $D(\alpha) = 2\theta(\alpha) + \alpha$ yields that $D(\alpha)$ satisfies

$$\dot{D}(\alpha)\cos D(\alpha) = f(\alpha)\cos\alpha, \quad \alpha \in (\bar{\alpha},\pi/2). \tag{3.2}$$

This equation may be formally solved (compare [7], Table I) to yield

$$\sin D(\alpha) = \sin D(\bar{\alpha}) + S(\bar{\alpha},\alpha), \tag{3.3}$$

where

$$S(\alpha_1,\alpha_2) = \int_{\alpha_1}^{\alpha_2} f(\tau)\cos\tau d\tau, \quad \alpha_1,\alpha_2 \in [\bar{\alpha},\tfrac{\pi}{2}],$$

and the initial condition is placed at $\alpha = \bar{\alpha}$. The reader should keep in mind that the equation for θ is singular where $\cos D(\alpha) = 0$. In particular, because of (2.2) we need to have

$$\theta(\pi/2) = 0, \tag{3.4}$$

and the equation is singular at $\alpha = \pi/2$ (alternatively, $\cos D(\alpha) = 0$ at $\alpha = \pi/2$).

3.2. Now we examine the singularity set of (α,θ) where the coefficients in (3.2) vanish. It is more convenient to use the set

$$\Delta \equiv \{(\alpha,\theta) \mid \tan^2\theta + 2\tan\theta\tan\alpha - 1 = 0\}.$$

The solutions of this equation in the rectangle $Q = [0,\pi/2) \times (-\pi/2,\pi/2)$ are given by the segments

$$\theta = -\frac{\pi}{4} - \frac{\alpha}{2}; \; \theta = \frac{\pi}{4} - \frac{\alpha}{2}.$$

Consequently the rectangle Q splits into three regions:

$$Q_1 = \left\{(\alpha,\theta) \middle| 0 \le \alpha < \frac{\pi}{2}, -\frac{\pi}{2} < \theta < -\frac{\pi}{4} - \frac{\alpha}{2}\right\},$$

$$Q_2 = \left\{(\alpha,\theta) \middle| 0 \le \alpha < \frac{\pi}{2}, -\frac{\pi}{4} - \frac{\alpha}{2} < \theta < \frac{\pi}{4} - \frac{\alpha}{2}\right\},$$

$$Q_3 = \left\{(\alpha,\theta) \middle| 0 \le \alpha < \frac{\pi}{2}, \frac{\pi}{4} - \frac{\alpha}{2} < \theta < \frac{\pi}{2}\right\}.$$

Within each of these regions the standard theorems [3] from the theory of ordinary differential equations guarantee existence and uniqueness of a solution of an initial value problem for (3.2).

3.3. The solution of (3.2), given by (3.3), is defined only for α such that the right hand side has absolute value less than or equal to one. Assume now that the integral $S(\bar{\alpha},\alpha)$ exists (perhaps as an improper integral) on the interval $[\bar{\alpha},\pi/2]$, and suppose the following relation

$$1 - \sin[2\theta(\bar{\alpha}) + \bar{\alpha}] = S(\bar{\alpha},\pi/2) \tag{3.5}$$

is satisfied. Then, since $S(\bar{\alpha},\alpha)$ on $[\bar{\alpha},\pi/2]$ is a monotone function of α (the integrand is positive),

$$|\sin D(\alpha)| = |\sin D(\bar{\alpha}) + S(\bar{\alpha},\alpha)| < 1, \ \alpha \in [\bar{\alpha},\tfrac{\pi}{2}),$$

and (3.3) defines a solution on $[\bar{\alpha},\pi/2]$. Furthermore, $\sin D(\frac{\pi}{2}) = \cos 2\theta(\frac{\pi}{2}) = 1$, which implies $\theta(\pi/2) = 0$, and (3.4) is satisfied. Also, $|\theta(\alpha)| < \pi/2$ when $\alpha \in [\bar{\alpha},\pi/2]$.

(3.5) is a compatibility condition to be satisfied if $\theta(\bar{\alpha})$ is to be chosen so as to satisfy (3.4). Since $(\pi/2,0) \notin \bar{Q}_1$, it can not support a solution satisfying (3.4). From this it follows that we have two solutions of (3.2), (3.4), - one in Q_2 and one in Q_3. The choice is determined by $\theta(\alpha)$.

3.4. We now need to consider the boundary condition (2.3). We rewrite it in the form

$$(\phi + \sin\bar{\alpha}) \tan^2\theta(\bar{\alpha}) - 2\tan\theta(\bar{\alpha})\cos\bar{\alpha} + (\phi - \sin\bar{\alpha}) = 0. \tag{3.6}$$

From this

$$\tan\theta(\bar{\alpha}) = \frac{\cos\alpha \pm \sqrt{1 - \phi^2}}{\phi + \sin\bar{\alpha}}.$$

Using proposition 2.3 we find the equation of $\partial\Omega$, namely,
$\partial\Omega = \{\tilde{\alpha} = -\pi/2 + \arccos\phi, \ 0 \le \beta \le 2\pi\}$. Then

$$\tan\theta(\bar{\alpha}) = \frac{\cos\bar{\alpha} \pm \sin(\frac{\pi}{2} + \tilde{\alpha})}{\cos(\frac{\pi}{2} + \tilde{\alpha}) + \sin\bar{\alpha}}$$

$$= \begin{cases} \cot \dfrac{\bar{\alpha} - \tilde{\alpha}}{2} & \text{if the sign "+" is taken,} \\ -\tan \dfrac{\bar{\alpha} + \tilde{\alpha}}{2} & \text{if the sign "-" is taken.} \end{cases}$$

Since $\bar{\alpha} \in (0,\pi/2)$, $\tilde{\alpha} \in (-\pi/2,0)$, and $\theta(\bar{\alpha})$ must be the ordinate of a point in $Q_2 \cup Q_3$, we conclude with the use of (3.5) that for a fixed $\tilde{\alpha}$ the value of $\bar{\alpha}$ should be chosen so that

$$1 + \sin\tilde{\alpha} = S(\bar{\alpha},\pi/2).$$

(It is easy to see that this condition is the condition (1.19) in radially symmetric case.) Then for $\theta(\bar{\alpha})$ we should take either

$$\theta(\bar{\alpha}) = -\frac{\bar{\alpha}}{2} - \frac{\tilde{\alpha}}{2} \text{ for } Q_2 \tag{3.7}$$

or

$$\theta(\bar{\alpha}) = \frac{\pi}{2} - \frac{\bar{\alpha}}{2} + \frac{\tilde{\alpha}}{2} \text{ for } Q_3. \tag{3.8}$$

We summarize the results of this section in the following theorem.

3.5. **Theorem**. Suppose $f(\alpha)$ is positive and continuous on $[0,\pi/2)$ and that the integral $S(0,\pi/2)$ exists. Further, let $\tilde{\alpha}$ be any number in the interval $(-\pi/2,0)$ and $\bar{\alpha}$ be the solution of the equation

$$1 + \sin\tilde{\alpha} = S(\bar{\alpha},\pi/2), \quad \bar{\alpha} \in (0,\pi/2).$$

Then for each choice of $\theta(\bar{\alpha})$ according to (3.7) or (3.8) there exists a unique solution of class $C^2([\bar{\alpha},\pi/2)) \cap C([\bar{\alpha},\pi/2])$ of the boundary value problem (3.2), (3.4), and (3.6).

3.5.1. **Remark**. An obvious sufficient condition on f for the equation $1 + \sin\tilde{\alpha} = S(\bar{\alpha},\pi/2)$ to have a solution $\bar{\alpha} \in (0,\pi/2)$ is $S(0,\pi/2) \geq 1$.

3.5.2. **Remark**. Observe that construction of the map $\gamma^{-1} = m(y)$ given by (1.10) requires the knowledge of $\dot{p}/p$ only, and, therefore, m can be constructed from $\tan\theta$. Furthermore, the two solutions correspond to maps for reflectors without real caustics or with two real caustics respectively. Details on the physics of this situation can be found in [10], Ch. 3.

3.6. Our ultimate objective in this section is to construct a function p satisfying (2.1), (2.2), and (2.3) and then show that a reflector surface indeed can be described in terms of the function $\bar{p}(\alpha,\beta) = p(\alpha)$, $\alpha \in [\bar{\alpha},\pi/2]$, $\beta \in [0,2\pi]$. For that it is necessary to show that p can be recovered from θ, that $p > 0$ in $\bar{\omega} = \{(\alpha,\beta)|\bar{\alpha} \leq \alpha \leq \pi/2,\ 0 \leq \beta \leq 2\pi\}$, $\bar{p} \in C^2(\omega) \cap C^1(\bar{\omega})$, and that the map m is a diffeomorphism. We begin by establishing further smoothness of θ.

3.7. **Lemma**. Suppose the conditions of the Theorem 3.5 are satisfied and $\theta(\alpha)$, $\alpha \in [\bar{\alpha}, \pi/2]$, is the function given there. Let $f \in C([\bar{\alpha},\pi/2])$. Then $\theta \in C^1([\bar{\alpha},\pi/2])$. Furthermore, $2\dot{\theta}(\pi/2) = -1 \pm \sqrt{f(\pi/2)}$.

3.7.1. **Proof**. From standard results on ordinary differential equations it follows that $\theta \in C^1[\bar{\alpha},\pi/2)$. Thus, we need to consider only the point $\alpha = \pi/2$. Since $\theta(\bar{\alpha})$ is given by (3.7) or (3.8), $\sin D(\bar{\alpha}) = 1 - S(\bar{\alpha},\pi/2)$. Then from (3.3)

$$\sin D(\alpha) = 1 - S(\alpha,\tfrac{\pi}{2}).$$

By the mean value theorem

$$S(\alpha,\frac{\pi}{2}) = f(\tau(\alpha))(1-\sin\alpha),$$

where $\tau(\alpha) \in [\bar{\alpha},\pi/2]$. Then evidently

$$\sin D(\alpha) = 1 - f(\tau(\alpha))(1 - \sin\alpha).$$

In a left hand side neighborhood of $\pi/2$ we have

$$1 - \sin\alpha = \frac{1}{2}(\frac{\pi}{2} - \alpha)^2 + o(A),$$

$$\sin D(\alpha) = 1 - \frac{1}{2} f(\tau(\alpha))(\frac{\pi}{2} - \alpha)^2 + o(A),$$

and

$$1 - \sin^2 D(\alpha) = f(\tau(\alpha)(\frac{\pi}{2} - \alpha)^2 + o(A),$$

where

$$A = O(|\frac{\pi}{2} - \alpha|^2).$$

From the equation (3.2), satisfied by $D(\alpha)$, we have for $\alpha < \pi/2$

$$|\dot{D}(\alpha)|^2 = f^2 \frac{\cos^2\alpha}{\cos^2 D(\alpha)} = f^2 \frac{(\frac{\pi}{2} - \alpha)^2 + o(A)}{f(\tau(\alpha))(\frac{\pi}{2} - \alpha)^2 + o(A)}.$$

Thus, $\dot{\theta}(\pi/2)$ exists, and evidently,

$$\dot{\theta}(\pi/2) = \frac{-1 \pm \sqrt{f(\pi/2)}}{2},$$

which corresponds to two possible solutions according to the Theorem 3.5.

3.8. To recover $p(\alpha)$, given $\theta(\alpha)$, one uses $\frac{p'(\alpha)}{p(\alpha)} = \tan\theta(\alpha)$ where $\theta(\alpha)$ is given in the Theorem 3.5. Since $\theta(\alpha)$ is $C^1[\bar{\alpha},\pi/2]$, $|\theta(\alpha)| < \pi/2$ when $\alpha \in [\bar{\alpha},\pi/2]$, and has a limit at $\alpha = \pi/2$, $p(\alpha)$ is defined (up to a multiplicative constant) and $p \in C^2[\bar{\alpha},\pi/2) \cap C^1[\bar{\alpha},\pi/2]$.

3.9. __Lemma__. Let the function p be as in section 3.8 and define $p(\alpha,\beta) = p(\alpha)$, $r : \bar{\omega} \to R^3$ the map defined by (1.10), and $m: \bar{\omega} \to \bar{\Omega}$ the map defined (1.12). Then r defines an imbedded surface which projects radially one-to-one onto $\bar{\Omega}$ and the map m is a diffeomorphism. In addition, $r(\bar{\omega})$ is a surface of revolution.

3.9.1. __Proof__. The Jacobian of the map r will have rank equal to two if $\det(g_{ij}) = \det(\langle r_i,r_j\rangle) \neq 0$ in $\bar{\omega}$. From (1.15) it follows that $\det(g_{ij}) \neq 0$ if and only if $M(p) \neq 0$ in $\bar{\omega}$. This is indeed the case, since $M(p) = f > 0$ in $\bar{\omega}$.

The same computation, in essence, shows that the Jacobian of the map m also has rank 2.

Let us show now that $r(\bar{\omega})$ is a surface which projects univalently on $\bar{\Omega}$. This will imply that m is a C^1 diffeomorphism. Suppose on the countrary that a ray of direction $d \in \bar{\Omega}$ intersects $r(\bar{\omega})$ at two or more points. Then, evidently, there exists a point $\bar{u} \in \bar{\omega}$ such that $\langle r(\bar{u}), n(\bar{u})\rangle = 0$, where $n(\bar{u})$ is the normal vector to the surface r at the point $\bar{u}$. But from (1.9), (1.10) and (1.11) we have

$$- \langle r,n\rangle = \frac{N^2 - p\rho}{\sqrt{|\nabla\rho|^2 + \rho^2}} = \frac{\sqrt{|\nabla p|^2 + p^2}}{2} \neq 0 \text{ in } \bar{\omega}.$$

The last statement follows from 3.4.2. The lemma is proved.

3.10. Combining the assertions of Theorem 3.5 and Lemmas 3.6, 3.9 we have the following

<u>**Theorem 3.10**</u>. Let $f(\alpha)$ be a positive continious function on $[0,\pi/2]$, and $\tilde{\alpha}$ any number in $(-\pi/2,0)$. Assume that there exists $\bar{\alpha}$ satisfying the equation

$$1 + \sin\tilde{\alpha} = S(\bar{\alpha},\pi/2), \ \bar{\alpha} \in (0,\pi/2).$$

Then there exist two reflector surfaces described by the maps r constructed from functions p_1 and p_2 defined as in section 3.8. These reflector surfaces solve the inverse reflector problem stated in 1.5 for the prescribed aperture $\partial\Omega$, target domain $\bar{\omega}$, and intensity f. The function p ($= p_1$ or p_2) of which the reflector surface $r: \bar{\omega} \to R^3$ is constructed is of class $C^2(\bar{\omega})$.

References

[1] F. Brickell, L. Marder, and B.S. Westcott, The geometrical optics design of reflectors using complex coordinates, J. Phys. A: Math. Gen., Vol 10 (1977), 245-260.

[2] L. Caffarelli, L. Nirenberg, J. Spruck, The Dirichlet problem for nonlinear second order elliptic equations, I. Monge-Ampère equation, Comm. on Pure and Appl. Math., 37 (3) (1984), 369-402.

[3] E.A. Coddington and N. Levinson, Theory of ordinary differential equations, McGraw Hill, New York, 1955.

[4] R. Courant and D. Hilbert, Methods of Mathematical Physics, Vol. II, Interscience Publishers, J. Wiley and Sons, New York, 1962.

[5] L. Eisenhart, Riemannian Geometry, Princeton Univ. Press, 1949.

[6] P. Hartman and A. Wintner, On the third fundamental form of a surface, American J. of Math. 75 (1953), 298-334.

[7] J.B. Keller, The inverse scattering problem in geometical optics and the design of reflectors, IRE Transactions on antennas and propagation. 1958, 146-149.

[8] L. Marder, Uniqueness in reflector mappings and the Monge-Ampère equation, Proc. R. Soc. London, A 378 (1981), 529-537.

[9] A.P. Norris and B.S. Westcott, Computation of reflector surfaces for bivariate beamshaping in the elliptic case, J. Phys. A: Math. Gen. Vol. 9, No. 12 (1976), 2159-2169.

[10] B.S. Westcott, Shaped Reflector Antenna Design, Research Studies Press Ltd., Letchworth, Hertfordshire, England, 1983.

[11] B.S. Westcott and A.P. Norris, Reflector synthesis for generalized far fields, J. Phys. A: Math. Gen. Vol. 8, No. 4, (1975), 521-532.

Department of Mathematics and Computer Science, Emory University, Atlanta, Georgia, 30322

SCATTERING THEORY FOR THE WAVE EQUATION

ON A HYPERBOLIC MANIFOLD

Ralph Phillips, Bettina Wiskott and Alex Woo

§1. Introduction

In this paper we sketch the main ideas of our approach to the spectral theory of the Laplace-Beltrami operator with a short range perturbation acting on an n-dimensional hyperbolic manifold M; full details can be found in [10]. We assume that M is representable as the fundamental domain of a discrete subgroup Γ of motions on the real hyperbolic space $\mathbb{H}^n$, that is $M = \Gamma \backslash \mathbb{H}^n$. We limit ourselves to subgroups Γ having the finite geometric property (i.e., the fundamental domains obtained by the polygonal method with 'center' j have a finite number of sides), but are otherwise unrestricted. The theory is vacuous if M is compact.

The unperturbed metric on M, inherited from $\mathbb{H}^n$, is given by

$$ds^2 = g^0_{ij}\, dx_i\, dx_j = (dx^2 + dy^2)/y^2; \tag{1.1}$$

here in the middle expression we have set $y = x_n$. The corresponding Laplace-Beltrami operator Δ_0 is

$$\Delta_0 = y^n\, \partial_i\, y^{2-n}\, \partial_i. \tag{1.2}$$

It is more convenient for our purposes to work with the operator

$$L_0 = \Delta_0 + \left(\frac{n-1}{2}\right)^2. \tag{1.3}$$

The perturbed operator is of the form

$$L = \frac{1}{\sqrt{g}}\, \partial_i\, \sqrt{g}\, g^{ij}\, \partial_j + q = \Delta + q, \tag{1.4}$$

which corresponds to the Laplace-Beltrami operator for the metric

$$ds^2 = g_{ij}\, dx_i\, dx_j. \tag{1.5}$$

Acknowledgment. The work of the first author was supported in part by the National Science Foundation under Grant DMS-85-03297 and a Ford Foundation grant.

We impose the following conditions on g_{ij} and q, which we state in terms of the non-Euclidean distance r from the 'center' j of the fundamental domain and the local coordinates of charts in M: q lies in L_p^{loc} where $p = n/2$ for $n > 4$, $p > 2$ for $n = 4$, $p = 2$ for $n = 3$ and $p = \infty$ for $n = 2$; g_{ij} lies in $C^{(1)}(M)$ and

$$\frac{1}{\mu}(g_{ij}) \leq (g_{ij}^0) \leq \mu(G_{ij}) \qquad \text{for some} \quad \mu \geq 1;$$

$$|g^{ij} - g_0^{ij}| = y^2 \quad O(1/r^\alpha)$$

$$\left| \frac{\partial_i \sqrt{g}\, g^{ij}}{\sqrt{g}} - \frac{\partial_i \sqrt{g_0}\, g_0^{ij}}{\sqrt{g_0}} \right| = y \quad O(1/r^\alpha) \qquad \text{for some} \quad \alpha > 1; \tag{1.6}$$

$$|q + \Delta P/P| = O(1/r^\beta) \qquad \text{for some} \quad \beta > 2;$$

here P is a solution of

$$\Delta_0 P + \left(\frac{n-1}{2}\right)^2 P = 0, \tag{1.7}$$

chosen as in (2.3). The condition on q can be stated more simply if $\alpha > 2$, namely as

$$|q - ((n-1)/2)^2| = O(1/r^\alpha).$$

We also require the unique continuation property for L.

The principal result of this paper is the existence and completeness of the wave operators $W_\pm$. Since the continuous part of the spectrum of L_0 is absolutely continuous and of uniform multiplicity on $\mathbb{R}_-$, the same will be true of the perturbed operator L.

Using stationary scattering theory techniques developed for the Schroedinger wave equation, Peter Perry (unpublished) has been able to establish similar results for long range potentials on a more limited class of manifolds.

We use the wave equation as our principal tool. The general pattern of the proof follows [8,9,11,12]. The perturbed wave equation is

$$u_{tt} = Lu, \tag{1.8}$$

with initial data

$$u(w,0) = f_1(w) \quad \text{and} \quad u_t(w,0) = f_2(w). \tag{1.9}$$

We denote the pair of data (f_1, f_2) by f and we denote the solution operator of the perturbed wave equation by $U(t)$. Our treatment relies heavily on the translation representations developed for the unperturbed wave equation in [2,3,4]. To simplify the presentation we shall assume throughout that both the perturbed and unperturbed generators have no null data.

We shall denote by B_ρ the ball in M of radius ρ about the 'center' j of the fundamental domain F of Γ. A superscript ρ on a norm will limit the integration of the form to the ball B_ρ.

§2. The perturbed system

In this section we present without proof the routine background material needed for the proof of the existence and completeness of the wave operators; the arguments can be found in [10].

The energy form for the perturbed system is

$$E(f,h) = -(Lf_1,h_1) + (f_2,h_2); \tag{2.1}$$

here $(\cdot,\cdot)$ denotes the $L_2(M,dg)$ inner product. An integration by parts transforms (2.1) into a more symmetric form:

$$E(f,h) = \int_M (g^{ij}\, \partial_i f_i\, \overline{\partial_j h_1} - qf_1\, \overline{h_1} + f_2\, \overline{h_2})\, \sqrt{g}\, dx. \tag{2.2}$$

Depending on q, E can be indefinite and hence cannot be used as the defining form for a Hilbert space. Instead we construct a locally positive definite form:

$$G = E + K \tag{2.3}$$

where K is chosen to be compact with respect to G. Our Hilbert space $\mathcal{H}$ is then obtained as the completion with respect to G of all $C_0^\infty(M)$ data.

In order to define G we cover M by a finite number of charts, which are either interior charts, regular charts at infinity, or neighborhoods of cusps at infinity. We then choose a finite partition of unity, subordinate to these charts. Denote the functions in this partition of unity by ω_j, ϕ_j and ψ_j, according as they are supported in interior, regular infinite or cusp charts, respectively. Finally for all charts except those corresponding to cusps of intermediate rank, we choose the solutions P of (1.7) equal to

$$P = y^{(n-1)/2} \tag{2.3}$$

in the local chart coordinates (see [10]).

For $\theta = \phi$ or ψ we now set

$$G^\theta(f) = \int \theta[P^2 \left|\nabla\left(\frac{f_1}{P}\right)\right|^2 + |f_2|^2]\, \sqrt{g}\, dx \tag{2.4}$$

and for interior charts we set

$$G^{\omega}(f) = \int \omega[P^2 \left|\nabla\left(\frac{f_1}{P}\right)\right|^2 + |f_1|^2 + |f_2|^2] \sqrt{g}\, dx. \qquad (2.4')$$

G itself is obtained by summing over all charts:

$$G(f) = \sum G_j^{\omega}(f) + \sum G_j^{\phi}(f) + \sum G_j^{\psi}(f). \qquad (2.5)$$

We now list some of the properties of G.

Property 1. *For any compact subset* S *of* M *there is a constant* c_S *such that*

$$\int_S |f_1|^2 \sqrt{g}\, dx \leq c_S\, G(f). \qquad (2.6)$$

Property 2. *The forms* G *and* G_0 *for the perturbed and unperturbed systems are equivalent.*

Property 3. *The form* $K = G - E$ *is compact with respect to* G.

Corollary. E *is positive on a subspace of finite codimension.*

Property 4. *There exists a constant* c *such that for every* $\theta = \phi$ *or* ψ

$$\int \theta |f_1/r|^2 \sqrt{g}\, dx \leq c\, G(f). \qquad (2.7)$$

Property 5. *If* E *is positive on a closed subspace* H'' *of* H, *then* E *and* G *are equivalent on* H''.

Property 6. *If* f *and* g *belong to* H *and* f *vanishes for* $r < \rho$, *then for* $\gamma' = \min(\alpha, \beta-2)$ *and all sufficiently large* ρ

$$|E(f,g) - E_0(f,g)| \leq \frac{c}{\rho^{\gamma'}} \|f\|_G \|g\|_G. \qquad (2.8)$$

The restriction L' of L to $L_2(M)$ plays a central role in our development. We define L' by means of the form

$$C(u,v) = \int_M (g^{ij} \partial_i u \overline{\partial_j v} - qu\bar{v} + cu\bar{v}) \sqrt{g}\, dx. \qquad (2.9)$$

Making use of the local constraints on q, it can be shown that

$$\left(\int |q u^2| \sqrt{g}\, dx\right)^{1/2} \leq c' \|u\| + \frac{1}{2} \|\nabla u\|. \qquad (2.10)$$

We can therefore choose c so that

$$C(u,u) \geq (\|u\|)^2. \qquad (2.11)$$

We now complete $C_0^{\infty}(M)$ with respect to the form C and denote the completion by W^1. Any u in W^1 for which there exists an f in $L_2(M)$ such that

$$C(u,v) = (f,v)$$

for all v in W^1 lies in the domain of L' and

$$L'u = -f + cu.$$

As so defined L' is selfadjoint. It also follows that if u belongs to W^1 and Lu, defined in the weak sense, lies in $L_2(M)$, then u is in $D(L')$ and $L'u = Lu$.

It is easy to prove from (2.10) that

$$\|\nabla u\| \leq c''(\|u\| + \|L'u\|). \tag{2.12}$$

It also follows from (2.10), (2.12) and elliptic theory that

$$\|\Delta u\|^{\rho} \leq c_{\rho} \sum_{|\alpha|\leq 2} \|\partial^{\alpha}u\|^{\rho+1} \leq c'_{\rho}(\|u\| + \|L'u\|). \tag{2.13}$$

Since q is bounded for large r and since $q\,u = L'u - \Delta u$, we conclude for u in $D(L')$ that

$$\|qu\| \leq c(\|u\| + \|L'u\|). \tag{2.14}$$

Making use of the corollary to Property 3, it is easy to show that

Lemma 2.1. *The nonnegative eigenspace of* L' *is finite dimensional.*

Let $\{\lambda_j^2;\ j = 1,\cdots,m\}$ denote the positive eigenvalues of L', if any, and let $\{\xi_j\}$ be the corresponding eigenfunctions:

$$L'\xi_j = \lambda_j^2\xi_j, \quad \|\xi_j\| = 1/\sqrt{2}, \quad \lambda_j > 0. \tag{2.15}$$

Then the data

$$q_j^{\pm} = (\xi_j, \pm\lambda_j\xi_j) \tag{2.16}$$

are eigendata of

$$A = \begin{pmatrix} 0 & I \\ L & 0 \end{pmatrix}; \tag{2.16'}$$

that is

$$A\, q_j^{\pm} = \pm\, \lambda_j\, q_j^{\pm}\,. \tag{2.16''}$$

We will denote the analogous eigendata of A_0 by $\{p_j^{\pm}\}$. The eigendata $\{q_j^{\pm}\}$ satisfy the following biorthogonality relations (see [5]):

$$E(q_j^+,q_k^+) = 0 = E(q_j^-,q_k^-)$$
$$E(q_j^+,q_k^-) = -\lambda_j^2\,\delta_{jk} \tag{2.17}$$

for all j, k.

Let P denote the span of the $\{q_j^\pm\}$. It is clear from (2.17) that E is nondegenerate on P. Hence if we denote the E-orthogonal complement of P by H_c, then every f in H has a unique decomposition of the form $f = g + q$, where g lies in H_c and q in P. We denote the projection $f \to g$ by Q. Q is E-orthogonal and because of (2.17) it can be expressed as

$$Qf = f + \sum a_j(f)\, q_j^+ + \sum b_j(f)\, q_j^-, \tag{2.18}$$

where

$$a_j(f) = E(f,q_j^-)/\lambda_j^2 \quad\text{and}\quad b_j(f) = E(f,q_j^+)/\lambda_j^2. \tag{2.18'}$$

Lemma 2.2. *The energy form* E *is nonnegative on* H_c.

In general E can have a nontrivial null space in H_c which coincides with the null space of A. However, as noted in the introduction, we shall assume that E is positive on H_c. This allows us to substantially simplify our exposition since it now follows from Property 5 above that E and G are equivalent forms on H_c.

Next we define the pregenerator A' of the solution operators by (2.16') with L replaced by L' and having $D(L') \times D(L')$ as its domain. A itself is defined as the closure of A'.

Lemma 2.3. A *is* E-*skewsymmetric.*

One proves by means of Lemma 2.3 and simple estimates on K that A satisfies the Hille-Yosida criterion and hence that

Theorem 2.4. A *generates a group of* E-*unitary operators on* H.

Lemma 2.5. *The following assertions hold:*

(a) $D(L') = D(L_0')$;

(b) $G(L')$ *and* $G(L_0')$ *are equivalent graphs;*

(c) *If* u *belongs to* $D(L')$ *and vanishes in the ball* B_ρ *of radius* ρ *about* j, *then for* $\gamma = \min(\alpha,\beta)$ *and* ρ *sufficiently large*

$$\| L'u - L_0'u\| \le \frac{c}{\rho^\gamma}\,(\|u\| + \|L'u\|). \tag{2.19}$$

Lemma 2.6. *If* f *belongs to* $D(A^2) \cap H_c$, *then* f_2 *lies in* $D(L')$ *and*

$$\|f_2\| + \|\nabla f_2\| + \|L'f_2\| \le c\,(\|f\|_E + \|A^2 f\|_E). \qquad (2.20)$$

Next we define the notion of incoming and outgoing subspaces. A solution $u(w,t)$ of the unperturbed wave equation, $u_{tt} = L_0 u$, is called outgoing if $u(w,t)$ vanishes in the ball B_t, centered at j and of radius $t > 0$. Incoming solutions are defined analogously with t replaced by $-t$ (see Section 8 of [4]). We call initial data of outgoing solutions outgoing data and denote the set of all such data by $\mathcal{D}_+$. Incoming data are defined analogously and denoted by $\mathcal{D}_-$. We say that data is eventually outgoing [incoming] if $U_0(t)f$ lies in $\mathcal{D}_+$ $[\mathcal{D}_-]$ for some t.

The incoming and outgoing translation representors $R^F_\pm$ are maps from the data to $L_2(\mathbb{R} \times B_F) \times L_2(\mathbb{R})^N$; here N is the number of inequivalent cusps of maximal rank in the fundamental domain F and B_F is the boundary at infinity of F. According to Theorem 2.3 of [2], $R^F_\pm$ transmutes the action of $U_0(t)$ into translation:

$$R^F_\pm U_0(t) = T_\pm(t)\, R^F_\pm, \qquad (2.21)$$

where $T_+(t)$ $[T_-(t)]$ is translation to the right [left] by t. According to Lemma 2.8 of [2], $R^F_\pm$ is an isometry on $\mathcal{D}_\pm$:

$$\|R^F_\pm\, d_\pm\| = \|d_\pm\|_{E_0} \qquad (2.22)$$

for $d_\pm$ in $\mathcal{D}_\pm$. According to Theorem 8.1 of [4] $R^F_+\mathcal{D}_+$ and $R^F_-\mathcal{D}_-$ are supported on $\mathbb{R}_\pm$. We shall introduce further properties of the translation representors later on.

Lemma 2.7. E *is equivalent to* G *on* $\mathcal{D}_\pm$.

This follows from Property 5 and (2.22).

Lemma 2.8. *For any* d_+ *in* $\mathcal{D}_+$

$$d_+ = Q_0\, d_+ + p_-, \qquad (2.23)$$

where p_- *is in the span of the* $\{p_j^-\}$. *Further*

$$\|Q_0\, d_+\|_{E_0} = \|d_+\|_{E_0}. \qquad (2.24)$$

The analogous assertions hold for d_- *in* $\mathcal{D}_-$.

§3. Existence and completeness of the wave operators

In this section we sketch the proof of the existence and completeness of the wave operators

$$W_{\pm} = \text{st.}\lim_{t\to\pm\infty} W(t), \tag{3.1}$$

defined on H_c^0 to H_c; here

$$W(t)f = Q\, U(-t)\, U_0(t)f. \tag{3.2}$$

To prove completeness we show that the range of $W_{\pm}$ fills out the continuous part of the spectrum of A, which is

$$H_1 = H_c \ominus \text{eigenspace of } A\big|_{H_c}. \tag{3.3}$$

The proof of completeness for the incoming and outgoing translation representors $R_{\pm}^F$ restricted to H_c^0 (see [3]) depended on the fact that for a certain smooth subspace of eventually outgoing [incoming] data, denoted by $D'_{\pm}$

$$\text{closure } Q_0 D'_{\pm} = H_c^0. \tag{3.4}$$

The subspace $D'_{\pm}$ was defined as the range of the inverse translation representor $J_{\pm}^F$ acting on C_0^∞ functions $\ell(s,\beta)$ of compact support in $\mathbb{R} \times B_F$ to H (see p. 322 of [2]).

Theorem 3.1. *The wave operators exist and are isometries on* H_c^0 *to* H_c.

Proof. We treat only the case W_+. Once we have established existence and isometry for a dense subspace of H_c^0, a continuity argument completes the proof.

We choose as our dense subspace of H_c^0 the set $Q_0 D''_+$, where D''_+ consists of the range of J_+^F acting on C_0^∞ functions $\ell(s,\beta)$ of compact support on $\mathbb{R} \times B_F$ and satisfying the added condition

$$\int \ell(s,\beta)\, ds = 0. \tag{3.5}$$

Obviously $D''_+ \subset D'_+$ and it is easy to show that D''_+ is dense in D'_+. Consequently (3.4) remains valid if we replace D'_+ by D''_+.

We now set

$$d = J_+^F \ell \quad\text{and}\quad d_0 = J_+^F \left(-\int_{-\infty}^{s} \ell(\sigma,\beta)\, d\sigma\right). \tag{3.6}$$

Because of (3.5) both d and d_0 lie in D'_+. The action of A_0 in the translation representation is $-\partial_s$ and it follows that

$$d = A_0 d_0. \tag{3.6'}$$

Thus

$$\mathcal{D}''_+ \subset D(A_0^\infty) \cap R(A_0).$$

Setting

$$V = A - A_0 = \begin{pmatrix} 0 & 0 \\ L - L_0 & 0 \end{pmatrix} \tag{3.7}$$

and following Cook's recipe, we have for f in $D(A_0)$

$$\begin{aligned} W(t)f - W(s)f &= -\int_s^t Q\, U(-\tau)\, V\, U_0(\tau) f \, d\tau \\ &= -\int_s^t U(-\tau)\, Q\, V\, U_0(\tau) f \, d\tau. \end{aligned} \tag{3.8}$$

Since U is E-isometric and Q is G-bounded

$$\begin{aligned} \|W(t)f - W(s)f\|_E &\le \int_s^t \|Q\, V\, U_0(\tau) f\|_E \, d\tau \\ &\le c \int_s^t \|V\, U_0(\tau) f\|_G \, d\tau \\ &\le c \int_s^t \|(L - L_0)[U_0(\tau) f]_1\| d\tau. \end{aligned} \tag{3.9}$$

Thus $\lim W(t)f$ exists if $\|(L - L_0)[U(\tau)f]_1\|$ is integrable. We now prove that this is so for data f of the form $f = Q_0 d$ with d in $\mathcal{D}''_+$. With d defined as in (3.6), we have

$$f = Q_0 d, \quad f_0 = Q_0 d_0 \quad \text{and} \quad f = A_0 f_0. \tag{3.10}$$

According to Lemma 2.8

$$\|f\|_{E_0} = \|d\|_{E_0} \quad \text{and} \quad \|f_0\|_{E_0} = \|d_0\|_{E_0} \tag{3.10'}$$

and

$$f = d + p_-, \tag{3.11}$$

where by (2.18) p_- is of the form

$$p_- = \sum b_j\, p_j^-, \quad b_j = E(d, p_j^+)/\lambda_j^2. \tag{3.11'}$$

Finally we note that

$$U_0(t) f = U_0(t) d + \sum b_j\, e^{-\lambda_j t}\, p_j^-. \tag{3.12}$$

Now $f_1 = [f_0]_2$ and hence by Lemma 2.6

$$\|f_1\| + \|\nabla_0 f_1\| + \|L_0' f_1\| \le c(\|f_0\|_{E_0} + \|A_0 f_0\|_{E_0} + \|A_0^2 f_0\|_{E_0}). \tag{3.13}$$

Using the fact that U_0 is E_0-isometric together with the exponential decay of the second term on the right in (3.12) we get a similar estimate for $v(t) = [U_0(t)d]_1$, namely

$$\|v(t)\| + \|\nabla_0 v(t)\| + \|L_0' v(t)\| \le c' \qquad \text{for all } t \ge 0. \tag{3.13'}$$

Since data in $\mathcal{D}_+''$ is eventually outgoing, $v(t)$ vanishes in a ball B_{t-a} for some real a. It therefore follows by Lemma 2.5c that

$$\|(L' - L_0')v(t)\| \le c/(t-a)^{\gamma} \tag{3.14}$$

for some $\gamma > 1$. Applying (3.12) once again we see that

$$\|(L' - L_0')[U_0(t)f]_1\| \le \|(L' - L_0')v(t)\| + \text{exponentially decaying terms.} \tag{3.14'}$$

Combining (3.14) and (3.14'), we see that $\lim W(t)f$ exists.

It remains to show that W_+ is an isometry for data $f = Q_0 d$ with d in $\mathcal{D}_+''$. For this it suffices to prove that asymptotically

$$\|W(t)f\|_E = \|Q\,U(-t)\,U_0(t)f\|_E = \|Q\,U_0(t)f\|_E \sim \|f\|_{E_0}. \tag{3.15}$$

By (3.10') $\|f\|_{E_0} = \|d\|_{E_0}$ and by (3.12)

$$\|U_0(t)f - U_0(t)d\|_G \to 0.$$

Hence it suffices to prove that

$$\|Q\,U_0(t)d\|_E \sim \|d\|_{E_0}. \tag{3.16}$$

Since $U_0(t)d$ vanishes in the ball B_{t-a} and since E is locally majorized by G, we see that

$$|E(U_0(t)d,g)| \le c\|U_0(t)d\|_G \|g\|_G^{r>t-a}.$$

Now by Lemma 2.7

$$\|U_0(t)d\|_G \le c\|U_0(t)d\|_{E_0} = c\|d\|_{E_0} \tag{3.17}$$

for $t \ge 0$. It follows that $E(U_0(t)d, q_j^{\pm}) \to 0$ and hence by (2.18) that

$$\|Q\,U_0(t)d - U_0(t)d\|_G \to 0. \tag{3.18}$$

Thus it remains to prove that

$$\|U_0(t)d\|_E \sim \|U_0(t)d\|_{E_0} = \|d\|_{E_0}. \tag{3.16'}$$

For this we appeal to Property 6, according to which

$$\left|\,\|U_0(t)d\|_E^2 - \|U_0(t)d\|_{E_0}^2\right| \le \frac{c}{(t-a)^{\gamma}}\,\|U\ (t)d\|_G^2.$$

Combining this with (3.17) we get (3.16'). This concludes the proof of Theorem 3.1.

We shall need a couple of technical lemmas which we state without proof (see [10]). The first is a weak form of local energy decay.

Lemma 3.2. *Suppose that* f *belongs to* $H_1 \cap D(A^5)$. *Then there exists a sequence* $\{t_n\}$, *tending to* ∞, *such that for all* $\rho > 0$,

$$\lim_{n\to\infty}[\|(U(t_n)f]_2\|^\rho + \sum_{j=1}^{4} \|U(t_n)A^j f\|_G^\rho = 0. \tag{3.19}$$

Lemma 3.3. *Suppose* f *belongs to* $H_c \cap D(A^2)$ *and*

$$\|f_2\|^\rho + \sum_{j=1}^{2} \{\|A^j f\|_G^\rho + \|[A^j f]_1\|^\rho\} \le \varepsilon. \tag{3.20}$$

Then for $|t| \le \rho^{1/2}$ *and* $\gamma = \min(\alpha,\beta)$

$$\|Q_0U(t)Af - Q_0U_0(t)Af\|_{E_0} \le c\varepsilon + \frac{c}{\rho^{\gamma-1/2}} \sum_{j=0}^{2} \|A^j f\|_E. \tag{3.21}$$

We are now in a position to prove the completeness of the wave operators. This amounts to showing that

$$R(W_\pm) = \text{Range of } W_\pm = H_1. \tag{3.22}$$

The intertwining property of W_+ together with the invariance of H_c^0 under the action of U_0 implies the invariance of the range of W_+ under the action of U. Since W_+ is an isometry the range of W_+ is a closed subspace and since the spectrum of A_0 is continuous on H_c^0, $R(W_+) \subset H_1$. Hence if (3.22) is not valid there will exist a nonzero f in H_1 orthogonal to the range of W_+; because of the invariance the same will be true of $U(t)f$ for all t. We can therefore choose a time smoothed version of f in $H_1 \cap D(A^\infty)$ with these same properties. We shall prove below that $A^2f = 0$ and since A has no null vectors in H_1 it will follow that $f = 0$, contrary to our choice of f. This will imply (3.22).

We begin by making use of Lemma 3.2 to obtain a sequence of t_n's, $t_n \to \infty$, such that for f chosen as above and all positive integers n,

$$\|[U(t_n)f]_2\|^{n^2} + \sum_{j=1}^{4} \|U(t_n)A^j f\|_G^{n^2} < \varepsilon_n, \tag{3.23}$$

where ε_n is chosen so that for $f_n = U(t_n)f$, $j = 1,2,3$ and $|t| \le n$ Lemma 3.3 gives

$$\|Q_0U(t)A^j f_n - Q_0U_0(t)A^j f_n\|_{E_0} < \frac{1}{n^{2\gamma-1}} \quad (\gamma = \min(\alpha,\beta) > 1). \tag{3.24}$$

Next we make use of the incoming and outgoing translation

representations (for the unperturbed system) of $Q_0\ Af_n$ (see [2,3]):

$$k_n = R_+^F\, Q_0Af_n \quad \text{and} \quad \ell_n = R_-^F\, Q_0Af_n. \tag{3.25}$$

Then

$$\|k_n\| = \|Q_0Af_n\|_{E_0} \le \|Q_0Af_n\|_G \le c\|Af_n\|_G \le c'\|Af_n\|_E \le c'', \tag{3.26}$$

and likewise

$$\|\ell_n\| = \|Q_0Af_n\|_{E_0} \le c\|Af_n\|_G \le c''. \tag{3.27}$$

Next we choose χ in $C_0(\mathbb{R})$ so that $0 \le \chi \le 1$ and

$$\chi(s) = \begin{cases} 0 & \text{for} \quad s < 0, \\ 1 & \text{for} \quad s > 1. \end{cases}$$

We then set

$$g_n = (R_+^F)^{-1}(\chi k_n) \quad \text{and} \quad h_n = (R_-^F)^{-1}(\chi\ell_n); \tag{3.28}$$

here $(R_\pm^F)^{-1}$ denotes the operator $Q_0J_\pm^F$. Thus g_n and h_n belong to H_c^0 and

$$\begin{aligned} \|g_n\|_{E_0} &= \|\chi k_n\| \le \|k_n\| \le c'', \\ \|h_n\|_{E_0} &= \|\chi\ell_n\| \le \|\ell_n\| \le c''. \end{aligned} \tag{3.29}$$

We see from (3.25) and (3.28) that

$$\begin{aligned} E_0(Q_0Af_n, g_n) &= (k_n, \chi k_n) \ge \|\chi k_n\|^2, \\ E_0(Q_0Af_n, h_n) &= (\ell_n, \chi\ell_n) \ge \|\chi\ell_n\|^2. \end{aligned} \tag{3.30}$$

We shall prove below that

$$\lim_{n\to\infty} E_0(Q_0Af_n, g_n) = 0 = \lim_{n\to\infty} E_0(Q_0Af_n, h_n). \tag{3.31}$$

It follows from this and (3.30) that

$$\|k_n\|^{s>1} + \|\ell_n\|^{s>1} \to 0. \tag{3.32}$$

The zeroth component of $R_\pm^F$ is defined as

$$R_\pm^0 f = \partial_s\, P\hat{f}_1 \mp P\hat{f}_2 \tag{3.33$_0$}$$

where P is in general an integral-differential operator (see [12]) and $\hat{u}$ is the Radon transform of u (see [2]). Each of the other components corresponds to a cusp of M of maximal rank; for the jth such cusp

$$R_\pm^j f(s) = \partial_s[\exp((1-n)s/2)\ \bar{f}_1(e^s)] \mp \exp((1-n)s/2)\ \bar{f}_2(e^s), \tag{3.33$_j$}$$

Now for $(v_1, v_2) = A^2 f_n$, we have

$$v_1 = [Af_n]_2, \quad v_2 = [A^2 f_n]_2 \quad \text{and} \quad L'v_1 = [A^3 f_n]_2$$

and since

$$\|A^2 f_n\|_E^2 = -(L'v_1, v_1) + (v_2, v_2)$$
$$= -([A^3 f_n]_2, [Af_n]_2) + \|[A^2 f_n]_2\|^2,$$

it follows that

$$\|A^2 f\|_E = \|A^2 f_n\|_E \to 0.$$

Finally since f lies in H_1 this implies that $A^2 f = 0$ and hence that $f = 0$. This proves the completeness of W_+ modulo (3.31).

Proof of (3.31). According to Theorem 8.3 of [4] for a given f in H_c^0 with $R_\pm^F f = 0$ for $s < 0$ there exists a $d_\pm$ in $\mathcal{D}_\pm$ and a $p_\pm$ in the span of $\{p_j^\pm\}$ such that $f = d_\pm + p_\mp$. Clearly $Q_0 d_\pm = f$ and by Lemma 2.8, $\|f\|_{E_0} = \|d_\pm\|_{E_0}$. In particular associated with g_n and h_n are $d_n^+ \in \mathcal{D}_+$ and $d_n^- \in \mathcal{D}_-$ such that

$$g_n = Q_0 d_n^+ \quad \text{and} \quad h_n = Q_0 d_n^- \tag{3.39}$$

with

$$\|g_n\|_{E_0} = \|d_n^+\|_{E_0} \quad \text{and} \quad \|h_n\|_{E_0} = \|d_n^-\|_{E_0}. \tag{3.40}$$

It follows from this and (3.29) that

$$\|U_0(t) d_n^\pm\|_{E_0} = \|d_n^\pm\|_{E_0} \le c''. \tag{3.40'}$$

Our proof of (3.31) now follows an argument due to Enss [1]. We conclude from (3.24) and (3.29) that

$$E_0(Q_0 Af_n, g_n) = E_0\left(Q_0 U_0(n) Af_n, U_0(n) g_n\right)$$
$$= E_0\left(Q_0 U(n) Af_n, U_0(n) g_n\right) + \varepsilon_n. \tag{3.41}$$

Since Q_0 is an E_0-orthogonal projection, we can bring Q_0 over to the right. Replacing $U_0(n) g_n$ by $Q_0 U_0(n) d_n^+$, (3.41) becomes

$$E_0(Q_0 Af_n, g_n) = E_0\left(U(n) Af_n, Q_0 U_0(n) d_n^+\right) + \varepsilon_n. \tag{3.41'}$$

Now

$$\|U(n) Af_n\|_G \le c\|U(n) Af_n\|_E = c\|Af\|_E \le c' \tag{3.42}$$

and by Lemma 2.7 and (3.40')

$$\|U_0(n) d_n^+\|_G \le c\|U_0(n) d_n^+\|_{E_0} = c\|d_n^+\|_{E_0} \le c'. \tag{3.42'}$$

where

$$\bar{f}_i(y) = \int f_i(x,y)\, dx, \qquad i = 1,2,$$

integrated over the fundamental domain of the stability group for this cusp.

Subtracting the outgoing representation of Q_0Af_n from the incoming representation, we see from (3.32) and (3.33) that

$$\|P[Q_0Af_n]_2^{\wedge}\|^{s>1} \to 0 \qquad (3.34)_0$$

and

$$\|\exp\big((1-n)s/2\big)[Q_0Af_n]_2^{-}\|^{s>1} \to 0. \qquad (3.34)_j$$

Now in parts (i) and (ii) of Theorem 3.2 in [3] the condition $R_+^F f = 0$ is used to establish the analogue of (3.34) for f and from this the assertions of these parts of the theorem follow. As a consequence given $\varepsilon > 0$ there exists a ρ_ε such that outside of a neighborhood of the cusps of intermediate rank

$$\overline{\lim_{n\to\infty}} \|[Q_0Af_n]_2\|^{r>\rho_\varepsilon} < \varepsilon. \qquad (3.35)$$

A similar but more elaborate argument shows that this result also holds for neighborhoods of cusps of intermediate rank (see [10]).

We also need the following lemma which we state without proof:

<u>Lemma 3.4</u>. *For* $\{f_n\}$ *chosen as above*

$$\lim_{n\to\infty} \|Q_0Af_n - Af_n\|_G = 0. \qquad (3.36)$$

Combining (3.35) and (3.36) we see that

$$\overline{\lim_{n\to\infty}} \|[Af_n]_2\|^{r>\rho_\varepsilon} < \varepsilon. \qquad (3.37)$$

On the other hand (3.23) shows that

$$\lim_{n\to\infty} \|[Af_n]_2\|^{\rho_\varepsilon} = 0. \qquad (3.37')$$

Since ε is arbitrarily small we conclude that

$$\lim_{n\to\infty} \|[Af_n]_2\| = 0. \qquad (3.38)$$

We can replace f by Af and A^2f in the previous development to obtain

$$\lim_{n\to\infty} \|[A^jf_n]_2\| = 0 \qquad \text{for } j = 1,2,3. \qquad (3.38')$$

By making use of the estimates (3.42) and (3.42'), we see from (3.12) that we can replace $Q_0U_0(n)d_n^+$ by $U_0(n)d_n^+$ in (3.41'):

$$E_0(Q_0Af_n, g_n) = E_0\big(U(n)Af_n, U_0(n)d_n^+\big) + \varepsilon_n'. \tag{3.43}$$

Since $U_0(n)d_n^+$ vanishes in the Ball B_n, we can apply Property 6 to obtain

$$|E\big(U(n)Af_n, U_0(n)d_n^+\big) - E_0\big(U(n)Af_n, U_0(n)d_n^+\big)| \leq c/n^{\gamma'}, \tag{3.44}$$

where $\gamma' = \min(\alpha, \beta-2)$. Thus we can write

$$\begin{aligned} E_0(Q_0Af_n, g_n) &= E\big(U(n)Af_n, U_0(n)d_n^+\big) + \varepsilon_n'' \\ &= E\big(Af_n,\ Q\ U(-n)\ U_0(n)d_n^+\big) + \varepsilon_n''. \end{aligned} \tag{3.45}$$

If d_n^+ were in the range of A_0 we could apply the argument used in the proof of Theorem 3.1 to show that

$$E_0(Q_0Af_n, g_n) = E(Af_n, W_+Q_0d_n^+) + \varepsilon_n'''. \tag{3.46}$$

Unfortunately d_n^+ is not in $R(A_0)$. Nevertheless the relation (3.46) can be justified by using the fact that Af_n is in $D(A)$. Finally since by choice Af_n is E-orthogonal to $R(W_+)$, it follows that

$$\lim_{n\to\infty} E_0(Q_0Af_n, g_n) = 0. \tag{3.31}$$

The argument for $E_0(Q_0Af_n, h_n)$ proceeds along similar lines.

$$\begin{aligned} E_0(Q_0Af_n, h_n) &= E_0\big(Q_0U_0(-n)Af_n, U_0(-n)h_n\big) \\ &= E_0\big(Q_0U(-n)Af_n, U_0(-n)h_n\big) + \varepsilon_n \\ &= E_0\big(U(-n)Af_n, U_0(-n)d_n^-\big) + \varepsilon_n' \\ &= E\big(U(-n)Af_n, U_0(-n)d_n^-\big) + \varepsilon_n'' \\ &= E(Af_n, W_-Q_0d_n^-) + \varepsilon_n'''. \end{aligned} \tag{3.47}$$

Since Af_n need not be E-orthogonal to $R(W_-)$, the rest of the proof requires a somewhat different strategy. Recall that $f_n = U(t_n)f$ so that if we make use of the intertwining property of W_-, we then get

$$\begin{aligned} E_0(Q_0Af_n, h_n) &= E\big(Af, W_-Q_0U_0(t_n)d_n^-\big) + \varepsilon_n''' \\ &= E_0\big(Q_0W_-^*Af, U_0(-t_n)d_n^-\big) + \varepsilon_n'''. \end{aligned} \tag{3.47'}$$

The analogue of (3.42') holds and since $U_0(-t_n)d_n^-$ vanishes in the ball B_n, we see that

$$|E_0(Q_0Af_n, h_n)| \leq c\|Q_0W_-^*Af\|^{r>t_n} + \varepsilon_n'''$$

and hence that

$$\lim_{n\to\infty} E_0(Q_0Af_n, h_n) = 0. \tag{3.31}$$

This concludes the proof of (3.31) in its entirety and establishes the completeness of W_+.

REFERENCES

1. Enss, V. "Asymptotic completeness for quantum mechanical potential scattering," Comm. Math. Phys., 61(1978), 285-291.
2. Lax, P. and Phillips, R. "Translation representations for automorphic solutions of the wave equation in non-Euclidean spaces,I" Comm. Pure and Appl. Math. 37(1984), 303-328.
3. _____, "Translation representations for automorphic solutions of the wave equation in non-Euclidean spaces, II," Comm. Pure and Appl. Math. 37(1984), 779-813.
4. _____, "Translation representations for automorphic solutions of the wave equation in non-Euclidean spaces, III," Comm. Pure and Appl. Math. 38(1985), 179-207.
5. _____, Scattering theory for automorphic functions, Annals of Math. Studies, 87, Princeton Univ. Press, 1976.
6. _____, "The asymptotic distribution of lattice points in Euclidean and non-Euclidean spaces," Journal of Functional Analysis 46(1982), 280-350.
7. Perry, Peter. "The Laplace operator on a hyperbolic manifold, I. Spectral and scattering theory," Journal of Functional Analysis, to appear.
8. Phillips, R. "Scattering theory for the wave equation with a short range perturbation, I," Indiana Univ. Math. Jr. 31(1982), 609-639.
9. _____, "Scattering theory for the wave equation with a short range perturbation, II," Indiana Univ. Math. Jr. 33(1984), 831-846.
10. Phillips, Ralph, Wiskott, Bettina and Woo, Alex. "Scattering theory for the wave equation on a hyperbolic manifold," Journal of Functional Analysis, to appear.
11. Wiskott, Bettina, "Scattering theory and spectral representation of short range perturbation in hyperbolic space." Dissertation, Stanford University, 1982.
12. Woo, A. C., "Scattering theory on real hyperbolic spaces and their compact perturbation." Dissertation, Stanford University, 1980.

Ralph Phillips
Department of Mathematics
Stanford University
Stanford, CA 94305

Bettina Wiskott
Digital Equipment Co.
Switzerland

Alex Woo
NASA Ames Research Center
Mountain View, CA 94035

ON UNSTEADY FLOW IN A TWO-DIMENSIONAL CASCADE WITH IN-PASSAGE SHOCKS

Kewal K. Puri
University of Maine
Orono, Maine 04469

Introduction. The phenomenon of dynamic aeroelastic instability, known as flutter, is one of the most serious problems encountered by a design engineer in the development of gas-turbine engines. It's accurate prediction entails the computation of unsteady aerodynamic loads on the blades in an incremental annulus The focus in the following discussion is on understanding this phenomenon by computing the forces and the moments on the blades in an infinite, rectilinear, two dimensional cascade which, as is customary, is regarded here as an adequate representation of the blade row in an annulus.

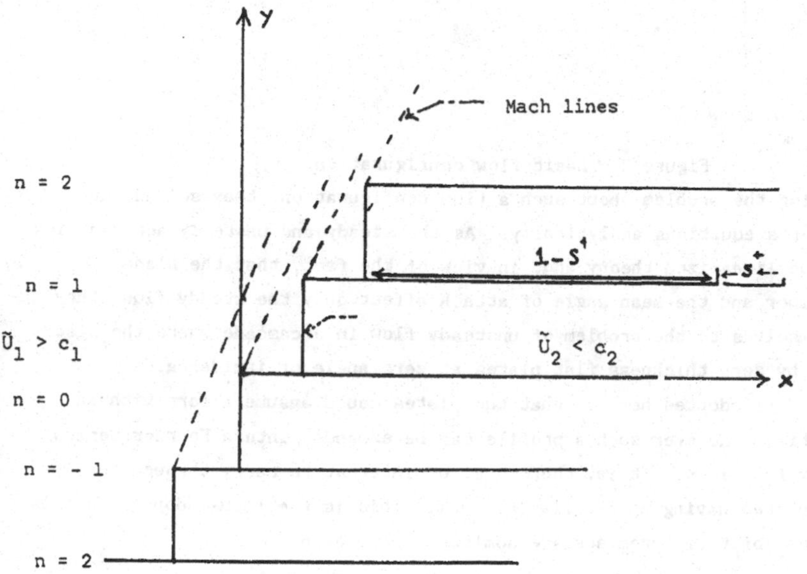

Figure 1 - Definition sketch of the rectilinear cascade

Modern fans and compressors operate with supersonic velocities relative to the blades. The axial velocities entering the blades, however, are usually subsonic. As such, the leading edge Mach waves extend all the way to infinity upstream, leaving no region ahead of the cascade undisturbed. Moreover, as is evidenced by the experimental investigations of Miller and Bailey [1] and Strazisar and Chima [2], these flow conditions exhibit an in-passage shock structure. Over most of the operating conditions, these shocks are nearly normal and are discernible in the tip region or in the trailing edge region or even in both. This implies that a viable model must admit a mixed supersonic-subsonic flow separated by in-passage shocks.

Over the past few years, such problems, both for a single isolated airfoil as well as for an infinite cascade, have been studied by several authors. These efforts are discussed in references [3]-[11]. This paper is based on the work of Goldstein, Braun and Adamiczyk [3] who defined the basic flow configuration as represented by a piece-wise constant step function, sketched in figure 2, below.

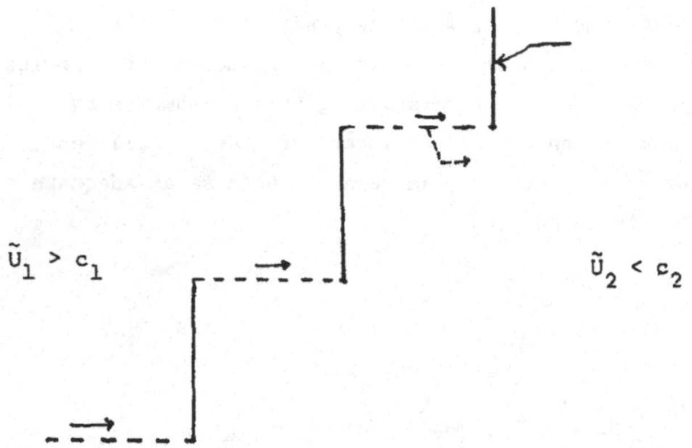

Figure 2 - Basic flow configuration

Linearizing the problem about such a flow configuration, they solved the resulting Euler's equations analytically. As the steady and unsteady aerodynamics uncouple in the linearized theory and, in view of the fact, that the blade thickness, camber and the mean angle of attack affect only the steady flow, they addressed themselves to the problem of unsteady flow in a cascade where the blades were replaced by zero thickness flat plates at zero angle of incidence.

The viewpoint adopted here is that the plates could assume a form with an arbitrary profile. However such a profile can be expanded into a Fourier Series of oscillatory functions. It is, therefore, of interest to carry through the analysis for plates having an oscillatory form. This is the focus adopted in this paper. The rest of their premises are admitted here also.

2. <u>Formulation</u>. Let the steady basic flow, defined above, be characterized by the densities ρ_i, local speeds of sound a_i and the flow velocities relative to the blade, $\tilde{U}_i$ where i = 1, 2 according as the subscript 'i' refers to the quantities related to the supersonic flow upstream or the subsonic flow downstream of the shocks. All variables appearing in the equations of motion have been non-

dimensionalized with reference to the chord length c, density ρ_1 and velocity $\tilde{U}_1$. The linearization of the equations describing the motion in the region R_1 is based on the quantities ρ_1, a_1, $\tilde{v}_1$ appropriate to the flow region R_1.

Upstream of the shock, the velocity potential,

$$\phi_1(x,y,t) = \Phi_1(x,y)\exp(-i\omega_1 t) \tag{2.1}$$

satisfies the equation,

$$\left(\frac{\partial^2}{\partial y^2} - \beta_1^2 \frac{\partial^2}{\partial x^2} + 2iM_1\beta_1^2 k \frac{\partial}{\partial x} + \beta_1^4 k_1^2\right)\Phi_1 = 0 \tag{2.2}$$

where $\omega_1 = \omega c/\tilde{U}_1$ is twice the reduced frequency, ω is the frequency of the motion, $M_1 = \tilde{U}_1/a_1 > 1$ is the free stream Mach number, $\beta_1^2 = M_1^2 - 1$ and $k_1 = \omega_1 M_1/\beta_1^2$. Following ϕ_1, the velocity vector $\vec{q}_1 = (u_1,v_1)$ is given by,

$$\vec{q}_1 = \nabla\phi_1 = \nabla\Phi_1\exp(-i\omega_1 t) = (U_1,V_1)\exp(-i\omega t) \tag{2.3}$$

and the pressure,

$$P_1 = P_1\exp(-i\omega_1 t) = (i\omega_1 - \frac{\partial}{\partial x})\Phi_1\exp(-i\omega_1 t) \tag{2.4}$$

For the subsonic flow, the velocity vector $\vec{q}_2 = (u_2,v_2) = (U_2,V_2)\exp(-i\omega_2 t)$ can be written as the sum of a solenoidal field characterized by the stream function $\psi_2(x,y,t)$ and an irrotational field represented by the velocity function $\phi_2(x,y,t)$ such that

$$\vec{q}_2 = \nabla\phi_2 + \nabla\times(0,0,\psi_2) \tag{2.5}$$

we define the functions $\Phi_2(x,y)$, $\Psi_2(x,y)$ to relate respectively to $\phi_2(x,y,t)$ and $\psi_2(x,y,t)$, as in equation (2.3).

It can be shown that the functions Φ_2,Ψ_2 are governed by the equations,

$$\left(\frac{\partial^2}{\partial y^2} + \beta_2^2 \frac{\partial^2}{\partial x^2} - 2iM_2\beta_2^2 k_2 \frac{\partial}{\partial x} + \beta_2^4 k_2^2\right)\Phi_2 = 0 \tag{2.6}$$

$$\left(i\omega_2 - \frac{\partial}{\partial x}\right)\Psi_2 = 0 \tag{2.7}$$

where $\beta_2^2 = 1 - M_2^2$, $\omega_2 = \omega c/\tilde{U}_2$ and $k_2 = \omega_2 M_2/\beta_2^2$, $M_2 = \tilde{U}_2/a_2 < 1$, is the mean steady flow Mach number, downstream of the shock, relating to M_1 by the equation,

$$M_2^2 = \left[(M_1^2 + \frac{2\mu}{\mu-1})/(\frac{2\mu}{\mu-1} M_1^2 - 1)\right] \tag{2.8}$$

and $\mu = c_p/c_v$ is the ratio of the specific heats.

$$\text{Also, } U_2 = \frac{\partial\Phi_2}{\partial x} + \frac{\partial\Psi_2}{\partial y},\quad V_2 = \frac{\partial\Phi_2}{\partial y} - \frac{\partial\Psi_2}{\partial x} \tag{2.9}$$

$$\text{and } P_2 = (i\omega_2 - \frac{\partial}{\partial x})\Phi_2 \tag{2.10}$$

The above equations are to be solved subject to certain boundary conditions on the blades. The latter are assumed to be infinitesimally thin and a typical one admits the profile:

$$y = B_0\sin\pi(x - d0) = \frac{1}{2i}(W(x,\pi) - W(x,-\pi)) \tag{2.11}$$

$$\text{with } W(x;a) = H_0(a)\exp(iax) \tag{2.12}$$

and $B_0 = H_0 \exp(i a d_0)$, H_0, d_0 are real constants. It is also assumed that the amplitude, $H_0 <<$ the amplitude of the fluid motion so that the boundary conditions may be applied at the mean position of the blades, viz.,

$y = ns$, $(n-1)s^+ < x < (n-1)s^+ \pm 1$, $n = 0, \pm 1, \pm 2, \ldots$ where s, s^+ are the gap and stagger distances measured, respectively, normal and parallel to the chord. The unsteady wakes, to the order of the linearized theory, are replaced by vortex sheets emanating from the trailing edges of the blades and lying along the lines,

$y = ns$, $x > s^+(n-1) + 1$, $n = 0, \pm 1, \pm 2, \ldots$ The mean shock wave positions are along the line segments,

$x = ns^+$, $ns < y < (n+1)s$, $n = 0, \pm 1, \pm 2, \ldots$ Finally, following Lane [6], we require that all blades oscillate harmonically with the same amplitude and a constant but arbitrary interblade phase angle σ. Thus,

$$V_i(x + ns^+, y + ns) = e^{i\sigma n} V_i(x,y) \tag{2.13}$$

for $-s^+ < x < 1 - s^+$, $y = 0^\pm$, $n = 0, \pm 1, \pm 2, \ldots$ where $y = 0^\pm$ denotes the limits as $y \to 0$ from above or below respectively. The above condition determines the upwash on the nth blade in terms of that on the zeroth blade. That, on the latter, it is determined from the kinematic conditioning.

$$V_2(x,y) = -(\tilde{U}_2/\tilde{U}_1)(i\omega_2 - \frac{\partial}{\partial x})W(x;a) \text{ for } \begin{cases} 0<x<1-s^+, y=0^+ \\ -s^+<x<1-s^+, y=0^- \end{cases} \tag{2.14}$$

$$V_1(x,y) = -(i\omega_1 - \frac{\partial}{\partial x})W(x;a) \text{ for } y = 0^+, \; -s^+<x<0 \tag{2.15}$$

The pressure and the upwash must be continuous across the wake. The discontinuities in the axial velocities are assumed to admit the Kutta conditions, which, when, applied at the trailing edges, render the solution unique.

The flow in the supersonic region is connected to that in the subsonic region by the jump conditions across the shock waves:

$$U_2 = -(1 + M_1^2)P_2/2M_1 + U_1 + \frac{1}{2}[1 + M_1^2 - B^4{}_1 M_1^{-2}(\mu-1)/(\mu+1)]P_1 \tag{2.16}$$

and

$$\frac{\partial P_2}{\partial y} = (M_1^2 + \frac{\mu-1}{\mu+1}\beta_1^2)\frac{\partial P_1}{\partial y} - (\mu+1)\left(\frac{M_1 M_2}{\beta_1\beta_2}\right)^2\left(\frac{\partial^2\Psi_2}{\partial y^2} - \omega_2^2\Psi_2\right) \tag{2.17}$$

$x = ns^+$, $sn < y < (n+1)s$, $n = 0, \pm 1, \pm 2, \ldots$. Finally we require that there be only outward propogating disturbances at large distances from the cascade so that no disturbance propogates upstream of the shock relative to the blade fixed coordinates.

For want of space, only an extended summary of the sdlution is presented below; the complete details will appear elsewhere [12].

3. <u>Solution of the problem in the supersonic regime</u>. In the sequel, the periodicity condition (2.13) is assumed to hold for all the variables. It will be a simple matter to verify that the final solution satisfies all the boundary conditions. Also, as is frequently done in boundary value problems posed in a

frictionless medium, we allow a small damping in the formulation by regarding k_1, k_2 and a as complex quantities with positive imaginary parts ϵ_1, ϵ_2, ϵ_3 respectively. They are set equal to zero in the final solution.

Since the disturbance cannot propogate upstream in a supersonic flow, the motion of the fluid in the region R_2 does not effect that in R_1. As such, we may solve for the supersonic part of the flow by solving the problem of the oscillating cascade of semi-infinite blades extending to $+\infty$. Solving equation (2.2) on the resulting domain, by using the method of separation of variables, we obtain,

$$\Phi_1 = \frac{1}{2\pi}\int_{-\infty+iM_1\epsilon_1}^{\infty+iM_1\epsilon_1} F_0(\alpha)\Lambda_1(\alpha,y)\exp\left\{-i(\alpha-M_1k_1)\tilde{x}_1\right\}d\alpha \tag{3.1}$$

where $\tilde{x}_1 = x_1 + s^+$,

$$\Lambda_1(\alpha,y) = \frac{1}{2i}\left[\frac{\exp\{i(\Delta_1^- + \beta_1\gamma_1 y)\}}{\sin\Delta_1^-} + \frac{\exp\{i(\Delta_1^+ - \beta_1\gamma_1 y)\}}{\sin\Delta_1^+}\right] \tag{3.2}$$

$$\Delta_1^{\pm} \equiv \frac{1}{2}(\sigma - M_1k_1s^+ \pm \beta_1\gamma_1 s) \qquad 0 < y < s \tag{3.3}$$

and, $F_0(\alpha)$ is an unknown function to be determined by using the condition that the jump, $[\Phi_1(x)]_n = \Phi(x,ns+0) - \Phi(x,ns-0)$, across the lines $L_n : y = ns$, $-\infty < x < (n-1)s^+$, $n = 0, \pm 1, \pm 2, \ldots$ be 0. This, in view of (2.13), amounts to requiring that,

$$[\Phi_1(x)]_0 = \int_{-\infty+iM_1k_1}^{\infty+iM_1k_1} F_0(\alpha)\exp\{-i(\alpha-M_1k_1)\tilde{x}_1\}d\alpha = 0 \text{ for } \tilde{x}_1 < 0 \tag{3.4}$$

Finally, the kinematic condition on the upwash velocity requires,

$$\frac{1}{2\pi}\int_{-\infty+iM_1k_1}^{\infty+iM_1k_1} F_0(\alpha)\kappa_1(\alpha,0)\exp\{-i(\alpha-M_1k_1)\tilde{x}_1\}d\alpha$$

$$= (-i_1\omega_1 + \frac{\partial}{\partial x})H_0\exp(iax) \qquad \text{for} \quad \tilde{x}_1 > 0 \tag{3.5}$$

Here $\kappa_1(\alpha,y) = \dfrac{\partial\Lambda_1}{\partial y}$. The equations (3.3), (3.4) are solved for $F_0(\alpha)$, by using Wiener-Hopf technique. This results into,

$$F_0^{(1)}(\alpha) = L_1\kappa_1^-(M_1k_1-a)/(a + \alpha - M_1k_1)\kappa_1^+(\alpha), \tag{3.6}$$

$$L_1 = H_0(\omega_1 - a)\exp(-ias^+) \tag{3.7}$$

$$\kappa_1(\alpha,0) = \frac{\kappa_1^+(\alpha}{\kappa_1^-(\alpha)} = \frac{\partial\Lambda_1}{\partial y}\Big|_{y=0} \tag{3.8}$$

$$\kappa_1^+(\alpha) = e^{\chi(\alpha)}\frac{\beta_1\gamma_1}{2}\sin(\beta_1\gamma_1 s)\left[\left(\frac{1}{2}d_1^+\right)^2\nu_0^+\nu_0^- \prod_{n=-\infty}^{\infty}\left(d_1^+/2n\pi\right)^2\nu_n^+\nu_n^-\right]^{-1}\left[\prod_{\pi=-\infty}^{\infty}(1-\alpha\_/\nu_n^-)\right]^{-1} \tag{3.9}$$

$$\kappa_1^-(\alpha) = e^{\chi(\alpha)}\prod_{n=-\infty}^{\infty}(1 - \alpha/\nu_n^+) \tag{3.10}$$

$$\chi(\alpha) = -\frac{1}{2} i\alpha(s^+ - \beta_1 s) \tag{3.11}$$

$$d_1^+ = (s_1^+ - \beta_1^2 s^2)^{1/2} \tag{3.12}$$

$$\nu_n^{\pm} = \Gamma_n^{(1)} s^+/d_1^+ \pm \frac{s\beta_1}{d_1^+}\left(\left(\Gamma_n^1\right)^2 - k_1^2\right)^{1/2}, \; n = 0, \pm 1, \pm 2, \ldots \tag{3.13}$$

$$\Gamma_n^{(1)} = (2n\pi + M_1 k_1 s^+ - \sigma)/d_1^+$$

The substitution of $F_0(\alpha)$ in the expression for Φ, yields,

$$\Phi_1 = \frac{L_1}{2\pi}\int_{-\infty+i\delta}^{\infty+i\delta} \frac{\kappa_1(M_1k_1-a)\Lambda_1(\alpha,y)\exp\{-i(\alpha-M_1k_1)\tilde{x}_1\}d\alpha}{(\alpha+a-M_1k_1)\kappa_1^+(\alpha)}$$

Our primary interest is to compute pressures $P_1(x,0\pm)$ on the blade surfaces as well as the axial velocity $U_1(0,y)$ and the pressure $P_1(0,y)$ on the shock. The latter two quantities are used to find the boundary conditions on the shock for the subsonic regime.

Using the residue theorem, with appropriate semicircular contours and the equation (2.4), it can be shown that

$$P_1(x,0^+) = s_1 + s_2 \tag{3.14}$$

where

$$s_1 = -iL_1 s\sum_{n=0}^{\infty} \frac{\kappa_1^-(M_1k_1-a)}{[\kappa_1^-(\nu_n^+)]'}\,\frac{\nu_n^+-\kappa_1/M_1}{\nu_n^++a-M_1k_1}\,\frac{\exp\{-i(\nu_n^+-M_1k_1)\tilde{x}_1\}}{d_1^+\Gamma_n^{(1)}-\nu_n^+s^+}$$

$$- \frac{i(1-\delta n,0)}{2\pi ns\beta_1}\,\frac{\kappa_1^-(M_1k_1-a)}{\kappa_1^-(\infty)}\exp\left[-i\left\{\frac{d_1^+\Gamma_n^{(1)}}{s^+-s\beta_1} - M_1k_1\right\}(x+s^+)\right]$$

$$- \frac{iL\kappa_1^-(M_1k_1-a)}{\beta_1\kappa_1^-(\infty)}\exp\left[-i\left\{\frac{\beta_1 sM_1k_1-\sigma}{s^+-s\beta_1}\right\}(x+s^+)\left(\frac{1}{2} - \frac{x+s^+}{s^+-\beta_1}\right)\right] -s^+<x<-s\beta_1 \tag{3.15}$$

The ' on the first term in the denominator denotes derivative with respect to α and

$$s_2 = \exp(iM_1k_1x)\sum_{n=0}^{\infty}\Big[\{T_n^+\exp(-i\lambda_n^{(1)}x) + T_n^-\exp(i\lambda_n^{(1)}x)\}$$

$$- \frac{2i}{\beta_1\pi}\left(1-\delta_{n,0}\right)\frac{\kappa_1^-(M_1k_1)}{\kappa^-(\infty)}L_1\frac{(-1)^n}{n}\exp(i\sigma)\sin\left(\frac{n\pi x}{\beta_1 s}\right)\Big]$$

$$- \frac{i}{\beta_1^2}\,\frac{\kappa_1^-(M_1k_1-a)}{\kappa_1^-(\infty)}L_1\frac{x}{s}\exp\left\{i(\sigma+M_1k_1x)\right\}$$

$$+ \frac{L_1(a-\omega_1)}{\tilde{\omega}_1\sin s\tilde{\omega}_1}\exp(iax)\{\exp(i\sigma)-\cos\tilde{\omega}_1 s\exp(is^+a)\} -s\beta_1<x<0 \tag{3.16}$$

where $\tilde{\omega}_1 = [M_1^2(\omega_1-a)^2 - a^2]^{1/2}$

$$T_n^{\pm} = - \left[\frac{(\pm\lambda_n^{(1)}-k_1/M_1)}{\pm\lambda_n^{(1)}-M_1k_1+a)}\right]Q_n^{\pm} \qquad - s\beta_1 < x < 0$$

$$Q_n^{\pm} = \frac{\kappa_1^-(M_1k_1-a)}{\kappa_1(\pm\lambda_n^{(1)})} \frac{[\exp\{i(\sigma-n\pi\}-\exp\{is^+(M_1k_1\mp\lambda_n^{(1)}\}]}{s(\pm\lambda_n^{(1)})(1+\delta n,0)\beta_1^2}$$

$$\Phi_1(0,y) = i\sum_{n=0}^{\infty}\left(\frac{Q_n^+}{\lambda_n^{(1)}-M_1k_1+a} + \frac{Q_n^-}{-\lambda_n^{(1)}-M_1k_1+a}\right)\cos\frac{n\pi y}{s}$$
$$+ L_1\frac{\cos\tilde{\omega}_1 y\exp(i\sigma)-\cos\tilde{\omega}(y-s)\exp(ias^+)}{\tilde{\omega}_1\sin s\tilde{\omega}_1} \qquad (3.17)$$

$$P_1(0,y) = -\left\{\sum_{n=0}^{\infty}\frac{Q_n^+\left(\lambda_n^{(1)}-\frac{k_1}{M_1}\right)}{\lambda_n^{(1)}-M_1k_1+a} + \frac{Q_n^-\left(\lambda_n^{(1)}+\frac{k_1}{M_1}\right)}{\lambda_n^{(1)}+M_1k_1-a}\right\}\cos\frac{n\pi y}{s}$$
$$+ L_1(a-\omega_1)\cdot\frac{\cos\tilde{\omega}_1 y\exp(i\sigma)-\cos\tilde{\omega}_1(y-s)\exp(ias^+)}{\tilde{\omega}_1\sin\tilde{\omega}_1 s}$$
$$= \sum_{n=0}^{\infty}R_n\cos\frac{n\pi y}{s} - L_1(\omega_1-a)\cdot\frac{e^{i\sigma}\cos\tilde{\omega}_1 y-e^{ias^+}\cos\tilde{\omega}_1(y,s)}{\tilde{\omega}_1\sin\tilde{\omega}_1 s} \qquad (3.18)$$

$$U_1(0,y) = \sum_{n=0}^{\infty}Q_n\cos\frac{n\pi y}{s} - La\frac{\cos\tilde{\omega}_1 y\exp(i\sigma)-\cos\omega_1(y,s)\exp(ias^+)}{\tilde{\omega}_1\sin s\tilde{\omega}_1} \qquad (3.19)$$

We are now in a position to construct the solution of the subsonic regime.

4. Solution of the problem in the subsonic regime. Here we are concerned with solving the equations (2.6), (2.7) subject to the jump conditions on the shocks given by (2.16), (2.17) together with (3.17) and (3.18), the kinematic conditions (2.14), (2.15) to be satisfied on the blades and the periodicity condition (2.13). Also the solution must satisfy the continuity of the pressure and the upwash velocity across the wakes which are approximated, to the leading order, by vortex sheets emanating from the trailing edges, the Kutta condition at the trailing edges together with the requirement that only the outward propogating disturbances exist for downstream.

The solution consists of two parts, namely, the accoustical part $\Phi_2(x,y)$ and the vortical part $\Psi_2(x,y)$. The latter can be easily seen to be given by

$$\Psi_2(x,y) = \Omega(y)\exp(i\omega_2 x) \qquad (4.1)$$
$$ns < y < (n+1)s,\ n = 0, \pm 1, \pm 2, \ldots$$

where the periodicity requirement entails

$$\Omega(y + ns) = \exp\{in(\sigma - \omega_2 s^+)\}\Omega(y) \tag{4.2}$$

$$0 < y < s, \; n = 0, \pm 1, \pm 2, \ldots$$

we augment the solution (4.1) by the function,

$$\tilde{\psi} = \frac{\Omega(s(n+1))\sinh \omega_2(y-ns) + \Omega(ns)\sinh \omega_2[s(n+1)-y]\exp(i\omega_2 x)}{\sinh \omega_2 s} \tag{4.3}$$

$$ns < y < (n + 1)s, \; n = 0, \pm 1, \pm 2, \ldots.$$

such that

$$\tilde{\Psi}_2 = \Psi_2 - \tilde{\psi} \tag{4.4}$$

satisfies the differential equation (2.7) together with vanishing upwash on the blades, has zero jump in pressure or upwash velocity across the wakes. Indeed it does not make any contribution to the pressure field at any point. Also we shall regard $\tilde{\Psi}_2$ as given by (4.4) as our required vortical solution. It follows then, that the boundary conditions on the blades and the jump condition across the wake must be satisfied entirely by the accoustical part, Φ_2 of the solution. The construction of this solution consists of two parts, $\Phi_2^{(1)}$, $\Phi_2^{(2)}$. The first one satisfies the kinematic boundary condition on the blades, the Kutta condition at the trailing edges and has the correct wake jump conditions. This represents the incident infinite duct waves and the reflected waves produced by the reflection of the incident wave on the open end of the cascade. The second part $\Phi_2^{(2)}$ has the right behavious at infinity, as at the trailing edges and at the wakes but otherwise has zero contribution to the boundary condition at the blades. It's determination shows that it has an arbitrarily specified downstream propogating wave field for upstream of the trailing edges. The composite solution $\Psi_2 + \Phi_2^{(1)} + \Phi_2^{(2)}$ then satisfies all the boundary contitions and the arbitrariness in Ψ_2 and Φ_2 is just sufficient to satisfy the condition at the shock.

The solution $\Phi_2^{(2)}$ is computed by Main and Harvey (7). In terms of the terminology of this paper, it is given as,

$$\Phi_2^{(2)} = -i \sum_{n=0}^{\infty} B_n \left[\frac{\exp(i\eta_n^+ x)}{k_2/M_2 - \lambda_n^{(2)}} \cos \frac{n\pi y}{s} - \frac{\exp(i\eta_n^+)}{4\pi i}\{\exp(in\pi - i\sigma) - \exp(-i\eta_n^+ s^+)\} \right.$$

$$\left. \times \int_{-\infty+i\delta_2}^{\infty+i\delta_2} \frac{\kappa_2^-(-\lambda_n^{(2)})}{\kappa_2^-(\alpha)} \, \frac{\Lambda_2(\alpha,y)\exp\{-i(\alpha+M_2k_2)\tilde{x}_2\}}{(\alpha+\lambda_n^{(2)})(\alpha+k2/M_2)} \, d\alpha \right] \quad \text{for } 0 < y < s \tag{4.5}$$

$$\text{where } x_2^+ = x + s^+ - 1 \tag{4.6}$$

$$\Lambda_2(\alpha,y) = \frac{1}{2i}\left\{ \frac{\exp(-\beta_2\gamma_2 y + i\Delta_2^+)}{\sin \Delta_2^+} + \frac{\exp(\beta_2\gamma_2 y + i\Delta_2^-)}{\sin \Delta_2^-} \right\} \quad \text{for } 0 < y < s \tag{4.7}$$

$$\Delta_n^{\pm} = \frac{1}{2}(\sigma + M_2k_2s^+ + \alpha s^+) \pm s\beta_2\gamma_2/2i \tag{4.8}$$

$$\eta_n^{\pm} \equiv -M_2k_2 \pm \lambda_n^{(2)} \tag{4.9}$$

$$\lambda_n^{(2)} \equiv i\left[\left(\frac{n\pi}{s\beta_2}\right)^2 - k_2^2\right]^{1/2} \tag{4.10}$$

$-\varepsilon_2 < \delta_2 < \varepsilon_2 \equiv \mathrm{Im}k_2$ and $\gamma_2 \equiv (\alpha^2 - k_2^2)^{1/2}$ with the branches of the roots are chosen as in figure (3).

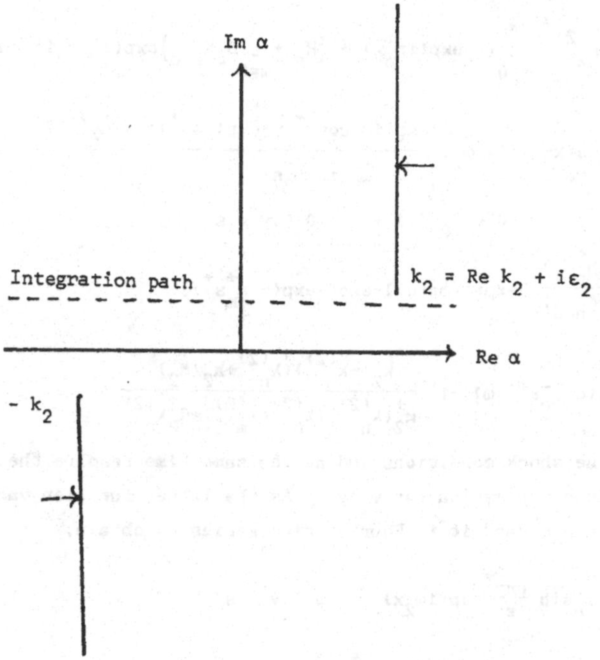

Fig 3 Branch cuts for $\gamma_2 = \sqrt{\alpha^2 - k_2^2}$

The unknown constants B_n are the amplitudes of the incident duct waves $[\exp(i\eta_n^+ x)\cos\frac{n\pi y}{s}]$ and the second terms, as mentioned above are the reflected waves.

The function $\kappa_1^-(\alpha)$ arises in the factorization $\kappa_1^-(\alpha)/\kappa_2^+(\alpha) = [-\frac{\partial}{\partial y}\Lambda_2(\alpha,y)]|_{y=0}/(\alpha+k_2/m_2)$, is required in the implementation of Wiener—Hopf technique to obtain the solution. Following the procedure, as shown in the supersonic case, one can compute the second part of the solution,

$$P_2^{(1)} = (i\omega_2 - \frac{\partial}{\partial x})\phi_2^{(1)} = \sum_{n=0}^{\infty} \frac{H_n \exp(i\eta_2^- \tilde{x}_2)(\omega_2+\lambda_n^{(2)}+M_2k_2)}{(\lambda_n^{(2)} + k_2/M_2)} \cos\frac{n\pi y}{s}$$

$$+ L_2 \frac{\exp\{ia(x_2-s^+)\}(\omega_2-a)[\exp(i\sigma)\cos\omega_2 y - \exp(ias^+)\cos\tilde{\omega}_2(y-s)]}{\tilde{\omega}_2 \sin\tilde{\omega}_2 s} \tag{4.11}$$

$$P_2^{(2)}(x,y) = (i\omega_2 - \frac{\partial}{\partial x})\phi_2^{(2)}(x,y)$$

$$= \sum_{n=0}^{\infty}\left[\left(\frac{B_n\exp(i\eta_n^+x)}{k_2/M_2-\lambda_n^{(2)}}\right)(\omega_2-\eta_n^+) + \sum_{m=0}^{\infty} B_m K_{m,n}\frac{\exp(i\eta_n^-\tilde{x}_2}{(\lambda_n^{(2)}+k_2/M_2)}(\omega_2-\eta_n^-)\right]\cos\frac{n\pi y}{s}$$

$$0 < x_2 < 1 - s^+,\ 0 < y < s \qquad (4.12)$$

As a result,

$$P_2 = P_2^{(1)} + P_2^{(2)} = \sum_{n=0}^{\infty}\{B_n\exp(i\eta_n^+x) + \left(H_n + \sum_{m=0}^{\infty} B_m K_{m,n}\right)\exp(i\eta_x^-x_2)\}\cos\frac{n\pi y}{s}$$

$$+ L_2\exp(ia(x-1))(a-\omega_2)\frac{\exp(i\sigma)\cos\tilde{\omega}_2 y-\exp(ias^+)\cos\tilde{\omega}_2(y-s)}{\tilde{\omega}_2\sin\tilde{\omega}_2 s}$$

$$0 < x_2 < 1 - s^+,\ 0 < y < s \qquad (4.13)$$

$$K_{m,n} = \frac{\exp(i\eta_{n-1}^+)}{-2(1+\delta_{n,0})}[\exp\{i(m\pi-\sigma)\}-\exp(-\exp(-i\eta_m^+s^+)]$$

$$\times[\exp\{i(\sigma-\eta_n^-s^+-n\omega)\}-1]\frac{\kappa_2^-(-\lambda_n^{(2)})(\lambda_n^{(2)}+k_2/M_2)}{\kappa_2^+(\lambda_n^{(2)})(\lambda_n^{(2)}+\lambda_m^{(2)})s\beta_2^2\lambda_n^{(2)}} \qquad (4.14)$$

We now satisfy the shock conditions and at the same time resolve the arbitrariness $\Omega(y)$ present in the expression for $\Psi_2(y)$. As the latter function vanishes at $y = 0$ and $y = s$, we may expand it in Fourier sine series to obtain,

$$\Psi_2(y) = \sum_{n=1}^{\infty} b_n\sin\frac{n\pi y}{s}\exp(i\omega_2 x) \qquad 0 \le y \le s. \qquad (4.15)$$

We now solve the for axial velocity,

$$U_2 = \frac{\partial\Phi_2^{(1)}}{\partial x} + \frac{\partial\Phi_2^{(2)}}{\partial x} + \frac{\partial\Psi_2}{\partial y} = \sum_{n=0}^{\infty}\frac{B_n\eta_n^+\exp(i\eta_n^+x)}{(k_2/M_2-\lambda_n^{(2)})} + \left(\sum_{m=0}^{\infty} B_m k_{m,n}\right)\frac{\eta_n^-\exp(i\eta_n^-\tilde{x}_2)}{(\lambda_n^{(2)}+k_2/M_2)}$$

$$- \frac{(\lambda_n^{(2)}+M_2k_2)H_n\exp(i\eta_n^-\tilde{x}_2)}{(\lambda_n^{(2)}+k_2/M_2)}\cos\frac{n\pi y}{s}$$

$$+ aL_2\exp\{ia(\tilde{x}_2-s^+)\frac{\exp(i\sigma)\cos\tilde{\omega}_2 y-\exp(ias^+)\cos\tilde{\omega}_2(y-s)}{\tilde{\omega}_2\sin\tilde{\omega}_2 s} + \sum_{n=1}^{\infty}\frac{n\pi}{s}b_n\cos\frac{n\pi y}{s}$$

Substituting the above results, we can now satisfy the shock conditions (2.16), (2.17). This leads us to a Fourier Series in terms of ($\cos\frac{n\pi}{s}$, $\sin\frac{n\pi}{s}$). Invoking the fact that they form a complete set, we deduce

$$a_n^+B_n+a_n^-\left(\sum_{m=0}^{\infty} B_m K_{m,n}\right)\exp\left\{-i\eta_n^-(s^+-1)\right\} = -\frac{2}{s}\frac{F_n}{1+\delta_{n,0}} \qquad (4.16)$$

where

$$F_n = -\frac{s}{(2-\sigma_{n,0})}\left[\frac{2M_2^2}{\beta_2^4}\left(\frac{n^2\pi^2}{s^2}+\omega_2^2\right]Q_n-a_n^-H_n\exp\{i\eta_n^-(s^+-1)\}$$

$$+ \frac{2M_1^2M_2^2-\beta_1^2}{\beta_2^4}\left(\frac{n^2\pi^2}{s^2} + \omega_2^2M_2^2r_1\right)R_n\Big]$$

$$+ c_n^{(2)}L_2\exp(ia)(\omega_2-a)\left\{a^2-4aM_2k_2-(k_2M_2^2(3 + \frac{1}{M_1^2M_2^2}\right\}$$

$$+ L_1(\omega_1-a)c_n^{(1)}\left\{\frac{(2M_1^2M_2^2-\beta_1^2)}{\beta_2^4}\left(\frac{n^2\pi^2}{s^2} + \omega_2^2M_2^2r_1\right) + \frac{2M_2^2a}{\beta_2^4}\,\frac{n^2\pi^2}{s^2}\right\}$$

where

$$a_n^{\pm} = (\lambda_n^{(2)})^2 \pm 2M_2k_2\lambda_n^{(2)} + \left(\frac{k_2}{M_1}\right)^2$$

$$c_n^{(j)} = \frac{\exp(ia)-\exp\{i(\sigma+n\pi)\}}{n^2\pi^2/s^2-M_j^2\pi_j^2} \qquad r_1 = M_1^2 + 1 \tag{4.17}$$

The infinite set of above equations are to be solved for the infinitely many unknowns B_n. This, in turn, allows us to determine the pressure on the upper surface of the blades. Invoking the periodicity condition, we can also calculate $P_2(x,0-)$ i.e. the pressure on the lower side of the blade. These expressions are rather unwidely and are given in the complete treatment [12] elsewhere.

5. The total fluctuating lift and moments on the surface of a blade consists of the contributions to these due to the unsteady surface pressures together with the direct contributions due to the motion of the shock. The amplitude of the shock-induced lift fluctuations, non- dimensionalized by $c\rho_1$, U_1^2 is given by,

$$L_s = \frac{i}{2k_1M_1}\,\frac{2M_1^2-\mu+1}{\mu+1}\,P_1(0,0^+) + \frac{\mu}{\mu+1}\,U_1(0,0^+) - P_2(0,0^+) \tag{5.1}$$

Assuming that the equilibrium position of the shock is slightly ahead of the blading edge of the upper blade, the contribution to the fluctuating lift from the motion of its foot-print is zero and the dimensionless moment coefficient about a point $x = d_0$ is

$$M = \int_{-s^+}^{1-s^+}(x-d_0)[P]_0dx + d_0\zeta_s \tag{5.2}$$

where $d_0\zeta_s$ is the moment of the shock-induced lift and $[P]_0$ is the pressure jump across the blade, n = 0. For the torsional motion, the work per cycle done on the flow by the blades is equal to $\pi A_0\rho_1\tilde{U}_1^2\text{Im } M$. When this quantity is positive, the blades receive energy from the flow and develop instability. In terms of dimensionless moment coefficient $(\frac{M}{-A_0})$ where A_0 is the instantaneous angle of attack, it follows that cascade will flutter when $\text{Im}(\frac{M}{-A_0}) < 0$. These parameters can easily be computed by using various expressions obtained above.

References

[1] Miller, G. R. and Bailey, E. E. Static pressure contours in the blade passage at the tip of several high Mach number ratios. N.A.S.A. Tech. Memo No. X-2170.

[2] Strazisar, A. J. and Chima, R. V. Comparison between optical measurements and a numerical solution of the flow field within a transonic axial-flow compressor rotor. Proc. of the AIAA/ASME/SAE 16th Joint Prop. Conf., Hartford, Conn. (1980).

[3] Goldstein, M. E., Braun, W., Adamczyk. Unsteady flow in supersonic cascades with strong in-passage shocks. J. Fluid Mech. 83, (1977), 596.

[4] Eckhaus, W. Two dimensional transonic unsteady flow with shock waves. Office Sci. Res. Tech. Note No. 59 (1959), 491.

[5] Williams, M. H. Linearization of unsteady transonic flows containing shocks. AIAA J. 17 (1979), 394.

[6] Lane, F. System mode shapes in the flutter of compressor blade rows. J. Aero. Sci. 23 (1956), 54.

[7] Mani, R. and Harvey, G. Sound transmission through blade power. J. Sound Vib. 12 (1970), 1159.

[8] Namba, M. and Minami, R. Effect of mean blade loading on supersonic cascade flutter. "Aeroelasticity in Turbomachines" P. Suter ed. Juris-Verlag, Zurich (1981).

[9] Ni, R. H. A rational analysis of periodic flow perturbation in supersonic two-dimensional cascade. Trans. ASME, J. Eng. Power 101, 3 pg. 431 (1979).

[10] Adamerzyk, J. J. et al. Supersonic Stall Flutter of High Speed Fans. Trans. ASME J. Engg. Power 104, 3 (1982).

[11] Gostelow, J. R. Cascade Aerodynamics, Pergamon Press (1984).

[12] Puri, K. Unmixed flow in a cascade with in-passage shocks. Submitted for publication in Int. J. of Comp. Mechanics.

ON THE ABSORPTION OF SINGULARITIES IN DISSIPATIVE NONLINEAR EQUATIONS

Jeffrey Rauch[1]
Department of Mathematics
University of Michigan
Ann Arbor, Michigan 48109

Michael C. Reed[2]
Department of Mathematics
Duke University
Durham, North Carolina 27706

In this talk I want to describe some recent work on the absorption of singularities by dissipation. We formulate the problem as follows: Let

$$L = \partial_t + A(x,t)\partial_x + B(x,t)$$

be a strictly hyperbolic operator with smooth coefficients in one space dimension. Consider the semi-linear initial value problem

$$\begin{aligned} L u^\epsilon &= f(x,t,u^\epsilon) \\ u^\epsilon|_{t=0} &= g + h^\epsilon \end{aligned} \tag{1}$$

where f is smooth, $g \in L^\infty$ and h^ϵ is a family of smooth functions which become singular as $\epsilon \to 0$. For example, imagine that h^ϵ approaches the delta function as $\epsilon \to 0$. If f is "dissipative" the solution u^ϵ should not become singular for $t > 0$ even though the initial data become singular, that is, the singularity should be smoothed out by the dissipation. To see what behavior to expect, consider a simple example:

$$\partial_t u^\epsilon(x,t) = -(u^\epsilon(x,t))^3$$

$$u^\epsilon(x,0) = j_\epsilon(x)$$

where $j_\epsilon(x)$ is the usual approximation to the delta function. The solution is

$$u^\epsilon(x,t) = j_\epsilon(x)(1 + 2t\, j_\epsilon(x)^2)^{-1/2} .$$

For each fixed $t > 0$, $u^\epsilon(x,t) \to 0$ in L^1 as $\epsilon \to 0$. Note however that $u^\epsilon \not\to 0$ in L^∞. Although u^ϵ converges uniformly to zero on compact sets away from zero, at $x = 0$ we have

$$u^\epsilon(0,t) \to (2t)^{-1/2}.$$

---

[1] Research partially supported by NSF Grant MCS-8301061.

[2] Research partially supported by NSF Grant DMS-8401590.

So, in the L^1 sense, nothing remains of the singularity as $\epsilon \to 0$, while in the L^∞ sense what remains is the singular line $\{x = 0\}$. This is the phenomenon; the problem is to show that for appropriate hypotheses this behavior persists in the system (1).

For a fixed $T > 0$, let R denote the domain of determinancy of $\{t = T, -N \le x \le N\}$ and by R_t the time slice $\{x | \langle t,x\rangle \in R\}$.

<u>Hypotheses on f</u>: We assume that $f(t,x,u)$ is measurable on $R \times \mathbb{R}^k$ and that

$$\frac{\partial f}{\partial u} \in L^\infty(K) \quad \text{for any compact } K \subset R \times \mathbb{R}^k . \tag{2}$$

In order to prevent blowup we assume that there is a constant c so that

$$\operatorname{sgn}(u_j) f_j(t,x,u) \le c\Big(1 + \sum_{i=1}^{k} |u_i|\Big) \tag{3}$$

holds for all t,x,u,j. One then has the differential inequality

$$(\partial_t + \lambda_j \partial_x)(|u_j|) \le c(1 + \Sigma |u_i|) .$$

It follows that

$$\sum_{i=1}^{k} \sup_{x\in R_t} |u_i(t,x)| \le c'\Big(1 + \sum_{i=1}^{k} \sup_{x\in R_0} |u_i(0,x)|\Big)$$

for all $0 \le t \le T$, so if the data are in L^∞ the solution is in L^∞. We refer to (3) as the <u>non-explosive hypothesis</u>.

The hypothesis of <u>superlinear</u> <u>dissipation</u> is

$$\lim_{|u_j|\to\infty} f_j(t,x,u)/u_j = -\infty \tag{4}$$

the limit holding uniformly for $(t,x,u_1,\dots,u_{j-1},u_{j+1},\dots,u_k)$ in each set $R \times K$ where K is compact in $\mathbb{R}^{k-1}$.

Notice that the Lipschitz hypothesis (4) is only required to hold on compact subsets of u and the dissipative hypothesis for each j is only required to hold for compact subsets of u_i, $i \ne j$. A sample f satisfying these hypotheses is:

$$f_j(u) = \sum_{i=1}^{k} a_{ij}\, u_i - u_j\Big(1 + \sum_{i=1}^{k} u_i^2\Big) .$$

Since these hypotheses are still satisfied if we add linear terms we may include the linear terms in f. By strict hyperbolicity a change of dependent variable diagonalizes A, so we need just study

$$(\partial_t + \Lambda\, \partial_x) u^\epsilon = f(t,x,u^\epsilon) \tag{5}$$

$$u^\epsilon|_{t=0} = g + h^\epsilon$$

where $\Lambda = \mathrm{diag}\{\lambda_1,\dots,\lambda_n\}$.

Hypotheses on $\{h^\epsilon\}$: The functions $\{h^\epsilon\}$ are smooth and:

$\{h^\epsilon\}_{0<\epsilon<1}$ is bounded in $L^1(R_0)$ (6)

$\{h^\epsilon\}$ converges to zero in measure, that is, if (7)

$S^{\epsilon,\eta} = \{x \in R_0 : |h^\epsilon(x)| > \eta\}$ then $\lim_{\epsilon\to 0} m(S^{\epsilon,\eta}) = 0$

for each $\eta > 0$, where m is Lebesgue measure.

There is a nested family of closed measurable sets $\{T^\epsilon\}$, (8)

$T^{\epsilon_1} \subseteq T^{\epsilon_2}$ if $\epsilon_1 \le \epsilon_2$, with $\lim_{\epsilon\to 0} m(T^\epsilon) = 0$, which satisfy:

For each fixed $\epsilon_2 > 0$ and $\eta > 0$, there exists an ϵ_1 so that $S^{\epsilon,\eta} \subset T^{\epsilon_2}$ if $\epsilon \le \epsilon_1$.

Let S_i denote the flow out of $\cap T^\epsilon$ under the vector field $\partial_t + \lambda_i\partial_x$ and $S = \cup S_i$. We can now state:

Theorem. Let g, f and h^ϵ satisfy the above hypotheses and suppose that

$$(\partial_t + \Lambda\partial_x)u^\epsilon = f(t,x,u^\epsilon), \quad u^\epsilon(0,x) = g + h^\epsilon$$

$$(\partial_t + \Lambda\partial_x)\bar{u} = f(t,x,\bar{u}), \quad \bar{u}(0,x) = g.$$

Then $u^\epsilon \to \bar{u}$ in L^1 on each time slice for $t > 0$ and uniformly for $t \in [\bar{t},T]$ for any $\bar{t} > 0$, $T < \infty$. Furthermore, if K is any compact set in the complement of S and if η is given, then

$$|u^\epsilon(t,x) - \bar{u}(t,x)| \le \eta, \qquad (t,x) \in K$$

for ϵ small enough.

Thus, this theorem recovers the two phenomena in the example: L^1 convergence on time slices; sup norm convergence off the flowout of the singular set.

The proof of the theorem begins by reducing to the case $g = 0$ and $f(t,x,0) = 0$. Let $\mathcal{S}_j^{\epsilon,\eta}(t)$ denote the time slice of the flow out of the set $\mathcal{S}^{\epsilon,\eta}$ under $\partial_t + \lambda_i\partial_x$ and set $\mathcal{S}^{\epsilon,\eta}(t) = \cup_j \mathcal{S}_j^{\epsilon,\eta}(t)$. Let $\bar{t} > 0$ be given. Through a series of bootstrap estimates one shows

$$|u^\epsilon(t,x) - \bar{u}(t,x)| \lesssim \eta, \qquad (t,x) \in R\backslash\mathcal{S}^{\epsilon,\eta}(t)$$

and

$$\int_{\mathcal{S}^{\epsilon,\eta}(t)} |u^\epsilon(t,x)|dx \lesssim c\eta \quad \text{if} \quad t > \bar{t}$$

from which the theorem follows. One already knows $|u_j^\epsilon| \lesssim cM(\epsilon)$ from the non-explosive hypothesis, where $M(\epsilon)$ is the sup of the data, and $\|u^\epsilon\|_{L^1(R)} \lesssim c$. An integration over backward characteristic regions then yields

$$\int_{\mathcal{C}_j} |v_j^\epsilon| \lesssim c \qquad i \neq j .$$

Using Gronwall's inequality on characteristics then gives

$$|u_j^\epsilon(p)| \lesssim c, \quad \text{for all} \quad p \notin \mathcal{S}_j^{\epsilon,\eta} \quad \text{for each} \quad j .$$

Now, one uses the superlinear dissipation to control the values of u_j in $\mathcal{S}_j^{\epsilon,\eta}$ showing that either

$$|u_j^\epsilon(p)| \lesssim \frac{\eta}{m(T^\epsilon)} c + c$$

or

$$|u_j^\epsilon(p)| \lesssim \eta(h_j^\epsilon(q) + c)$$

where q is the point where $\mathcal{C}_j$ intersects $\{t = 0\}$. This implies

$$\int_{\mathcal{C}_i \cap \mathcal{S}_j^{\epsilon,\eta}} |u_j^\epsilon| \lesssim c\eta$$

from which the crucial estimates described above follow by an appropriate use of Gronwall's inequality.

It may seem at first that in the presence of the condition of superlinear dissipation one shouldn't need the non-explosive hypothesis at all. But actually the dissipation condition is not a strong hypothesis since it is only required to hold for large u_j if the other u_i are in a compact set. Nothing is said in the hypothesis about what happens if all the variables are big. The following example ($\Box u + 8u_t^3 = 0$ written as a system) makes this situation clear:

$$(\partial_t + \partial_x)w = -(w + v)^3$$

$$(\partial_t - \partial_x)v = -(w + v)^3 .$$

The non-linearity is easily seen to satisfy the superlinear dissipation hypotheses but it does not satisfy the non-explosive condition. Note that $\partial_t(w - v) + \partial_x(v + w) = 0$ so $\int (w - v)dx$ is a conserved quantity. Therefore if $v^\epsilon(0,x) \equiv 0$

and $w^{\epsilon}(0,x) = j_{\epsilon}(x)$ we will have $\int w^{\epsilon}(t,x) - v^{\epsilon}(t,x)dx = 1$, so we can't have $(w^{\epsilon},v^{\epsilon}) \to 0$ in L^1 for $t > 0$.

The analogous problem in higher space dimensions is open.

Reference

[1] Rauch, J. and M. Reed, "Nonlinear superposition and absorption of delta waves in one space dimension", J. Functional Analysis, to appear, 1987.

FEEDBACK CONTROL FOR AN ABSTRACT PARABOLIC EQUATION

Rouben Rostamian and Thomas I. Seidman

Department of Mathematics
University of Maryland Baltimore County
Catonsville, Maryland 21228, U.S.A.

Takao Nambu

Department of Mathematics
Kumamoto University
Kumamoto, Japan

1. Introduction

In our previous paper [6] we considered the problem of the exponential stabilization of the heat equation with Neumann boundary conditions

$$\begin{aligned} \frac{\partial u}{\partial t} &= \Delta u && \text{in } \Omega \times (0,\infty) \\ \frac{\partial u}{\partial \nu} &= 0 && \text{on } \partial\Omega \times (0,\infty) \\ u(x,0) &= u_0(x) && x \text{ in } \Omega \end{aligned} \tag{1.1}$$

in the smooth and bounded domain Ω in $\mathbf{R}^n$, where $\partial/\partial\nu$ denotes the differentiation in the direction normal to the boundary $\partial\Omega$. The solutions of this problem, although stable (in the sense of Liapunov, considered for any L^p space, say) are not asymptotically stable, because spatially homogeneous functions do not decay. A practical way to induce asymptotic stability is to introduce a feedback mechanism which observes the temperature at a patch of the boundary $\partial\Omega$ and causes an appropriate exchange of heat through the boundary or through an interior subset of Ω. A feedback mechanism of the latter type may be modeled by introducing a term of the form

$$\Phi(x,t) := -\epsilon\left(\int_{\partial\Omega} u(y,t)\varphi(y)\,dy\right)\sigma(x) \tag{1.2}$$

on the right hand side of the equation in (1.1). The functions φ and σ, prescribed on $\partial\Omega$ and Ω, respectively, are indicators of the sites where the observation and the feedback take place. The coefficient ϵ is the 'gain factor' of the feedback. Feedback mechanisms of the former type may also be treated by combining the method of section 6 of [6] and the estimates developed in the present paper.

More generally, we may consider a more complex diffusion processs governed by a general (autonomous) linear elliptic operator A, and several feedback mechanisms of the type above that act simultaneously and independently, so that the governing equation becomes

$$\frac{\partial u}{\partial t} = Au - \epsilon\sum_{k=1}^{p}\sum_{j=1}^{q}\gamma_{jk}\left(\int_{\partial\Omega} u(y,t)\varphi_k(y)\right)\sigma_j(x). \tag{1.3}$$

For instance,

$$Au = \sum_{i,j=1}^{n}\frac{\partial}{\partial x_i}\left(a_{ij}(x)\frac{\partial u}{\partial x_j}\right) + \sum_{i=1}^{n} b_i(x)\frac{\partial u}{\partial x_i}$$

may correspomd to a heat conduction problem in an anisotropic and inhomogeneous medium.

The method introduced in [6] for the stability analysis extends to this and more general situations. In this paper we study an abstract evolution equation which includes the problem in [6] and the problem corresponding to equation (1.3) as special cases. Our main result, in the context of the latter problem is, roughly, that if certain hypotheses are satisfied and if ϵ is sufficiently small then the system is exponentially asymptotically stable. Studies of similar kind, albeit with quite more complicated sufficient stability conditions, appear in various places in the literature; we refer the reader to the articles in [1], [5], [7], and [8].

The smallness of the feedback gain factor ϵ is essential for stability. In [6] we give an example where for a large ϵ the solution grows exponentially as $t \to \infty$. For a comprehensive study of the spectral properties and the spectral resolution of elliptic operators of the type occuring in (1.3) see the recent article of van Harten[3].

2. The Abstract Problem

Let A be a sectorial operator (cf. [4]) on a Hilbert space H, and denote by e^{tA} the analytic semigroup generated by A. Suppose that the null-space H_0 of A is finite-dimensional and non-trivial, and that the following hypotheses hold:

Hypotheses on A:

(i) There exists a continuous projection operator Q, not necessarily orthogonal, of H onto H_0, which commutes with A; thus $AQ = QA = 0$.

(ii) If $P = I - Q$, where I is the identity mapping in H, and $H_1 = \text{range}(P) = \text{null space}(Q)$, then there exist positive constants a and α such that

$$\left|e^{tA}v\right| \leq ae^{-\alpha t}|v|, \qquad v \in H_1. \tag{2.1}$$

Thus, in the dynamical system corresponding to e^{tA}, all orbits in H_1 are asymptotically exponentially stable. The points in H_0, on the other hand, are equilibria and remain stationary. We now introduce a feedback process, motivated by the concrete application in [6], which makes *all* orbits in H tend exponentially to zero as $t \to \infty$.

We suppose we are given q elements $\{\sigma_1, \cdots, \sigma_q\}$ of H and p linear (possibly unbounded) functionals $\{l_1, \cdots, l_p\}$ on H, for which there exists a $p \times q$ real matrix γ_{jk} such that the feedback operator B, formally defined by

$$B(u) = -\sum_{k=1}^{p}\sum_{j=1}^{q}\gamma_{jk}l_k(u)\sigma_j, \tag{2.2}$$

satisfies the following hypotheses:

Hypotheses on B:

(i) For each $t > 0$ and $k = 1, 2, \cdots, p$, the expression $l_k\left(e^{tA}\right)$ defines a *bounded* linear functional on H.

(ii) There exists a $\delta \in (0,1)$ such that

$$\left|l_k\left(e^{tA}u\right)\right| \leq at^{-\delta}e^{-\alpha t}|u|, \qquad u \in H_1, \tag{2.3}$$

and

$$|l_k(u)| \leq a|u|, \qquad u \in H_0. \tag{2.4}$$

(iii) If we define the operator $B_0 : H_0 \to H_0$ by

$$B_0 w = -\sum_{k=1}^{p}\sum_{j=1}^{q} \gamma_{jk} l_k(w) Q\sigma_j, \qquad w \in H_0, \tag{2.5}$$

then there exist positive constants b and β such that

$$\left|e^{tB_0}w\right| \leq be^{-\beta t}|w|, \qquad w \in H_0. \tag{2.6}$$

Remark: The inequalities (2.4) and (2.6) imply the estimate

$$\left|l_k\left(e^{tB_0}w\right)\right| \leq abe^{-\beta t}|w|, \qquad w \in H_0. \tag{2.7}$$

Note that, formally, $B_0 = QBQ$, and that the operator Be^{tA} is bounded in H for each $t > 0$. In view of this we are able to show that the equation

$$\begin{aligned} \dot{u} &= Au + \epsilon Bu \\ u(0) &= u_0 \end{aligned} \tag{2.8}$$

makes sense in H. We prove the

Stabilization Theorem: *There exists a number $\epsilon_0 > 0$, depending on the constants $a, \alpha, b, \beta, \delta$ entering in estimates in (2.1), (2.3), and (2.6), as well as on p, q, γ_{jk}, and the norms of the σ_j's, such that for any $\epsilon \in (0, \epsilon_0)$ the solution of (2.8) decays exponentially as $t \to \infty$.*

We defer the proof to the next section but first make some remarks:

(i) It is possible to obtain an *explicit* upper bound ϵ_0 in terms of the quantities mentioned above. The computation proceeds along the lines described in section 4 of [6].

(ii) It is worthwhile to observe that in our computation we use the estimates in (2.1), (2.3), (2.4), (2.6) very sparingly. In fact, the bulk of the computation in the following section, up to the derivation of the system of integral equations (3.7), is in terms of equalities rather than estimations. This has the advantage of keeping the results as sharp as possible; moreover, the general theorem on integral equations developed in [6] applies here directly.

(iii) The inequality in (2.1) applies in situations where the initial condition is in the space H. This assumption is of course unnecessarily strong; in concrete cases the initial conditions are often allowed to be outside H. For instance, H may be $L^2(\Omega)$ whereas u_0 is a distribution. The regularizing effect of the analytic semigroup puts the orbit in H in positive times. The estimate (2.1) can be modified to apply to such irregular data; see [6] for details.

(iv) The estimate in (2.6) is an assumption of the positivity of the operator $-B_0$ on the space H_0. One might also consider a possible order-preserving property of $-B_0$, in the form

$$w \geq 0 \Rightarrow -B_0 w \geq 0, \quad \forall w \in H_0$$

where the inequalities are understood in the sense of a partial order on H for which one has a maximum principle for A. It seems likely, but we have not verified this, that under this stronger condition the equation (2.8) is asymptotically stable with no restrictions on the magnitude of the coefficient ϵ. See [6] for further discussion and practical impications.

(v) It is possible to show, although we do not do it here, that the operator $A + \epsilon B$ entering in (2.8) is sectorial for arbitrary ϵ. Hence it generates an analytic semigroup on H.

3. The Proof of the Stabilization Result

We apply the projections P and Q to the evolution equation (2.8) and use the notation $v = Pu$, $w = Qu$, to arrive at the following coupled system of differential equations:

$$\dot{v} = Av - \epsilon \sum_{k=1}^{p}\sum_{j=1}^{q} \gamma_{jk} l_k(u) P\sigma_j \tag{3.1}$$

$$\dot{w} = -\epsilon \sum_{k=1}^{p}\sum_{j=1}^{q} \gamma_{jk} l_k(u) Q\sigma_j \tag{3.2}$$

We also set $v_0 = v(0) = Pu(0)$ and $w_0 = w(0) = Qu(0)$. Note that the equation (3.2) may be re-written, in view of the definition of the operator B_0 in (2.5) and the identity $u = v + w$, as:

$$\dot{w} + \epsilon B_0 w = -\epsilon \sum_{k=1}^{p}\sum_{j=1}^{q} l_k(v) Q\sigma_j.$$

Applying the variation of constants formula, we get

$$v(t) = e^{tA} v_0 - \epsilon \sum_{k=1}^{p}\sum_{j=1}^{q} \int_0^t \gamma_{jk} \left[e^{(t-s)A} P\sigma_j\right] l_k(v(s) + w(s))\, ds, \tag{3.3}$$

$$w(t) = e^{\epsilon t B_0} w_0 - \epsilon \sum_{k=1}^{p}\sum_{j=1}^{q} \int_0^t \gamma_{jk} \left[e^{\epsilon(t-s)B_0} Q\sigma_j\right] l_k(v(s))\, ds. \tag{3.4}$$

Now define

$$\rho_i(t) = l_i(v(t)),$$
$$\lambda_{ik}(t) = \sum_{j=1}^{q} \gamma_{jk} l_i\left(e^{tA} P\sigma_j\right),$$
$$\eta_{ik}(t) = \sum_{j=1}^{q} \gamma_{jk} l_i\left(e^{\epsilon t B_0} Q\sigma_j\right),$$

and apply l_i to both of equations (3.3) and (3.4) to obtain

$$\rho_i(t) = l_i\left(e^{tA} v_0\right) - \epsilon \sum_{k=1}^{p} \int_0^t \lambda_{ik}(t-s)\big(\rho_k(s) + l_k(w(s))\big)\, ds,$$

$$l_i(w(t)) = l_i\left(e^{\epsilon t B_0} w_0\right) - \epsilon \sum_{k=1}^{p} \int_0^t \eta_{ik}(t-s) \rho_k(s)\, ds.$$

Substitute the second of these equations in the first, interchange the order of integration in the resulting double-integral (justified by the estimates in (2.1) and (2.3),) and simplify, to arrive at

$$\begin{aligned}\rho_i(t) &= l_i\left(e^{tA}v_0\right) - \epsilon\sum_{k=1}^{p}\int_0^t \lambda_{ik}(t-s)l_k\left(e^{\epsilon sB}w_0\right)ds \\ &-\epsilon\sum_{k=1}^{p}\int_0^t \lambda_{ik}(t-s)\rho_k(s)\,ds \\ &+\epsilon^2\sum_{k,k'=1}^{p}\int_0^t\int_0^s \lambda_{ik}(t-s)\eta_{kk'}(s-\tau)\rho_{k'}(\tau)\,d\tau\,ds. \qquad (3.5)\end{aligned}$$

Further simplification is achieved by letting $\rho_i^*(t)$ denote the sum of the first two terms on the right hand side of (3.5) (i.e., the terms which contain the initial data) and setting

$$\mu_{ij}(t) = \lambda_{ij}(t) - \epsilon\sum_{k=1}^{p}\int_0^t \lambda_{ik}(r)\eta_{kj}(t-r)\,dr. \qquad (3.6)$$

After a bit of manipulation, (3.5) then yields the following system of integral equations:

$$\rho_i(t) = \rho_i^*(t) - \epsilon\sum_{j=1}^{p}\int_0^t \mu_{ij}(t-\tau)\rho_j(\tau)\,d\tau \qquad (3.7)$$

for $i = 1, 2, \cdots, p$.

It is easy to see that by the assumption $0 < \delta < 1$ in (2.3), the functions ρ_i^* and μ_{ik} have *integrable* singularities, hence the system of Volterra equations (3.7) has a unique solution. By a computation similar to the one in [6], it is possible to show that this solution decays exponentially as $t \to \infty$, if $\epsilon > 0$ is sufficiently small. Going back to (3.4), recalling that $l_k(v(t)) = \rho_k(t)$, and using the estimations in (2.1) and (2.3), we conclude that the norm of $w(t)$ decays. Now since

$$l_k(u(t)) = l_k(v(t)) + l_k(w(t)) = \rho_k(t) + l_k(w(t))$$

we get exponential decay for $l_k(u(t))$. Finally, using this in (3.3) we obtain exponential decay for $v(t)$ and, consequently, for $u(t) = v(t) + w(t)$. This completes the proof of the stabilization result.

Remark: A variant of the control problem treated in this paper is that of the boundary observation and *boundary control*. In this situation the feedback mechanism involves interaction with Ω entirely through the boundary of Ω. The corresponding mathematical model is:

$$\frac{\partial u}{\partial t} = \sum_{i,j=1}^{n} \left(a_{ij}(x) \frac{\partial u}{\partial x_j} \right) \qquad \text{in } \Omega \times (0,\infty)$$

$$\frac{\partial u}{\partial \nu} = -\epsilon \sum_{j,k}^{p} \gamma_{jk} \left(\int_{\partial\Omega} u(y,t)\varphi_k(y)\, dy \right) \sigma_j \qquad \text{on } \partial\Omega \times (0,\infty)$$

$$u(x,0) = u_0(x) \qquad x \in \Omega.$$

Abstractly, this corresponds to having $e^{tA}\sigma_j \in H$ for $t > 0$ rather than $\sigma_j \in H$; we must then assume an estimate like (2.3) for $\left|e^{tA}\sigma_j\right|$, corresponding to having $\sigma_j \in \mathcal{D}(A^\delta)$. It is possible to show that this system is also asymptotically exponentially stable provided that the coefficient ϵ is sufficiently small. The proof is along the lines of the argument in section 6 of [6], combined with the generalized frmework that we have developed in the resent paper.

Acknowledgments: The research of R. Rostamian was partially supported by grants DMS–8602006 from the National Science Foundation and N00014–86–K–0498 from the Office of Naval Research. The research of T. I. Seidman was partially supported by grants AFOSR–82–0271 from the Air Force Office of Scientific Research and CDR–85–00108 from the National Science Foundation. T. Nambu, while visiting UMBC, was on leave from Kumamoto University and was also partially supported by grant AFOSR–82–0271.

References:

1. R. F. Curtain, Finite dimensional compensators for parabolic distributed systems with unbounded control and observations, SIAM J. Control Optim. **22** (1984), pp. 255–276.

2. D. Fujiwara, Concrete characterization of the domains of fractional powers of some elliptic differential operators of second order, Proc. Japan Acad. **43**(1967), pp. 82–86.

3. A. van Harten, On the spectral properties of a class of elliptic functional differential operators arising in feedback control theory for diffusion processes, SIAM J. Math. Anal. **17**(1986), pp. 352–383.

4. D. Henry, *Geometric Theory of Semilinear Parabolic Equations*, Lecture Notes in Mathematics, no. 840, Springer-Verlag, 1981.

5. T. Nambu, On the stabilization of diffusion equations: Boundary observation and feedback, J. Diff. Euations **52**(1984), pp. 204–233.

6. T. Nambu, R. Rostamian and T. I. Seidman, Feedback stabilization of a marginally stable heat conduction problem, to appear in Applicable Analysis.

7. Y. Sakawa and T. Matsushita, Feedback stabilization of a class of distributed system and construction of a state estimator, IEEE Trans. Automatic Control AC-20(1975), pp. 748–753.

8. Y. Sakawa, Feedback stabilization of linear differential systems, SIAM J. Control Optim. **21**(1983), pp. 667–676.

APPROXIMATE SOLUTION OF RANDOM DIFFERENTIAL EQUATION*

M. Sambandham
Department of Mathematics
Morehouse College/Atlanta University
Atlanta, GA 30314

and

Negash Medhin
Department of Mathematics and Computer Science
Atlanta University
Atlanta, GA 30314

Abstract

Chebyshev method for solving random differential equation is presented. The convergence of the random coefficients of the Chebyshev series is established. Statistical properties of the random coefficients are discussed.

1. INTRODUCTION

In recent years, increasing interest in the numerical solution of random differential equations has led to the progressive development of several numerical methods. A large number of papers have appeared in the literature containing approximate solutions of random differential equations. For Newton's method, successive approximations, perturbation methods, method of moments, finite element methods and other methods on the approximate solution of random equations we refer to Bharucha-Reid [1]. Numerical methods of random polynomials can be found in Bharucha-Reid and Sambandham [2]. For a short and elegant note on several of these methods we also refer to Lax [9, 10]. Some analytical and numerical estimates on error estimates of stochastic differential equations are presented in Ladde et al. [6-8]. For other interesting numerical techniques we refer to Boyce [3] and [11]. Numerical treatment of Ito equations can be found in Klauder and Petersen [4], Rumelin [12] and Taley [15]. We notice that most of these numerical methods are successful when applied to specific problems and have restrictions to apply to general problems. However, we remark that Chebyshev methods are proved to be more efficient and useful to obtain numerical solutions of random

*Research supported by U.S. Army Research Office, Grant No. DAAG-29-85-G-0109.

Fredholm equations and random singular integral equations [13, 14]. Therefore, the purpose of this paper is to introduce Chebyshev methods to random differential equations. We restrict ourselves to random initial value problems.

The organization of the paper is as follows. Section 2 presents a brief outline on the application of Chebyshev polynomials to random differential equations and special advantages and usefulness of properties of Chebyshev polynomials. In Section 3 we discuss the convergence of the coefficients of Chebyshev series. Section 4 contains two examples and finally in Section 5 we discuss the numerical procedure and present few figures and a table to illustrate some statistical properties of the solution of the examples in Section 4.

2. CHEBYSHEV METHOD

Among various strategies and methods developed for approximating deterministic equations, Chebyshev method plays a major role. The main reason is that Chebyshev polynomials have special properties which are easy to implement and very much useful for computational purposes. Here we develop a computationally feasible method to obtain approximate solution of random differential equations.

Suppose we have random differential equations with polynomial coefficients as follows:

$$g_1(x,\omega)y' + g_2(x,\omega)y = 0, \quad y(0,\omega) = y_0(\omega), \tag{2.1}$$

$$g_1(x,\omega)y'' + g_2(x,\omega)y' + g_3(x,\omega)y = 0, \quad \begin{aligned} y(0,\omega) &= y_0(\omega) \\ y'(0,\omega) &= 0, \end{aligned} \tag{2.2}$$

where $(x,\omega) \in I \times \Omega$, $I = [t_0,T)$ and (Ω,F,P) is the underlying probability space. We assume that the solutions of (2.1) and (2.2) exist with probability one (w.p.1.) and can be expressed as a Chebyshev series, namely,

$$y(x,\omega) = \frac{1}{2} a_0(\omega)T_0(x) + \sum_{j=1}^{\infty} a_j(\omega)T_j(x), \tag{2.3}$$

where $a_i(\omega)$ are random coefficients and $T_j(x)$ are Chebyshev polynomials of the first kind. We notice that if

$$y^{(s)}(x,\omega) = \frac{1}{2} a_0^{(s)}(\omega)T_0(x) + \sum_{j=1}^{\infty} a_j^{(s)}(\omega)T_j(x), \tag{2.4}$$

where $y^{(s)}(x,\omega)$ is the s-th derivative of $y(x,\omega)$, then

$$\begin{aligned} 2ra_r^{(s)}(\omega) &= a_{r-1}^{(s+1)}(\omega) - a_{r+1}^{(s+1)}(\omega) \\ 2xT_r(x) &= T_{r+1}(x) + T_{|r-1|}(x) \\ T_r(x)T_s(x) &= \frac{1}{2}(T_{s+r}(x) + T_{|s-r|}(x)). \end{aligned} \tag{2.5}$$

Substituting (2.3) and (2.4) in (2.1) and (2.2) and separating the coefficients of $T_j(x)$, we get backward random difference equation, which can be solved to get $a_j(\omega)$. The sample space Ω of $a_j(\omega)$ thus obtained can be used later for further discussions on statistical properties of $a_j(\omega)$ and $y(x,\omega)$. The mean and covariance of $y(x,\omega)$ can be stated as follows:

$$\langle y(x,\omega)\rangle = \frac{1}{2}\langle a_0(\omega)\rangle T_0(x) + \sum_{j=1}^{\infty}\langle a_j(\omega)\rangle T_j(x), \tag{2.6}$$

$$\langle y(x,\omega),y(t,\omega)\rangle = \langle \frac{1}{2} a_0(\omega)T_0(x) + \sum_{j=0}^{\infty} a_j(\omega)T_j(x), \frac{1}{2} a_0(\omega)T_0(t) + \sum_{j=0}^{\infty} a_j(\omega)T_j(t)\rangle \tag{2.7}$$

We remark that in (2.7) the right hand side expression can be simplified by an application of (2.5). In the following sections we assume that all equalities and inequalities are true w.p.l.

3. REMARK ON THE RATE OF CONVERGENCE

In this section we establish the rate of convergence of the random coefficients of (2.3). It is well known that, if $g_1(x,\omega) > 0$, $0 \le x \le 1$ w.p.l., we can transform (2.2) into

$$y'' + b(x,\omega)y = 0, \quad y(0,\omega) = y_0(\omega), \quad y'(0,\omega) = y_1(\omega), \tag{3.1}$$

where $b(\cdot,\omega)$ is at least continuous on [0,1]. Let us write

$$y(x,\omega) = \frac{1}{2} a_0(\omega)T_0^*(x) + \sum_{k=1}^{\infty} a_k(\omega)T_k^*(x), \tag{3.2}$$

where T_k^* are Chebyshev polynomials on [0,1]. We remark that (3.1) is equivalent to the integral equation

$$y(x,\omega) = y_0(\omega) + y_1(\omega)x + \int_0^x (t-x)b(x,\omega)y(t,\omega)dt. \tag{3.3}$$

Let

$$\beta(x,\omega) = 1 + \int_0^x (1-t)|b(t,\omega)|\exp\{\int_t^x (1-s)|b(s,\omega)|ds\}dt. \tag{3.4}$$

We note that $\beta(x,\omega) \le \beta(1,\omega)$ w.p.l. Using Gronwall's inequality, from (3.3) and (3.4), we have

$$|y(x,\omega)| \le \beta(x,\omega)(|y_0(\omega)| + |y_1(\omega)|x) \text{ w.p.l.} \tag{3.5}$$

Further, if $y(\cdot,\omega)$ is infinitely differentiable on (0,1) it is well known that w.p.l.

$$a_n(\omega) \sim \frac{d^n y(\frac{1}{2},\omega)/dx^n}{n!\,2^{n-1}}. \tag{3.6}$$

If $b(\cdot,\omega)$ is continuous on [0,1], let us write

$$y'' = \frac{1}{2} d_0(\omega) T_0^*(x) + \sum_{k=1}^{\infty} d_k(\omega) T_k^*(x). \qquad (3.7)$$

Using the properties of Chebyshev polynomials we have

$$a_k(\omega) = \frac{1}{16k(k-1)} (d_{k-2}(\omega) - d_k(\omega)) - \frac{1}{16k(k-1)} (d_k(\omega) - d_{k-2}(\omega)). \qquad (3.8)$$

Since $y'' = -b(x,\omega)y$, from (3.7) we have

$$-b(x,\omega)y = \frac{1}{2} d_0(\omega) T_0^*(x) + \sum_{k=1}^{\infty} d_k(\omega) T_k^*(x). \qquad (3.9)$$

Therefore from the properties of Chebyshev polynomial we get

$$d_k(\omega) = \frac{2}{\pi} \int_{-1}^{1} \frac{-b(\frac{x+1}{2}, \omega) y(\frac{x+1}{2}, \omega) T_k(x)}{\sqrt{1 - x^2}} dx \qquad (3.10)$$

From (3.5) and (3.10) we obtain

$$|d_k(\omega)| \le 2\beta(1,\omega)(|y_0(\omega)| + |y_1(\omega)|) \|b(\cdot,\omega)\|_\infty. \qquad (3.11)$$

From (3.8), by an application of (3.11) we have

$$a_k(\omega) \sim \frac{1}{k^2} \quad \text{w.p.1.} \qquad (3.12)$$

We remark that (3.6) and (3.12) establish the rate of convergence of the coefficients of Chebyshev series. To obtain the approximate solution of (2.1) or (2.2) we truncate (2.3) as follows.

$$y(x,\omega) = \frac{1}{2} a_0(\omega) T_0(x) + \sum_{j=1}^{n} a_j(\omega) T_j(x) + \varepsilon_n. \qquad (3.13)$$

By an application of (3.12) and the properties of Chebyshev polynomials we can estimate ε_n, which is the error if we omit ε_n in (3.13).

4. EXAMPLES

In this section we illustrate Chebyshev method with two examples.

Example 4.1. Let

$$g_1(x,\omega)y' + g_2(x,\omega)y = 0, \quad y_0(\omega) = y_0(\omega),$$
$$g_1(x,\omega) = 2(1+x), \quad g_2(x,\omega) = \nu(\omega). \qquad (4.1)$$

By an application of the method in Section 2 we get

$$(2r+\nu(\omega))a_r^{(0)}(\omega) = -3a_r^{(1)}(\omega) - a_{r+1}^{(1)}(\omega), \quad r = 0,1,2,\ldots, \qquad (4.2)$$

$$\frac{1}{2} a_0^0(\omega) - a_1^{(0)}(\omega) + \ldots = y_0(\omega) \qquad (4.3)$$

Example 4.2. Let

$$g_1(x,\omega)y'' + g_2(x,\omega)y' + g_3(x,\omega)y = 0, \quad y(0,\omega) = y_0(\omega)$$
$$y'(0,\omega) = 0$$

$$g_1(x,\omega) = x\nu_1(\omega), \quad g_2(x,\omega) = \nu_2(\omega), \quad g_3(x,\omega) = x \tag{4.4}$$

By an application of the method in Section 2 we obtain

$$a^{(0)}_{|r-2|}(\omega) = a^{(0)}_{r+2} - \frac{1}{8}(\nu_2(\omega)(r-1) + \nu_1(\omega))(a^{(1)}_{r+1} + a^{(1)}_{r-1}) \tag{4.5}$$

$$r = 0,1,2,\ldots$$

$$\frac{1}{2}a^{(0)}_0 - a^{(0)}_2 + \ldots = y_0(\omega). \tag{4.6}$$

We notice that (4.2) and (4.5) are backward random difference equations and $a_i(\omega)$ can be obtained from them and later these coefficients are normalized using (4.3) and (4.6).

5. RESULTS AND DISCUSSIONS

In this section we consider different forms of Examples 4.1 and 4.2. In the following we take

$$y(x,\omega) = \frac{1}{2}a_0(\omega) + \sum_{j=1}^{n} a_j(\omega)T_j(x)$$

and $N(m,\sigma^2)$ represents normal distribution with mean m and variance σ^2.

Example 4.1 (a). $\nu_1(\omega) = 1$, $y_0(\omega) \in N(1,0.04)$, $n = 4$.

Example 4.1 (b). $\nu_1(\omega)$ is uniform in $1 \le x \le 2$, $y_0(\omega) = 1$, $n = 4$.

Example 4.1 (c). $\nu_1(\omega)$ is uniform in $1 \le x \le 2$, $y_0(\omega) \in N(1,0.04)$, $n = 4$.

Example 4.2 (a). $\nu_1(\omega) = 1 = \nu_2(\omega)$, $y_0(\omega) \in N(1,0.04)$, $n = 10$.

Example 4.2 (b). $\nu_1(\omega), \nu_2(\omega)$ are uniform in $1 \le x \le 2$, $y_0(\omega)$ $N(0,0.04)$ and $n = 10$.

In our simulation, we have taken $a_n = 1$, $a_n^{(1)} = 0$. Each example is simulated for 100 sample values and the sample of $a_i(\omega)$ are plotted in Figures 1-5, which are the frequency distributions of $a_i(\omega)$. Figures 1-5 highlight the behavior of $a_i(\omega)$ and its rapid convergence to zero. Table I illustrates the behavior of the expectations of the coefficients $a_i(\omega)$. From the sample space of $a_i(\omega)$ we can obtain other statistical properties of $a_i(\omega)$ and the solution properties of $y(x,\omega)$.

The foregoing analysis demonstrates a fairly simple and yet useful procedure for the approximate solution of random differential equations. These results highlight that without much restrictions Chebyshev method can be applied to a wide variety of random differential equations.

DISTRIBUTION OF $a_i(\omega)$

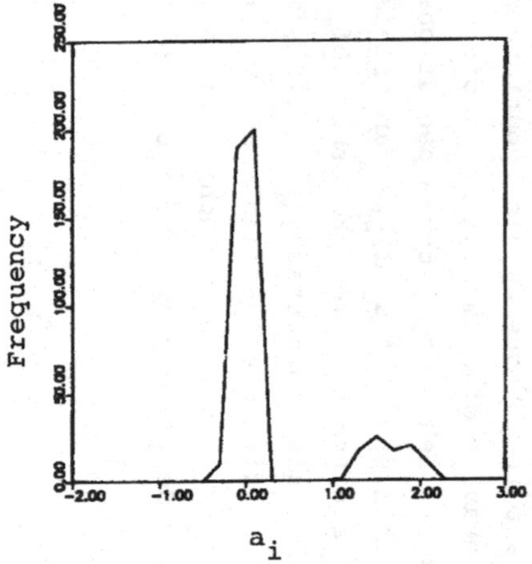

Fig. 1: Example 4.1 (a).

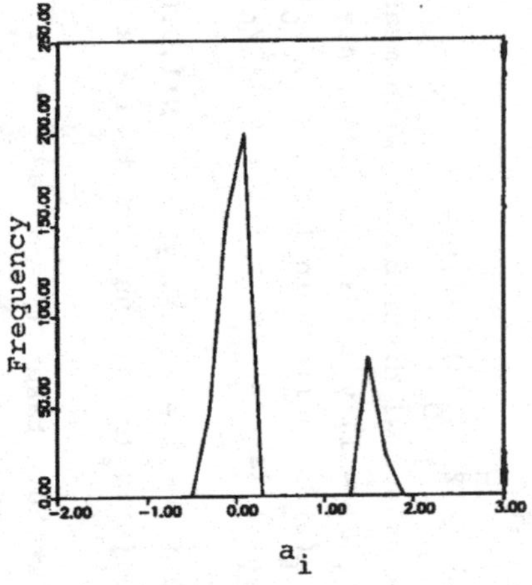

Fig. 2: Example 4.1 (b).

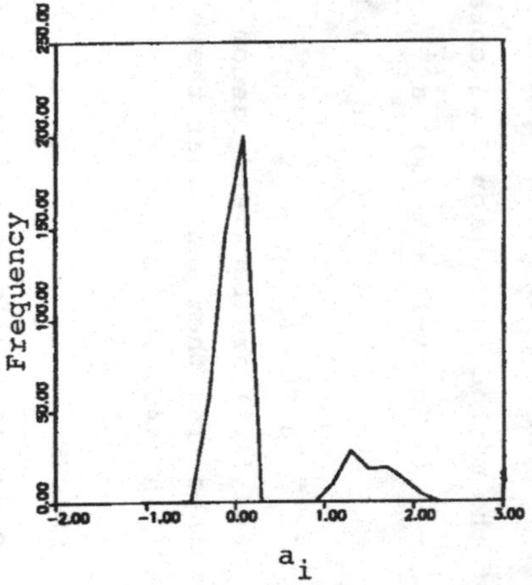

Fig. 3: Example 4.1 (c).

DISTRIBUTION OF $a_{2i}(\omega)$

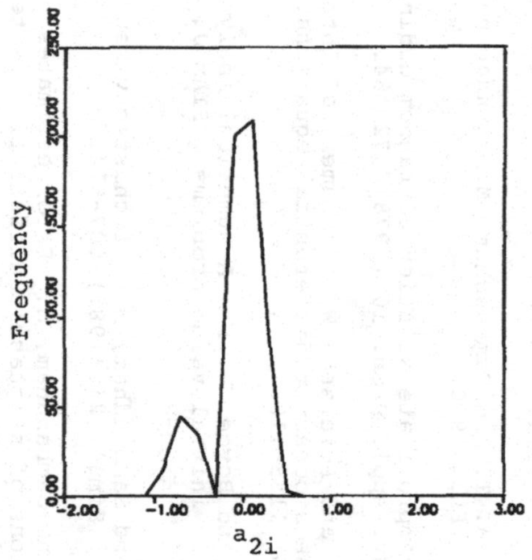

Fig. 4: Example 4.2 (a).

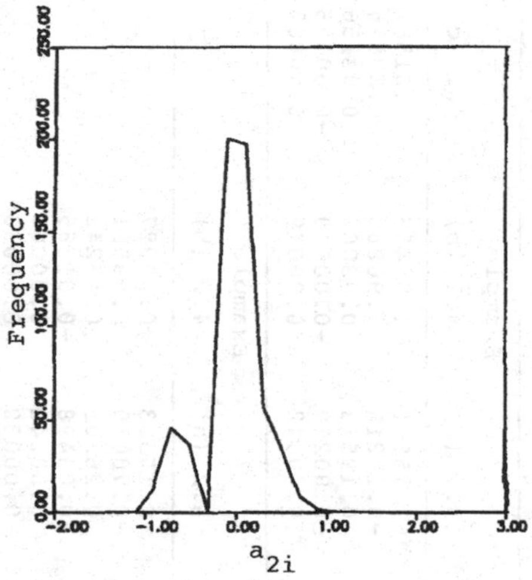

Fig. 5: Example 4.2 (b).

TABLE I. Expected Values of $a_j(\omega)$

j	Example 4.1 (a)	Example 4.1 (b)	Example 4.1 (c)
0	1.75681	1.53456	1.61581
1	-0.15218	-0.96863	-0.20696
2	0.19513	0.03005	0.03156
3	-0.00289	-0.00500	-0.00525
4	0.00042	0.00079	0.00083

j	Example 4.2 (a)	Example 4.2 (b)
0	0.10553	0.39987
2	-0.70019	-0.68581
4	0.26205	0.15242
6	-0.03498	-0.01365
8	0.00242	0.00067
10	-0.00032	-0.00030

REFERENCES

1. Bharucha-Reid, A. T. *Approximate solution of random equations.* North Holland, NY 1979.

2. Bharucha-Reid, A. T. and Sambandham, M. *Random polynomials.* Academic Press, FL, 1986.

3. Boyce, W. E. Approximate solution of random ordinary differential equations. Adv. Appl. Prob. 10 (1978) 172-184.

4. Klauder, J. R. and Petersen, W. P. Numerical integration of multi plicative-noise stochastic differential equations. SIAM J. Numer. Anal. 22 (1985) 1153-1166.

5. Kohler, W. E. and Boyce, W. E. A numerical analysis of some first order stochastic initial value problems. SIAM J. Appl. Math. 27 (1974) 167-179.

6. Ladde, G. S. and Sambandham, M. Stochastic versus deterministic. Math. Computer. Simul. 24 (1982) 507-514.

7. Ladde, G. S. and Sambandham, M. Error estimates of solutions and mean of solutions of stochastic differential systems. J. Math. Physics. 24 (1983) 815-822.

8. Ladde, G. S., Lakshmikantham, V. and Sambandham, M. Comparison theorems and error estimates of stochastic differential systems. Stoch. Anal. Appl. 3 (1985) 23-62.

9. Lax, M. D. Numerical solution of random nonlinear equations. Stoch. Anal. Appl. 3 (1985) 163-169.

10. Lax, M. D. "Approximate solution of random differential and integral." *Applied Stochastic Processes*, G. Adomian, ed. Academic Press, NY 1980.

11. Nina-Mao Xia and Boyce, W. E. The density function of the solution of a random initial value problem containing small stochastic process. SIAM J. Appl. Math. 44 (1984) 1192-1209.

12. Rumelin, W. Numerical treatment of stochastic differential equations. SIAM J. Numer. Anal. 19 (1982) 604-613.

13. Sambandham, M., Srivatsan, T. and Bharucha-Reid, A. T. Numerical solution to singular integral equation in crack problems. *Integral Methods in Science and Engineering*, ed. F. Payne, Hemisphere Publishers, NY, 1986, 130-148.

14. Sambandham, M., Christensen, M. J. and Bharucha-Reid, A. T. Numerical solution to random integral equation III: Chebyshev polynomial and Fredholm equation of the second kind. Stoch. Anal. Appl. 3 (1985) 467-484.

15. Talay, D. "How to discretize stochastic differential equations." *Nonlinear Filtering and Stochastic Control*. Lecture Notes in Mathematics. Springer Verlag, NY (1980) Vol. 972, 276-292.

GEOMETRIC PROPERTIES AND BOUNDS FOR POSITIVE SOLUTIONS OF SEMILINEAR ELLIPTIC EQUATIONS

by

Klaus Schmitt*
University of Utah
Department of Mathematics
Salt Lake City, UT 84112

1. Introduction

In this paper I shall present a collection of results about a priori lower bounds for positive solutions of Dirichlet problems of a semilinear elliptic equation. The motivation for this study in two fold. On the one hand the results given here show how certain a priori lower bounds for positive solutions of the one-dimensional case (i.e., the case of ordinary differential equations) carry over to the partial differential equations case, either for special types of solutions in the case of arbitrary domains, or else for all solutions in a certain class in the case the domain satisfies certain geometric properties. On the other hand, these results allow for distinguishing procedures between numerically "relevant" and "irrelevant" solutions obtained by finite difference approximations of the problem at hand (see [PS], [PSS], [Sch2]).

I shall not strive for generality, but restrict myself to the following model problem

$$(1)\qquad \begin{cases} \Delta u + \lambda f(u) = 0, & x \in \Omega \\ u = 0, & x \in \Omega, \end{cases}$$

where Ω is a bounded domain in $\mathbb{R}^n$ with smooth boundary and f is a C^1 nonlinearity as given in the following figure 1, and λ is a positive real parameter.

*Research supported by NSF grant DMS-8501311

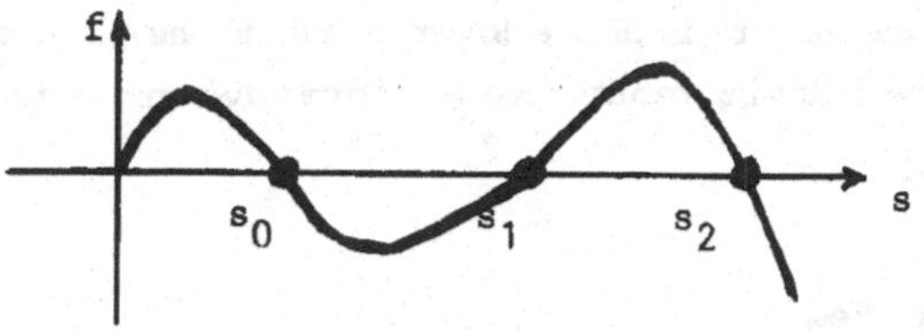

Figure 1

And I will be interested mainly in positive solution of (1) satisfying

(2) $$s_1 < u_{max} < s_2.$$

2. Motivation - The case of an ODE.

If one considers the one-dimensional case of (1), one sees, using phase plane arguments, that a necessary condition for the existence of a positive solution u of (1) satisfying (2) is that

(3) $$\int_{s_0}^{s_2} f(s)ds > 0.$$

(This fact will be demonstrated below also for the higher dimensional case). Furthermore in this case one may demonstrate, again using phase plane arguments (see e.g. [BB]) the following result.

Theorem 1 Let (3) hold, then there exists λ^* such that for all $\lambda > \lambda^*$ problem (1) has at least two solutions u_1, u_2 with

(4) $$r < \max u_i(x) < s_2, \quad i = 1,2,$$

where r, $s_1 < r < s_2$, is such that

(5) $$\int_{s_0}^{r} f(s)ds = 0.$$

The question now arises whether r is also a lower bound, in the p.d.e. case, for solutions of (1) satisfying (2). The following results give an affirmative answer for several special situations.

3. A consequence of a result of Hess.

Associate with f the following mapping f_1 given by:

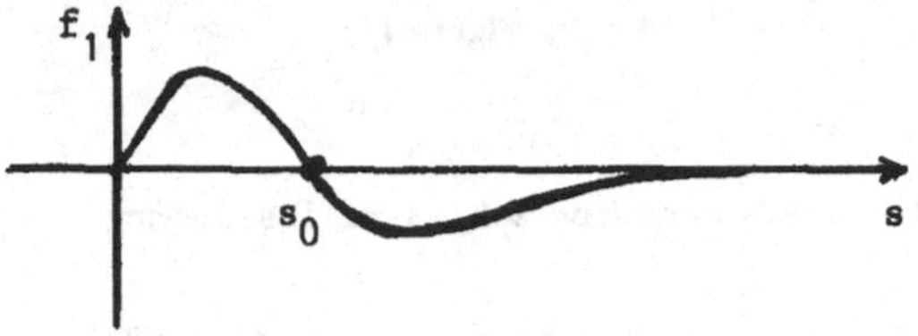

Figure 2

i.e., $f_1(s) = f(s)$, $0 < s < s_0$. Consider the problem

$$(6) \qquad \begin{cases} \Delta u + \lambda f_1(u) = 0, & x \in \Omega \\ u = 0, & x \in \partial\Omega. \end{cases}$$

Then critical points of the functionals

$$\phi_1(\lambda,u) = \frac{1}{2}\int_\Omega |\nabla u|^2 dx - \lambda \int_\Omega F_1(u)dx$$

and

$$(7) \qquad \phi(\lambda,u) = \frac{1}{2}\int_\Omega |\nabla u|^2 dx - \lambda \int_\Omega F(u)dx,$$

$(F_1(s) = \int_{s_0}^{s} f_1(\tau)d\tau,\ F(s) = \int_{s_0}^{s} f(\tau)d\tau)$ such that $0 < u(x) < s_0$ are solutions of (1). Hess [H] (see also [DF]) shows that for all $\lambda > 0$, $\phi_1(\lambda,u)$ has a minimum u such that $0 < u(x) < s_0$, hence showing that (1) has a positive solution u_0 with $0 < u_0(x) < s_0$. He then shows that there exists λ^* and $w \in H_0^1(\Omega)$, $\max_{x\in\Omega} w(x) > s_1$, such that for $\lambda > \lambda^*$

$$\phi(\lambda,w) - \phi_1(\lambda,u) < 0,$$

for all $u \in H_0^1(\Omega)$ with $0 \le u(x) \le s_0$, hence producing the existence of a minimum u_1 of $\phi(\lambda,\cdot)$ such that

$$\max_{x\in\Omega} u_1(x) > s_1,$$

and hence a positive solution of (1) satisfying (2), and

$$\phi(\lambda,u_1) - \phi_1(\lambda,u) < 0,$$

for all $u \in H_0^1(\Omega)$, $0 \le u(x) \le s_0$. Letting $\Omega_{s_0} = \{x \in \Omega: u_1(x) > s_0\}$ and

$$\bar{u}(x) = \begin{cases} s_0, & x \in \bar{\Omega}_{s_0} \\ u_1(x), & x \notin \Omega_{s_0} \end{cases},$$

then $\bar{u} \in H_0^1(\Omega)$ and $0 \le \bar{u}(x) \le s_0$, thus

$$\phi(\lambda,u_1) - \phi_1(\lambda,\bar{u}) < 0,$$

i.e.,

$$\frac{1}{2}\int_\Omega(|\nabla u|^2 - |\nabla\bar{u}|^2)dx - \lambda\int_\Omega(F(u_1) - F_1(\bar{u}))dx < 0.$$

The latter expression, on the other hand, may be computed as

$$\frac{1}{2}\int_{\Omega_{s_0}} |\nabla u_1|^2 dx - \lambda\int_{\Omega_{s_0}} (F(u_1) - F_1(s_0))dx = \frac{1}{2}\int_{\Omega_{s_0}} |\nabla u_1|^2 dx - \int_{\Omega_{s_0}} \int_{s_0}^{u_1(x)} f(x)ds < 0$$

and hence

$$\int_{\Omega_{s_0}} \int_{s_0}^{u_1(x)} f(s)ds > 0.$$

Hence there exists $x \in \Omega_{s_0}$ such that

$$\int_{s_0}^{u_1(x)} f(s)ds > 0,$$

i.e., there exists $x \in \Omega_{s_0}$, such that $u_1(x) > r$. We hence have the following result.

Theorem 2: Let (3) hold, then for any $n > 1$, there exists $\lambda^* > 0$ such that for all $\lambda > \lambda^*$ (1) has a positive solution u satisfying (2) and

$$\max_{x\in\overline{\Omega}} u(x) > r, \tag{8}$$

where r is given by (5). Such a solution may be obtained as a minimum of the functional defined in (7).

Using degree theoretic arguments one may show that for $\lambda > \lambda^*$ one has in fact at least two positive solutions of (1) satisfying (2). Whether the second solution also satisfies (8) is not known for general domains Ω. In the next section we shall establish this fact for special kinds of domains.

4. Convex domains.

Let Ω be a domain which is convex, $0 \in \Omega$, and let u be a positive solution of (1) satisfying (2), $u \in C^3(\overline{\Omega})$. Consider the function $P(x)$ given by

$$P(x) = \frac{1}{2}\,|\nabla u(x)|^2 + \lambda \int_{s_0}^{u(x)} f(x)ds, \tag{9}$$

it then follows from a maximum principle of Payne (see [S]) that P can only assume its maximum where $\nabla u(x) = 0$ (convexity of Ω is important here!).

Theorem 3: Let Ω be convex and let u be a positive solution of (1) satisfying (2), then

$$\max_{x\in\overline{\Omega}} u(x) > r,$$

where r is given by (5).

Proof: Let $x_0 \in \overline{\Omega}$ be such that

$$P(x_0) = \max_{x\in\overline{\Omega}} P(x) = \lambda \int_{s_0}^{u(x_0)} f(s)ds$$

then $x_0 \in \Omega$ and $\nabla u(x_0) = 0$, and $P(x_0) > 0$, since $\{x : u(x) = s_0\} \neq \emptyset$. Hence either $u(x_0) > r$, or else $u(x_0) = s_0$. If $u(x_0) = s_0$, then $P(x_0) = 0$ and $P(x) = 0$ for all x such that $u(x) = s_0$, i.e. $|\nabla u(x)| = 0$, for all such x. Letting

$F(u(x)) = \int_{s_0}^{u(x)} f(s)ds$, we have, using an identity of Rellich (see[B]) on the set $\Omega_{s_0} = \{x : u(x) > s_0\}$, that

$$\frac{1}{2}\int_{\partial\Omega_{s_0}} \nu\cdot x\left(\frac{\partial u}{\partial\nu}\right)^2 ds + \lambda F(s_0)\int_{\partial\Omega_{s_0}} \nu\cdot x ds = \tag{10}$$

$$\frac{2-n}{2}\int_{\Omega_{s_0}} |\nabla u|^2 dx + n\int_{\Omega_{s_0}} \lambda F(u)dx,$$

Hence, since $n > 2$, there exists $x \in \Omega_{s_0}$ such that $\int_{s_0}^{u(x)} f(s)ds > 0$. But $x \in \Omega_{s_0}$ implies $u(x) > s_0$ and thus $\max_{x\in\overline{\Omega}} u(x) > r$.

The above result was first established in [CS1].

5. Symmetric domains.

Let Ω be a domain with $0 \in \Omega$. Let $\{\gamma_1, \dots, \gamma_n\}$ be n linearly independent unit vectors such that Ω satisfies the Gidas-Ni-Nirenberg symmetry conditions (see [GNN], [CS1], [CS2]) with respect to each of the hyperplanes

$$P(\gamma_i) = \{x : x \cdot \gamma_i = 0\}, 1 \le i \le n.$$

Using some considerations from the theory of reflection groups (see [GB]) it was proved in [CS2] that all level sets

$$\Omega_c = \{x : u(x) > c\}$$

of a positive solution u of (1) are starlike with respect to the origin. I.e. if ν is the outward normal vector field to Ω_c, then $\nu \cdot x \ge 0$, $x \in \partial\Omega_c$. Letting $f(u(x)) = \int_{s_0}^{u(x)} F(s)ds$, we have using again Rellich's identity (10), that

$$n\lambda\int_{\Omega_{s_0}} F(u)dx = \frac{1}{2}\int_{\partial\Omega_{s_0}} \nu \cdot x\left(\frac{\partial u}{\partial\nu}\right)^2 ds + \frac{n-2}{2}\int_{\Omega_{s_0}} |\nabla u|^2 dx,$$

and hence (because of starlikeness) that

$$\int_{\Omega_{s_0}} F(u)dx > 0.$$

The following theorem summarizes the above considerations.

Theorem 4: Let Ω be a domain which satisfies the Gidas-Ni-Nirenberg symmetry conditions with respect to n-linearly independent unit vectors $\{\gamma_1, \dots, \gamma_n\}$ and let u be a positive solution of (1) satisfying (2), then $\max_{x\in\bar\Omega} u(x) > r$, where r is given by (5).

Remark: It follows from the results in [GNN] that if u is a positive solution of (1) on a domain Ω satisfying the conditions of theorem 4, then in fact it achieves a unique maximum at the origin and $\nabla u(x) = 0$, $x \in \Omega$, if and only if $x = 0$. Hence the above considerations in fact imply that $\int_{\Omega_{s_0}} F(u)dx > 0$ and therefore there exists $x \in \Omega_{s_0}$ such that $\int_{s_0}^{u(x)} f(s)ds > 0$. This observation will be important in the next section

6. Necessary condition for existence.

We observed in section 2 that in the case of ordinary differential equations a necessary condition for the existence of a positive solution u satisfying (2) is that the nonlinearity f satisfy condition (3). We shall demonstrate here that this is also the case for partial differential equations. This fact was also observed by Dancer [D], using different considerations.

Theorem 6: Assume $\int_{s_0}^{s_2} f(s)ds < 0$, then for any $\lambda > 0$, (1) has no positive solution u satisfying

$$s_1 < \max_{x\in\Omega} u(x) < s_2.$$

(This result is valid for arbitrary domains with the smooth boundary.)

Proof: Assume there exists $\lambda > 0$ such that (1) has a positive solution u, with $s_1 < \max_{x\in\bar\Omega} u(x) < s_2$. Let B be an open ball, centered at the origin, with $\bar\Omega \subset B$. Let

$$\alpha(x) = \begin{cases} u(x), & x \in \bar\Omega \\ 0, & x \in \bar B \setminus \Omega, \end{cases}$$

and consider the boundary value problem

$$(11)\qquad \begin{cases} \Delta v + \lambda f(v) = 0, & x \in B \\ v = 0, & x \in \partial B. \end{cases}$$

It follows from the results in [Sch1] that α is a lower solution for (11) and that $\beta(x) \equiv s_2$ is an upper solution. Hence (11) has a solution $v \in C^2(\overline{B})$ such that

$$\alpha(x) < v(x) < s_2,$$

and hence by the maximum principle $v(x) > 0$, $x \in B$. The results in [GNN] imply that v must be radially symmetric and $\nabla v(x) = 0$, $x \in B$, if and only if $x = 0$. We may therefore apply Rellich's identity (10) on

$$B_{s_0} = \{x \in B\colon v(x) > s_0\}$$

and conclude that

$$\int_{B_{s_0}} F(v)dx > 0,$$

i.e., there exists $x \in B_{s_0}$ such that

$$(12)\qquad \int_{s_0}^{v(x)} f(s)ds > 0 .$$

But $x \in B_{s_0}$ implies $v(x) > s_0$, and by (12), $v(x) > s_1$, a contradiction.

REFERENCES

[B] C. Bandle, Isoperimetric Inequalities and Applications, Pitman, Boston, 1980.

[BB] K. Brown and H. Budin, On the existence of positive solution for a class of semilinear elliptic boundary value, SIAM J. Math. Anal. 10(1979), 875-883.

[CS1] C. Cosner and K. Schmitt, A priori bounds for positive solutions of semilinear elliptic equations, Proc. Amer. Math. Soc. 95(1985), 47-50.

[CS2] C. Cosner and K. Schmitt, On the geometry of level sets of positive solutions of semilinear elliptic equations, Rocky Mtn. J. Math., to appear.

[D] E. Dancer, Private communication, 1986.

[DF] D. DeFigueiredo, On the existence of multiple ordered solutions of nonlinear eigenvalue problems, preprint, 1986.

[GB] L. Grover and C. Benson, Finite reflection groups, 2nd ed., Springer-Verlag, Berlin, 1985.

[GNN] B. Gidas, W. Ni and L. Nirenberg, Symmetry and related properties via the maximum principle, Comm. Math. Phys. 68(1979), 209-243.

[H] P. Hess, On multiple positive solutions of nonlinear elliptic eigenvalue problems, Comm. PDE 6(1981), 951-961.

[PS] H. Peitgen and K. Schmitt, Positive and spurious solutions of nonlinear eigenvalue problems, Springer Lecture Notes in Math. 878 (1981), 275-324.

[PSS] H. Peitgen, D. Saupe and K. Schmitt, Nonlinear elliptic boundary value problems versus their finite difference approximations: numerically irrelevant solutions, J. Reine Angew. Mathematik 322(1981), 74-117.

[S] R. Sperb, Maximum Principles and Their Appliations, Acad. Press, New York, 1981.

[Sch1] K. Schmitt, Boundary value problems for quasilinear second order elliptic equations, Nonlinear Anal. 2(1978), 263-309.

[Sch2] K. Schmitt, A Study of Eigenvalue and Bifurcation Problems for Nonlinear Elliptic Partial Differential Equations via Topological Continuation Methods, CABAY, Louvain-la-Neuve, 1982.

Asymptotic Completeness of Multiparticle Scattering

I.M. Sigal A. Soffer

Introduction

In this talk I present a new approach to scattering theory developed in collaboration with A. Soffer [SigSof1,2]. We test this approach on the problem of asymptotic completeness for N-body short-range systems. We have also made some progress in understanding long-range and spin-wave scattering.

Consider an N-body system in the center-of-mass frame and let H be its self-adjoint Schrödinger operator (or Hamiltonian, see the next section for the definitions). The scattering theory studies large time behaviour of the orbits $e^{-iHt}\psi$, for the states (≡ vectors) ψ orthogonal to the bound states (≡ eigenfunctions of H). One expects that the probability that at time t the system is not broken into independently moving, stable subsystems vanishes as $t \to \pm\infty$. Cast in precise mathematical terms this problem is called asymptotic completeness. The distinguishing feature of many-body systems, compared to 2-body ones, is in the geometry of their potentials. The potential V(x) of a many-body system does not vanish as $|x| \to \infty$ along certain planes X_a, where a labels different break-ups of the system in question into subsystems. In this language the statement of asymptotic completeness expresses the fact that as $t \to +\infty$ (or $t \to -\infty$), $e^{-iHt}\psi$ approaches a superposition of waves propagating freely (classically) along the planes X_a while committed to a bounded (quantum) motion in the perpendicular directions.

In this talk I sketch a proof of asymptotic completeness for short-range systems. The result presented here was announced in [SigSof1] and full proofs have appeared in [SigSof2]. (Here I will use a different phase-space partition of unity.) Reviews of earlier works and references to them can be found in [G, RSIII, Sig2, SigSof2].

Our approach integrates several recent developments in the rigorous QM. These are geometrical methods [Z, E1, Sim, DMSV, Sig1], phase-space analysis [E1-7] and commutator estimates [K1,2, L1-7, M1-3]. Our first step is phase-space (micro-local) analysis of the Schrödinger states, $\psi_t = e^{-iHt}\psi$. In other words, we change the point of view, and instead of following evolution of the system in question in the physical space we analyse the dynamics of its phase-space distribution. We show that, in a certain micro-local sense, the evolution ψ_t with the total energy E (i.e. with ψ in a small spectral interval around E), which is away from the thresholds of H, concentrates as $|t| \to \infty$ on a certain set, called the propagation set at the energy E, PS_E. This set is a collection of classical trajectories passed by free stable clusters resulting from various break-ups of the system. Our propagation estimates show that the probability for ψ_t to be in a phase-space region disjoint from the propagation set vanishes in a certain sense as $|t| \to \infty$. This reflects the intuitive picture that after the collision a system in question disintegrates into a number of stable clusters whose centers-of-mass follow the classical trajectories. We establish this picture for a certain time averaging and it probably breaks down if uniform (in time) estimates on decay of probabilities outside of the propagation region are required. We expect that in some regions of the phase-space the quantum phenomena take over and for some arbitrary remote time intervals there exists probabilities, bounded away from 0, for the system to move into certain classically forbidden regions. As a result the uniform estimates suggested by the classical considerations break down and should be replaced by some sort of time average estimates.

To connect the phase-space picture of scattering with the usual, physical-space picture we construct a phase-space partition of unity $\{j_{a,E}(x,p)\}$ (here a runs through all break-ups and $p = -i\nabla_x$). The members of this partition are supported in phase-space regions where the system is broken into clusters with mutual distances proportional to the total size of the system. Moreover, the x-boundaries of these regions, $\text{supp}\,\nabla_x j_{a,E}(x,p)$, lie outside of the propagation set (i.e. in the region into which no propagation takes place). Thus each member of the partition is associated with a certain break-up, a, of the system into well separated clusters which is dynamically decoupled from other break-ups (decoupling of channels). We expect that on $j_a(x,p)$ the total evolution e^{-iHt} behaves as the cluster evolution $e^{-iH_a t}$, where $H_a = H$ - (inter-cluster interaction) (i.e. H_a describes independently moving clusters associated with the break-up a). To compare these two evolutions we introduce the Deift-Simon wave operators, $W^{\pm}_{a,E} = \text{s-}\lim_{t\to\pm\infty} e^{iH_a t} j_{a,E}(x,p) e^{-iHt}$. (Such operators for configuration space partitions of unity were introduced and applied to study of asymptotic completeness in [DS]. This work, however, did not have the essential notion of propagation set. This notion is necessary for the geometric decoupling of channels.) The phase-space partition and Deift-Simon wave operators relate the phase-space space analysis of scattering with the usual, physical-space analysis. The existence of $W^{\pm}_{a,E}$ is equivalent to the statement of asymptotic completeness. To prove the existence of $W^{\pm}_{a,E}$ we use, via the Kato argument, the result of [PSS] on the local decay of ψ_t and the propagation estimates discussed above.

Finally we mention that the conditions on the potentials (e.g. strength of local singularities) can be relaxed if we exercise more care in our estimation (e.g. in controlling various commutators of H). Since our priority in this work is exposition of the method, we use conditions more restrictive than necessary. Note also that our methods are applicable to a more general class of potentials, considered in [Ag]. These potentials are defined by the condition that they vanish as $|x| \to \infty$ in all directions except along a certain system of planes.

1. Hamiltonians

Consider an N-body system in the physical space R^ν. The configuration space in the center-of-mass frame is

$$X = \{x \in R^{\nu N} \mid \Sigma m_i x_i = 0\} \tag{1.1}$$

with the inner product $\langle x,y \rangle = 2\Sigma m_i x_i \cdot y_i$. Here $m_i > 0$ are masses of the particles in question. The Schrödinger operator of such a system is

$$H = -\Delta + V(x) \quad \text{on} \quad L^2(X) . \tag{1.2}$$

Here Δ is the Laplacian on X and $V(x) = \Sigma V_{ij}(x_i - x_j)$, where (ij) runs through all the pairs satisfying $i < j$. We assume that the potentials V_{ij} are real and obey: $V_{ij}(y)$ are Δ_y-compact.

Now we describe the decomposed systems. Denote by a,b,..., partitions of the set $\{1,...,N\}$ into non-empty disjoint subsets, called clusters. The relation $b \subset a$ means that b is a refinement of a. Then a_{min} is the partition into N clusters $(1),...,(N)$. Usually, we assume that partitions have at least two clusters. $\#(a)$ denotes the number of clusters in a. We also identify pairs $l = (ij)$ with partitions on N-1 clusters: $(ij) \longleftrightarrow \{(ij)(1) \cdots (\hat{i}) \cdots (\hat{j}) \cdots (N)\}$.

We define the intercluster interaction for a partition a as

$$I_a = \text{sum of all potentials linking different clusters in } a . \tag{1.3}$$

For each a we introduce the truncated Hamiltonian:

$$H_a = H - I_a . \tag{1.4}$$

These operators are clearly self-adjoint. They describe the motion of the original system broken into non-interacting clusters of particles.

For each cluster decomposition a, define the configuration space of the relative motion of the clusters in a:

$$X_a = \{x \in X \mid x_i = x_j \text{ if } i \text{ and } j \text{ belong to same cluster of } a\}$$

and the configuration space of the internal motion within those clusters:

$$X^a = \{x \in X \mid \sum_{j \in C_i} m_j x_j = 0 \text{ for all } C_i \in a\} .$$

Clearly that X_a and X^a are orthogonal (in our inner product) and they span X:

$$X = X^a \oplus X_a .$$

Given generic vector $x \in X$ its projections on X^a and X_a will be denoted by x^a and x_a, respectively.

If i and j belong to some cluster of a, then $x_i - x_j \in X^a$. This elementary fact, yields the following decomposition:

$$H_a = H^a \otimes 1 + 1 \otimes T_a \text{ on } L^2(X) = L^2(X^a) \otimes L^2(X_a). \quad (1.5)$$

Here H^a is the Hamiltonian of the non-interacting a-clusters with their centers-of-mass fixed at the origin, acting on $L^2(X^a)$, and $T_a = -$(Laplacian on X_a), the kinetic energy of the center-of-mass motion of those clusters.

The eigenvalues of H^a, whenever they exist, will be denoted by ε_α, where $\alpha = (a,m)$ with m, the number of the eigenvalue in question counting the multiplicity. For $a = a_{min}$, we set $\varepsilon_\alpha = 0$. The set $\{\varepsilon_\alpha , \text{ all } \alpha\}$ is called the threshold set of H and ε_α are called the thresholds of H. (For more details see [Sig2].)

Our method is based on localization of operators in the phase- space $F = X \times X'$. Here and henceforth, the prime stands for taking dual of the space in question. The dual (momentum) space X' is identified with

$$X' = \{k \in R^{\nu N} \mid \Sigma k_i = 0\} \text{ with the inner product } \langle k,u\rangle = \Sigma \frac{1}{2m_i} k_i \cdot u_i . \quad (1.6)$$

Thus $|k|^2$ is the symbol of $-\Delta$ and $-\Delta = |p|^2$ with $p = -i \nabla_x$. We use extensively the natural bilinear form on $X \times X'$: $\langle x,k\rangle = \Sigma x_i \cdot k_i$. Given generic vector

$k \in X'$, its projections on X'_a and $(X^a)'$ will be denoted by k_a and k^a, correspondingly. Accordingly, the momenta canonically conjugate to x_a and x^a and corresponding to k_a and k^a will be denoted by p_a and p^a, respectively. Thus $T_a = |p_a|^2$.

Below $P_\Delta(H)$ stands for the spectral projection for H.

2. Asymptotic Completeness

In this section we cast the intuitive notion of asymptotic completeness into precise mathematical terms and discuss different aspects of this notion. To this end we need the notion of channel. The channel is a scenario according to which the scattering process develops. Different scenarios are labeled by an asymptotic state of the system in question at $t = -\infty$ or $t = +\infty$. Thus a channel is described by a pair: $\alpha = (a,m)$, where a is a decomposition of the system into subsystems and m specifies a stable motion (i.e. a bound state) within each subsystem of this decomposition.

We say that the many-body system in question is <u>short-range</u> if the pair potentials $V_{ij}(y)$ vanish at ∞ faster than $|y|^{-1}$ (the precise definition will be given in the next section). We say that the short-range many-body system is <u>asymptotically complete</u> if for any $\psi \in L^2(X)$ orthogonal to all eigenfunctions of H and for any $\varepsilon > 0$ there exist a finite subset, A_ε, of channels and for each α from this subset, vectors $u^{\pm}_{\alpha,\varepsilon} \in L^2(X_a)$ s.t.

$$\lim_{t\to\pm\infty} \|e^{-iHt}\psi - \sum_{\alpha} \psi^{\alpha} \otimes e^{-iH_a t} u^{\pm}_{\alpha}\| \le \varepsilon . \tag{2.1}$$

where ψ^α is the eigenfunction of H^a corresponding to ε_α. Our definition is equivalent to the standard one (see [SigSof2]).

Our next goal is to reduce the statement of asymptotic completeness to a certain "geometric" statement which does not involve the dynamical notion of bound state (for subsystems). We say that a short-range many-body system described by H is asymptotically clustering at an energy E, if there exist an interval Δ around E s.t. for any $\psi \in \text{Ran}\, P_\Delta(H)$, orthogonal to the eigenfunctions of H there exist vectors $\psi^{\pm}_{a,E} \in L^2(X)$ s.t.

$$\|e^{-iHt}\psi - \sum_{\#(a)>1} e^{-iH_a t}\psi^{\pm}_{a,E}\| \to 0 \quad \text{as} \quad t \to \pm\infty . \tag{2.2}$$

In other words the asymptotic clustering describes the fact that as $t \to \pm\infty$, the system breaks down into independently moving subsystems. However, it does not say anything whether the resulting subsystems are stable or not. Nevertheless, we expect that these subsystems which are not stable split further into smaller fragments and so on till only stable clusters remain

Proposition 2.1. Suppose the system in question and all of its subsystems are asymptotically clustering for all non-threshold energies. Then it is asymptotically complete.

Proof. See [SigSof2]. □

3. Results

In this section we formulate our main result and outline its proof. We begin with the assumptions on potentials which we use in this work.

(A) $V_{ij}(y)$ are Δ_y-compact

(B) $\langle y\rangle^{1+\theta}\,|\nabla V_{ij}(y)|$ are Δ_y-bounded for some $\theta > 0$

(C) $|y|^2 \dfrac{\partial^2}{\partial y_k \partial y_l} V_{ij}(y)$ are Δ_y-bounded

(D) $V_{ij}(y)\langle y\rangle^{\mu}$ are Δ_y-bounded.

By a short range system we understand a system obeying (A) and (D) with $\mu > 1$. The main result of this paper is

Theorem 3.1. Assume an N-body system is described by potentials obeying conditions (A)-(D) with $\mu > 1$. Then asymptotic completeness holds for this system.

The proof of Theorem 3.1 is sketched in sections 4-9.

4. Propagation Set

In this section we introduce the notion of propagation set and state our main result in this direction.

Fix energy $E \in \mathbf{R}$. A channel α is said to be open if $E - \varepsilon_\alpha \geq 0$. This condition is suggested by the law of conservation of energy applied to the channel α:

$$\underbrace{E}_{\text{total energy}} = \underbrace{\varepsilon_\alpha}_{\substack{\text{internal energy}\\ \text{of stable clusters}}} + \underbrace{|p_a|^2}_{\substack{\text{kinetic energy of}\\ \text{c-of-m motion of clusters}}} \tag{4.1}$$

Thus $E - \varepsilon_\alpha$ is the kinetic energy available to the clusters of α. For an open channel α, introduce $\kappa_\alpha = \sqrt{E-\varepsilon_\alpha}$. Furthermore, let $\Sigma_E = \{\pm\kappa_\alpha \text{ all open } \alpha\}$.

Below we use the symbol $u\|v$ to signify that the vectors u and v are parallel (multiple of each other). Also a cluster decomposition corresponding to a channel α is denoted by $a(\alpha)$. Given E, define the set

$$PS_E = PS_E^+ \cup PS_E^- , \tag{4.2}$$

where

$$PS_E^{\pm} = \bigcup_{\text{open } \alpha} \{(x,k) \in F \mid x^a = 0, x^c \neq 0 \ \forall c \supsetneq a, x_a \| \pm \nabla |k_a|^2 ,$$
$$|k_a|^2 = \kappa_\beta , \text{ where } a(\beta) \subseteqq a\} . \tag{4.3}$$

The sets $PS_E^{\pm}$ describe different possibilities of outgoing/incoming free motion of non-interacting stable clusters of the total energy E. This free motion develops along classical trajectories: the coordinates and momenta of the cluster centers-of-mass are parallel or antiparallel depending on whether the clusters are outgoing or incoming. The restriction on the kinetic energy of this free classical motion stems from the energy conservation law and the fact that the stable internal motion of the clusters is described by a bound state of their internal Hamiltonian and, consequently, the energy of the internal motion is given by the corresponding eigenvalue. This mixture of classical and quantum-mechanical pictures is the unique feature of the many-body scattering theory. The next theorem shows that asymptotically the quantum evolution $\psi_t = e^{-iHt}\psi$ with ψ having the energy near E becomes localized in the phase-space region $PS_E^{\pm}$.

To formalize notion of phase-space localization we introduce a class of operators which appears naturally in our analysis. Those are operators of the form

$$J \equiv \Sigma j_i(x) f_i(p) , \tag{4.4}$$

where the sum is finite, all functions are smooth and bounded and, in addition, j_i obey the estimates $|D^\alpha j_i(x)| \le C_\alpha \langle x \rangle^{-|\alpha|}$ for all α. Such operators will be called the <u>phase-space operators.</u> The function

$$j(x,k) = \Sigma j_i(x) f_i(k) \tag{4.5}$$

will be called the <u>symbol</u> of the operator J. We say also that the phase-space operator, J, is <u>supported</u> in the phase-space region (which is denoted by f-supp J) supp j.

Given E, we say that a subset of the phase-space $F = X \times X'$ <u>is a propagation set at the energy</u> E if for any phase-space operator J supported outside of the set in question there is a small interval Δ around E s.t.

$$\int_{-\infty}^{\infty} \|J \frac{1}{\sqrt{\langle x \rangle}} e^{-iHt}\psi\|^2 dt \le C\|\psi\|^2 \tag{4.6}$$

for any $\psi \in \mathrm{Ran}\, P_\Delta(H)$ and with $C < \infty$ and independent of ψ.

Remark 4.1. This definition is equivalent to one in which the phase-space operators are replaced by more general pseudodifferential operators. However, since, as was mentioned above, the phase-space operators arise naturally in our approach we restrict our definitions to such operators.

Much more importantly, we can refine the definition of the propagation set by breaking it into two pieces each defined through one of the inequalities

$$\pm \int_{0}^{\pm\infty} \|J \frac{1}{\sqrt{\langle x \rangle}} e^{-iHt}\psi\|^2 dt \le C\|\psi\|^2 . \tag{4.7}$$

Respectively, PS_E^+ (resp. PS_E^-) is a candidate for the future (resp. past) propagation set. The further refinement of the notion of the propagation set would be to introduce the coefficient of proportionality in the relation $x_a \,|\pm\nabla\,|k_a|^2$ - time. These notions and their ramifications will be discussed elsewhere ([SigSof3]).

Henceforth, we always assume that E is inside of the continuous spectrum of H: $E \ge \min\,\{\varepsilon_\alpha \,|\, \#(a(\alpha)) > 1\}$.

Theorem 4.2. Assume conditions (A) - (C) are satisfied (note: condition (D) is not required!). Let E be away from the thresholds and eigenvalues of H. Then PS_E is a propagation set for H at energy E.

Proof of theorem (i.e. of propagation estimates of type (4.6)) is based on positivity estimates of commutators of H with appropriate bounded operators which we call the propagation observables. These observable "live" in certain regions of phase-spaces whose boundaries are vacated by the flow generated by H.

In our analysis we slice the phase-space into two pieces $\{\hat{x}\cdot k \notin \Sigma_E\}$ and $\{\hat{x}\cdot k \in \Sigma_E\}$. Note that $\hat{x}\cdot k|_{PS_E} = \Sigma_E$. This suggests that the most difficult should be the analysis of the second region. However, though the analysis of the second region requires some subtlety, most of the work goes into proving that there is no propagation into the region $\{\hat{x}\cdot k \notin \Sigma_E\}$. Once this is done we localize $\hat{x}\cdot k$ close to a point $\kappa \in \Sigma_E$ and then apply hand-made phase-space analysis to show that there is no propagation into the region $(F\backslash PS_E) \cap \{\hat{x}\cdot k = \kappa\}$.

5. Geometry of N-Body Systems

The geometry of the N-body potential $V(x) = \Sigma V_{ij}(x_j - x_j)$ is determined by the system of planes X_a, where a runs through all cluster decompositions. These are exactly the planes along which V(x) does not vanish. The main property of these planes is the relation

$$X_a \cap X_b = X_{a\cup b} . \tag{5.1}$$

This relation follows directly from the definition of X_a. Note that $X_{a_{max}} = \{0\}$. In this section we discuss geometry of these planes. This discussion is adopted to needs of the next section.

Let ε_a be small, positive numbers obeying

$$\varepsilon_a \le \frac{1}{20\alpha}\,\varepsilon_f \quad \text{for} \quad \#(f) > \#(a) , \tag{5.2}$$

where the number $\alpha \ge 1$ depending only on the masses (and the number of particles) is chosen so that

$$|x^f| \le \alpha(|x^a| + |x^b|) \quad \text{with} \quad f = a \cup b . \tag{5.3}$$

Define the open homogeneous sets

$$B_\varepsilon(X_a) = \{x \in X \mid |x^a| < \varepsilon |x|\}.$$

Then eqns. (5.2) and (5.3) imply

$$B_{\frac{3}{2}\varepsilon_a}(X_a) \cap B_{\frac{3}{2}\varepsilon_b}(X_b) \subset B_{\frac{1}{8}\varepsilon_f}(X_f) \quad \text{with } f = a \cup b, \tag{5.4}$$

provided $f \neq a,b$. Here the r.h.s. is empty if $\#(f) = 1$. This for instance is always the case, if $\#(a) = \#(b) = 2$ and $a \neq b$. Eqn. (5.4) will play an important rôle in our construction.

Define the open (conical) sets

$$\Gamma_{a,\varepsilon} = B_{\frac{10}{9}\varepsilon_a}(X_a) \setminus \bigcup_{\substack{b \supset a \\ \neq}} \overline{B_{\frac{1}{2}\varepsilon_b}(X_b)}, \tag{5.5}$$

where ε stands for the multi-index with components ε_a.

We use the following standard notation

$$|x|_a = \min\{|x_i - x_j| \mid i \text{ and } j \text{ belong to different clusters of } a\}.$$

This is the <u>intercluster distance</u> in the decomposition a.

Lemma 5.1. (i) The homogeneous sets $\Gamma_{a,\varepsilon}$ form an open covering of X.
(ii) <u>There is a number</u> $\delta > 0$ s.t.

$$\Gamma_{a,\varepsilon} \subset \{x \in X \mid |x|_a > \delta |x|\}. \tag{5.6}$$

Proof. (i) This follows from the following relation, which is implied by (5.1):

$$X = \bigcup_{a \neq a_{max}} [X_a \setminus \bigcup_{b \supset a} X_b]. \tag{5.7}$$

(ii) Let a pair $l \not\subset a$. Let $b = a \cup l$. Then $b \underset{\neq}{\supset} a$. Due to eqn. (5.4)

$$B_{\varepsilon_l}(X_l) \cap B_{\frac{3}{2}\varepsilon_a}(X_a) \subset B_{\frac{1}{8}\varepsilon_b}(X_b).$$

Since $b \underset{\neq}{\supset} a$, we have that

$$\Gamma_{a,\varepsilon} \cap B_{\varepsilon_l}(X_l) \subset [B_{\varepsilon_l}(X_l) \cap B_{\frac{3}{2}\varepsilon_a}(X_a)] \setminus B_{\frac{1}{2}\varepsilon_b}(X_b) = \emptyset .$$

In other words

$$\Gamma_{a,\varepsilon} \subset \{x \in X \mid |x^l| > \varepsilon_l |x|\} .$$

Since this is true for all $l \not\leq a$, (5.7) follows with $\delta = \min_{l \not\leq a} \varepsilon_l$. □

Observe for future references that eqns. (5.4) and (5.6) imply

$$B_{\varepsilon_a}(X_a) \cap \Gamma_{b,\frac{1}{2}\varepsilon} = \emptyset \quad \text{if } a \cup b \neq a,b \tag{5.8}$$

This lemma implies the following important fact about the geometry of X:

The configuration space can be covered by open sets in each of which the distances between certain clusters are proportional to the total size of the system.

6. Phase-Space Partition of Unity

The main result of this section is a construction of a phase-space partition of unity which decouples different channels (in the sense stated in Introduction). We begin with some definitions. Fix the energy E away from the thresholds and eigenvalues of H. Introduce

$$\Sigma_{E,\varepsilon} = \{\kappa_\varepsilon \mid \kappa \in \Sigma_E\} , \quad \text{where } \kappa_\varepsilon = \sqrt{1-\varepsilon^2}\, |\kappa| . \tag{6.1}$$

Pick ε_a's satisfying (5.2) - (5.3) and s.t.

$$\Sigma_{E,\varepsilon_a} \cap \Sigma_{E,\frac{1}{2}\varepsilon_a} = \emptyset \quad \text{for all } a .$$

This is possible since for E non threshold, Σ_E is a bounded, closed subset of R^+ of the form $\Sigma_E = \bigcup_{k=1}^{n} \Sigma^{(k)}$ for some $n < \infty$, where $\Sigma^{(k)}$ is a countable set whose accumulation points belong to $\Sigma^{(k-1)}$ and $\Sigma^{(1)}$ is finite. (We will omit an elementary proof of this fact.) Pick now ε_0 s.t.

$$\varepsilon_0 < \frac{1}{10} \min_a \left(\text{dist}(\Sigma_{E,\varepsilon_a}, \Sigma_{E,\frac{1}{2}\varepsilon_a})^2, \varepsilon_a^5\right) \tag{6.2}$$

Then we have for all a's the relation

$$\sqrt{\varepsilon_0} - \text{nbhd}\,(\Sigma_{E,\varepsilon_a}) \cap \sqrt{\varepsilon_0} - \text{nbhd}(\Sigma_{E,\frac{1}{2}\varepsilon_a}) = \phi\,. \tag{6.3}$$

Henceforth we will use the following notation

$$r^a = \frac{\langle x^a \rangle}{\langle x \rangle} \quad \text{and} \quad r_a = \frac{\langle x_a \rangle}{\langle x \rangle}\,.$$

We use the following cut-off functions:

$$F_\varepsilon(s \in \Omega) = \begin{cases} 1 & \text{dist}(s,\Omega^c) \geq 2\varepsilon \\ 0 & \text{dist}(s,\Omega) \geq 2\varepsilon \end{cases} \tag{6.4a}$$

if $|\Omega| > 0$ and $|\Omega^c| > 0$, and

$$F_\varepsilon(s \in \Omega) = \begin{cases} 1 & \text{dist}(s,\Omega) \leq \varepsilon \\ 0 & \text{dist}(s,\Omega) \geq 2\varepsilon \end{cases} \tag{6.4b}$$

if $|\Omega| = 0$, and

$$F_\varepsilon(s \in \Omega) = 1 - F_\varepsilon(s \in \Omega^c) \tag{5.4c}$$

if $|\Omega^c| = 0$. Here $\Omega \subset R$ (usually, an interval), $\Omega^c = R \backslash \Omega$ and the relation $s \notin \Omega$ is treated as $s \in \Omega^c$. Besides, F_ε are assumed to be smooth, $0 \leq F_\varepsilon \leq 1$ and for Ω a semi-finite interval,

$$F'_\varepsilon(s \in \Omega) = \text{const} \quad \text{for} \quad \text{dist}(s,\partial\Omega) \leq \varepsilon\,.$$

In particular, F_ε can be adjusted in such a way as to satisfy for Ω, a semi-finite interval,

$$F'_\varepsilon(s \in \Omega) = \text{const} \cdot F_\varepsilon(s \in \partial\Omega)\,. \tag{6.5}$$

Next we define our basic cut-off functions on the phase-space:

$$\tilde{j}_{a,E}(x,k) = F_{\varepsilon_0}(r^a < \varepsilon_a)[1 - F_{\varepsilon_0}(r^a > \frac{3}{4}\varepsilon_a)]F_{\sqrt{\varepsilon_0}}[|k_a| \in \Sigma_{E,\varepsilon_a}]] \tag{6.6a}$$

if $\#(a) < N$ and

$$\tilde{j}_{a,E}(x,k) = \prod_{\text{all } l} F_{\varepsilon_0}(r^l > \varepsilon_a) \tag{6.6b}$$

if $\#(a) = N$. Here, recall, l denotes a pair of indices and $x^l = x_i - x_j$ for $l = (ij)$. (Remember, also, that l is identified with an $(N-1)$-cluster decomposition.) We define

$$j_{a,E}(x,k) = \tilde{j}_{a,E}(x,k) \prod_{n=2}^{\#(a)-1} [1 - \sum_{\#(b)=n} \tilde{j}_{b,E}(x,k)], \tag{6.7}$$

where $\prod_{n=2}^{\#(a)-1}(\ldots) = 1$ for $\#(a) = 2$.

Henceforth we will use an abbreviated notation for subsets of the phase-space $F = X \times X'$, displaying only inequalities defining these subsets. Denote

$$\Omega_{a,E} = \bigcup_{\substack{b \supseteq a \\ \alpha = \frac{3}{4}, 1}} \left\{ |r^b - \alpha\varepsilon_b| \le \varepsilon_0, \ \mathrm{dist}(|k_b|, \Sigma_{E,\alpha\varepsilon_b}) \ge \sqrt{\varepsilon_0} \right\}. \tag{6.8}$$

It can be shown (though it is not used directly in our analysis) that

$$\Omega_{a,E} \cap PS_E = \emptyset \tag{6.9}$$

Theorem 6.1.

(i) $j_{a,E}(x,k)$ *are symbols and are independent of* k^a,

(ii) $\Sigma j_{a,E}(x,k) = 1$,

(iii) supp $j_{a,E} \subset \{|x|_a \ge \delta|x|\}$ *for some* $\delta > 0$,

(iv) $\mathrm{supp}(\nabla_x j_{a,E}) \subset \Omega_{a,E}$.

Discussion. Relation (iii) shows that $j_{a,E}$ is supported in the region where the clusters in a move away from each other and relation (iv) shows that the x-"boundary", $\mathrm{supp}(\nabla_x j_{a,E})$, of $j_{a,E}$ lives in the region where no propagation takes place.

Proof. (i) The fact that $j_{a,E}(x,k)$ are symbols follows readily from the definition (6.6)-(6.7). The fact that $j_{a,E}(x,k)$ is independent of k^a follows from (for the proof see [SigSof2])

Lemma 6.2. *The terms with* $b \not\supset a$ *can be dropped from the r.h.s. of* (6.7).

(ii) We begin with

Lemma 6.3

$$\prod_{n=2}^{N} [1 - \sum_{\#(b)=n} \tilde{j}_{b,E}(x,k)] = 0 . \tag{6.10}$$

Proof. Due to the definition (6.3)

$$\mathrm{supp}\,[1 - \tilde{j}_{a_{min},E}] \subset \bigcup_{\text{all } l} B_{\varepsilon_{a_{min}}}(X_l) . \tag{6.11}$$

Hence eqn. (6.8) and the inequalities on the ε_a's imply

$$\mathrm{supp}\,[1 - \tilde{j}_{a_{min},E}] \subset \bigcup_{\#(d)\le N-1} \Gamma_{d,\frac{1}{2}\varepsilon} . \tag{6.12}$$

Using eqns. (5.4), (5.8) and the equation $B_{c_f} = \bigcup_{d \supset f} \Gamma_{d,\varepsilon}$ one can show that

$$1 - \sum_{\#(c)=\#(d)} \tilde{j}_{c,E}(x,k) = 0 \quad \text{on } \Gamma_{d,\frac{2}{3}\varepsilon} \tag{6.13}$$

The last two equations imply

$$\prod_{n=2}^{N-1} [1 - \sum_{\#(c)=n} \tilde{j}_{c,E}(x,k)] = 0 \quad \text{on} \quad \text{supp}[1 - \tilde{j}_{a_{min},E}] . \quad \square$$

Now we derive (ii) from Lemma 7.3. Expanding the product, we find

$$\begin{aligned}
0 &= \prod_{n=2}^{N} (1 - \Sigma\tilde{j}_{a,E}) \\
&= \prod_{n=2}^{N-1} (1 - \Sigma\tilde{j}_{a,E}) - \tilde{j}_{a_{min},E} \prod_{n=2}^{N-1} (1 - \Sigma\tilde{j}_{a,E}) \\
&= \prod_{n=2}^{N-2} (1 - \Sigma\tilde{j}_{a,E}) - [\sum_{\#(b)=N-1} \tilde{j}_{b,E}] \prod_{n=2}^{N-2} (1 - \Sigma\tilde{j}_{a,E}) - j_{a_{min},E} \\
&= \cdots \\
&= 1 - \Sigma j_{a,E} .
\end{aligned}$$

This implies (ii).

(iii) Eqn. (6.13) yields

$$\prod_{n=2}^{s} [1 - \sum_{\#(c)=n} \tilde{j}_{c,E}(x,k)] = 0 \quad \text{on} \quad \Gamma_{b,\frac{2}{3}\varepsilon} \quad \text{if} \ \#(b) \le s . \tag{6.14}$$

Next, the definition (5.5) of $\Gamma_{a,\varepsilon}$ implies

$$B_{\varepsilon_a}(X_a) = \Gamma_{a,\varepsilon} \bigcup_{b \supsetneq a} \Gamma_{b,\frac{2}{3}\varepsilon} . \tag{6.15}$$

The last two equations imply

$$j_{a,E}(x,k) = F(\Gamma_{a,\varepsilon}) \prod_{n=2}^{\#(a)-1} [1 - \sum_{\#(c)=n} \tilde{j}_{a,E}(x,k)] , \tag{6.16}$$

where $F(\Gamma_{a,\varepsilon})$ is an appropriate smooth function living in $\Gamma_{a,\varepsilon}$. The latter equation yields: $\text{supp } j_{a,E} \subset \Gamma_{a,\varepsilon}$ which, due to (5.6), implies (iii).

(iv) As in Lemma 6.3 we show that $\nabla_x \tilde{j}_{a,E}(x) \prod_{n=2}^{N-1} (1 - \sum_{\#(b)=n} \tilde{j}_{b,E}(x,k)) = 0$ for

$\#(a) = N$. Hence it suffices to consider $\nabla_x \tilde{j}_{a,E}(x,k)$ with $\#(a) < N$. We differentiate

$$\nabla_x \tilde{j}_{a,E}(x,k) = \Big\{F'_{\varepsilon_0}(r^a < \varepsilon_a)[1 - F_{\varepsilon_0}(r^a > \tfrac{1}{2}\varepsilon_a)F_{\sqrt{\varepsilon_0}}(|k_a| \in \Sigma_{E,\varepsilon_a})]$$
$$- F_{\varepsilon_0}(r^a < \varepsilon_a)F'_{\varepsilon_a}(r^a > \tfrac{1}{2}\varepsilon_a)F_{\sqrt{\varepsilon_0}}[|k_a| \in \Sigma_{E,\varepsilon_a}]\Big\}\nabla_x(r^a)\,. \tag{6.17}$$

Using eqns (6.4c) and (6.5) and using that

$$F_{\varepsilon_0}(r^a \geq \tfrac{1}{2}\varepsilon_a) = 1 \quad \text{on supp } F'_{\varepsilon_0}(r^a \leq \varepsilon_a)$$

$$F_{\varepsilon_0}(r^a \leq \varepsilon_a) = 1 \quad \text{on supp } F'_{\varepsilon_0}(r^a \geq \tfrac{1}{2}\varepsilon_a)\,,$$

we obtain

$$\nabla_x \tilde{j}_{a,E}(x,k) = \text{const}[F_{\varepsilon_0}(r^a = \varepsilon_a)F_{\sqrt{\varepsilon_0}}[|k_a| \notin \Sigma_{E,\varepsilon_a}]$$
$$+ F_{\varepsilon_0}(r^a = \tfrac{1}{2}\varepsilon_a)F_{\sqrt{\varepsilon_0}}[|k_a| \in \Sigma_{E,\varepsilon_0}]\nabla_x(r^a)\,. \tag{6.18}$$

This relation together with definition (6.6) and eqn.(6.3) yields (iv). □

Now we quantize the partition of unity $j_{a,E}(x,k)$ defined above. Using the standard procedure of pseudodifferential calculus we define $J_{a,K} \equiv j_{a,K}(x,p)$ by the formula

$$j_{a,E}(x,p)\psi = \frac{1}{(2\pi)^{n/2}} \int j_{a,E}(x,k)\hat{\psi}(k)e^{ik\cdot x}dx\,, \tag{6.19}$$

where $\hat{\psi}$ is the Fourier transform of ψ and $n = \nu(N-1)$, the dimension of X.

Theorem 6.4. The operators $j_{a,E}(x,p)$ form a partition of unity on $L^2(X)$:

$$\sum_a j_{a,E}(x,p) = 1 + O(|x|^{-1})\,. \tag{6.20}$$

Moreover, they have the following properties:

(i) $j_{a,E}(x,p)$ are phase-space operators commuting with x^a,

(ii) $j_{a,E}(x,p)$ are supported in $\{|x|_a > \delta|x|\}$ for some $\delta > 0$,

(iii) $\nabla_x j_{a,E}(x,p)$ are supported in $\Omega_{a,\kappa}$.

A proof of this theorem follows easily from Theorem 6.1.

7. Propagation Estimates

In this section we discuss and sketch (in varied detail) the proof of Theorem 4.2 (Propagation Theorem). We assume that conditions (A) - (C) are satisfied. We begin with a special kind of propagation estimate.

First, we introduce some operators which play an important rôle in our phase-space analysis

$$\gamma = \frac{1}{2}(\hat{x}\cdot p + p\cdot\hat{x}) \quad \text{with} \quad \hat{x} = \frac{x}{\langle x\rangle}, \quad \langle x\rangle = \sqrt{1+|x|^2},$$

$$\gamma_a = \frac{1}{2}(\hat{x}\cdot p_a + p_a\cdot\hat{x}_a) \quad \text{and} \quad \gamma^a = \frac{1}{2}(\hat{x}^a\cdot p^a + p^a\cdot\hat{x}^a)$$

These operators are self-adjoint on their natural domains and the set $s(X)$ and $D(|p|^n)$ are their cores.

Theorem 7.1. Let E be a fixed energy which is not a threshold or an eigenvalue of H. Then for any smooth bounded function F(s) supported away from Σ_E there is a small interval Δ around E s.t.

$$\int_{-\infty}^{\infty} \left\|F(\gamma)\frac{1}{\sqrt{\langle x\rangle}}\, e^{-iHt}\psi\right\|^2 dt \le C\|\psi\|^2 \tag{7.1}$$

for any $\psi \in \operatorname{Ran} P_\Delta(H)$ and with $C < \infty$ independent of ψ.

First observe that due to the relation

$$\frac{d}{dt}\langle\phi\psi_t,\psi_t\rangle = \langle i[H,\phi]\psi_t,\psi_t\rangle \tag{7.2}$$

where ϕ is a bounded operator and $\psi_t = e^{-iHt}\psi$, the propagation estimates can be derived from certain commutator estimates. Namely, we derive them from the following result. Let $F(s \geq s_0)$ denote smooth cut-off function, $0 \leq F \leq 1$, satisfying

$$F(s \geq s_0) = \begin{matrix} 1 \text{ for } s \geq s_0+\delta_1 \\ 0 \text{ for } s \leq s_0-\delta_1 \end{matrix}$$

with δ_1 specified in an appropriate place. For an interval Δ, let $F_\Delta = F_\Delta(H)$, where $F_\Delta(s)$ is a smooth function s.t. $0 \leq F_\Delta(s) \leq 1$, $F_\Delta(s) = 1$ for $s \in \Delta$, $F_\Delta(s) = 0$ for $\text{dist}(s,\Delta) > \delta$ with $\delta < \frac{1}{10}\ |\Delta|$ sufficiently small.

Proposition 7.2. Let E be away from the thresholds and eigenvalues of H and let $\gamma_0 \notin \Sigma_E$, $|\gamma_0| < \max \kappa_\alpha$. Then there is a small interval Δ around E s.t. the estimate

$$F_\Delta i[H,F(\gamma \geq \gamma_0)]F_\Delta \geq \theta F_\Delta \frac{1}{\sqrt{<x>}} F'(\gamma \geq \gamma_0) \frac{1}{\sqrt{<x>}} F_\Delta\,, \tag{7.3}$$

holds for some $\theta > 0$ and for any $F(s \geq \gamma_0)$ with derivative $F'(s \geq \gamma_0)$, supported in $[\gamma_0-\delta_1, \gamma_0+\delta_1]$ with $\delta_1 < \frac{1}{10}\ \frac{1}{10+|\gamma_0|}$ dist (γ_0,Σ_E).

First we mention basic ingredients in the proof of proposition.

For any bounded smooth function F with a C_0^∞ derivative

$$\begin{aligned} F_\Delta i[H,F(\gamma)]F_\Delta &= F_\Delta F(\gamma)^{\frac{1}{2}} F_{\Delta'} i[H,\gamma]F_{\Delta'} F(\gamma)^{\frac{1}{2}} F_\Delta \\ &\quad + O(|x|^{-2})\,, \end{aligned} \tag{7.4}$$

where $\Delta \subset \Delta'$ and $F_{\Delta'}F_\Delta = F_\Delta$. A similar relation holds if H and γ are replaced by H_a and γ_a. The proof of this estimate is given in [SigSof2]. Next we use

$$i[H,\gamma] = \frac{1}{\sqrt{<x>}} (i[H,A] - \gamma^2) \frac{1}{\sqrt{<x>}} + O(|x|^{-2}) \tag{7.5}$$

Note that the last two estimates show that the commutator of any two factors in the first term on the r.h.s. of (7.4) is $O(|x|^{-2})$ so the order in which they stand is, in fact, inessential. We will estimate the first term in the r.h.s. of (7.5). This estimate is based on the following result, which, though somewhat different in spirit, is closely related to the Mourre estimate ([M1, PSS, FH]).

Let P_α be the projection on the channel bound state ψ^α: $P_\alpha f = \psi^\alpha \langle f, \psi^\alpha \rangle$ and $P_\alpha = 1$ for the free channel, let p_α be the channel momentum: $p_\alpha = p_a$ for $a = a(\alpha)$ and let $\gamma_\alpha = \gamma_a$ for $a = a(\alpha)$. Denote by S the tuples of the form $S = (a_2, a_3, \ldots, a_k)$, where $\#(a_i) = i$, $a_i \supset a_{i+1}$. We let $\alpha(S) = \alpha$ if $a(\alpha) = a_k$.

Theorem 7.3. For any $\varepsilon, \delta > 0$ and an energy $E \neq \varepsilon_\alpha$ for all α, there are smooth functions F_α with $0 \leq F_\alpha \leq 1$, $\operatorname{supp} F_\alpha \subset (-\frac{\delta}{2}, \frac{\delta}{2})$, a finite subset, A_ε, of channels and a small interval Δ around E s.t. the following estimates hold

$$F_\Delta i[H,A] F_\Delta \geq F_\Delta \sum (2\kappa_\alpha - \varepsilon) \phi_{\alpha,E} F_\Delta \tag{7.6}$$

and

$$(F_\Delta)^2 \gtrless (1 \mp \varepsilon) F_\Delta \Sigma \phi_{\alpha,E} F_\Delta \tag{7.7}$$

with the sums extending over all open channels from A_ε and

$$\phi_{\alpha,E} = \sum_{\alpha(S)=\alpha} j_S P_\alpha F_\alpha(|p_\alpha| - \kappa_\alpha) j_S^* , \tag{7.8}$$

where j_S are bounded operators which are finite sums of finite products of the P_α's and bounded smooth cut-off functions. We do not state here the properties of these cut-off functions, though we use them later in a technical estimate (eqn (7.16)).

Proof. See [SigSof2].

Finally, we mention that we use the following conventions

$$B = O(|x|^{-\alpha}) <=> \langle x\rangle^{\beta} B \langle x\rangle^{\alpha-\beta} \text{ is bounded for all } |\beta| \leq \max(\alpha,2) .$$

$$B = O_1(|x|^{-\alpha}) <=> (H+i)^{-1+k} B (H+i)^{-k} \text{ is } O(|x|^{-\alpha}) \text{ for } k=0,1 .$$

$$B = O_2(|x|^{-\alpha}) <=> (H+i)^{-1} B (H+i)^{-1} \text{ is } O(|x|^{-\alpha})$$

$$B \doteq 0 <=> B = O(|x|^{-1-\epsilon}), \text{ similarly, } B \dot{\geq} 0 .$$

We will use the following property of these symbols

$$f(B) O(|x|^{-\alpha}) = O(|x|^{-\alpha}) = O(|x|^{-\alpha}) f(B) \tag{*}$$

where f has C_0^∞ derivative and B stands for one of the following operators $H_a, \gamma_a, |p_a|$ (any a).

Sketch of the proof of Proposition 7.2. Take Δ' slightly larger than Δ so that $F_{\Delta'} F_\Delta = F_\Delta$ and fix it. Eqns. (*), (7.4) and (7.5) imply

$$F_\Delta i[H, F(\gamma \geq \gamma_0)] F_\Delta \doteq$$
$$F_\Delta F'(\gamma \geq \gamma_0)^{\frac{1}{2}} F_{\Delta'} \frac{1}{\sqrt{\langle x\rangle}} (i[H,A] - \gamma^2) \frac{1}{\sqrt{\langle x\rangle}} F_{\Delta'} F'(\gamma \geq \gamma_0)^{\frac{1}{2}} F_\Delta . \tag{7.9}$$

Next, using the relations

$$[F_{\Delta'}, \frac{1}{\sqrt{\langle x\rangle}}] = O(|x|^{-\frac{3}{2}}) \quad \text{and} \quad [F'(\gamma \geq \gamma_0)^{\frac{1}{2}}, \frac{1}{\sqrt{\langle x\rangle}}] = O(|x|^{-\frac{3}{2}}),$$

we move $\frac{1}{\sqrt{\langle x\rangle}}$ outside of $F'(\gamma \geq \gamma_0)^{\frac{1}{2}}$. After that we estimate the γ^2-term. First, we observe that

$$[F'(\gamma \geq \gamma_0)^{\frac{1}{2}}, F_{\Delta'}] = O(|x|^{-1}), \tag{7.10}$$

Next we use the localization of $F'(\gamma \geq \gamma_0)^{\frac{1}{2}}$ and the spectral theorem to deduce

$$\gamma^2 F'(\gamma \ge \gamma_0)^{\frac{1}{2}} \ge (\gamma_0^2 - \varepsilon) F'(\gamma \ge \gamma_0)^{\frac{1}{2}} \tag{7.11}$$

with $\varepsilon = 4|\gamma_0|\delta_1$. The result of these manipulations is

$$F_\Delta i[H,F]F_\Delta \,\dot{\ge}$$

$$F_\Delta \frac{1}{\sqrt{<x>}} (F')^{\frac{1}{2}} F_{\Delta'}(i[H,A] - 2\gamma_0^2 - 2\varepsilon) F_{\Delta'}(F')^{\frac{1}{2}} \frac{1}{\sqrt{<x>}} F_\Delta ,$$

where we have used the obvious abbreviation. Applying to this inequality expansions (7.6) and (7.7) with ε given above and $\delta = 10\delta_1$ we arrive at

$$F_\Delta i[H,F]F_\Delta \,\dot{\ge}\, 2F_\Delta \frac{1}{\sqrt{<x>}} (F')^{\frac{1}{2}} F_{\Delta'} \sum_{\alpha\text{-open}} (\kappa_\alpha^2 - \gamma_0^2 - 2\varepsilon)\phi_\alpha F_{\Delta'}(F')^{\frac{1}{2}} \frac{1}{\sqrt{<x>}} F_\Delta \tag{7.12}$$

with ϕ_α given in (7.8).

Lemma 7.4. Assume f is smooth and

$$\operatorname{supp} f \subset \{s \mid |s - \gamma_0| < \frac{1}{4}\delta\} \tag{7.13}$$

with δ given in Theorem 7.3. Then for any ε_1

$$\|f(\gamma)\phi_\alpha - O(|x|^{-1})\| \le \varepsilon_1 \text{ for } \alpha \text{ s.t. } |\gamma_0| > \kappa_\alpha + 2\delta \tag{7.14}$$

Here $O(|x|^{-1})$ depends on ε_1.

Proof. Routine commutator estimates yield (see [SigSof2]) that for f as above and any ε_1

$$\|f(\gamma)P_\alpha - P_\alpha f(\gamma_\alpha) - O(|x|^{-1})\| < \varepsilon_1 \tag{7.15}$$

and

$$\|f(\gamma)j_S - j_S f(\gamma) - O(|x|^{-1})\| \le \varepsilon_1 , \tag{7.16}$$

where $\alpha(S) = \alpha$ and $O(|x|^{-1})$ depends on ε_1.

To deduce (7.14) we recall the definition of ϕ_α and using (7.15) and (7.16) pull

$f(\gamma)$ through j_s and P_α inside of ϕ_α next to $F_\alpha(|p_\alpha| - \kappa_\alpha)$. Next we use the following localization estimate

$$f(\gamma)F_\alpha(|p_\alpha| = \kappa_\alpha) = O(|x|^{-1}) \text{ for } \alpha \text{ s.t. } |\gamma_0| > \kappa_\alpha + 2\delta,$$

which is just a quantum expression of an elementary classical inequality: $|\hat{x}_\alpha \cdot k_\alpha| \le |k_\alpha|$. This yields (7.14). □

Let $\kappa(\varepsilon)$ be the number of terms on the r.h.s. of (7.12). Take $\varepsilon_1 < \varepsilon[10\kappa(\varepsilon)\max\kappa_\alpha^2]^{-1}$ Pull $F_1(\gamma)$ with $F_1(s)$ satisfying (7.13) out of $F'(\gamma \ge \gamma_0)^{\frac{1}{2}}$ (i.e. $F_1 = 1$ on supp. F') and use eqn. (7.14) to obtain

$$\begin{aligned}(F')^{\frac{1}{2}}F_{\Delta'} &\sum_{\alpha \text{ open}} (\kappa_\alpha^2 - \gamma_0^2 - 2\varepsilon)\phi_\alpha F_{\Delta'}(F')^{\frac{1}{2}} \\ \ge (F')^{\frac{1}{2}}F_{\Delta'} &\sum_{\substack{\alpha \text{ open} \\ |\gamma_0|<\kappa_\alpha+2\delta}} (\kappa_\alpha^2 - \gamma_0^2 - 2\varepsilon)\phi_\alpha F_{\Delta'}(F')^{\frac{1}{2}} \\ &- \frac{\varepsilon}{10}(F')^{\frac{1}{2}}F_{\Delta'}^2(F')^{\frac{1}{2}} + O(|x|^{-1})\end{aligned} \tag{7.17}$$

Now we use that $\gamma_0 \notin \Sigma_E$. Since $\delta < \frac{1}{10}\,\mathrm{dist}(\gamma_0, \Sigma_E)$, we have

$$|\gamma_0| \le \kappa_\alpha + 2\delta \longleftrightarrow |\gamma_0| < \kappa_\alpha \tag{7.18}$$

Since $\varepsilon < \frac{1}{10}\theta$, where $\theta = \min_{\kappa_\alpha > |\gamma_0|} (\kappa_\alpha^2 - \gamma_0^2)$, we have

$$\begin{aligned}&\text{1st term on the r.h.s. of (4.25)} \\ &\ge \frac{2}{3}\theta(F')^{\frac{1}{2}}F_{\Delta'} \sum_{\alpha:\,|\gamma_0|<\kappa_\alpha} \phi_\alpha F_{\Delta'}(F')^{\frac{1}{2}}.\end{aligned} \tag{7.19}$$

Now using (7.14) again we restore the summation on the r.h.s. of (7.19) to over all open α:

$$\text{r.h.s. of (7.19)} \geq \frac{2}{3}\theta\,(F')^{\frac{1}{2}}F_{\Delta'}\sum_{\alpha\text{ open}}\phi_\alpha F_{\Delta'}(F')^{\frac{1}{2}}$$
$$+\,O(|x|^{-1}) - \frac{\varepsilon}{10}(F')^{\frac{1}{2}}F_{\Delta'}^2(F')^{\frac{1}{2}} \tag{7.20}$$

Folding back the sum on the r.h.s. by using (7.7) and remembering that $\varepsilon < \frac{\theta}{10}$, we obtain finally

$$(F')^{\frac{1}{2}}F_{\Delta'}\sum_\alpha(\kappa_\alpha^2 - \gamma_0^2 - 2\varepsilon)\phi_\alpha F_{\Delta'}(F')^{\frac{1}{2}}$$
$$\geq \frac{\theta}{2}(F')^{\frac{1}{2}}(F_{\Delta'})^2(F')^{\frac{1}{2}} + O(|x|^{-1}) \tag{7.21}$$

Combining this inequality with (7.12) and pulling $(F_{\Delta'})^2$ through $(F')^{\frac{1}{2}}$ and absorbing it into one of the F_Δ, we arrive at (7.7). □

Now we observe that $F_0(\gamma = \gamma_0) \equiv [F'(\gamma \geq \gamma_0)]^{\frac{1}{2}}$ is, in fact, an arbitrary positive function supported in a small but fixed neighbourhood of γ_0. Hence, inequality (7.7) implies

$$\langle i[H,F(\gamma \geq \gamma_0)]\psi_t,\psi_t\rangle \geq \theta\left\|\frac{1}{\sqrt{\langle x\rangle}}F_0(\gamma=\gamma_0)\psi_t\right\|^2 \tag{7.22}$$

for $\theta > 0$ and for any $\psi \in \operatorname{Ran} P_\Delta(H)$. This together with eqn. (7.2) yields the statement of Theorem 5.2 with an additional restriction that $F(\gamma)$ is supported in $(-\max\kappa_\alpha, \max\kappa_\alpha)$.

It remains to prove (7.1) for a bounded smooth function supported $\{s \mid |s| > \max\kappa_\alpha\}$. Let $M > \max_\alpha \kappa_\alpha$ and let $F(s \geq M)$ be supported away from $[\max\kappa_\alpha, \infty)$. It is a simple exercise to show (see [SigSof2]) that

$$\int_{-\infty}^{\infty}\left\|F(|\gamma| \geq M)\frac{1}{\sqrt{\langle x\rangle}}\psi_t\right\|^2 dt \leq c\|\psi\|^2 \tag{7.23}$$

for all $\psi \in \text{Ran } P_\Delta$, which completes the proof of Theorem 7.1. □

Now we consider the case $\gamma \in \Sigma_E$. First, we observe that the proof of Prop. 8.1 shows that it suffers to prove the propagation estimates for the operators $j_{a,i}$ introduced there. The proof of Theorem 6.1(iv) shows that these operators are of the form (for simplicity we consider only the case of two-cluster a)

$$J_{a,E,\varepsilon} = F_{\varepsilon_0}(r^a = \varepsilon)F_{\sqrt{\varepsilon_0}}(|p_a| \notin \Sigma_{E,\varepsilon}), \text{ where } \Sigma_{E,\varepsilon} = \{\sqrt{1-\varepsilon^2}\kappa \mid \kappa \in \Sigma_E\} . \quad (7.24)$$

The main difference between the 2-cluster and general cases is in the geometric fact that $r^a \leq \varepsilon$, for ε sufficiently small, implies in the two-cluster case that $|x|_a > \delta|x|$ for some $\delta > 0$. This is not in general true for the a's with more than 2 clusters.

Theorem 7.4. Let E be away from the thresholds and eigenvalues of H. Let $J_{a,E,\varepsilon}$ be an operator defined in (7.24) (for two-cluster a). There exists a small interval Δ around E s.t.

$$\int_{-\infty}^{\infty} \|J_{a,E,\varepsilon} \frac{1}{\sqrt{\langle x\rangle}} \psi_t\|^2 dt \leq C\|\psi\|^2 \quad (7.25)$$

for all $\psi \in \text{Ran } P_\Delta(H)$ and with C independent of ψ.

A generalization of Theorem 7.4 to the multicluster a's yields the essential part of Theorem 4.2 (propagation theorem).

Recall that eqn. (7.2) reduces the proof of propagation estimates of the form (7.25) to estimates of commutators of H with suitable propagation observables, restricted to an energy shell. Now we use propagation observables of the form

$$\phi = j(x)f(p_a)g(\gamma_a)\phi(\gamma) , \quad (7.26)$$

where all the functions are real smooth and bounded. Besides j satisfies the estimate $|D^\alpha j(x)| \leq C_\alpha \langle x\rangle^{-|\alpha|}$ and

$$\text{supp } j \subset \{x \varepsilon X \mid |x|_a > \delta |x| \text{ and } |x_a| > \delta |x|\} \text{ for some } \delta > 0$$

f, g' and ϕ have compact supports

The factors on the r.h.s. of (5.26) commute modulo $O(|x|^{-1})$. This fact follows from simple commutator estimates (see [SigSof2]) and the relation

$$<x_a>^{-1} F\left(\frac{|x_a|}{|x|} > \delta\right) = <x>^{-1}$$

(this relation is needed only for the commutators of $g(\gamma_a)$ with $f(p_a)$). This implies, for instance, that

$$\phi = \text{Re } \phi + O(|x|^{-1}), \tag{7.27}$$

where $\text{Re } \phi = \frac{1}{2} (\phi + \phi^*)$ is a self-adjoint operator.

Moreover, since different factors on the r.h.s. of (7.26) are positive operators, whose square roots commute modulo $O(|x|^{-1})$, we have

$$\phi = \text{positive oper.} + O(|x|^{-1}). \tag{7.28}$$

Indeed, the difference between a symmetrized expression,

$$\phi^{sym} = \phi_1^* \phi_1 \text{, where } \phi_1 = j(x)^{\frac{1}{2}} f(p_a)^{\frac{1}{2}} g(\gamma_a)^{\frac{1}{2}} \phi(\gamma)^{\frac{1}{2}} .$$

and ϕ is $O(|x|^{-1})$.

The philosophy of the analysis presented below hinges on the fact that terms of higher orders in $|x|^{-1}$ can be dropped. They lead to contributions into the time integrals which are convergent by a result of [PSS] (see eqn (7.29) below). This allows us to treat factors entering into the r.h.s. of (7.26) as commuting, reducing our analysis, essentially, to classical phase-space estimates. Below we identify, without mentioning it separately, operators of the form (7.26) with their self-adjoint realizations ($\phi \sim \text{Re } \phi$).

We use the following eqivalence relation:

$$B \doteq O \iff B = (H+i)\frac{1}{\sqrt{<x>}}F(\gamma\notin\Sigma_E)(\text{bndd opr})F(\gamma\notin\Sigma_E)\frac{1}{\sqrt{<x>}}(H+i) + O_2(|x|^{-1-\varepsilon}),$$

where $F(s \notin \Sigma_E)$ is a C_0^∞ function supported away from Σ_E. Similarly we define $B \gtrsim 0$ etc. Recall, that due to Thm 5.1,

$$\int_{-\infty}^{\infty} \|F(\gamma\in\Sigma_E)\frac{1}{\sqrt{<x>}}e^{-iHt}\psi\|^2 dt \le C\|(H+i)^{-1}\psi\|^2 \tag{7.29}$$

for all $\psi \in \operatorname{Ran} P_\Delta(H)$ for sufficiently small Δ around E. Moreover, due to a result of [PSS] (see also [M1])

$$\int_{-\infty}^{\infty} \|<x>^{-\frac{1}{2}-\varepsilon}e^{-iHt}\psi\|^2 dt \le C\|(H+i)^{-1}\psi\|^2 \tag{7.30}$$

for all $\psi \in \operatorname{Ran} P_\Delta(H)$ for sufficiently small $\Delta \ni E$. Hence

$$B \doteq O \iff \int_{-\infty}^{\infty} |<B\psi_t, \psi_t>| dt \le C\|\psi\|^2$$

for all $\psi \in \operatorname{Ran} P_\Delta(H)$ sufficiently small interval around E.

<u>Sketch of the proof of Theorem 7.4.</u> Since we have already obtained the propagation estimates for $\gamma \notin \Sigma_E$, it suffices therefore to consider only the operators

$$F_{\varepsilon_0}(r^a = \varepsilon)F_{\sqrt{\varepsilon_0}}(|p_a| \neq \kappa_\varepsilon)F(\gamma = \kappa)$$

for some $\kappa \in \Sigma_E$ (remember a is 2-cluster). Here $\kappa_\varepsilon = \sqrt{1-\varepsilon^2}\kappa$ and $F(\gamma = \kappa)$ is chosen as follows: F is smooth, $0 \le F \le 1$,

$F(s = \kappa)$ is supported in $\{s \mid |s-\kappa| < \varepsilon_0\}$

and

$$F'(s = \kappa) \text{ is supported away from } \Sigma_E . \tag{7.31}$$

The cut-off function $F(\gamma = \kappa)$ is chosen in such a way that on one hand it localizes γ around κ and on the other hand

$$[H, F(\gamma{=}\kappa)] \doteq 0 \tag{7.32}$$

The latter relation follows from eqn (7.31) and equations related to (7.4) and (7.5) but with H replaced by $(H{+}n)^{-1}$ with n s.t. $H + n \geq 1$.

Next, we introduce into consideration one observable-γ_a - and consider separately the cases $\gamma_a > \kappa_\varepsilon$, $\gamma_a < \kappa_\varepsilon$ and $\gamma_a = \kappa_\varepsilon$. Since the main ideas of our analysis are similar in these three cases we demonstrate them on one of the case - the case $\gamma_a > \kappa_\varepsilon$. We introduce the observable

$$\phi_{\varepsilon,\lambda} = F_{\varepsilon_0}(r^a \leq \varepsilon)F_{\varepsilon_0}(\gamma_a \geq \mu)F(\gamma = \kappa)F_{\sqrt{\varepsilon_0}}(|p_a| = \lambda)\ , \tag{7.33}$$

where $\mu > \kappa_\varepsilon$ and $\mu \neq \lambda$. Define the cut-off operator

$$\tilde{\phi}_{\varepsilon,\lambda} = F_{\varepsilon_0}(r^a = \varepsilon)F_{\varepsilon_0}(\gamma_a \geq \mu)F(\gamma = \kappa)F_{\sqrt{\varepsilon_0}}(|p_a| = \lambda)\ . \tag{7.34}$$

Lemma 7.5. For some $\theta > 0$ and $i = 1,2$,

$$i[H,\phi_{\varepsilon,\lambda}] \ \dot{\geq}\ \theta \frac{1}{\sqrt{<x>}}\, \phi_{\varepsilon,\lambda} \frac{1}{\sqrt{<x>}} \tag{7.35}$$

Proof: We drop all the indices at the F, ϕ, $\tilde{\phi}$ and μ. Commutator estimates and some geometry ([SigSof2]) yield

$$[f(B),I_a]F(|x|_a > \delta|x|) = O(|x|^{-1-\varepsilon})(H + i) \tag{7.36}$$

for any bounded function f with a C_0^∞ derivative. Here B stands for one of the operators p_a^2, γ_a and γ. Hence

$$[I_a,\phi] = O_2(|x|^{-1-\varepsilon}) \text{ and therefore } [H,\phi] = [H_a,\phi] + O_2(|x|^{-1-\varepsilon}) \tag{7.37}$$

To compute the commutators of H_a with the first two factors in ϕ we use the relations:

$$[H_a, f(r^a)] = [p^2, f(r^a)]$$
$$= \frac{1}{\langle x \rangle} f'(r^a)(\gamma^a - \gamma r^a) + O_1(|x|^{-2}) \tag{7.38}$$

and

$$[H_a, f(\gamma_a)] = [p_a^2, f(\gamma_a)]$$
$$= \frac{2}{\langle x \rangle} f'(\gamma_a)(p_a^2 - \gamma_a^2) + O_2(|x|^{-2}) \tag{7.39}$$

Collecting these estimates, remembering (5.32) and observing that $F'_{\varepsilon_0}(s \le \varepsilon) \le 0$ we obtain

$$i[H,\phi] \gtrsim \frac{1}{\sqrt{\langle x \rangle}} [(-F')(r^a\gamma - \gamma^a)FFF$$
$$+ FF'(p_a^2 - \gamma_a^2)FF] \frac{1}{\sqrt{\langle x \rangle}} \tag{7.40}$$

First we estimate the second term on the r.h.s. If $\lambda < \mu$, then

$$F'_{\varepsilon_0}(\gamma_a \ge \mu)F_{\sqrt{\varepsilon_0}}(|p_a| = \lambda) = O(|x|^{-1}), \tag{7.41}$$

provided $\sqrt{\varepsilon_0} < \frac{1}{10}(\mu - \lambda)$. This is a quantum version of the classical fact that $|\hat{x}_a \cdot k_a| \le |k_a|$. If $\lambda > \mu$, then we claim

$$FF'(p_a^2 - \gamma_a^2)FF \ge \frac{1}{2}(\lambda^2 - \mu^2)FF'FF + O_2(|x|^{-2}) \tag{7.42}$$

with the same abbreviations as in (7.40). Indeed, using the Fourier transform (i.e. the spectral theorem for $|p_a|$), we obtain

$$p_a^2 F_{\sqrt{\varepsilon_0}}(|p_a| = \lambda) \ge (\lambda - 2\sqrt{\varepsilon_0})^2 F_{\sqrt{\varepsilon_0}}(|p_a| = \lambda)$$

Furthermore, straightforward commutator estimates (see [SigSof2]) give

$$(F'_{\varepsilon_0})^{\frac{1}{2}} F_{\sqrt{\varepsilon_0}} (F'_{\varepsilon_0})^{\frac{1}{2}} = F'_{\varepsilon_0} F_{\sqrt{\varepsilon_0}} + O(|x|^{-1})$$

The last three estimates imply

$$F'_{\varepsilon_0} p_a^2 F_{\sqrt{\varepsilon_0}} > (\lambda - 2\sqrt{\varepsilon_0})^2 F'_{\varepsilon_0} F_{\sqrt{\varepsilon}} \tag{7.43}$$

where, remember, both sides should be understood in terms of their self-adjoint realizations. Similarly

$$\begin{aligned} &F'_{\varepsilon_0}(\gamma_a \geq \mu)\gamma_a^2 F_{\sqrt{\varepsilon_0}}(|p_a| = \lambda) \\ &= (F_{\sqrt{\varepsilon_0}})^{\frac{1}{2}} F'_{\varepsilon_0}\gamma_a^2 (F_{\sqrt{\varepsilon_0}})^{\frac{1}{2}} + O(|x_a|^{-1}) \\ &\leq (\mu + 2\varepsilon_0)^2 (F_{\sqrt{\varepsilon_0}})^{\frac{1}{2}} F'_{\varepsilon_0} (F_{\sqrt{\varepsilon_0}})^{\frac{1}{2}} + O(|x_a|^{-1}) \\ &= (\mu + 2\varepsilon_0)^2 F'_{\varepsilon_0} F_{\sqrt{\varepsilon_0}} + O(|x_a|^{-1}) \end{aligned} \tag{7.44}$$

The last two equations together yield

$$\begin{aligned} &F'_{\varepsilon_0}(\gamma_a \geq \mu)(p_a^2 - \gamma_a^2) F_{\sqrt{\varepsilon_0}}(|p_a| = \lambda) \\ &\geq \frac{1}{2}(\lambda^2 - \mu^2) F'_{\varepsilon_0}(\gamma_a \geq \mu) F_{\sqrt{\varepsilon_0}}(|p_a| = \lambda) + O(|x|^{-1}|x_a|^{-1}), \end{aligned}$$

where, recall, $F'(p_a^2 - \gamma_a^2)F$ and $F'F$ are identified with their self-adjoint realizations. Taking into account that $O(|x_a|^{-1})F(r^a \leq \varepsilon) = O(|x|^{-1})$ and that $[F(\gamma = \kappa), F_{\sqrt{\varepsilon_0}}(|p_a| = \lambda)] = O(|x|^{-1})$ we arrive at (7.42).

Now we analyze the first term on the r.h.s. We use the equation

$$\gamma = r^a\gamma^a + r_a\gamma_a + \frac{i(n-1)}{2\langle x\rangle}. \tag{7.45}$$

Using that $r_a = \sqrt{1 - (r^a)^2}$, we obtain

$$\begin{aligned} r^a\gamma - \gamma^a &= \frac{\sqrt{1-(r^a)^2}}{r^a}(\gamma_a - \sqrt{1-(r^a)^2}\gamma) \\ &+ \frac{i(n-1)}{2\langle x^a\rangle} \end{aligned} \tag{7.46}$$

On the phase-space support of the operator $F'_{\varepsilon_0}(r^a \leq \varepsilon)F_{\varepsilon_0}(\gamma^a > \mu)F(\gamma = \kappa)F_{\sqrt{\varepsilon_0}}(|p_a| = \lambda)$ we compute

$$r^a\gamma - r^a \geq \frac{\sqrt{1-\varepsilon^2}}{2\varepsilon}(\mu - \kappa_\varepsilon) = \theta \tag{7.47}$$

since $\varepsilon_0 \ll \varepsilon$. Thus, proceeding as above we derive $(-F')(r^a\gamma - \gamma^a)FFF \gtrsim \theta(-F')FFF$ (remember also our agreement about the symmetrization).

Putting these estimates together and choosing θ to be the minimum of $\frac{1}{2}(\lambda^2 - \mu^2)$ if $\lambda > \mu$ and θ_1, gives

$$i[H,\phi] \gtrsim -\theta \frac{1}{\sqrt{<x>}} F'FFF \frac{1}{\sqrt{<x>}} \tag{7.48}$$

This inequality together with the inequality

$$C(-F'FFF) \geq \bar{\phi} + O(|x|^{-1}) \tag{7.49}$$

yields (7.34). □

Write now $\tilde{\phi}_{\varepsilon,\lambda} = \Lambda^*_{\varepsilon,\lambda}\Lambda_{\varepsilon,\lambda} + O(|x|^{-1})$ with

$$\Lambda_{\varepsilon,\lambda} = F_{\varepsilon_0}(r^a = \varepsilon)^{\frac{1}{2}} F_{\varepsilon_0}(\gamma_a > \mu)^{\frac{1}{2}} F(\gamma = \kappa)^{\frac{1}{2}} F_{\sqrt{\varepsilon_0}}(|p_a| = \lambda)^{\frac{1}{2}} \tag{7.50}$$

Then Lemma 7.5 and eqn. (7.2) yield

$$\begin{aligned} &\frac{d}{dt} <\phi_{\varepsilon,\lambda}\psi_t,\psi_t> + C\|<x>^{-1}\psi_t\|^2 + \|F(\gamma \notin \Sigma_E) \frac{1}{\sqrt{<x>}} \psi_t\|^2 \\ &\qquad \geq \theta\|\Lambda_{\varepsilon,\lambda} \frac{1}{\sqrt{<x>}} \psi_t\|^2 , \end{aligned} \tag{7.51}$$

provided $\psi \in \text{Ran}\, P_\Delta(H)$ with compact Δ. Taking Δ, sufficiently small interval around E (remember $E \neq \varepsilon_\alpha$ for all α) and using results (7.29) and (7.30) we obtain that

$$\int_{-\infty}^{\infty} \|\Lambda_{\varepsilon,\lambda} \frac{1}{\sqrt{<x>}} \psi_t\|^2 dt \leq C\|\psi\|^2 \tag{7.52}$$

This completes the proof of propagation estimate for the $\gamma_a > \kappa_\varepsilon$ case. As was mentioned before other cases are treated similarly. □

8. Deift-Simon Wave Operators

Our contention is that each member $j_{a,E}(x,p)$ of the phase-space partition of unity constructed in section 7 is supported in a group of geometrically identical channels (i.e. channels with the same cluster decomposition a). To test this contention we compare on $j_a(x,p)$ dynamics generated by H and H_a. To this end we introduce the Deift-Simon wave operators:

$$W_{a,E}^{\pm} = \text{s-lim}\, e^{iH_a t} j_{a,E}(x,p) e^{-iHt}, \tag{8.1}$$

if the limits exist on Ran $P_\Delta(H)$. (Deift and Simon [DS] have introduced such operators for a partition of unity depending only on x.)

Proposition 8.1. Let (A)-(D), with $\mu > 1$, hold. Let the energy E be away from the thresholds and eigenvalues of H. Then the Deift-Simon wave operators $W_{a,E}^{\pm}$ exist on Ran P_Δ, provided Δ is sufficiently small interval around E.

Proof. Let (omit the subindex E from the partition of unity)

$$W_a(t) = e^{iH_a t} j_a(x,p) e^{-iHt}. \tag{8.2}$$

By the Fundamental theorem of Calculus

$$W_a(t) = j_a + i\int_0^t e^{iH_a s}(H_a j_a - j_a H)e^{-iHs}ds. \tag{8.3}$$

We have

$$H_a j_a - j_a H = \underbrace{[H_a, j_a]}_{\substack{\text{use propag. est.} \\ \text{and [PSS]}}} - \underbrace{j_a I_a}_{\text{use [PSS]}}. \tag{8.4}$$

Due to Theorem 7.4(ii) and condition (D) on the potentials

$$j_a I_a = O_1(|x|^{-\mu}). \tag{8.5}$$

(This is the only place where condition (D) is used!)

Now we analyze $[H_a, j_a]$. We compute

$$\begin{aligned} H_a j_a(x,p)] &= [p^2 j_a(x,p)] \\ &= 2p \cdot \nabla_x j_a(x,p) + O(|x|^{-2}), \end{aligned} \tag{8.6}$$

since due to Theorem 7.4(i), $j_a(x,p)$ commutes with $V_a = V - I_a$ (the latter depends only on x^a) and since j_a is a symbol.

Collecting these estimates and writing the operators $j_{a,E}$ as $\nabla_x j_{a,E} = \frac{1}{\sqrt{<x>}} j_{a,i}^* j_{a,i} \frac{1}{\sqrt{<x>}} + O(|x|^{-2})$, we obtain

$$\begin{aligned} i[H_a, j_a] &= \frac{1}{\sqrt{<x>}} \Sigma\, j_{a,i}^* p_i j_{a,i} \frac{1}{\sqrt{<x>}} \\ &\quad + O(|x|^{-2}), \end{aligned} \tag{8.7}$$

where $j_{a,i}$ are phase-space operators supported away from PS_E. Let $g \in \operatorname{Ran} P_\Delta(H)$ and let $\phi \in C_0^\infty$ obey $\phi \equiv 1$ on Δ. Consider $<[H_a, j_a]g, f>$. Using eqn. (8.7), pulling $\phi(H)$ from g and commuting it through $j_{a,i}<x>^{-\frac{1}{2}}$ (the commutator picks up an extra $<x>^{-1}$) and using that the operators $p_i\phi(H)$ are bounded, we obtain

$$\begin{aligned} |<[H_a, j_a]g, f>| &\le C\Sigma\Big(\|j_{a,i} \frac{1}{\sqrt{<x>}} g\|\, \|j_{a,i} \frac{1}{\sqrt{<x>}} f\| \\ &\quad + \|\frac{1}{<x>} g\|\, \|\frac{1}{<x>} f\|\Big). \end{aligned} \tag{8.8}$$

Collecting (8.5) and (8.7) and using (8.4) we obtain

$$\begin{aligned} |<(H_a j_a - j_a H)g, f>| &\le C\Sigma\Big(\|j_{a,i} \frac{1}{\sqrt{<x>}} g\|\|j_{a,i} \frac{1}{\sqrt{<x>}} f\| \\ &\quad + \|<x>^{-\min(\mu/2,1)} g\|\|<x>^{-\min(\mu/2,1)} f\|\Big). \end{aligned}$$

Let Δ' be slightly larger than Δ so that $\Delta' \backslash \bar{\Delta}$ is open. Let $\psi \in \operatorname{Ran} P_\Delta(H)$ and $u \in \operatorname{Ran} P_{\Delta'}(H_a)$. We have

$$\int_0^{|t|} |\langle (H_a j_a - j_a H) e^{-iHs}\psi, e^{-iH_a s} u\rangle| \, ds$$

$$\leq C[\Sigma \int_0^{|t|} \| j_{a,i} \frac{1}{\sqrt{\langle x\rangle}} e^{-iHs}\psi\|^2]^{\frac{1}{2}} \times [\Sigma \int_0^{|t|} \| j_{a,i} \frac{1}{\sqrt{\langle x\rangle}} e^{-iH_a s} u\|^2 ds]^{\frac{1}{2}} \tag{8.9}$$

$$+ C[\int_0^{|t|} \|\langle x\rangle^{-\alpha} e^{-iHs}\psi\|^2 ds]^{\frac{1}{2}} \times [\int_0^{|t|} \|\langle x\rangle^{-\alpha} e^{-iH_a s} u\|^2 ds]^{\frac{1}{2}} ,$$

where $\alpha = \min(1, \mu/2)$. Now due to Thm 4.2 and the fact that $j_{a,i}$ are supported away from PS_E, the first product on the r.h.s. of (8.9) is bounded by $\text{const}\|\psi\| \, \|u\|$, provided Δ and Δ' are sufficiently small. Furthermore, a Perry-Sigal-Simon result ([PSS], see eqn. (7.30)) implies that $\langle x\rangle^{-\alpha}$ with $\alpha > \frac{1}{2}$ is locally $H-$ (and therefore also H_a-) smooth provided Δ' avoids the thresholds and eigenvalues of H and, consequently, also of H_a). (This is the only place where the condition $\mu > 1$ is used!) Hence the second product on the r.h.s. of (8.9) is bounded by the same $\text{const}\|\psi\| \, \|u\|$. This proves the estimate

$$\int_0^{|t|} |\langle (H_a j_a - j_a H) e^{-iHs}\psi, e^{-iH_a s} u\rangle| \, ds \leq \text{const}\|\psi\| \, \|u\| \tag{8.10}$$

for all $\psi \in \text{Ran}\, P_\Delta(H)$ and $u \in \text{Ran}\, P_{\Delta'}(H_a)$. Hence the integral on the r.h.s. of (8.3), multiplied from the left by $P_{\Delta'}(H_a)$, converges strongly on $\text{Ran}\, P_\Delta(H)$, which proves that $\underset{t\to\pm\infty}{\text{s-lim}}\, P_{\Delta'}(H_a) W_a(t)$ exists on $\text{Ran}\, P_\Delta(H)$.

Finally, using a standard technique, one shows that $\chi(H_a) W_a(t) \xrightarrow{s} 0$ as $|t| \to \infty$ for any smooth $\chi(s)$ supported away from $\overline{\Delta}$. This, together with the previous result implies the existence of s-lim $W_a(t)$ on $\text{Ran}\, P_\Delta(H)$. □

9. Proof of Asymptotic Completeness

In this section we combine the results of previous sections in order to prove asymptotic completeness. A simple argument we use below was first suggested by Deift and Simon [DS].

Theorem 9.1. Assume conditions (A)-(D) with $\mu > 1$ hold. Let E be a non-threshold and non-bound state energy. Then the system in question is asymptotically clustering at the energy E.

Proof. Let ψ be orthogonal to the bound state subspace of H. Write

$$e^{-iHt}\psi = \Sigma j_{a,E}e^{-iHt}\psi = \Sigma e^{-iH_at}e^{iH_at}j_{a,E}e^{-iHt}\psi ,$$

where we have omitted the term $O(|x|^{-1})e^{-iHt}\psi$ which tends to 0 by [PSS]. Introducing $\phi_a^\pm = W_{a,E}^\pm\psi$ we rewrite this as

$$e^{-iHt}\psi = \Sigma e^{-iH_at}\phi_a^\pm + R^\pm(t) , \tag{9.1}$$

where

$$R^\pm(t) = \Sigma e^{-iH_at}[e^{iH_at}j_{a,E}e^{-iHt} - W_{a,E}^\pm]\psi .$$

Since $W_{a,E}^\pm$ exist by Prop. 8.1, we have $\|R^\pm(t)\| \to 0$ as $t \to \pm\infty$. Hence the system is asymptotically clustering. □

Since this result is valid for any number of particles, it implies, due to Prop. 2.1, that the system in question is asymptotically complete (Theorem 3.1). □

References

[AG] S. Agmon, Lectures on Exponential Decay of Solutions of Second-Order Elliptic Equations, Math. Notes, Princeton Univ. Press, 1982.

[CFKS] H. Cycon, R. Froese, W. Kirsch and B. Simon, Lectures on the Schrödinger Equation (to appear).

[DS] P. Deift and B. Simon, Comm. Pure Appl. Math 30 (1977), 573-583.

[DMSV] P. Deift, W. Hunziker, B. Simon and E. Vock, Comm. Math. Phys. 64 (1978), 1-34.

[E1] V. Enss, Comm. Math. Phys. 52 (1977), 233-238.

[E2] V. Enss, Comm. Math. Phys. 61 (1978), 285-291.

[E3] V. Enss, J. Funct. Anal. 52 (1983), 219-268.

[E5] V. Enss, in: Differential Equations, I.W. Knowles and R.T. Lewis, eds., North Holland Math. Studies 92 (1984), 173-204.

[E6] V. Enss, Physica 124 A (1984), 269-292.

[E7] V. Enss, in: Proceedings of the Como Summer School on the Schrödinger Equation, Como 1984 (to be published).

[FH] R. Froese and I. Herbst, Duke Math. J. 49 (1982), 1975.

[G] J. Ginibre, in: "Schrödinger Equation", W. Thirring and P. Urban, eds., Springer-Verlag, 1977.

[Ka1] T. Kato, Math. Anal. 162 (1966), 258-279.

[Ka2] T. Kato, Studio Math xxxi (1968), 535-546.

[L1] R. Lavine, Proc. AMS 22 (1969), 55-60.

[L2] R. Lavine, J. Funct. Anal. 5 (1970), 368-382.

[L3] R. Lavine, Math. Phys. 20 (1971), 301-323.

[L4] R. Lavine, Indiana Univ. Math. J. 21 (1972), 643-656.

[L5] R. Lavine, J. Funct. Anal. 12 (1973), 30-54.

[L6] R. Lavine, J. Math. Phys. 14 (1973), 376-379.

[L7] R. Lavine, in: Scattering Theory in Mathematical Physics, J.A. Lavita and J.-P. Marchard, eds., D. Reidel Publ. Co.

[M1] E. Mourre, Comm. Math. Phys. 78 (1981), 391-408.

[M2] E. Mourre, Comm. Math. Phys. 91 (1983), 279-300.

[M1] E. Mourre, Preprint, Marseille 1983.

[PSS] P. Perry, I.M. Sigal and B. Simon, Ann. of Math. 144 (1981), 519-567.

[RS] M. Reed and B. Simon, Methods of Modern Mathematical Physics III, Academic Press.

[Sig1] I.M. Sigal, Comm. Math. Phys. 85 (1982), 309-324.

[Sig2] I.M. Sigal, Scattering Theory for Many-Body Quantum Mechanical Systems, Springer Lecture Notes in Math. N1011, 1983.

[SigSof1] I.M. Sigal and A. Soffer, Bulletin AMS 14 N1 (1986).

[SigSof2] I.M. Sigal and A. Soffer, preprint 1985.

[SigSof3] I.M. Sigal and A. Soffer, in preparation.

[Sim] B. Simon, Comm. Math. Phys. 55 (1977), 259-274.

I.M. Sigal* **
Department of Mathematics
University of California, Irvine, CA 92717

A. Soffer †
Division of Mathematics, Physics and Astronomy
CALTECH
Pasadena, CA 91125

* Supported by NSF Grant No. DMS8507040

** Present Address: Department of Mathematics, University of Toronto, Toronto, Canada M5S 1A1

† Weizmann Fellow of Mathematics and Physics

ON A MULTI-DIMENSIONAL INVERSE PROBLEM RELATED TO THE GEL'FAND-LEVITAN THEORY

Takashi Suzuki
Department of Mathematics, Faculty of Science, University of Tokyo
Hongô, Tokyo, 113 Japan

1. Problem For $I = (0,1)$ and $S^1 = [0,1]/0\sim 1 = \{e^{i2\pi s} | 0 \le s < 1\}$, let $\Omega = I \times S^1$. Then, $\partial\Omega = \gamma_0 \cup \gamma_1$, where $\gamma_0 = \{0\} \times S^1$ and $\gamma_1 = \{1\} \times S^1$. For $p = p(z) \in C^\infty(\bar{\Omega})$ and $F = F(\xi,t) \in C^\infty(\partial\Omega \times [0,T])$, let us consider the parabolic equation

$$(1.1) \qquad \frac{\partial u}{\partial t} = \Delta u - pu \qquad (z=(x,\theta) \in \Omega,\ 0 < t < T)$$

with the boundary condition

$$(1.2) \qquad \frac{\partial u}{\partial \nu}\Big|_{\partial\Omega} = F \qquad (0 < t < T)$$

and with the initial condition

$$(1.3) \qquad u|_{t=0} = 0 \qquad (z=(x,\theta) \in \Omega),$$

where $\Delta = \frac{\partial^2}{\partial x^2} + \frac{\partial^2}{\partial \theta^2}$ and ν denotes the outer unit normal vector on $\partial\Omega$. The problem which we study is to determine the coefficient p from two functions $F \not\equiv 0$ and $f = u|_{\partial\Omega}$ $(0 \le t \le T)$, which may be regarded as the input and the output of the system, respectively. We shall show a uniqueness theorem obtained in [4] and its underlying idea.

A one-dimensional version of our problem has been studied by A. Pierce ([2]). Namely, for the coefficient $P = (p,h,H) \in C^1(\bar{I}) \times \mathbf{R} \times \mathbf{R}$ and the input $F \in L^2(0,T)$, let $(E_{p,F})$ be the parabolic equation

$$(1.1') \qquad \frac{\partial u}{\partial t} = \frac{\partial^2}{\partial x^2} u - pu \qquad (x \in I,\ 0 < t < T)$$

with the boundary condition

$$(1.2') \qquad -\frac{\partial u}{\partial x} + hu\Big|_{x=0} = 0, \quad \frac{\partial u}{\partial x} + Hu\Big|_{x=1} = F \qquad (0 < t < T)$$

and with the initial condition

$$(1.3') \qquad u|_{t=0} = 0 \qquad (x \in I).$$

Furthermore, let $u = u(x,t;P,F)$ be the solution of $(E_{P,F})$. Then

Theorem 1. The equality

(1.4) $$u(\cdot,t;Q,F)|_{\cdot=1} = u(\cdot,t;P,F)|_{\cdot=1} \qquad (0 \le t \le T)$$

for some $Q = (q,j,J) \in C^1(\bar{I}) \times \mathbf{R} \times \mathbf{R}$ implies

(1.5) $$Q \equiv P,$$

provided that $F \neq 0$. □

The proof is based on the fact that $F \neq 0$ and $f = u|_{x=1}$ determine the "spectral characteristics" of $P = (p,h,H)$. Then, the conclusion follows immediately from the Gel'fand-Levitan theory.

2. Gel'fand-Levitan theory. We have arrived at the point to discuss the Gel'fand-Levitan theory. It is the inverse theory of Titchmarsh - Kodaira's general expansion theorem. Both of them are constructive and applies to the singular boundary value problem. Thus, the latter constructs the spectral density function $\rho = \rho(\lambda)$ from the differential operator $-\frac{d^2}{dx^2} + p(x)$ on $[0,+\infty)$ with $(-\frac{d}{dx} + h)\cdot|_{x=0} = 0$, while the former recovers (p,h) from ρ.

To fix the idea, let us restrict ourselves to the regular case and to the uniqueness aspect. Namely, for $P = (p,h,H) \in C^1(\bar{I}) \times \mathbf{R} \times \mathbf{R}$, let A_P be the Sturm-Liouville operator $-\frac{d^2}{dx^2} + p(x)$ on $[0,1]$ with the boundary condition $(-\frac{d}{dx} + h)\cdot|_{x=0} = (\frac{d}{dx} + H)\cdot|_{x=1} = 0$. The set $\sigma(A_P) = \{\lambda_n(P)\}_{b=0}^{\infty}$ $(-\infty < \lambda_0(P) < \lambda_1(P) < \dots \to +\infty)$ denotes its eigenvalues, and $\phi_n(\cdot;P)$ denotes its eigenfunction corresponding to A_P normalized by $\phi_n(\cdot;P)|_{x=0} = 1$. The norming constant $\rho_n = \rho_n(P)$ is defined by

$$\rho_n = \int_0^1 \phi_n(x;P)^2 dx,$$

and the set of two sequences

$$\mathscr{C} = \mathscr{C}(P) = \{\lambda_n(P),\ \rho_n(P)\}_{n=0}^{\infty}$$

is referred to as the spectral characteristics of A_P. Then, the Gel'fand-Levitan theory says

Theorem 2. The relation $\mathscr{C}(P) = \mathscr{C}(Q)$ implies $P \equiv Q$. □

We wish to go into details. The key idea is to connect two eigen functions through an integral transformation. Namely, for $\lambda \in \mathbf{R}$, let $\phi = \phi(\cdot;p,h;\lambda)$ be the solution of

$$(2.1) \qquad \left(-\frac{d^2}{dx^2} + p(x)\right)\phi = \lambda\phi \quad (0 \le x \le 1), \quad \phi|_{x=0} = 1, \quad \phi'|_{x=0} = h.$$

Then, we have the identity

$$(2.2') \qquad \phi(x;p,h;\lambda) = \cos\sqrt{\lambda}x + \int_0^x K(x,y)\cos\sqrt{\lambda}y\, dy$$

for $0 \le x \le 1$ and $\lambda \in \mathbf{R}$. Here $K(x,y) = K(x,y;p,h)$ is independent of λ and solves the hyperbolic equation

$$(2.3') \qquad K_{xx} - K_{yy} = p(x)K \qquad (\text{in } D)$$

with

$$(2.4') \qquad K(x,x) = h + \frac{1}{2}\int_0^x p(s)ds \quad \text{and} \quad K_y(x,0) = 0 \quad (0 \le x \le 1),$$

where $D = \{(x,y) | 0 < y < x < 1\}$. Next, the integral equation

$$(2.5) \qquad F(x,y) + \int_0^x K(x,z)F(z,y)dz + K(x,y) = 0 \qquad (\text{in } D)$$

is derived from the Parseval's relation and the identity (2.2'), the function $F = F(x,y)$ being represented by the spectral characteristics $\mathscr{G}(P)$ of A_P. Finally, the unique solvability with repsect to K of (2.5) is verified by the Fredholm theory, from which (p,h) is recovered as

$$p(x) = 2\frac{d}{dx}K(x,x) \qquad \text{and} \qquad h = K(0,0).$$

The constant H is recovered as

$$H = -\phi'(1;p,h;\lambda_n(P))/\phi(1;p,h;\lambda_n(P)),$$

which is shown to be independent of n. Thus in particular, Theorem 2 follows.

A more direct proof of Theorem 2 is obtained by modifying (2.2') as

$$(2.2) \qquad \phi(x;q,j;\lambda) = \phi(x;p,h;\lambda) + \int_0^x K(x,y)\phi(y;p,h;\lambda)dy.$$

This time, $K(x,y) = K(x,y;q,j;p,h)$ solves

$$(2.3) \qquad K_{xx} - K_{yy} + p(y)K = q(x)K \qquad (\text{in } D)$$

with

(2.4) $K(x,x)=(j-h)+\frac{1}{2}\int_0^x\{q(s)-p(s)\}ds$ and $K_y(x,0)=hK(x,0)$ $(0\leq x\leq 1)$.

Again, we shall arrive at the integral equation (2.5), $F = F(x,y)$ being represented by both $\mathscr{C}(P)$ and $\mathscr{C}(Q)$, this time. Actually, the relation $\mathscr{C}(P) = \mathscr{C}(Q)$ implies $F \equiv 0$, so that $K \equiv 0$ and $(p,h) = (q,j)$ follow in turn. Finally, then $H = -\phi'(1;p,h;\lambda_n(P))/\phi(1;p,h;\lambda_n(P)) = J$ is obvious.

A crucial point in extending the above argument to multi-dimensions lies in the identity (2.2). In fact it may be regarded as

$$\phi(\cdot;q,j;\lambda) = \phi(\cdot;p,h;\lambda) + \int_0^1 H(\cdot-y)K(\cdot,y)\phi(y;p,h;\lambda)dy, \tag{2.6}$$

where $H = H(z)$ is the Heaviside function. There seems to be no natural way to extend the Heaviside function to multi-dimensions. However, under the assumption of iso-spectral:

$$\lambda_n(P) = \lambda_n(Q) \qquad (n \in \mathbf{N} \equiv \{0,1,2,\ldots\}), \tag{2.7}$$

(2.2) may be read as

$$\phi_n(\cdot;Q) = (I + \mathscr{K})\phi_n(\cdot;P), \tag{2.8}$$

$\mathscr{K}$ being the integral operator with the kernel $H(x-y)K(x,y)$. We note, conversely, that the operator $\mathscr{K}$ can be defined through (2.8):

$$\mathscr{K} : \phi_n(\cdot;P) \mapsto \phi_n(\cdot;Q) - \phi_n(\cdot;P) \qquad (n \in \mathbf{N}).$$

The corresponding kernel $\tilde{K} = \tilde{K}(x,y)$ is formally given as

$$\tilde{K}(x,y) = \sum_{n=0}^{\infty} \{\phi_n(x;Q) - \phi_n(x;P)\}\phi_n(y;P)/\rho_n(P). \tag{2.9}$$

At this standpoint, the identity (2.2) is nothing but

$$\tilde{K}(x,y) = H(x-y)K(x,y). \tag{2.10}$$

We note that the formal expression (2.9) has meaning even in multi-dimensions, though (2.10) does not. That idea is made use of in our study. Incidentally, the function F in (2.5) corresponds to the kernel $\tilde{F}(x,y) = H(x-y)F(x,y)$, with $\tilde{F}(x,y) = \sum_{n=0}^{\infty}\{\phi_n(x;P) - \phi_n(x;Q)\}\phi_n(y;Q)/\rho_n(Q)$.

3. Back to the Problem. Let $(E_{p,F})$ be the equation (1.1) with (1.2) and (1.3), and let $u = u(z,t;p)$ be its solution. We assume

$$u(\cdot,t;q)|_{\partial\Omega} = u(\cdot,t;p)|_{\partial\Omega} \qquad (0 \leq t \leq T) \tag{3.1}$$

for some q. Let A_p denote the elliptic operator $-\Delta + p$ in Ω with $\frac{\partial}{\partial\nu}\cdot|_{\partial\Omega} = 0$. Its eigenvalues and eigenfunctions are denoted by $\{\lambda_n(p)\}_{n=0}^{\infty}$ $(-\infty < \lambda_0(p) < \lambda_1(p) \leq \dots \to +\infty)$ and $\{\phi_n(\cdot;p)\}_{n=0}^{\infty}$ $(\|\phi_n(\cdot;p)\|_{L^2} = 1)$, respectively. We assume the following, for the sake of simplicity:

(A1) Each $\lambda_n(p)$ is simple.

Furthermore, for the input $F = F(\xi,t)$, we suppose that

(A2) $F = g(t)h(\xi)$ $(\xi \in \partial\Omega,\ 0 \leq t \leq T)$, with $g \neq 0$,

$$\int_{\partial\Omega} h(\xi)\phi_n(\xi;p)d_\xi \neq 0 \quad \text{and} \quad \int_{\partial\Omega} h(\xi)\phi_n(\xi;q)d\sigma_\xi \neq 0.$$

The solution $u = u(z,t;p)$ is given as

$$u(z,t;p) = \int_0^t d\tau \int_{\partial\Omega} d\sigma_\xi G(z,\xi;t-\tau;p)F(\tau,\xi),$$

where G is the Green's function of $-\frac{\partial}{\partial t} + A_p$: $G(z,w;t;p) = \sum_{n=0}^{\infty} e^{-t\lambda_n(p)}\phi_n(z;p)\phi_n(w;p)$. Therefore, (3.1) reads

$$\int_0^t s(\xi,t-\tau;q)g(\tau)d\tau = \int_0^t s(\xi,t-\tau;p)g(\tau)d\tau \quad (\xi \in \partial\Omega)$$

with

$$s(z,t;p) = \sum_{n=0}^{\infty} e^{-t\lambda_n(p)}\phi_n(z;p)\int_{\partial\Omega}\phi_n(\xi;p)h(\xi)d\sigma_\xi.$$

From $g \neq 0$ follows

$$s(\xi,t;q) = s(\xi,t;p) \quad (\xi \in \partial\Omega,\ 0 \leq t \leq T),$$

which holds for $0 \leq t < +\infty$ by analytic continuation. Comparing the asymptotic behavior as $t \to +\infty$ of both-hand sides, we have

(3.2) $$\lambda_n \equiv \lambda_n(q) = \lambda_n(p) \quad (n \in \mathbf{N})$$

and

(3.3') $$\phi_n(\cdot;q)\int_{\partial\Omega}\phi_n(\xi;q)h(\xi) = \phi_n(\cdot;p)\int_{\partial\Omega}\phi_n(\xi;p)h(\xi) \quad (n\in \mathbf{N},\ \cdot\in \partial\Omega)$$

by (A.2) and $\phi_n(\cdot;p)|_{\partial\Omega} \neq 0$. The relation (3.3') implies that

(3.3) $$\phi_n(\cdot;q)|_{\partial\Omega} = c_n\phi_n(\cdot;p)|_{\partial\Omega} \quad \text{with} \quad c_n^2 = 1 \quad \text{for} \quad n \in \mathbf{N}.$$

At this point, we introduce the function

(3.4) $$K_s(z,w;\lambda) = \sum_{n=0}^{\infty} \{c_n\phi_n(z;q) - \phi_n(z;p)\}\phi_n(w;p)(\lambda_n+\lambda)^{-s},$$

taking $\lambda > -\lambda_0$ and $s > 0$ sufficiently large. Then, it becomes a C^2-function and by (3.2) satisfies the ultra-hyperbolic equation

$$(\Box - c(z,w))K_s(z,w;\lambda) = c(z,z)G_s(z,w;p,\lambda),$$

where $\Box = -\Delta_z + \Delta_w$, $c(z,w) = -q(z) + p(w)$ and

$$G_s(z,w;p,\lambda) = \sum_{n=0}^{\infty} \phi_n(z;p)\phi_n(w;p)(\lambda_n+\lambda)^{-s},$$

which is nothing but the Green's function of $(A_p + \lambda)^s$. Furthermore,

$$K_s|_{\Gamma_0} = \frac{\partial}{\partial\nu} K_s|_{\Gamma_0} = 0$$

holds by (3.3) and $c_n^2 = 1$, where $\Gamma_0 = \partial\Omega \times \Omega \subset \partial(\Omega \times \Omega)$.

Henceforth, we suppose that

(A.3) p and q are real analytic.

Then, G_s is real analytic on $\bar{\Omega} \times \bar{\Omega} \setminus \bar{D}$, where $D = \{(z,z)|z \in \Omega\}$. By Holmgren's theorem, K_s is also real analytic in a neighborhood U_1 of Γ_1 in $\Omega \times \Omega \setminus \bar{D}$. Actually, U_1 can contain all points in $\bar{\Omega} \times \bar{\Omega} \setminus \bar{D}$ which are reached by deforming a portion of the initial surface Γ_1 analytically through noncharacteristic surfaces with respect to $\Box$, having the same boundary. Therefore, $K = K(z,w)$ $(-\Delta_w + p(w) + \lambda)^s K_s(z,w;\lambda) \in \mathcal{D}'(\Omega \times \Omega)$ is real analytic in U_1 and satisfies.

(3.5) $$(\Box - c(z,w))K = c(z,z)\delta(z - w)$$

in $\Omega \times \Omega$ with $K|_{\Gamma_1} = \frac{\partial}{\partial\nu} K|_{\Gamma_1} = 0$. Again by Holmgren's theorem, we obtain $K = 0$ in $U_1 \subset \bar{\Omega} \times \bar{\Omega} \setminus \bar{D}$.

Now, we set

$$F_s(x,w;\lambda) = \sum_{n=0}^{\infty} \phi_n(z;q)\{c_n\phi_n(w;p) - \phi_n(w;q)\}(\lambda_n + \lambda)^{-s}.$$

Then, for $F(z,w) = (-\Delta_z + q(z) + \lambda)^s F_s(z,w;\lambda) \in \mathcal{D}'(\Omega \times \Omega)$ we have $F = 0$ in a neighborhood U_2 of $\Gamma_2 = \Omega \times \partial\Omega$, similarly. However, $F = K$ is verified by a standard argument. Indeed, formally it is obvious: $F(z,w) = \sum_{n=0}^{\infty} c_n\phi_n(z;q)\phi_n(w;p) - \sum_{n=0}^{\infty} \phi_n(z;q)\phi_n(w;q)$ $= \sum_{n=0}^{\infty} c_n\phi_n(z;q)\phi_n(w;p) - \delta(z-w) = \sum_{n=0}^{\infty} c_n\phi_n(z;q)\phi_n(w;p) - \sum_{n=0}^{\infty} \phi_n(z;p)\phi_n$ $(w;p) = K(z,w)$. Therefore, $K = 0$ in $U_1 \cup U_2$, from which we can conclude that

$$\mathrm{Supp}\ K \subset \{y = x\} \cup \{y = 1 - x\} \qquad (z = (x,\theta),\ w = (y,\omega))$$

and hence

$$(3.6) \qquad K(z,w) = \sum_{\ell=0}^{m} \{a_\ell(x,\omega) \otimes \delta^{(\ell)}(x-y) + b_\ell(z,\omega) \otimes \delta^{(\ell)}(1-x-y)\},$$

$a_\ell(z,\cdot)$, $b_\ell(z,\cdot)$ $\mathcal{D}'(S^1)$ being $w^* - C^2$ in z. By (3.4)-(3.6), we can show that $k \equiv 0$ and hence $q \equiv p$. In this way we can state

Theorem 3. The equality (3.1) implies $q \equiv p$, provided that (A.1)-(A.3). □

Concluding the present paper, we wish to point out that our method will apply to other multi-dimensional inverse problems.

Acknowledgement: This work was supported partly by Inoue Foundation for Science.

References

[1] Gel'fand, I.M., Levitan, B.M., On the determination of a differential equation from its spectral function (English translation), Amer. Math. Soc. Transl., (2) 1 (1955) 253-304.

[2] Pierce, A., Unique identification of eigenvalues and coefficients in a parabolic problem, SIAM J. Control & Optim., 17 (1979) 494-499.

[3] Suzuki, T., Gel'fand-Levitan's theory, deformation formulas and inverse problems, J. Fac. Sci. Univ. Tokyo, Sec. IA, 32 (1985) 223-271.

[4] Suzuki, T., On a multi-dimensional inverse parabolic problem, Proc. Japan Acad., Ser. A, 62 (1986) 83-86.

Fundamental Solution of the Poisson-Boltzmann Equation

Kyril Tintarev*
School of Mathematics
University of Minnesota
127 Vincent Hall
206 Church Street SE
Minneapolis, MN 55455

1. The main results.

The Poisson-Boltzmann equation is used for the study of electric potential in polyelectrolytes (vis. e.g. [1]). Recently a cylindrical case was studied by [2] and [3]. Both papers considered the equation

$$\frac{1}{2\pi}\Delta u_a = f_a(u_a) \quad \text{in } \Omega ,$$

$$f_a(u_a) = \frac{e^{\alpha u_a}}{\int_{\Omega_a} e^{\alpha u_a} dx} + \frac{Ne^{\beta u_a}}{\int_{\Omega_a} e^{\beta u_a} dx} - \frac{Ne^{-\beta u_a}}{\int_{\Omega_a} e^{-\beta u_a} dx} , \tag{1.1}$$

where α,β,N are positive constants, $\Omega_a = \Omega \{|x| \leq a\}$ and Ω is a 2-dimensional domain containing the ball $\{|x| \leq a\}$. In addition, u_a satisfies

$$\frac{\partial u_a}{\partial |x|} = -\frac{1}{a} \quad \text{on } \{|x| = a\} , \tag{1.2}$$

$$u_a = 0 \quad \text{on } \partial\Omega . \tag{1.3}$$

The paper [3] has studied this equation in the radial case. The paper [2] considered arbitrary Ω. Uniqueness, existence of a classical solution and asymptotics of u at $|x| = a$ for $a \to 0$ were established.

This paper deals with the formal limit of the problem (1.1)-(1.3):

$$\frac{1}{2\pi}\Delta u = f(u) - \delta(x) , \quad x \in \Omega ,$$

$$f(u) = \frac{e^{\alpha u}}{\int_\Omega e^{\alpha u} dx} + \frac{Ne^{\beta u}}{\int_\Omega e^{\beta u} dx} - \frac{Ne^{-\beta u}}{\int_\Omega e^{-\beta u} dx} , \tag{1.4}$$

---

*Research supported by Chaim Weizmann Postdoctoral Fellowship and by National Science Foundation Grant DMS 8501397.

with the boundary condition

$$u|_{\partial\Omega} = 0 \tag{1.5}$$

The boundary $\partial\Omega$ is assumed C^2 piecewise. A solution of (1.4)-(1.5) will be called a fundamental solution of the Poisson-Boltzmann equation. The present paper establishes existence, uniqueness and asymptotic behaviour of solution near the origin.

Set

$$\gamma = \max\{\alpha,\beta\} \tag{1.6}$$

THEOREM 1.1. There exists a unique, weak solution u of (1.4), (1.5) in $W^1_{2-\epsilon}(\Omega)$, $\epsilon > 0$. The function u is smooth in Ω_a for any $a > 0$.

THEOREM 1.2. Let u_a be a solution of (1.1)-(1.3). There exists a function $\hat{u}_a \in W^1_{2-\epsilon}(\Omega)$ such that $\hat{u}_a = u_a$ in Ω_{3a},

$$u_a \to u \text{ in } W_2^{1-\delta}(\Omega) \text{ for any } \delta > 0 . \tag{1.7}$$

THEOREM 1.3. The fundamental solution u satisfies the following relations:

$$|u - \log \frac{1}{|x|}| \le c \text{ in } \Omega , \ \gamma < 2 , \tag{1.8a}$$

$$|u - \frac{2}{\gamma} \log \frac{1}{|x|} + \frac{2}{\gamma} \log \log \frac{1}{|x|}| \le c \text{ in } \Omega , \ \gamma \ge 2 , \tag{1.8b}$$

for some $c > 0$, dependent on γ and Ω.

2. Preliminary estimates.

In this section we prove some uniform estimates for u_a. Let φ_a be the solution of the problem:

$$\begin{aligned} &\Delta\varphi_a = Me^{\gamma\varphi_a} \text{ in } \{a < |x| < R\} \supset \Omega_a \\ &\varphi|_{|x|=R} = 0 \\ &\frac{\partial\varphi}{\partial|x|}\Big|_{|x|=a} = -\frac{1}{a} \end{aligned} \tag{2.1}$$

with some constant $M > 0$. The paper [3] gives an explicit formula for φ_a.

LEMMA 2.1. The solution of (1.1)-(1.3) satisfies the following estimate:

$$\tilde{\varphi}_a - c \le u_a \le \varphi_a + c , \ c > 0 \text{ independent of } a \tag{2.2}$$

where $\tilde{\varphi}_a$ differs from φ_a by value of parameter M.

The upper bound in (2.2) is given in [2] by Lemma 4.6, and the lower bound in the proof of Lemma 4.4 therein.

LEMMA 2.2. The solution of (2.1) satisfies the following relations:

$$\varphi_a \to \varphi \quad \text{in} \quad W^1_{2-\epsilon}(\Omega)\ ,\ a\to 0\ ,\ \epsilon>0 \tag{2.3}$$

$$|\varphi_a - \frac{2}{\gamma}\log\frac{1}{|x|}| \le c\ ,\ \gamma<2 \tag{2.4}$$

$$|\varphi_a - \frac{2}{\gamma}\log\frac{1}{|x|} + \frac{2}{\gamma}\log\log\frac{1}{|x|}| \le c\ ,\ \gamma\ge 2\ , \tag{2.5}$$

where

$$\varphi = \frac{2}{\gamma}\log\frac{R}{|x|} - \frac{1}{\gamma}\log[K^{-1}\sinh^2(\frac{2-\gamma}{2}\log\frac{R}{|x|} - \text{arc sinh } K)]$$

for $\gamma<2$,

$$\varphi = \frac{2}{\gamma}\log\frac{R}{|x|} - \frac{1}{\gamma}\log[K_1^{-1}(\frac{1}{\sqrt{2}}\log\frac{R}{|x|} + K_1)]$$

for $\gamma\ge 2$,

$c, K, K_1 > 0$.

The lemma trivially follows from the explicit expression of φ_a in [3].

LEMMA 2.3. Let φ_a be solution of (2.1). Then

$$\|u_a-\varphi_a\|W^1_2(\Omega) \le C \quad \text{uniformly in} \quad a\ . \tag{2.6}$$

Proof: By (1.1), (2.1)

$$\Delta(u_a-\varphi_a) = f_a(u_a) - Me^{\gamma\varphi_a}\ . \tag{2.7}$$

Multiplying (2.7) by $(u_a-\varphi_a)$ and integrating by parts one obtains:

$$\|u_a-\varphi_a\|W^1_2(\Omega) \le \|u_a-\varphi_a\|L^\infty(\Omega)(\|f_a(u_a)\|L^1(\Omega) + M\|e^{\gamma\varphi_a}\|L^1(\Omega))\ . \tag{2.8}$$

The term $\|u_a-\varphi_a\|L^\infty(\Omega)$ is bounded by (2.2),(2.4),(2.5). The term $\|f_a(u_a)\|L^1(\Omega)$ by (1.2) does not exceed $2N+1$. The term $\|e^{\varphi_a}\|L^1(\Omega)$ is bounded by (2.4),(2.5).

LEMMA 2.4. There exist a function $u_a \in C^1_0(\Omega)$, such that $\hat{u}_a = u_a$ in Ω_{3a} ,

$$\|\hat{u}_a\|W^1_{2-\epsilon}(\Omega) \le C \quad \text{uniformly in} \quad a\ , \tag{2.9}$$

$$\|\hat{u}_a - u_a\|W^1_{2-\epsilon}(\Omega_a) \to 0 \quad \text{for} \quad a\to 0\ . \tag{2.10}$$

Proof: Let $\chi \in C^\infty(R_+)$, $\chi=0$ for $t<2$, $\chi=1$ for $t>3$ and $0\le\chi\le 1$ for $2\le t\le 3$. Set

$$\hat{u}_a = \chi\left(\frac{|x|}{a}\right) u_a \tag{2.11}$$

Then,

$$\|u'_a - \hat{u}'_a\|L^{2-\epsilon}(\Omega_a) \le \|(u'_a - \varphi'_a)(1-\chi)\|L^{2-\epsilon}(\Omega_a) + \|\varphi'_a(1-\chi)\|L^{2-\epsilon}(\Omega_a) + \|\chi' u_a\|L^{2-\epsilon}(\Omega_a) \; . \tag{2.12}$$

By (2.6) and Cauchy inequality,

$$\|(u'_a - \varphi'_a)(1-\chi)\|L^{2-\epsilon}(\Omega_a) \le \int_{2a\le|x|\le 3a} |u'_a - \varphi'_a|^{2-\epsilon} \le O(1)\|u'_a - \varphi'_a\|^{2-\epsilon/2}_{L^2(\Omega_a)} \to 0 \; , \; a \to 0 \tag{2.13}$$

By (2.3) and Cauchy inequality,

$$\|\varphi'_a(1-\chi)\|L^{2-\epsilon}(\Omega_a) \le \|\varphi'_a\|L^{2-\epsilon/2} O(1) \to 0 \; , \quad a \to 0 \; . \tag{2.14}$$

Finally, by (2.2),(2.4),(2.5),

$$\|\chi' u_a\|L^{2-\epsilon}(\Omega_a) \le c\left(\frac{1}{a}\int_{2a\le|x|\le 3a} |u_a|^{2-\epsilon} dx\right)^{1/2-\epsilon} \to 0 \; , \; a \to 0 \; . \tag{2.15}$$

Substituting (2.13),(2.14) and (2.15) into (2.12), one obtains (2.10). In order to prove (2.9) it suffices now (in view of (2.15)) to show that

$$\|u_a\|W^1_{2-\epsilon}(\Omega_a) < c \; . \tag{2.16}$$

Relation (2.16) follows immediately from (2.3) and (2.6). The lemma is proved.

3. Proofs of main theorems.

Proof of existence: Let $w \in C_0^\infty(\Omega)$. By multiplication of (1.1) by w , one obtains:

$$-\frac{1}{2\pi}\int_{\Omega_a} \nabla u_a \nabla w = -\frac{1}{2\pi a}\int_{|x|=a} w + \frac{1}{2\pi}\int_{|x|=a} u_a \frac{\partial w}{\partial \nu} + \int_{\Omega_a} f_a(u_a) w \; . \tag{3.1}$$

By Lemma 2.4 there is $u \in W^1_{2-\epsilon}(\Omega)$, such that $\hat{u}_a \to u$ in $W_2^{1-\delta}(\Omega)$ and a.e. in Ω on a subsequence of $a \to 0$. By an imbedding theorem, $u = 0$ on $\partial\Omega$. Remark now, that by (2.10),(2.2),(2.4),(2.5),

$$\frac{1}{2\pi}\int_{\Omega_a} \nabla(u_a - \hat{u}_a)\nabla w + \frac{1}{2\pi}\int_{|x|=a} (u_a - \hat{u}_a)\frac{\partial w}{\partial \nu} + \int_{\Omega_a} [f_a(u_a) - f_a(\hat{u}_a)] w \to 0 \; \text{ for } \; a \to 0 \; . \tag{3.2}$$

Therefore,

$$\frac{1}{2\pi}\int_\Omega \nabla\hat{u}_a\nabla w - \frac{1}{2\pi a}\int_{|x|=a} w + \frac{1}{2\pi}\int_{|x|=a} u_a\frac{\partial w}{\partial \nu} + \int_\Omega f_a(\hat{u}_a)w$$
$$\to 0 \quad \text{for} \quad a \to 0 . \tag{3.3}$$

By (2.2),(2.4),(2.5) one may apply the Lebesgue theorem to (3.3) and obtain:

$$\frac{1}{2\pi}\int_\Omega \nabla u_a\nabla w + \int_\Omega (f(u)-\delta)w = 0 \ , \ w\in C_0^\infty(\Omega) \ ,$$
$$u\in W^1_{2-\epsilon}(\Omega) \tag{3.4}$$

Proof of Theorem 1.3: Let u be a solution satisfying (3.4). Then from (2.2),(2.4), (2.5) the estimates (1.8) are immediate.

Corollary 3.1: There exists a constant $C>0$, such that for any two solutions of (1.4),(1.5), u and v, there holds:

$\|u-v\|L^\infty(\Omega) < C$.

This follows from (1.8).

Proof of uniqueness: Let u,v solve (1.4),(1.5) and w_ϵ be the Sobolev average of $u-v$. By (3.5),

$$\|w_\epsilon\|L^\infty(\Omega) < c \tag{3.6}$$

and since $f(u),f(v)\in L^1(\Omega)$, one gets from (1.4):

$$\|\Delta w_\epsilon\|L^1(\Omega) < c . \tag{3.7}$$

Multiplying (1.4) for u and for v by w_ϵ and taking the difference one has:

$$\frac{1}{2\pi}\int_\Omega (u-v)\Delta w_\epsilon = \int_\Omega (f(u)-f(v))w_\epsilon \tag{3.8}$$

The right-hand side of (3.8) has a limit for $\epsilon\to 0$ since $f(u)-f(v)\in L^1$, which equals

$$A = \int_\Omega (f(u)-f(v))(u-v) . \tag{3.9}$$

The proof of uniqueness for u_a in [2] shows that $A>0$ unless $u=v$. Let us estimate the upper limit of the left-hand side when $\epsilon\to 0$. One has:

$$\int_\Omega (u-v)\Delta w_\varepsilon + \int_\Omega |\nabla w_\varepsilon|^2 = \int_\Omega (u-v-w_\varepsilon)\Delta w_\varepsilon$$

$$= \int_\Omega (u-v-w_\varepsilon)(f(u)-f(v)) + \int_\Omega (u-v-w_\varepsilon)(\Delta w_\varepsilon - f(u) + f(v)$$

$$\leq \int_\Omega (u-v)(f(u)-f(v)) - \int_\Omega w_\varepsilon (f(u)-f(v)) \tag{3.10}$$

$$+ c \int_\Omega |(f(u)-f(v))_\varepsilon - (f(u)-f(v))| \ .$$

The last term in the final expression in(3.10) clearly tends to zero. Obtain also that

$$\int w_\varepsilon (f(u)-f(v)) = \int w(f(u)-f(v))_\varepsilon \underset{\varepsilon\to 0}{\to} \int w(f(u)-f(v)) \tag{3.11}$$

when the Sobelev averages are computed with a symmetric kernel. Therefore

$$\limsup_{\varepsilon\to 0} \int (u-v)\Delta w_\varepsilon = \limsup_{\varepsilon\to 0} - \int_\Omega |\nabla w_\varepsilon|^2 \leq 0$$

and therefore (3.8) implies with necessity that $u = v$.

By uniqueness the boundness of u_a in $W^1_{2-\varepsilon}$, $\varepsilon > 0$ provides now (1.7). By standard techniques of iterated interior estimates one can proof smoothness of u in every Ω_a .

This completes the proofs of Theorems 1.1 and 1.2.

REFERENCES

[1] H. Fixman, The Poisson-Boltzmann equation and its application to polyelectrolytes, J. Chem. Phys. 70 (1979), 4995-5005.

[2] A Friedman, K. Tintarev, Boundary asymptotics for solutions of the Poisson-Boltzmann equation, J. Diff. Eq. (to appear).

[3] I. Rubinstein, Counterion condensation as an exact limiting property of solutions of Poisson-Boltzmann equation, SIAM J. Appl. Math. (to appear).

EXAMPLES OF EXPONENTIAL DECAY OF EIGENFUNCTIONS OF MAGNETIC SCHRÖDINGER OPERATORS

Jun Uchiyama
Department of Mathematics
Kyoto Institute of Technology
Matsugasaki Sakyoku Kyoto 606 Japan

Let $u(x)$ satisfy

$$(*)\begin{cases} -\sum_{j=1}^{n}\left(\partial_j+\sqrt{-1}\,b_j(x)\right)^2u(x)+\left(q_1(x)+q_2(x)\right)u(x)=0 \quad \text{in} \quad \Omega, \\ u\in H^2_{loc}(\Omega), \\ \operatorname{supp}[u] \text{ is not a compact set in } \overline{\Omega}, \end{cases}$$

where $\partial_j=\partial/\partial x_j$.

Assumptions

(1) there exists some constant $R_0>1$ such that $\Omega\supset\{\,x\in R^n \mid |x|>R_0\,\}$,

(2) $q_1(x)$ is a real-valued function and $q_2(x)$ is a complex-valued function,

(3) each $b_j(x)$ is a real-valued function,

(4) for any $w\in H^1_{loc}(\Omega)$ and $1\le i,\ j\le n$ we have

$$q_1|w|^2,\quad (\partial_r q_1)|w|^2,\quad q_2|w|^2,\quad |b_j|^2|w|^2,\quad |\partial_i b_j|^2|w|^2\ \in L^1_{loc}(\Omega),$$

where $\partial_r=\partial/\partial r$,

(5) there exist some constants β, γ, δ satisfying

$$\beta>0,\quad 2-2\beta<\gamma<2,\quad 0<\delta<\min\{2-\gamma,\ \gamma-2+2\beta\},$$

$$M^{(\delta)}\equiv\{\,\mu>0 \mid \limsup_{r\to\infty}[\,r^{2-2\beta}(r\partial_r q_1+\gamma q_1)+$$

$$+(4\beta\mu)^{-1}r^{4-3\beta}|q_2|^2+(2-\gamma-\delta)^{-1}r^{2-2\beta}|B(x)x|^2]$$

$$<(\beta\mu)^2(\gamma-2+2\beta)\ \}\neq\emptyset,$$

where $B(x)=\left(\partial_i b_j(x)-\partial_j b_i(x)\right)$ is an $n\times n$ matrix,

(6) there exist some constants $0 < a_1 < 1$, $a_2 > 0$, $\nu^* \geq 0$, $C > 0$ such that for any $w \in H^1(\Omega)$ having a compact support we have

$$\int_\Omega (q_1 + \mathrm{Re}[q_2])_- |w|^2 dx \leq a_1 \int_\Omega |\nabla w|^2 dx + C\int_\Omega \exp(\nu^* r^\beta)|w|^2 dx,$$

$$\int_\Omega (q_1)_- |w|^2 dx \leq a_2 \int_\Omega |\nabla w|^2 dx + C\int_\Omega \exp(\nu^* r^\beta)|w|^2 dx,$$

where $(f)_-(x) = \max\{0, -f(x)\} \geq 0$ for a real-valued function $f(x)$, and $(\mathrm{Re}[q_2])(x)$ means the real part of $q_2(x)$.

Then by [5, Theorem 1.2] we have

Theorem 1. If we assume (1) ∿ (5), then for any u satisfying (*) and for any $\mu > \mu^* \equiv \liminf_{\eta \downarrow 0} M^{(\eta)}$ we have

$$\lim_{R\to\infty} \exp(2\mu R^\beta) \int_{|x|=R} [|\partial_r u + \sqrt{-1}(\sum_{j=1}^n b_j(x)x_j)r^{-1}u|^2 + (1+(q_1)_-)|u|^2]dS = \infty.$$

If we assume (1) ∿ (6), then for any u satisfying (*) and for any $\mu > \mu^* + 2^{-1}\nu^*$ we have

$$\lim_{R\to\infty} \exp(2\mu R^\beta) \int_{R<|x|<R+1} |u|^2 dx = \infty.$$

Remark 2. In [5] we assumed that each $b_j(x) \in C^1(\Omega)$ instead of the conditions given in (4). But by (4) we have $(\partial_j + \sqrt{-1}\, b_j(x))v(x) \in H^1_{loc}(\Omega)$ for $v(x) \in H^2_{loc}(\Omega)$. Then Lemma 2.1 in [4] and Lemma 2.3 in [5] are also valid, since for w satisfying $w, \partial_j w \in L^1_{loc}(\Omega)$ we have by integration by parts

$$\int_{s<|x|<t} \partial_j w \, dx = \int_{|x|=t} x_j r^{-1} w \, dS - \int_{|x|=s} x_j r^{-1} w \, dS.$$

See also Appendices 1 and 3 in [1]. We use this weaker condition in Examples 10 and 11.

In Examples 3, 5, 7, 10 and 11 we apply Theorem 1 to $u(x)$ satisfying (*), and we have for any $\mu > \mu^*$

$$(\#) \quad \lim_{R\to\infty} \exp(2\mu R^\beta) \int_{R<|x|<R+1} |u|^2 dx = \infty,$$

where β and μ^* are given in each example respectively.

Example 3. Let $c > 0$ and $\lambda \in R$ be constants and let

$$q_1(x) + q_2(x) = cr^2 + o(r) - \lambda \quad \text{as} \quad r \to \infty,$$

$$b_j(x) \equiv 0 \quad (1 \le j \le n),$$

then we have $\beta = 2$ and $\mu^* = 2^{-1}\sqrt{c}$ in (#).

<u>Remark 4.</u> Let $c > 0$ be a constant, $k \ge 0$ be an integer. And let

$$q_1(x) = cr^2 - (2k+n)\sqrt{c},$$

$$q_2(x) = b_j(x) \equiv 0 \quad (1 \le j \le n).$$

Then following $u_k(x)$ satisfies (*).

$$u_k(x) = H_{\ell_1}(c^{1/4}x_1)\cdots H_{\ell_n}(c^{1/4}x_n)\exp(-2^{-1}\sqrt{c}\ r^2),$$

where $x = (x_1,\ldots,x_n)$, $\ell_j \ge 0$ is an integer satisfying $\ell_1 + \ldots + \ell_n = k$, and $H_\ell(t)$ is the Hermite polynomial of degre ℓ. Then for any $\varepsilon > 0$ there exists some constant $C > 0$ such that for any $r > R_0$ we have

$$|u_k(x)| \le C \exp\left(-(1-\varepsilon)2^{-1}\sqrt{c}\ r^2\right).$$

<u>Example 5.</u> Let $c \neq 0$, $\delta > 0$ and $\lambda < 0$ be constants, and let

$$q_1(x) + q_2(x) = cr^{-\delta} + o(r^{-1/2}) - \lambda \quad \text{as} \quad r \to \infty,$$

$$b_j(x) \equiv 0 \quad (1 \le j \le n).$$

Then we have $\beta = 1$ and $\mu^* = \sqrt{-\lambda}$ in (#).

<u>Remark 6.</u> Let $n = 3$, $c > 0$ be a constant, k be a positive integer and $\lambda = -(2k)^{-2}c^2$. And let

$$q_1(x) = -cr^{-1} - \lambda, \quad \text{and} \quad q_2(x) = b_j(x) \equiv 0 \quad (1 \le j \le 3).$$

Then the following $u_{k\ell m}(x)$ satisfies (*).

$$u_{k\ell m}(x) = \exp(-\sqrt{-\lambda}\ r)r^\ell L_{k+\ell}^{2\ell+1}(2\sqrt{-\lambda}\ r)P_\ell^m(\cos\theta)\exp(\sqrt{-1}\ m\phi),$$

where $0 \le \ell \le k-1$ and $-\ell \le m \le \ell$ are integers, (r, θ, ϕ) is a polar co-ordinate of $x \in R^3$, $L_k^\ell(t) = (d^\ell/dt^\ell)L_k(t)$ ($L_k(t)$ is the Laguerre polynomial of degree k), and $P_\ell^m(t) = (1-t^2)^{m/2}(d^m/dt^m)P_\ell(t)$ ($P_\ell(t)$ is the Legendre polynomial of degree ℓ). Then for any $\varepsilon > 0$ there exists some constant $C > 0$ such that for any $r > R_0$

$$|u_{k\ell m}(x)| \le C \exp(-(1-\varepsilon)\sqrt{-\lambda}\, r).$$

Example 7. Let $n = 2$, $b_0 \neq 0$, c, λ, α, σ, ρ be real constants. And let

$$q_1(x) + q_2(x) = cr^\sigma + o(r^\rho) - \lambda, \quad \text{as } r \to \infty,$$

$$b_1(x) = -2^{-1}b_0x_2f(r), \quad b_2(x) = 2^{-1}b_0x_1f(r),$$

where $f(r) = r^\alpha$ or $f(r) = (1+r)^\alpha$. Noting

$$|B(x)x|^2 = 4^{-1}b_0^2(rf'+2f)^2r^2,$$

we have the followings :

(7) If $\alpha > -1$, $\sigma < 2(\alpha+1)$ and $\rho = 1+(3\alpha)/2$, then we have $\beta = 2+\alpha$ and $\mu^* = 2^{-1}(2+\alpha)^{-1}|b_0|$ in (#). The choice $\gamma = -\alpha$ in (5) is the best one for the determination of μ^*.

(8) If $\alpha > -1$, $\sigma = 2(\alpha+1)$, $b_0^2+4c < 0$ and $\rho = 4^{-1}(\alpha+2)\left(1-\sqrt{1+(4c)^{-1}b_0^2}\right)+\alpha$, then we have $\beta > 2^{-1}(\alpha+2)\{1-(1+(4c)^{-1}b_0^2)^{1/2}\}$ is arbitrary and $\mu^* = 0$ in (#). Here we must choose γ in (5) to satisfy $\max\{2-2\beta, -(\alpha+2)\sqrt{1+(4c)^{-1}b_0^2}-\alpha\} < \gamma < (\alpha+2)\sqrt{1+(4c)^{-1}b_0^2}-\alpha$ (< 2), since we have $c(2\alpha+2+\gamma) + (2-\gamma)^{-1}4^{-1}b_0^2(\alpha+2)^2 < 0$ for this γ.

(9) If $\alpha > -1$, $\sigma = 2(\alpha+1)$, $b_0^2+4c \ge 0$ and $\rho = 1+(3\alpha)/2$, then we have $\beta = 2+\alpha$ and $\mu^* = 2^{-1}(2+\alpha)^{-1}(b_0^2+4c)^{1/2}$ in (#). Here we choose $\gamma = -\alpha$ in (5).

(10) If $\alpha = -1$, $\lambda \le 4^{-1}b_0^2$, $\sigma < 0$ and $\rho = -1/2$, then we have $\beta = 1$ and $\mu^* = 2^{-1}(b_0^2-4\lambda)^{1/2}$ in (#). The choice $\gamma = 1$ is the best one in (5).

(11) If $\alpha = -1$, $\lambda > 4^{-1}b_0^2$, $\sigma < 0$ and $\rho = -4^{-1}\{3+(1-(4\lambda)^{-1}b_0^2)^{1/2}\}$, then we have $\beta > 2^{-1}\{1-(1-(4\lambda)^{-1}b_0^2)^{1/2}\}$ is arbitrary and $\mu^* = 0$ in (#). We must choose γ in (5) to satisfy $\max\{2-2\beta, 1-(1-(4\lambda)^{-1}\times b_0^2)^{1/2}\} < \gamma < 1+(1-(4\lambda)^{-1}b_0^2)^{1/2}$ (< 2), since we have $(2-\gamma)^{-1}4^{-1}b_0^2-\gamma\lambda < 0$ for this γ.

(12) If $\alpha < -1$, $\lambda > 0$, $\sigma < 0$ and $\rho = -1$, then we have $\beta > 0$ is arbitrary and $\mu^* = 0$ in (#). We must choose γ in (5) to satisfy $\max\{0, 2-2\beta\} < \gamma < 2$.

Remark 8. Let $b_0 \neq 0$ and α be real constants, and let

$$H = -(\partial_1 - \sqrt{-1}\, 2^{-1} b_0 (1+r)^{\alpha} x_2)^2 - (\partial_2 + \sqrt{-1}\, 2^{-1} b_0 (1+r)^{\alpha} x_1)^2,$$

$$D(H) = C_0^{\infty}(R^2).$$

[3] shows the followings : If $\alpha < 0$, H is essentially selfadjoint in $L^2(R^2)$ and $\sigma_{ess}(H) = [0, \infty)$. If $\alpha < -1$, H has no eigenvalues in $(0,\infty)$. If $\alpha = -1$, H has no eigenvalues in $(4^{-1}b_0^2, \infty)$ and $\sigma_p(H)$ is dence in $[0, 4^{-1}b_0^2]$. If $-1 < \alpha < 0$, H has dence point spectrum in $[0,\infty)$.

Remark 9. The following equation in $\Omega \subset R^2$

$$[-(\partial_1 - \sqrt{-1}\, 2^{-1} b_0 x_2)^2 - (\partial_2 + \sqrt{-1}\, 2^{-1} b_0 x_1)^2 - (2k+1)|b_0|]u(x_1,x_2) = 0$$

has a solution

$$u_{k\ell}(x_1,x_2) = (\partial_1 \pm \sqrt{-1}\partial_2 \mp 2^{-1} b_0 (x_1 \pm \sqrt{-1} x_2))^k \cdot$$
$$\cdot\{(x_1 \mp \sqrt{-1} x_2)^{\ell} \exp(\mp 4^{-1} b_0 (x_1^2 + x_2^2))\},$$

where $k \geq 0$ and $\ell \geq 0$ are integers, and $\pm b_0 > 0$ is a constant. Then for any $\varepsilon > 0$ there exists some constant $C > 0$ such that for any $r > R_0$ we have

$$|u_{k\ell}(x_1,x_2)| \leq C \exp(-(1-\varepsilon)4^{-1}|b_0|r^2).$$

Compare with (7).

Example 10. Let $n = 3$, and $b_0 \neq 0$, $\alpha' > -1$, α, c, σ, ρ and λ be real constants. And let

$$q_1(x) + q_2(x) = cr^{\sigma} + o(r^{\rho}) - \lambda \quad \text{as} \quad r \to \infty,$$

$$b_1(x) = -2^{-1} b_0 x_2 f(x), \quad b_2(x) = 2^{-1} b_0 x_1 f(x), \quad b_3(x) \equiv 0,$$

where $f(x) = r^{\alpha}(x_1^2+x_2^2)^{\alpha'/2}$. The condition $\alpha' > -1$ is imposed to satisfy (4). Noting $|B(x)x|^2 = 4^{-1}b_0^2(x_1^2+x_2^2)(r\partial_r f + 2f)^2 = 4^{-1}b_0^2 \times (\alpha+\alpha'+2)^2 r^{2\alpha}(x_1^2+x_2^2)^{\alpha'+1}$, we have the same results as (7) $\sim$ (12), where we replace α with $\alpha+\alpha'$.

<u>Example 11.</u> Let $N \geq 1$ be integer, c_j, c_{jk}, c, $b_0 \neq 0$, α, α', σ, ρ and λ are real constants. And let

$$q_1(x) + q_2(x) = -\sum_{j=1}^{N} c_j r_j^{-1} + \sum_{1\leq j<k\leq N} c_{jk} r_{jk}^{-1} + cr^{\sigma} + o(r^{\rho}) - \lambda,$$

$$b_{3j-2}(x) = -2^{-1} b_0 x_{3j-1} f_j(x), \quad b_{3j-1}(x) = 2^{-1} b_0 x_{3j-2} f_j(x),$$

$$b_{3j}(x) \equiv 0 \qquad (1 \leq j \leq N),$$

where $f_j(x) = r^{\alpha} r_j^{\alpha'}$ ($\alpha' \geq -1$) or $f_j(x) = r^{\alpha}(x_{3j-2}^2 + x_{3j-1}^2)^{\alpha'/2}$ ($\alpha' > -1$), $x = (x^{(1)}, \ldots, x^{(N)}) \in R^{3N}$, $x^{(j)} = (x_{3j-2}, x_{3j-1}, x_{3j}) \in R^3$, $r_j = |x^{(j)}|$ and $r_{jk} = |x^{(j)} - x^{(k)}|$. (See [1, Example 5.3.3, p.134] and [2, Example 1, p.162]). Since $\max \{ s_1^{\theta} + \cdots + s_N^{\theta} \mid s_1^2 + \cdots + s_N^2 = 1, s_j \geq 0, 1 \leq j \leq N \} = \max \{ N^{1-\theta/2}, 1 \}$ for $\theta \geq 0$, we have

$$|B(x)x|^2 = 4^{-1} b_0^2 \sum_{j=1}^{N} (x_{3j-2}^2 + x_{3j-1}^2)(r\partial_r f_j + 2f_j)^2$$

$$= 4^{-1} b_0^2 (\alpha+\alpha'+2)^2 \sum_{j=1}^{N} (x_{3j-2}^2 + x_{3j-1}^2) f_j^2,$$

$$\limsup_{r\to\infty} r^{-2(\alpha'+1)} \sum_{j=1}^{N} (x_{3j-2}^2 + x_{3j-1}^2) r_j^{2\alpha'} = \max \{ N^{-\alpha'}, 1 \},$$

$$\limsup_{r\to\infty} r^{-2(\alpha'+1)} \sum_{j=1}^{N} (x_{3j-2}^2 + x_{3j-1}^2)^{\alpha'+1} = \max \{ N^{-\alpha'}, 1 \}.$$

Noting that we must choose $\gamma = 1$ in (5), we have the followings :

(13) If $\alpha+\alpha' > -1$, $\sigma < 2(\alpha+\alpha'+1)$ and $\rho = 1+3(\alpha+\alpha')/2$, then we have $\beta = 2+\alpha+\alpha'$ and $\mu^* = 2^{-1}|b_0|(3+2\alpha+2\alpha')^{-1}$ in (#).

(14) If $\alpha+\alpha' = -1$, $\sigma < 0$, $\lambda > 4^{-1} b_0^2 \max\{N^{-\alpha'}, 1\}$ and $\rho = -3/4$, then we have $\beta > 1/2$ is arbitrary and $\mu^* = 0$ in (#).

(15) If $\alpha+\alpha' = -1$, $\sigma < 0$, $\lambda < 4^{-1} b_0^2 \max\{N^{-\alpha'}, 1\}$ and $\rho = -1/2$, then we have $\beta = 1$ and $\mu^* = (4^{-1} b_0^2 \max\{N^{-\alpha'}, 1\} - \lambda)^{1/2}$ in (#).

(16) If $\alpha+\alpha' < -1$, $\sigma < 0$, $\lambda > 0$ and $\rho = -3/4$, then we have $\beta > 1/2$ is arbitrary and $\mu^* = 0$ in (#).

(17) If $\alpha+\alpha' < -1$, $\sigma < 0$, $\lambda < 0$ and $\rho = -1/2$, then we have $\beta = 1$ and $\mu^* = \sqrt{-\lambda}$ in (#).

<u>References.</u>

[1] Eastham, M.S.P. and H. Kalf, *Schrödinger-type operators with continuous spectra*, Research notes in mathematics 65, Pitman Advanced Publishing Program, 1982.

[2] Kalf, H., Non-existence of eigenvalues of Schrödinger operators, *Proc. Roy. Edinburgh*, 79 A (1977), 152-172.

[3] Miller, K. and B. Simon, Quantum magnetic Hamiltonians with remarkable spectral properties, *Phys. Rev. Letters*, 44 (1980), 1706-1707.

[4] Uchiyama, J., Lower bounds of decay order of eigenfunctions of second-order elliptic operators, *Publ. RIMS, Kyoto Univ.*, 21 (1985), 1281-1297.

[5] Uchiyama, J., Decay order of eigenfunctions of second-order elliptic operators in an unbounded domain, and its applications, to appear on *Publ. RIMS, Kyoto Univ.*

SPATIALLY LOCALIZED FREE VIBRATIONS OF CERTAIN SEMILINEAR WAVE EQUATIONS ON $\mathbb{R}^2$: RECENT RESULTS AND OPEN PROBLEMS

Pierre-A. Vuillermot
Mathematics Department, The University of Texas
Arlington, Texas 76019 U.S.A.

I. INTRODUCTION

In this short expository article, we analyze and compare three recent results concerning the existence of spatially localized free vibrations of semilinear wave equations of the form

$$u_{tt} = u_{xx} - g(u) \tag{I.1}$$

on $\mathbb{R}^2$. Nonlinear Klein-Gordon equations such as (I.1) occur in various physical contexts (see for instance [1], [2] and the numerous references therein), and it is of considerable interest to determine the class of nonlinearities $u \to g(u)$ in (I.1) for which equations (I.1) possesses nontrivial T-periodic solutions in time (with preassigned period $T > 0$) which are simultaneously strongly localized in space (the so-called breather-solutions). A celebrated example of such a nonlinearity is $g(u) \equiv \sin(u)$, since it has been known for a long time that the Sine-Gordon equation

$$u_{tt} = u_{xx} - 2\sin(u) \tag{I.2}$$

on $\mathbb{R}^2$ possesses the particular breather-solution

$$u_{SGB}(x;t) = 4\,\mathrm{arctang}\left\{\frac{\sin(t)}{\cosh(x)}\right\} \tag{I.3}$$

with 2π - periodicity in time and exponential fall-off in the spatial direction [2]. Because of the above example, it has become customary to associate the existence of breather-solutions with those nonlinearities for which equation (I.1) is completely integrable. It has even been recently speculated by Brezis in [3] that periodic solutions to (I.1) with an asymptotic behaviour similar to that of (I.3) as $|x| \to \infty$ should be an intrinsic feature of the Sine-Gordon equation. While the three results that we analyze in the following sections do not yet (dis)prove the above conjecture, they shall already reveal that the class of nonlinearities in (I.1) for which breather-solutions may exist is indeed surprisingly small. In section II, we state, compare and discuss two theorems which pertain to the above questions; the first one is due to J.M. Coron [4] and the second one to A. Weinstein [5]. In section III, we state a recent nonexistence result of the author concerning breather-solutions with polynomial decay in the spatial direction [6]. Section IV is devoted to the discussion of several examples and open problems. For an announcement of the results, see [21].

II. THE SIGNIFICANCE OF THE PERIOD CONDITIONS OF CORON AND WEINSTEIN

The first result essentially asserts that the requirement of periodicity in time is incompatible with the requirement of sufficiently fast decay of the wave profile in the spatial direction, in that every classical solution to (I.1) which satisfies both requirements is necessarily trivial in a certain sense - unless the preassigned period of vibration is chosen sufficiently large. Indeed, we have the following theorem, which contrasts sharply with other known results concerning periodic solutions to equation (I.1) in which x runs over a finite interval (see for instance [3] and its numerous references, in particular our references [7] and [8]).

Theorem II.1 (Coron, [4]). Consider equation (I.1) on $\mathbb{R}^2$ in which $g \in C^{(2)}(\mathbb{R},\mathbb{R})$ and $g(0) = 0$. Let $T > 0$ and let $u \in C^{(2)}(\mathbb{R}^2;\mathbb{R})$ be a classical time-periodic solution to equation (I.1) of period T. Assume moreover that the following conditions hold:

$$(C_1) \quad g'(0) < (\tfrac{2\pi}{T})^2$$

$$(C_2) \quad \int_0^\infty dx \int_0^T dt\,|u(x;t)| < \infty$$

$$(C_3) \quad \lim_{x\to\infty} \int_0^T dt(u_t^2(x;t) + u_x^2(x,t)) = 0$$

$$(C_4) \quad \lim_{x\to\infty} \max_{t\in[0,T]} |u(x;t)| = 0$$

Then u is independent of t.

While condition (C_1) already emerges from the physical arguments of [9], a natural question that arises is to know what happens if condition (C_1) is replaced by the reversed period condition $g'(0) > (\frac{2\pi}{T})^2$. The case of the Sine-Gordon equation suggests that nontrivial breather-solutions should then exist, since u_{SGB} obviously satisfies (C_2), (C_3) and (C_4) and since $g'(0) = 2$, $T = 2\pi$. In his analysis of the above question, Weinstein recently proved that exponentially localized breathers indeed do exist, under the crucial additional assumption that equation (I.1) be considered on the half-plane $\mathbb{R}_0^+ \times \mathbb{R}$ instead of $\mathbb{R}^2$. Indeed he proved the following

Theorem II.2 (Weinstein, [5]). Consider equation (I.1) on $\mathbb{R}_0^+ \times \mathbb{R}$ in which $g \in C^{(2)}(\mathbb{R};\mathbb{R})$ and $g(0) = 0$; let $T > 0$ and assume that the following period condition holds:

$$(W_1) \quad g'(0) > (\tfrac{2\pi}{T})^2.$$

Then there exist nontrivial classical time-periodic solutions to (I.1) with period T which decay exponentially fast along the spatial direction, in that condition

$$(W_2) \quad \int_0^T dt(u^2(x;t) + u_t^2(x;t) + u_x^2(x;t)) < Ae^{-Bx}$$

holds for some $A, B > 0$.

It is instructive to observe that condition (W_2) implies conditions (C_2), (C_3) and (C_4) and that the results of Coron and Weinstein are thus complementary. The remaining question is to investigate the existence of breather-solutions for equation (I.1) on the whole of $\mathbb{R}^2$, when condition (W_1) holds. We propose a solution in the next section.

III. NONEXISTENCE OF BREATHER-SOLUTIONS ON $\mathbb{R}^2$ WITH POLYNOMIALLY DECAYING PROFILES

In this section, we prove the nonexistence of free vibrations of arbitrary period with polynomially decreasing profiles for a large class of semilinear wave equations such as (I.1) on $\mathbb{R}^2$. Some of the examples of the next section will show that our class of admissible wave equations contains both completely integrable equations (like the Sinh-Gordon equation) and nonintegrable ones (like the nonlinear scalar Higgs equation of classical field theory). They will also reveal that the breather-phenomenon on $\mathbb{R}^2$ is entirely different from the corresponding phenomenon on $\mathbb{R}_0^+ \times \mathbb{R}$. In view of these applications, our class of admissible wave equations must allow for both polynomial and exponential nonlinearities in (I.1); according to the physical picture of a breather being a bound state made out of two solitons which dissociates as the solitons travel far apart from each other [2], our class of admissible breathers must contain solutions to (I.1) which converge to constant solutions $u = u_0 \in \mathbb{R}$ as $|x| \to \infty$. These remarks motivate the following definitions.

Definition III.1. Let $u_0 \in \mathbb{R}$; the function $g : \mathbb{R} \to \mathbb{R}$ is said to be an admissible nonlinearity for equation (I.1) if the following hypotheses hold:

(H_1) $g \in C(\mathbb{R},\mathbb{R})$

(H_2) $\limsup_{|u|\to\infty} \frac{e^{-(u-u_0)^2} |g(u)|}{u - u_0} < \infty$

(H_3) $\limsup_{u\to u_0} \frac{|g(u)|}{|u - u_0|} < \infty$

We note that hypotheses (H_1) and (H_3) imply $g(u_0) = 0$, hence that $u = u_0$ solves (I.1) as requested. For u smooth on $\mathbb{R}^2$, we now define $(u)_j = u, u_x, u_t$ for $j = 0,1,2$ respectively, and write $u_{ij} = ((u)_i)_j$. We still need the following

Definition III.2. Let $T > 0$, $u_0 \in \mathbb{R}$. We denote by $B(T;u_0)$ the set of all breather-solutions with polynomial decay to equation (I.1), that is the set of all $u \in C^{(2)}(\mathbb{R}^2,\mathbb{R})$ such that the following conditions hold:

(V_1) u is a time-periodic solution to (1.1) of period T, that is $u(x;t + T) = u(x,t)$ for each $(x,t) \in \mathbb{R}^2$.

(V_2) $u(x;0) = u_0$ for each $x \in \mathbb{R}$

(V_3) $\sup_{x\in\mathbb{R}} \left| x^m \max_{t\in[0,T]} |(u(x;t) - u_0)_{ij}| \right| < \infty$

for each $m\in\{0;1;2\}$, each $i \in(0;1)$ and each $j \in\{0;1;2\}$ with $j \geq i$. It follows immediately from the above definitions that $u_0 \in B(T;u_0)$; also, for the Sine-Gordon case, $B_{SG}(T;0) = \{0\}$ for $T \in (0,\sqrt{2}\pi)$, but $|B_{SG}(2\pi;0)| \geqslant 2$.

This shows that the notion of admissible nonlinearity of definition (III.1) is not sufficient to obtain a nonexistence result valid for each $T > 0$. To achieve this, we must reinforce our assumptions on g; it turns out that a sufficient condition is a certain convexity property of the corresponding potential $G_{u_0}(u) = \int_{u_0}^{u} d\xi g(\xi)$ around the asymptotic solution u_0. To make this idea precise, we need the following.

Definition III.2. G_{u_0} is said to be convex in $(u-u_0)^2$ if there exists a function $H \in C^{(1)}(\mathbb{R}^+;\mathbb{R})$ such that $G_{u_0}(u) = H(y)$ with $y = (u-u_0)^2$ and H convex in y.

Remark. Example (IV.4) of Section IV will show that G_{u_0} need not be convex in $u-u_0$ in order to be convex in $(u-u_0)^2$. For other applications of this notion, see [12]-[16].

Our main result is the following

Theorem III.1. Let $T > 0$ and $u_0 \in \mathbb{R}$; let g be an admissible nonlinearity for equation (I.1) on $\mathbb{R}^2$. If G_{u_0} is convex in $(u-u_0)^2$, then $B(T;u_0) = \{u_0\}$.

Proof. Let $\Omega_T = \mathbb{R} \times (0,T)$ and consider the Sobolev space $\overset{\circ}{H}{}^1(\Omega_T;\mathbb{R})$. Define $g_{u_0}:\mathbb{R} \to \mathbb{R}$ by $g_{u_0}(u) = g(u+u_0)$ and write $\hat{G}_{u_0}(u) = \int_0^u d\xi g_{u_0}(\xi)$; it is easy to see that, because of hypotheses (H_1), (H_2) and (H_3), the indefinite functional defined by

$$S_{\Omega_T,u_0}[u] = \int_{\Omega_T} dxdt\{\tfrac{1}{2}\, u_t^2(x;t) - \tfrac{1}{2}\, u_x^2(x,t) - \hat{G}_{u_0}(u(x,t))\} \qquad \text{(III.1)}$$

is a real-valued, $C^{(1)}$-Fréchet differentiable functional on $\overset{\circ}{H}{}^1(\Omega_T;\mathbb{R})$, with Fréchet derivative

$$S'_{\Omega_T,u_0}[u](v) = \int_{\Omega_T} dxdt\{u_t(x;t)v_t(x;t) - u_x(x;t)v_x(x;t) - g_{u_0}(u(x;t))v(x;t)\} \qquad \text{(III.2)}$$

Now let $u \in B(T;u_0)$; conditions (V_2) and (V_3) readily imply that $u - u_0 \in \overset{\circ}{H}{}^1(\Omega_T;\mathbb{R})$, $xu_x \in \overset{\circ}{H}{}^1(\Omega_T;\mathbb{R})$. Furthermore, $\hat{u} \equiv u-u_0$ solves the wave equation $\hat{u}_{tt} = \hat{u}_{xx} - g_{u_0}(\hat{u})$. It then follows from the continuity of S'_{Ω_T,u_0} and standard considerations in the calculus of variations that $u - u_0$ is a critical point of S_{Ω_T,u_0} on $\overset{\circ}{H}{}^1(\Omega_T;\mathbb{R})$, namely $S'_{\Omega_T,u_0}[u](v) = 0$ for each $v \in \overset{\circ}{H}{}^1(\Omega_T;\mathbb{R})$. Choosing first $v = u-u_0$ leads to the identity

$$\int_{\Omega_T} dxdt(u_t^2(x,t) - u_x^2(x,t)) = \int_{\Omega_T} dxdt\; g(u(x,t))(u(x,t) - u_0). \qquad \text{(III.3)}$$

Choosing next $v = xu_x$ gives

$$\int_{\Omega_T} dxdt(u_t^2(x,t) + u_x^2(x,t)) = 2 \int_{\Omega_T} dxdt\; G_{u_0}(u(x;t)) \qquad \text{(III.4)}$$

upon integrating by parts and using (V_3). Combining (III.3) and (III.4) then gives

$$\int_{\Omega_T} dxdt\, u_x^2(x;t) = \int_{\Omega_T} dxdt (G_{u_0}(u(x;t)) - \frac{1}{2} g(u(x;t))(u(x;t) - u_0)) \qquad \text{(III.5)}$$

where G_{u_0} is now the potential of Theorem III.1. By the convexity of G_{u_0} in $(u-u_0)^2$, we now have

$$G_{u_0}(u) - \frac{1}{2} g(u)(u-u_0) = H(y) - y\,H'(y) \leq 0 \quad \text{for each } u \in \mathbb{R}, \text{ hence}$$

$u_x \equiv 0$ on Ω_T by (III.5). This implies $u \equiv u_0$ on $\mathbb{R}^2$ by (V_1) and (V_3). ■

IV. SOME TYPICAL EXAMPLES AND OPEN PROBLEMS

In the following examples, we write $G_0 = G$ and $B(T;0) = B(T)$ if $u_0 = 0$. We begin with the following

Examples IV.1 : The Polynomial Case. Let $T > 0$ and consider the wave equation

$$u_{tt} = u_{xx} - \sum_{j=1}^{N} c_j\, u^{2j-1} \qquad \text{(IV.1)}$$

on $\mathbb{R}^2$; here, $g(u) = \sum_{j=1}^{N} c_j\, u^{2j-1}$, $G(u) = \sum_{j=1}^{N} \hat{c}_j\, u^{2j}$.

Case (A): $c_1 \leq 0$, c_j arbitrary for $j \in \{2,\ldots N\}$. Then, $g'(0) < 0$ and since condition (V_3) implies (C_2), (C_3) and (C_4), we get $B(T) = \{0\}$ from Theorem (II.1) and (V_2).

Case (B): $c_j \geq 0$ for each $j \in \{1,2,\ldots N\}$. Then, $B(T) = \{0\}$ since G is convex in u^2.

Problem IV.1: The EKNS-Conjecture ([10], [11]). It has been conjectured in [10] that equation (I.1) has no breather-solutions whenever g is a polynomial with $g(0) = 0$; while example (IV.1) proves the conjecture in a significant number of cases, the structure of $B(T)$ in the remaining case $c_1 > 0$, c_j arbitrary for $j \in \{2,\ldots N\}$ is still unknown. The conjecture is puzzling in that if $B(T) = 0$ for every such polynomial g, one could for instance destroy the breather-solution u_{SGB} to (I.2) in replacing $g(u) = 2\sin(u)$ by one of its Taylor polynomials of arbitrary large degree.

The result of case (B) with $c_1 > 0$ contrasts sharply with the following, which is an immediate consequence of Theorem (II.2).

Example IV.2. Consider equation (IV.1) on the half-plane $\mathbb{R}_0^+ \times \mathbb{R}$ with $c_1 > 0$ and c_j arbitrary for $j \in \{2,\ldots N\}$. Let $T > 2\pi c_1^{-1/2}$; then there exist nontrivial solutions to (I.1), T-periodic in time, which decay exponentially fast in the x-direction.

The sharp contrast between Weinstein's results on $\mathbb{R}_0^+ \times \mathbb{R}$ and ours on $\mathbb{R}^2$ can be explained in two different ways. On the one hand, if one associates with (I.1) a dynamical system for which $x \geq 0$ plays the role of time, the problem is that of the intersection of the stable manifold with the unstable one [5]. This intersection is a very unlikely event in the corresponding phase space, which prevents one to carry over Weinstein's results to the whole plane. On the other hand, if one attempts to define a class of breather-solutions to (I.1) on $\mathbb{R}_0^+ \times \mathbb{R}$ in replacing condition (V_3) by

$$(V_3^+) \qquad \sup_{x\in\mathbb{R}_0^+} (x^m \max_{t\in[0,T]} |(u(x;t) - u_0)_{ij}|) < \infty$$

our argument breaks down since membership of $u - u_0$ in $\overset{\circ}{H}{}^1(\mathbb{R}_0^+ \times (0;T);\mathbb{R})$ does not hold. This is because condition (V_3^+) does not necessarily imply $u(0;t) - u_0 = 0$ for $t \in [0,T]$.

Another contrast of this kind is provided by the following

Example IV.3: The Exponential Case. Let $T > 0$ and consider

$$u_{tt} = u_{xx} - \sinh(u) \tag{IV.2}$$

on $\mathbb{R}^2$; it is well known that (IV.2) is completely integrable (see for instance [17]). However, $G(u) = \cosh(u) - 1$, hence $B(T) = \{0\}$ since $H(y) = \cosh(\sqrt{y}) - 1$ is convex in y.

Example IV.4. The Nonlinear Higgs Equation. Let $T > 0$ and consider

$$u_{tt} - u_{xx} - 2u + 3u^2 - u^3 \tag{IV.3}$$

on $\mathbb{R}^2$; we have $g(u) = (u-1)^3 - (u-1)$ and $u_0 = 1$ is an admissible breather-solution. Hence $G_1(u) = \frac{(u-1)^4}{4} - \frac{(u-1)^2}{2}$, which is convex in $(u-1)^2$ (but not in $u-1$); hence $B(T;1) = \{1\}$.

Problem IV.2. Are there breather-solutions to the translated equation $u_{tt} = u_{xx} + u - u^3$ on $\mathbb{R}^2$ which converge to $u_0 = \pm 1$ as $|x| \to \infty$? The above argument certainly does not proclude them since the potential $G_{\pm 1}(u) = \frac{u^4}{4} - \frac{u^2}{2} + 1/4 = 1/4(u-1)^2(u+1)^2$ is neither convex in $(u-1)^2$ nor convex in $(u+1)^2$. In fact, some numerical and asymptotic evidence for the existence of such solutions has recently been exhibited in [18]. This problem is currently of tremendous interest [20].

We hope to shed more light on these problems in the near future and to understand what is so special about the Sine-Gordon nonlinearity [19].

ACKNOWLEDGEMENTS. The author should like to thank several friends and colleagues: Ian Knowles and Yoshimi Saito for their invitation to this conference; David Campbell and Basil Nicolaenko for their invitation to the Center for Nonlinear Studies and to the Theoretical Division of the Los Alamos National Laboratory, where this work was begun. Alan Weinstein, Harvey Segur and Martin Kruskal for discussions, correspondence, and for having pointed out to him references [10], [11] and [17]. The generous supports of the Los Alamos National Laboratory under contract COL-2335 and of the University of Texas under a Summer grant are also gratefully acknowledged.

BIBLIOGRAPHICAL REFERENCES

[1] Dodd, R.K., Eilbeck, T.C., Gibbon, T.D., Morris, H.C., "Solitons and Nonlinear Wave Equations". Academic Press (1982).

[2] Lamb, G.L., "Elements of Soliton Theory", Wiley-Interscience Series in Pure and Applied Mathematics, J. Wiley and Sons, New York (1980).

[3] Brézis, H., "Periodic Solutions of Nonlinear Vibrating Stringa and Duality Principles", Bull. Am. Math. Soc. 8, 409-426 (1983).

[4] Coron, J.M., "Période Minimale pour une Corde Vibrante de Longueur Infinie", C.R. Acad. Sci. Paris, A294, 127-129 (1982).

[5] Weinstein, A., "Periodic Nonlinear Waves on a Half-Line", Commun. Math. Phys. 99, 385-388 (1985).

[6] Vuillermot, P.A., Nonexistence of Spatially Localized Free Vibrations for a Class of Nonlinear Wave Equations", Comment. Math. Helv., to appear (1987).

[7] Rabinowitz, P., "Free Vibrations for a Semilinear Wave Equation", Commun. Pure Appl. Math. 31, 31-68 (1978).

[8] Brézis, H., Coron, J.M., Nirenberg, L., "Free Vibrations for a Nonlinear Wave Equation and a Theorem of P. Rabinowitz", Commun. Pure Appl. Math. 33, 667-689.

[9] Dashen, R., Hasslacher, B., Neveu, A., "Particle Spectrum in Model Field Theories from Semiclassical Functional Integral Techniques", Phys. Rev. D11, 3424-3450 (1975).

[10] Eleonskii, V.M., Kulagin, N.E., Novoshilova, H.S., Silin, V.P., "Asymptotic Expansions and Qualitative Analysis of Finite-Dimensional Models in Nonlinear Field Theory", Teoreticheskaya i Mathematichskaya Fizika, 60, 3, 395-403 (1984).

[11] Eleonskii, V.M., Kulagin, N.E., Novoshilova, N.S., Silin, V.P., "The Asymptotic Expansion Method and a Qualitative Analysis of Finite-Dimensional Models in nonlinear Field Theory", in "Nonlinear and Turbulent Processes in Physics", Gordon and Breach Press, Harwood Acad. Publ., 1333-1336 (1984).

[12] Vuillermot, P.A., "A Class of Orlicz-Sobolev Spaces with Applications to Variational Problems Involving Nonlinear Hill's Equations", Jour. Math. Anal. Appl. 89, 1, 327-349 (1982).

[13] Vuillermot, P.A., "A Class of Elliptic Partial Differential Equations with Exponential Nonlinearities", Math. Ann. 268, 497-518 (1984).

[14] Vuillermot, P.A., "Hölder-Regularity for the Eigenfunctions of strongly nonlinear Eigenvalue Problems on Orlicz-Sobolev Spaces", Houston Journal of Mathematics, to appear (1986).

[15] Vuillermot, P.A., "Elliptic Regularization for a Class of Strongly Nonlinear Degenerate Eigenvalue Problems on Orlicz-Sobolev Spaces. I: The ODE Case", Houston Journal of Mathematics, to appear (1986).

[16] Vuillermot, P.A., "Existence and Regularity Theory for Isoperimetric Variational Problems on Orlicz-Sobolev Spaces: A Review", in "Nonlinear Systems of Partial Differential Equations in Applied Mathematics", AMS Series "Lectures in Applied Mathematics, part 2," 23, 109-122, (1986).

[17] Mikhailov, A.V., "Integrability of a Two-Dimensional Generalization of the Toda Chain", JETP Lett. 30, 7, 414-418 (1979).

[18] Campbell, D.K., Private communications concerning work in preparation by Campbell, D.K., Negele, J., Peyrard, M. and their collaborators, (1986).

[19] Vuillermot, P., "A Critical Point Theory for a Class of Nonlinear Wave Equations with Applications to the Breather Problem", in preparation.

[20] Kruskal, M., Segur, H., "Φ^4 has no Small Breathers", Preprint, (1986).

[21] Vuillermot, P., "Non-Existence de Vibrations Libres Localisées pour Certaines Equations des Ondes Semilineaires sur $\mathbb{R}^2$", C.R. Acad.Sci.Paris, to appear,(1986).

LECTURE NOTES IN MATHEMATICS
Edited by A. Dold and B. Eckmann

Some general remarks on the publication of proceedings of congresses and symposia

Lecture Notes aim to report new developments - quickly, informally and at a high level. The following describes criteria and procedures which apply to proceedings volumes.

1. One (or more) expert participant(s) of the meeting should act as the responsible editor(s) of the proceedings. They select the papers which are suitable (cf. points 2, 3) for inclusion in the proceedings, and have them individually refereed (as for a journal). It should not be assumed that the published proceedings must reflect conference events faithfully and in their entirety. Contributions to the meeting which are not included in the proceedings can be listed by title. The series editors will normally not interfere with the editing of a particular proceedings volume - except in fairly obvious cases, or on technical matters, such as described in points 2, 3. The names of the responsible editors appear on the title page of the volume.

2. The proceedings should be reasonably homogeneous (concerned with a limited area). For instance, the proceedings of a congress on "Analysis" or "Mathematics in Wonderland" would normally not be sufficiently homogeneous.

   One or two longer survey articles on recent developments in the field are often very useful additions to such proceedings - even if they do not correspond to actual lectures at the congress. An extensive introduction on the subject of the congress would be desirable.

3. The contributions should be of a high mathematical standard and of current interest. Research articles should present new material and not duplicate other papers already published or due to be published. They should contain sufficient information and motivation and they should present proofs, or at least outlines of such, in sufficient detail to enable an expert to complete them. Thus resumes and mere announcements of papers appearing elsewhere cannot be included, although more detailed versions of a contribution may well be published in other places later.

   Surveys, if included, should cover a sufficiently broad topic, and should in general not simply review the author's own recent research. In the case of surveys, exceptionally, proofs of results may not be necessary.

   The editors of a volume are strongly advised to inform contributors about these points at an early stage.

.../...

4. Proceedings should appear soon after the meeeting. The publisher should, therefore, receive the complete manuscript within nine months of the date of the meeting at the latest.

5. Plans or proposals for proceedings volumes should be sent to one of the editors of the series or to Springer-Verlag Heidelberg. They should give sufficient information on the conference or symposium, and on the proposed proceedings. In particular, they should contain a list of the expected contributions with their prospective length. Abstracts or early versions (drafts) of some of the contributions are very helpful.

6. Lecture Notes are printed by photo-offset from camera-ready typed copy provided by the editors. For this purpose Springer-Verlag provides editors with technical instructions for the preparation of manuscripts and these should be distributed to all contributing authors. Springer-Verlag can also, on request, supply stationery on which the prescribed typing area is outlined. Some homogeneity in the presentation of the contributions is desirable.

   Careful preparation of manuscripts will help keep production time short and ensure a satisfactory appearance of the finished book. The actual production of a Lecture Notes volume normally takes 6 -8 weeks.

   Manuscripts should be at least 100 pages long. The final version should include a table of contents.

7. Editors receive a total of 50 free copies of their volume for distribution to the contributing authors, but no royalties. (Unfortunately, no reprints of individual contributions can be supplied.) They are entitled to purchase further copies of their book for their personal use at a discount of 33 1/3%, other Springer mathematics books at a discount of 20% directly from Springer-Verlag.

   Commitment to publish is made by letter of intent rather than by signing a formal contract. Springer-Verlag secures the copyright for each volume.

Vol. 1173: H. Delfs, M. Knebusch, Locally Semialgebraic Spaces. XVI, 329 pages. 1985.

Vol. 1174: Categories in Continuum Physics, Buffalo 1982. Seminar. Edited by F.W. Lawvere and S.H. Schanuel. V, 126 pages. 1986.

Vol. 1175: K. Mathiak, Valuations of Skew Fields and Projective Hjelmslev Spaces. VII, 116 pages. 1986.

Vol. 1176: R.R. Bruner, J.P. May, J.E. McClure, M. Steinberger, H_∞ Ring Spectra and their Applications. VII, 388 pages. 1986.

Vol. 1177: Representation Theory I. Finite Dimensional Algebras. Proceedings, 1984. Edited by V. Dlab, P. Gabriel and G. Michler. XV, 340 pages. 1986.

Vol. 1178: Representation Theory II. Groups and Orders. Proceedings, 1984. Edited by V. Dlab, P. Gabriel and G. Michler. XV, 370 pages. 1986.

Vol. 1179: Shi J.-Y. The Kazhdan-Lusztig Cells in Certain Affine Weyl Groups. X, 307 pages. 1986.

Vol. 1180: R. Carmona, H. Kesten, J.B. Walsh, École d'Été de Probabilités de Saint-Flour XIV – 1984. Édité par P.L. Hennequin. X, 438 pages. 1986.

Vol. 1181: Buildings and the Geometry of Diagrams, Como 1984. Seminar. Edited by L. Rosati. VII, 277 pages. 1986.

Vol. 1182: S. Shelah, Around Classification Theory of Models. VII, 279 pages. 1986.

Vol. 1183: Algebra, Algebraic Topology and their Interactions. Proceedings, 1983. Edited by J.-E. Roos. XI, 396 pages. 1986.

Vol. 1184: W. Arendt, A. Grabosch, G. Greiner, U. Groh, H.P. Lotz, U. Moustakas, R. Nagel, F. Neubrander, U. Schlotterbeck, One-parameter Semigroups of Positive Operators. Edited by R. Nagel. X, 460 pages. 1986.

Vol. 1185: Group Theory, Beijing 1984. Proceedings. Edited by Tuan H.F. V, 403 pages. 1986.

Vol. 1186: Lyapunov Exponents. Proceedings, 1984. Edited by L. Arnold and V. Wihstutz. VI, 374 pages. 1986.

Vol. 1187: Y. Diers, Categories of Boolean Sheaves of Simple Algebras. VI, 168 pages. 1986.

Vol. 1188: Fonctions de Plusieurs Variables Complexes V. Séminaire, 1979–85. Edité par François Norguet. VI, 306 pages. 1986.

Vol. 1189: J. Lukeš, J. Malý, L. Zajíček, Fine Topology Methods in Real Analysis and Potential Theory. X, 472 pages. 1986.

Vol. 1190: Optimization and Related Fields. Proceedings, 1984. Edited by R. Conti, E. De Giorgi and F. Giannessi. VIII, 419 pages. 1986.

Vol. 1191: A.R. Its, V.Yu. Novokshenov, The Isomonodromic Deformation Method in the Theory of Painlevé Equations. IV, 313 pages. 1986.

Vol. 1192: Equadiff 6. Proceedings, 1985. Edited by J. Vosmansky and M. Zlámal. XXIII, 404 pages. 1986.

Vol. 1193: Geometrical and Statistical Aspects of Probability in Banach Spaces. Proceedings, 1985. Edited by X. Fernique, B. Heinkel, M.B. Marcus and P.A. Meyer. IV, 128 pages. 1986.

Vol. 1194: Complex Analysis and Algebraic Geometry. Proceedings, 1985. Edited by H. Grauert. VI, 235 pages. 1986.

Vol. 1195: J.M. Barbosa, A.G. Colares, Minimal Surfaces in $\mathbb{R}^3$. X, 124 pages. 1986.

Vol. 1196: E. Casas-Alvero, S. Xambó-Descamps, The Enumerative Theory of Conics after Halphen. IX, 130 pages. 1986.

Vol. 1197: Ring Theory. Proceedings, 1985. Edited by F.M.J. van Oystaeyen. V, 231 pages. 1986.

Vol. 1198: Séminaire d'Analyse, P. Lelong – P. Dolbeault – H. Skoda. Seminar 1983/84. X, 260 pages. 1986.

Vol. 1199: Analytic Theory of Continued Fractions II. Proceedings, 1985. Edited by W.J. Thron. VI, 299 pages. 1986.

Vol. 1200: V.D. Milman, G. Schechtman, Asymptotic Theory of Finite Dimensional Normed Spaces. With an Appendix by M. Gromov. VIII, 156 pages. 1986.

Vol. 1201: Curvature and Topology of Riemannian Manifolds. Proceedings, 1985. Edited by K. Shiohama, T. Sakai and T. Sunada. VII, 336 pages. 1986.

Vol. 1202: A. Dür, Möbius Functions, Incidence Algebras and Power Series Representations. XI, 134 pages. 1986.

Vol. 1203: Stochastic Processes and Their Applications. Proceedings, 1985. Edited by K. Itô and T. Hida. VI, 222 pages. 1986.

Vol. 1204: Séminaire de Probabilités XX, 1984/85. Proceedings. Edité par J. Azéma et M. Yor. V, 639 pages. 1986.

Vol. 1205: B.Z. Moroz, Analytic Arithmetic in Algebraic Number Fields. VII, 177 pages. 1986.

Vol. 1206: Probability and Analysis, Varenna (Como) 1985. Seminar. Edited by G. Letta and M. Pratelli. VIII, 280 pages. 1986.

Vol. 1207: P.H. Bérard, Spectral Geometry: Direct and Inverse Problems. With an Appendix by G. Besson. XIII, 272 pages. 1986.

Vol. 1208: S. Kaijser, J.W. Pelletier, Interpolation Functors and Duality. IV, 167 pages. 1986.

Vol. 1209: Differential Geometry, Peñíscola 1985. Proceedings. Edited by A.M. Naveira, A. Ferrández and F. Mascaró. VIII, 306 pages. 1986.

Vol. 1210: Probability Measures on Groups VIII. Proceedings, 1985. Edited by H. Heyer. X, 386 pages. 1986.

Vol. 1211: M.B. Sevryuk, Reversible Systems. V, 319 pages. 1986.

Vol. 1212: Stochastic Spatial Processes. Proceedings, 1984. Edited by P. Tautu. VIII, 311 pages. 1986.

Vol. 1213: L.G. Lewis, Jr., J.P. May, M. Steinberger, Equivariant Stable Homotopy Theory. IX, 538 pages. 1986.

Vol. 1214: Global Analysis – Studies and Applications II. Edited by Yu.G. Borisovich and Yu.E. Gliklikh. V, 275 pages. 1986.

Vol. 1215: Lectures in Probability and Statistics. Edited by G. del Pino and R. Rebolledo. V, 491 pages. 1986.

Vol. 1216: J. Kogan, Bifurcation of Extremals in Optimal Control. VIII, 106 pages. 1986.

Vol. 1217: Transformation Groups. Proceedings, 1985. Edited by S. Jackowski and K. Pawalowski. X, 396 pages. 1986.

Vol. 1218: Schrödinger Operators, Aarhus 1985. Seminar. Edited by E. Balslev. V, 222 pages. 1986.

Vol. 1219: R. Weissauer, Stabile Modulformen und Eisensteinreihen. III, 147 Seiten. 1986.

Vol. 1220: Séminaire d'Algèbre Paul Dubreil et Marie-Paule Malliavin. Proceedings, 1985. Edité par M.-P. Malliavin. IV, 200 pages. 1986.

Vol. 1221: Probability and Banach Spaces. Proceedings, 1985. Edited by J. Bastero and M. San Miguel. XI, 222 pages. 1986.

Vol. 1222: A. Katok, J.-M. Strelcyn, with the collaboration of F. Ledrappier and F. Przytycki, Invariant Manifolds, Entropy and Billiards; Smooth Maps with Singularities. VIII, 283 pages. 1986.

Vol. 1223: Differential Equations in Banach Spaces. Proceedings, 1985. Edited by A. Favini and E. Obrecht. VIII, 299 pages. 1986.

Vol. 1224: Nonlinear Diffusion Problems, Montecatini Terme 1985. Seminar. Edited by A. Fasano and M. Primicerio. VIII, 188 pages. 1986.

Vol. 1225: Inverse Problems, Montecatini Terme 1986. Seminar. Edited by G. Talenti. VIII, 204 pages. 1986.

Vol. 1226: A. Buium, Differential Function Fields and Moduli of Algebraic Varieties. IX, 146 pages. 1986.

Vol. 1227: H. Helson, The Spectral Theorem. VI, 104 pages. 1986.

Vol. 1228: Multigrid Methods II. Proceedings, 1985. Edited by W. Hackbusch and U. Trottenberg. VI, 336 pages. 1986.

Vol. 1229: O. Bratteli, Derivations, Dissipations and Group Actions on C*-algebras. IV, 277 pages. 1986.

Vol. 1230: Numerical Analysis. Proceedings, 1984. Edited by J.-P. Hennart. X, 234 pages. 1986.

Vol. 1231: E.-U. Gekeler, Drinfeld Modular Curves. XIV, 107 pages. 1986.